W9-AVF-368

Undergraduate Texts in Mathematics

Editors

S. Axler
F.W. Gehring
K.A. Ribet

Undergraduate Texts in Mathematics

Abbott: Understanding Analysis.

Anglin: Mathematics: A Concise History and Philosophy.
Readings in Mathematics.

Anglin/Lambek: The Heritage of Thales.
Readings in Mathematics.

Apostol: Introduction to Analytic Number Theory. Second edition.

Armstrong: Basic Topology.

Armstrong: Groups and Symmetry.

Axler: Linear Algebra Done Right. Second edition.

Beardon: Limits: A New Approach to Real Analysis.

Bak/Newman: Complex Analysis. Second edition.

Banchoff/Wermer: Linear Algebra Through Geometry. Second edition.

Berberian: A First Course in Real Analysis.

Bix: Conics and Cubics: A Concrete Introduction to Algebraic Curves.

Brémaud: An Introduction to Probabilistic Modeling.

Bressoud: Factorization and Primality Testing.

Bressoud: Second Year Calculus.
Readings in Mathematics.

Brickman: Mathematical Introduction to Linear Programming and Game Theory.

Browder: Mathematical Analysis: An Introduction.

Buchmann: Introduction to Cryptography.

Buskes/van Rooij: Topological Spaces: From Distance to Neighborhood.

Callahan: The Geometry of Spacetime: An Introduction to Special and General Relativity.

Carter/van Brunt: The Lebesgue–Stieltjes Integral: A Practical Introduction.

Cederberg: A Course in Modern Geometries. Second edition.

Chambert-Loir: A Field Guide to Algebra

Childs: A Concrete Introduction to Higher Algebra. Second edition.

Chung/AitSahlia: Elementary Probability Theory: With Stochastic Processes and an Introduction to Mathematical Finance. Fourth edition.

Cox/Little/O'Shea: Ideals, Varieties, and Algorithms. Second edition.

Croom: Basic Concepts of Algebraic Topology.

Curtis: Linear Algebra: An Introductory Approach. Fourth edition.

Daepp/Gorkin: Reading, Writing, and Proving: A Closer Look at Mathematics.

Devlin: The Joy of Sets: Fundamentals of Contemporary Set Theory. Second edition.

Dixmier: General Topology.

Driver: Why Math?

Ebbinghaus/Flum/Thomas: Mathematical Logic. Second edition.

Edgar: Measure, Topology, and Fractal Geometry.

Elaydi: An Introduction to Difference Equations. Third edition.

Erdős/Surányi: Topics in the Theory of Numbers.

Estep: Practical Analysis in One Variable.

Exner: An Accompaniment to Higher Mathematics.

Exner: Inside Calculus.

Fine/Rosenberger: The Fundamental Theory of Algebra.

Fischer: Intermediate Real Analysis.

Flanigan/Kazdan: Calculus Two: Linear and Nonlinear Functions. Second edition.

Fleming: Functions of Several Variables. Second edition.

Foulds: Combinatorial Optimization for Undergraduates.

Foulds: Optimization Techniques: An Introduction.

Franklin: Methods of Mathematical Economics.

(continued after index)

Theodore W. Gamelin

Complex Analysis

With 184 Illustrations

 Springer

Theodore Gamelin
Department of Mathematics
UCLA
Box 951555
Los Angeles, CA 90095-1555
USA

Mathematics Subject Classification (2000): 28-01, 30-01

Library of Congress Cataloging-in-Publication Data
Gamelin, Theodore W.
   Complex analysis / Theodore W. Gamelin
     p. cm. — (Undergraduate texts in mathematics)
   Includes bibliographical references and index.
    ISBN 0-387-95069-9 (alk. paper) – ISBN 0-387-95069-9 (softcover : alk. paper)
    1. Mathematical analysis.   2. Functions of complex variables.   I. Title.
  II. Series
  QA300 .G25 2000
  515—dc21   00-041905

ISBN-10: 0-387-95069-9
ISBN-13: 978-0387-95069-3

Printed in the United States of America.   (EB)

9 8 7 6 5 4

springeronline.com

To the many wonderful students at Perugia who took my course, and to the others there who contributed to an enjoyable experience

Preface

This book provides an introduction to complex analysis for students with some familiarity with complex numbers from high school. Students should be familiar with the Cartesian representation of complex numbers and with the algebra of complex numbers, that is, they should know that $i^2 = -1$. A familiarity with multivariable calculus is also required, but here the fundamental ideas are reviewed. In fact, complex analysis provides a good training ground for multivariable calculus. It allows students to consolidate their understanding of parametrized curves, tangent vectors, arc length, gradients, line integrals, independence of path, and Green's theorem. The ideas surrounding independence of path are particularly difficult for students in calculus, and they are not absorbed by most students until they are seen again in other courses.

The book consists of sixteen chapters, which are divided into three parts. The first part, Chapters I–VII, includes basic material covered in all undergraduate courses. With the exception of a few sections, this material is much the same as that covered in Cauchy's lectures, except that the emphasis on viewing functions as mappings reflects Riemann's influence. The second part, Chapters VIII–XI, bridges the nineteenth and the twentieth centuries. About half this material would be covered in a typical undergraduate course, depending upon the taste and pace of the instructor. The material on the Poisson integral is of interest to electrical engineers, while the material on hyperbolic geometry is of interest to pure mathematicians and also to high school mathematics teachers. The third part, Chapters XII–XVI, consists of a careful selection of special topics that illustrate the scope and power of complex analysis methods. These topics include Julia sets and the Mandelbrot set, Dirichlet series and the prime number theorem, and the uniformization theorem for Riemann surfaces. The final five chapters serve also to complete the coverage of all background necessary for passing PhD qualifying exams in complex analysis.

Note to the instructor
There is a glut of complex analysis textbooks on the market. It is a beautiful subject, so beautiful that a large number of experts have been moved to

write their own accounts of the area. In spite of the plethora of textbooks, I have never found an introduction to complex analysis that is completely suitable for my own teaching style and audiences.

The students in each of my various audiences have begun the course with a wide range of backgrounds. Teaching to students with disparate backgrounds and preparations has posed a major teaching challenge. I respond by including early some topics that can be treated in an elementary way and yet are usually new and capture the imagination of students with already some background in complex analysis. For example, the stereographic projection appears early, the Riemann surface of the square root function is explained early at an intuitive level, and both conformality and fractional linear transformations are treated relatively early. Exercises range from the very simple to the quite challenging, in all chapters. Some of the exercises that appear early in the book can form the basis for an introduction to a more advanced topic, which can be tossed out to the more sophisticated students. Thus for instance the basis is laid for introducing students to the spherical metric already in the first chapter, though the topic is not taken up seriously until much later, in connection with Marty's theorem in Chapter XII.

The second problem addressed by the book has to do with flexibility of use. There are many routes through complex analysis, and many instructors hold strong opinions concerning the optimal route. I address this problem by laying out the material so as to allow for substantial flexibility in the ordering of topics. The instructor can defer many topics (for instance, the stereographic projection, or conformality, or fractional linear transformations) in order to reach Cauchy's theorem and power series relatively early, and then return to the omitted topics later, time permitting.

There is also flexibility with respect to adjusting the course to undergraduate students or to beginning graduate students. The bulk of the book was written with undergraduate students in mind, and I have used various preliminary course notes for Chapters I-XI at the undergraduate level. By adjusting the level of the lectures and the pace I have found the course notes for all sixteen chapters appropriate for a first-year graduate course sequence.

One of my colleagues wrote in commenting upon the syllabus of our undergraduate complex analysis course that "fractional powers should be postponed to the end of the course as they are very difficult for the students." My philosophy is just the reverse. If a concept is important but difficult, I prefer to introduce it early and then return to it several times, in order to give students time to absorb the idea. For example, the idea of a branch of a multivalued analytic function is very difficult for students, yet it is a central issue in complex analysis. I start early with a light introduction to the square root function. The logarithm function follows soon, followed by phase factors in connection with fractional powers. The basic idea is returned to several times throughout the course, as in the applications of residue theory to evaluate integrals. I find that by this time most students are reasonably comfortable with the idea.

A solid core for the one-semester undergraduate course is as follows:

Chapter I

Chapter II

Sections III.1-5

Sections IV.1-6

Sections V.1-7

Sections VI.1-4

Sections VII.1-4

Sections VIII.1-2

Sections IX.1-2

Sections X.1-2

Sections XI.1-2

To reach power series faster I would recommend postponing I.3, II.6-7, III.4-5, and going light on Riemann surfaces. Sections II.6-7 and III.4-5 should be picked up again before starting Chapter IX.

Which additional sections to cover depends on the pace of the instructor and the level of the students. My own preference is to add more contour integration (Sections VII.5 and VII.8) and hyperbolic geometry (Section IX.3) to the syllabus, and then to do something more with conformal mapping, as the Schwarz reflection principle (Section X.3), time permitting. To gain time, I mention some topics (as trigonometric and hyperbolic functions) only briefly in class. Students learn this material as well by reading and doing assigned exercises. Finishing with Sections XI.1-2 closes the circle and provides a good review at the end of the term, while at the same time it points to a fundamental and nontrivial theorem (the Riemann mapping theorem).

Note to the student

You are about to enter a fascinating and wonderful world. Complex analysis is a beautiful subject, filled with broad avenues and narrow backstreets leading to intellectual excitement. Before you traverse this terrain, let me provide you with some tips and some warnings, designed to make your journey more pleasant and profitable.

Above all, give some thought to strategies for study and learning. This is easier if you are aware of the difference between the "what," the "how," and the "why," (as Halmos calls them). The "what" consists of definitions, statements of theorems, and formulae. Determine which are most important and memorize them, at least in slogan form if not precisely. Just as one maintains in memory the landmark years 1066, 1453, and 1776 as markers in the continuum of history, so should you maintain in memory the definition of analytic function, the Cauchy-Riemann equations, and the residue formula. The simplest of the exercises are essentially restatements of "what."

The "how" consists in being able to apply the formulae and techniques to solve problems, as to show that a function is analytic by checking the Cauchy-Riemann equations, or to determine whether a polynomial has a zero in a certain region by applying the argument principle, or to evaluate a definite integral by contour integration. Before determining "how" you must know "what." Many of the exercises are "how" problems. Working these exercises and discussing them with other students and the instructor are an important part of the learning process.

The "why" consists in understanding why a theorem is true or why a technique works. This understanding can be arrived at in many different ways and at various levels. There are several things you can do to understand why a result is true. Try it out on some special cases. Make a short synopsis of the proof. See where each hypothesis is used in the proof. Try proving it after altering or removing one of the hypotheses. Analyze the proof to determine which ingredients are absolutely essential and to determine its depth and level of difficulty. The slogan form of the Jordan curve theorem is that "every closed curve has an inside and an outside" (Section VIII.7). What is the level of difficulty of this theorem? Can you come up with a direct proof? Try it.

Finally, be aware that there is a language of formal mathematics that is related to but different from common English. We all know what "near" means in common English. In the language of formal mathematics the word carries with it a specific measure of distance or proximity, which is traditionally quantified by $\varepsilon > 0$ or a "for every neighborhood" statement. Look also for words like "eventually," "smooth," and "local." Prepare to absorb not only new facts and ideas but also a different language. Developing some understanding of the language is not easy – it is part of growing up and becoming mathematically sophisticated.

Acknowledgments

This book stems primarily from courses I gave in complex analysis at the Interuniversity Summer School at Perugia (Italy). Each course was based on a series of exercises, for which I developed a computer bank. Gradually I deposited written versions of my lectures in the computer bank. When I finally decided to expand the material to book form, I also used notes based on lectures presented over the years at several places, including UCLA, Brown University, Valencia (Spain), and long ago at the university at La Plata (Argentina). I have enjoyed teaching this material. I learned a lot, both about the subject matter and about teaching, through my students. I would like to thank the many students who contributed, knowingly or unknowingly, to this book.

The origins of many of the mathematical ideas have been lost in the thickets of the history of mathematics. Let me mention the source for one item. The treatment of the parabolic case of the uniformization theorem follows a line of proof due to D. Marshall, and I am grateful for his sharing his work. As far as I know, everything else is covered by the bibliography, and I apologize for any omissions.

Each time I reread a segment of the book manuscript I found various mathematical blunders, grammatical infringements, and stylistic travesties. Undoubtedly mistakes have persisted into the printed book. I would appreciate receiving your email about any egregious errors you come across, together with your comments about any passages you perceive to be particularly dense or unenlightening. My email address, while I am around, is twg@math.ucla.edu. I thank you, dear reader, in advance.

Julie Honig and Mary Edwards helped with the preparation of class notes that were used for parts of the book, and for this I thank them. Finally, I am happy to acknowledge the skilled assistance of the publishing staff, who turned my doodles into figures and otherwise facilitated publication of the book.

T.W. Gamelin
Pacific Palisades February 2001

Contents

Preface   vii
Introduction   xvii

FIRST PART

Chapter I   The Complex Plane and Elementary Functions   **1**
  1. Complex Numbers   1
  2. Polar Representation   5
  3. Stereographic Projection   11
  4. The Square and Square Root Functions   15
  5. The Exponential Function   19
  6. The Logarithm Function   21
  7. Power Functions and Phase Factors   24
  8. Trigonometric and Hyperbolic Functions   29

Chapter II   Analytic Functions   **33**
  1. Review of Basic Analysis   33
  2. Analytic Functions   42
  3. The Cauchy-Riemann Equations   46
  4. Inverse Mappings and the Jacobian   51
  5. Harmonic Functions   54
  6. Conformal Mappings   58
  7. Fractional Linear Transformations   63

Chapter III   Line Integrals and Harmonic Functions   **70**
  1. Line Integrals and Green's Theorem   70
  2. Independence of Path   76
  3. Harmonic Conjugates   83
  4. The Mean Value Property   85
  5. The Maximum Principle   87
  6. Applications to Fluid Dynamics   90
  7. Other Applications to Physics   97

Chapter IV   Complex Integration and Analyticity   **102**
  1. Complex Line Integrals   102
  2. Fundamental Theorem of Calculus for Analytic Functions   107
  3. Cauchy's Theorem   110
  4. The Cauchy Integral Formula   113
  5. Liouville's Theorem   117
  6. Morera's Theorem   119
  7. Goursat's Theorem   123
  8. Complex Notation and Pompeiu's Formula   124

Chapter V   Power Series   **130**
  1. Infinite Series   130
  2. Sequences and Series of Functions   133
  3. Power Series   138
  4. Power Series Expansion of an Analytic Function   144
  5. Power Series Expansion at Infinity   149
  6. Manipulation of Power Series   151
  7. The Zeros of an Analytic Function   154
  8. Analytic Continuation   158

Chapter VI   Laurent Series and Isolated Singularities   **165**
  1. The Laurent Decomposition   165
  2. Isolated Singularities of an Analytic Function   171
  3. Isolated Singularity at Infinity   178
  4. Partial Fractions Decomposition   179
  5. Periodic Functions   182
  6. Fourier Series   186

Chapter VII   The Residue Calculus   **195**
  1. The Residue Theorem   195
  2. Integrals Featuring Rational Functions   199
  3. Integrals of Trigonometric Functions   203
  4. Integrands with Branch Points   206
  5. Fractional Residues   209
  6. Principal Values   212
  7. Jordan's Lemma   216
  8. Exterior Domains   219

SECOND PART

Chapter VIII The Logarithmic Integral   **224**
  1. The Argument Principle   224
  2. Rouché's Theorem   229
  3. Hurwitz's Theorem   231
  4. Open Mapping and Inverse Function Theorems   232
  5. Critical Points   236
  6. Winding Numbers   242

Contents

7. The Jump Theorem for Cauchy Integrals   246
8. Simply Connected Domains   252

Chapter IX   The Schwarz Lemma and Hyperbolic Geometry   **260**
1. The Schwarz Lemma   260
2. Conformal Self-Maps of the Unit Disk   263
3. Hyperbolic Geometry   266

Chapter X   Harmonic Functions and the Reflection Principle   **274**
1. The Poisson Integral Formula   274
2. Characterization of Harmonic Functions   280
3. The Schwarz Reflection Principle   282

Chapter XI   Conformal Mapping   **289**
1. Mappings to the Unit Disk and Upper Half-Plane   289
2. The Riemann Mapping Theorem   294
3. The Schwarz-Christoffel Formula   296
4. Return to Fluid Dynamics   304
5. Compactness of Families of Functions   306
6. Proof of the Riemann Mapping Theorem   311

THIRD PART

Chapter XII   Compact Families of Meromorphic Functions   **315**
1. Marty's Theorem   315
2. Theorems of Montel and Picard   320
3. Julia Sets   324
4. Connectedness of Julia Sets   333
5. The Mandelbrot Set   338

Chapter XIII Approximation Theorems   **342**
1. Runge's Theorem   342
2. The Mittag-Leffler Theorem   348
3. Infinite Products   352
4. The Weierstrass Product Theorem   358

Chapter XIV Some Special Functions   **361**
1. The Gamma Function   361
2. Laplace Transforms   365
3. The Zeta Function   370
4. Dirichlet Series   376
5. The Prime Number Theorem   382

Chapter XV   The Dirichlet Problem   **390**
1. Green's Formulae   390
2. Subharmonic Functions   394
3. Compactness of Families of Harmonic Functions   398
4. The Perron Method   402
5. The Riemann Mapping Theorem Revisited   406

6. Green's Function for Domains with Analytic Boundary   407
7. Green's Function for General Domains   413

Chapter XVI Riemann Surfaces   **418**
1. Abstract Riemann Surfaces   418
2. Harmonic Functions on a Riemann Surface   426
3. Green's Function of a Surface   429
4. Symmetry of Green's Function   434
5. Bipolar Green's Function   436
6. The Uniformization Theorem   438
7. Covering Surfaces   441

Hints and Solutions for Selected Exercises   447
References   469
List of Symbols   471
Index   473

Introduction

Complex analysis is a splendid realm within the world of mathematics, unmatched for its beauty and power. It has varifold elegant and oftentimes unexpected applications to virtually every part of mathematics. It is broadly applicable beyond mathematics, and in particular it provides powerful tools for the sciences and engineering.

Already in the eighteenth century Euler discovered the connection between trigonometric functions and exponential functions through complex analysis. (It was he who invented the notation $e^{i\theta}$.) However, it was not until the nineteenth century that the foundations of complex analysis were laid. Among the many mathematicians and scientists who contributed, there are three who stand out as having influenced decisively the course of development of complex analysis. The first is A. Cauchy (1789-1857), who developed the theory systematically along the lines we shall follow, with the complex integral calculus, Cauchy's theorem, and the Cauchy integral formula playing fundamental roles. The other two are K. Weierstrass (1815-1897) and B. Riemann (1826-1866), who appeared on the mathematical scene about the middle of the nineteenth century. Weierstrass developed the theory from a starting point of convergent power series, and this approach led towards more formal algebraic developments. Riemann contributed a more geometric point of view. His ideas had a tremendous impact not only on complex analysis but upon mathematics as a whole, though his views took hold only gradually.

In addition to the standard undergraduate material, we shall follow several strands and obtain several poster theorems, which together with the more elementary material cover what might be called the "complex analysis canon," the part of complex analysis included in the syllabus of the typical PhD qualifying exam.

One of the strands we shall follow culminates in the prime number theorem. Already Euler in the eighteenth century had written down an infinite product for the zeta function, connecting the prime numbers to complex analysis. In the 1830's Dirichlet used variants of the zeta function to prove the existence of infinitely many primes in arithmetic progressions. Riemann did fundamental work connecting the zeta function to the distribution of prime numbers. And finally just before the close of the nineteenth century J. Hadamard and C.J. de la Vallée Poussin independently proved the prime number theorem using techniques of complex analysis.

Another strand we shall follow is the conformal mapping of domains in the plane and more generally of Riemann surfaces. We shall aim at two poster results: the Riemann mapping theorem and the uniformization theorem for Riemann surfaces. The definitive version of the Riemann mapping theorem, which one finds in all complex analysis textbooks today, was proved by W. Osgood in 1900. The uniformization theorem for Riemann surfaces was proved independently in 1907 by P. Koebe and H. Poincaré, thereby solving Hilbert's 22nd problem from his famous address to the International Mathematical Congress in 1900.

The first quarter of the twentieth century was one of rapid development of the foundations of complex analysis. P. Montel put his finger on the notion of compactness in spaces of meromorphic functions and developed the theory of normal families. P. Fatou and G. Julia used Montel's theorem in their seminal work around 1914-1921 on complex iteration theory. On another front, O. Perron developed in 1923 a powerful method for solving the Dirichlet problem.

By the end of the first quarter of the twentieth century, the complex analysis canon had been established, and nearly all the main results constituting the undergraduate and first-year graduate courses in complex analysis had been obtained. Nevertheless, throughout the twentieth century there has been much exciting progress on the frontiers of research in complex analysis, and meanwhile proofs of the most difficult foundational results have been gradually simplified and clarified. While the complex analysis canon has remained relatively static, the developments at the frontier have led to new perspectives and shifting emphases. For instance, the current research interest in dynamical systems and the advent of computer graphics contributed to elevating the work of Fatou and Julia to a more prominent position.

What lies before you is the distillation of the essential, the useful, and the beautiful, from two centuries of labor. Enjoy!

I

The Complex Plane and Elementary Functions

In this chapter we set the scene and introduce some of the main characters. We begin with the three representations of complex numbers: the Cartesian representation, the polar representation, and the spherical representation. Then we introduce the basic functions encountered in complex analysis: the exponential function, the logarithm function, power functions, and trigonometric functions. We view several concrete functions $w = f(z)$ as mappings from the z-plane to the w-plane, and we consider the problem of describing the inverse functions.

1. Complex Numbers

A **complex number** is an expression of the form $z = x + iy$, where x and y are real numbers. The component x is called the **real part** of z, and y is the **imaginary part** of z. We will denote these by

$$x = \operatorname{Re} z,$$
$$y = \operatorname{Im} z.$$

The set of complex numbers forms the **complex plane**, which we denote by $\mathbb{C}$. We denote the set of real numbers by $\mathbb{R}$, and we think of the real numbers as being a subset of the complex plane, consisting of the complex numbers with imaginary part equal to zero.

The correspondence

$$z = x + iy \longleftrightarrow (x, y)$$

is a one-to-one correspondence between complex numbers and points (or vectors) in the Euclidean plane $\mathbb{R}^2$. The real numbers correspond to the x-axis in the Euclidean plane. The complex numbers of the form iy are called **purely imaginary numbers**. They form the **imaginary axis** $i\mathbb{R}$

1

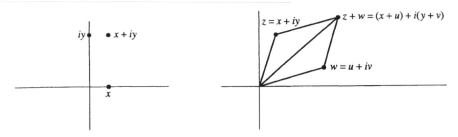

in the complex plane, which corresponds to the y-axis in the Euclidean plane.

We add complex numbers by adding their real and imaginary parts:

$$(x + iy) + (u + iv) = (x + u) + i(y + v).$$

Thus $\text{Re}(z + w) = \text{Re}(z) + \text{Re}(w)$, and $\text{Im}(z + w) = \text{Im}(z) + \text{Im}(w)$ for $z, w \in \mathbb{C}$. The addition of complex numbers corresponds to the usual componentwise addition in the Euclidean plane.

The **modulus** of a complex number $z = x + iy$ is the length $\sqrt{x^2 + y^2}$ of the corresponding vector (x, y) in the Euclidean plane. The modulus of z is also called the **absolute value** of z, and it is denoted by $|z|$:

$$|z| = \sqrt{x^2 + y^2}.$$

The triangle inequality for vectors in the plane takes the form

$$|z + w| \leq |z| + |w|, \qquad z, w \in \mathbb{C}.$$

By applying the triangle inequality to $z = (z - w) + w$, we obtain $|z| \leq |z - w| + |w|$. Subtracting $|w|$, we obtain a very useful inequality,

$$(1.1) \qquad\qquad |z - w| \geq |z| - |w|, \qquad z, w \in \mathbb{C}.$$

Complex numbers can be multiplied, and this is the feature that distinguishes the complex plane $\mathbb{C}$ from the Euclidean plane $\mathbb{R}^2$. Formally, the multiplication is defined by

$$(x + iy)(u + iv) = xu - yv + i(xv + yu).$$

One can check directly from this definition that the usual laws of algebra hold for complex multiplication:

$$
\begin{aligned}
(z_1 z_2) z_3 &= z_1 (z_2 z_3), && \text{(associative law)}\\
z_1 z_2 &= z_2 z_1, && \text{(commutative law)}\\
z_1 (z_2 + z_3) &= z_1 z_2 + z_1 z_3. && \text{(distributive law)}
\end{aligned}
$$

With respect to algebraic operations, complex numbers behave the same as real numbers. Algebraic manipulations are performed on complex numbers using the usual laws of algebra, together with the special rule $i^2 = -1$.

Every complex number $z \neq 0$ has a multiplicative inverse $1/z$, which is given explicitly by

$$\frac{1}{z} = \frac{x - iy}{x^2 + y^2}, \qquad z = x + iy \in \mathbb{C}, \ z \neq 0.$$

Thus for instance, the multiplicative inverse of i is $1/i = -i$.

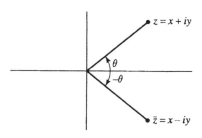

The **complex conjugate** of a complex number $z = x + iy$ is defined to be $\bar{z} = x - iy$. Geometrically, $\bar{z}$ is the reflection of z in the x-axis. If we reflect twice, we return to z,

$$\bar{\bar{z}} = z, \qquad z \in \mathbb{C}.$$

Some other useful properties of complex conjugation are

$$\overline{z + w} = \bar{z} + \bar{w}, \qquad z, w \in \mathbb{C},$$
$$\overline{zw} = \bar{z}\bar{w}, \qquad z, w \in \mathbb{C},$$
$$|z| = |\bar{z}|, \qquad z \in \mathbb{C},$$
$$|z|^2 = z\bar{z}, \qquad z \in \mathbb{C}.$$

Each of these identities can be verified easily using the definition of $\bar{z}$ and $|z|$. The last formula above allows us to express $1/z$ in terms of the complex conjugate $\bar{z}$:

$$1/z = \bar{z}/|z|^2, \qquad z \in \mathbb{C}, z \neq 0.$$

The real and imaginary parts of z can be recovered from z and $\bar{z}$, by

$$\operatorname{Re} z = (z + \bar{z})/2, \qquad z \in \mathbb{C},$$
$$\operatorname{Im} z = (z - \bar{z})/2i, \qquad z \in \mathbb{C}.$$

From $|zw|^2 = (zw)(\overline{zw}) = (z\bar{z})(w\bar{w}) = |z|^2|w|^2$, we obtain also

$$|zw| = |z||w|, \qquad z, w \in \mathbb{C}.$$

A **complex polynomial of degree** $n \geq 0$ is a function of the form

$$p(z) = a_n z^n + a_{n-1} z^{n-1} + \cdots + a_1 z + a_0, \qquad z \in \mathbb{C},$$

where $a_0, \ldots, a_n$ are complex numbers, and $a_n \neq 0$. A key property of the complex numbers, not enjoyed by the real numbers, is that any polynomial with complex coefficients can be factored as a product of linear factors.

Fundamental Theorem of Algebra. *Every complex polynomial $p(z)$ of degree $n \geq 1$ has a factorization*

$$p(z) = c(z - z_1)^{m_1} \cdots (z - z_k)^{m_k},$$

where the z_j's are distinct and $m_j \geq 1$. This factorization is unique, up to a permutation of the factors.

We will not prove this theorem now, but we will give several proofs later. Some remarks are in order.

The uniqueness of the factorization is easy to establish. The points $z_1, \ldots, z_k$ are uniquely characterized as the **roots** of $p(z)$, or the **zeros** of $p(z)$. These are the points where $p(z) = 0$. The integer m_j is characterized as the unique integer m with the property that $p(z)$ can be factored as $(z - z_j)^m q(z)$ where $q(z)$ is a polynomial satisfying $q(z_j) \neq 0$.

For the proof of the existence of the factorization, one proceeds by induction on the degree n of the polynomial. The crux of the matter is to find a point z_1 such that $p(z_1) = 0$. With a root z_1 in hand, one easily factors $p(z)$ as a product $(z - z_1)q(z)$, where $q(z)$ is a polynomial of degree $n - 1$. (See the exercises.) The induction hypothesis allows one to factor $q(z)$ as a product of linear factors, and this yields the factorization of $p(z)$. Thus the fundamental theorem of algebra is equivalent to the statement that every complex polynomial of degree $n \geq 1$ has a zero.

Example. The polynomial $p(x) = x^2 + 1$ with real coefficients cannot be factored as a product of linear polynomials with real coefficients, since it does not have any real roots. However, the complex polynomial $p(z) = z^2 + 1$ has the factorization

$$z^2 + 1 = (z - i)(z + i),$$

corresponding to the two complex roots $\pm i$ of $z^2 + 1$.

Exercises for I.1

1. Identify and sketch the set of points satisfying:
   (a) $|z - 1 - i| = 1$ (f) $0 < \operatorname{Im} z < \pi$
   (b) $1 < |2z - 6| < 2$ (g) $-\pi < \operatorname{Re} z < \pi$
   (c) $|z - 1|^2 + |z + 1|^2 < 8$ (h) $|\operatorname{Re} z| < |z|$
   (d) $|z - 1| + |z + 1| \leq 2$ (i) $\operatorname{Re}(iz + 2) > 0$
   (e) $|z - 1| < |z|$ (j) $|z - i|^2 + |z + i|^2 < 2$

2. Verify from the definitions each of the identities (a) $\overline{z + w} = \bar{z} + \bar{w}$, (b) $\overline{zw} = \bar{z}\bar{w}$, (c) $|\bar{z}| = |z|$, (d) $|z|^2 = z\bar{z}$. Draw sketches to illustrate (a) and (c).

3. Show that the equation $|z|^2 - 2\,\mathrm{Re}(\bar{a}z) + |a|^2 = \rho^2$ represents a circle centered at a with radius ρ.

4. Show that $|z| \leq |\mathrm{Re}\,z| + |\mathrm{Im}\,z|$, and sketch the set of points for which equality holds.

5. Show that $|\mathrm{Re}\,z| \leq |z|$ and $|\mathrm{Im}\,z| \leq |z|$. Show that
$$|z + w|^2 = |z|^2 + |w|^2 + 2\,\mathrm{Re}(z\bar{w}).$$
Use this to prove the triangle inequality $|z + w| \leq |z| + |w|$.

6. For fixed $a \in \mathbb{C}$, show that $|z - a|/|1 - \bar{a}z| = 1$ if $|z| = 1$ and $1 - \bar{a}z \neq 0$.

7. Fix $\rho > 0$, $\rho \neq 1$, and fix $z_0, z_1 \in \mathbb{C}$. Show that the set of z satisfying $|z - z_0| = \rho|z - z_1|$ is a circle. Sketch it for $\rho = \frac{1}{2}$ and $\rho = 2$, with $z_0 = 0$ and $z_1 = 1$. What happens when $\rho = 1$?

8. Let $p(z)$ be a polynomial of degree $n \geq 1$ and let $z_0 \in \mathbb{C}$. Show that there is a polynomial $h(z)$ of degree $n - 1$ such that $p(z) = (z - z_0)h(z) + p(z_0)$. In particular, if $p(z_0) = 0$, then $p(z) = (z - z_0)h(z)$.

9. Find the polynomial $h(z)$ in the preceding exercise for the following choices of $p(z)$ and z_0: (a) $p(z) = z^2$ and $z_0 = i$, (b) $p(z) = z^3 + z^2 + z$ and $z_0 = -1$, (c) $p(z) = 1 + z + z^2 + \cdots + z^m$ and $z_0 = -1$.

10. Let $q(z)$ be a polynomial of degree $m \geq 1$. Show that any polynomial $p(z)$ can be expressed in the form
$$p(z) = h(z)q(z) + r(z),$$
where $h(z)$ and $r(z)$ are polynomials and the degree of the remainder $r(z)$ is strictly less than m. *Hint.* Proceed by induction on the degree of $p(z)$. The resulting method is called the **division algorithm**.

11. Find the polynomials $h(z)$ and $r(z)$ in the preceding exercise for $p(z) = z^n$ and $q(z) = z^2 - 1$.

2. Polar Representation

Any point $(x, y) \neq (0, 0)$ in the plane can be described by polar coordinates r and θ, where $r = \sqrt{x^2 + y^2}$ and θ is the angle subtended by (x, y) and the x-axis. The angle θ is determined only up to adding an integral multiple of 2π. The Cartesian coordinates x, y are recovered from the polar coordinates r, θ by
$$\begin{cases} x = r\cos\theta, \\ y = r\sin\theta. \end{cases}$$

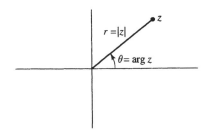

If we write the polar representation in complex notation, we obtain

(2.1) $$z = x + iy = r(\cos\theta + i\sin\theta).$$

Here $r = |z|$ is the modulus of z. We define the **argument** of z to be the angle θ, and we write

$$\theta = \arg z.$$

Thus $\arg z$ is a multivalued function, defined for $z \neq 0$. The **principal value of** $\arg z$, denoted by $\text{Arg } z$, is specified rather arbitrarily to be the value of θ that satisfies $-\pi < \theta \leq \pi$. The values of $\arg z$ are obtained from $\text{Arg } z$ by adding integral multiples of 2π:

$$\arg z = \{\text{Arg } z + 2\pi k \,:\, k = 0, \pm 1, \pm 2, \ldots\}, \qquad z \neq 0.$$

Example. The principal value of $\arg i$ is $\text{Arg } i = \pi/2$. The principal value of $\arg(1 - i)$ is $\text{Arg}(1 - i) = -\pi/4$.

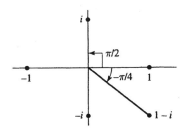

It will be convenient to introduce the notation

(2.2) $$e^{i\theta} = \cos\theta + i\sin\theta.$$

From (2.1) we obtain

$$z = re^{i\theta}, \qquad\qquad r = |z|, \ \theta = \arg z.$$

This representation is called the **polar representation** of z. The sine and cosine functions are 2π-periodic, that is, they satisfy $\sin(\theta + 2\pi m) = \sin\theta$, $\cos(\theta + 2\pi m) = \cos\theta$. Thus the various choices of $\arg z$ yield the same value for $e^{i\theta}$,

$$e^{i(\theta + 2\pi m)} = e^{i\theta}, \qquad m = 0, \pm 1, \pm 2, \ldots .$$

Example. Some common complex exponentials are

$$e^{i\pi} = -1, \quad e^{i\pi/2} = i, \quad e^{i\pi/3} = \frac{1 + \sqrt{3}i}{2}, \quad e^{i\pi/4} = \frac{1 + i}{\sqrt{2}}.$$

Also note that

$$e^{2\pi mi} = 1, \quad m = 0, \pm 1, \pm 2, \ldots .$$

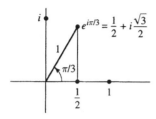
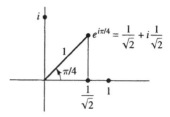

Several useful identities satisfied by the exponential function are

(2.3)
$$|e^{i\theta}| = 1,$$

(2.4)
$$\overline{e^{i\theta}} = e^{-i\theta},$$

(2.5)
$$1/e^{i\theta} = e^{-i\theta}.$$

The identity (2.3) is equivalent to the trigonometric identity $\cos^2\theta + \sin^2\theta = 1$, while (2.4) follows from $\cos(-\theta) = \cos\theta$ and $\sin(-\theta) = -\sin\theta$.

One of the most important properties of the exponential function is the **addition formula**

(2.6)
$$e^{i(\theta + \varphi)} = e^{i\theta}e^{i\varphi}, \quad -\infty < \theta, \varphi < \infty.$$

In view of the definition (2.2), this is equivalent to

$$\cos(\theta + \varphi) + i\sin(\theta + \varphi) = (\cos\theta + i\sin\theta)(\cos\varphi + i\sin\varphi).$$

Multiplying out the right-hand side and equating real and imaginary parts, we obtain the equivalent pair of identities

(2.7)
$$\begin{cases} \cos(\theta + \varphi) = \cos\theta\cos\varphi - \sin\theta\sin\varphi, \\ \sin(\theta + \varphi) = \cos\theta\sin\varphi + \sin\theta\cos\varphi, \end{cases}$$

which are the addition formulae for sine and cosine. Thus the addition formula (2.6) for the complex exponential is a compact form of the addition formulae (2.7) for the sine and cosine functions, and it is much easier to remember!

The properties (2.4), (2.5), (2.6) of the exponential function correspond respectively to the following properties of the argument function:

(2.8)
$$\arg\bar{z} = -\arg z,$$

(2.9)
$$\arg(1/z) = -\arg z,$$

(2.10)
$$\arg(z_1 z_2) = \arg z_1 + \arg z_2,$$

where each formula is understood to hold modulo adding integral multiples of 2π. To establish (2.8) and (2.9), note that if the polar representation of z is $re^{i\theta}$, then the polar representation of $\bar{z}$ is $re^{-i\theta}$, and that of $1/z$ is $(1/r)e^{-i\theta}$. For (2.10), write $z_1 = r_1 e^{i\varphi_1}$, $z_2 = r_2 e^{i\theta_2}$, and use the addition formula to obtain the polar form of $z_1 z_2$,

$$z_1 z_2 \;=\; r_1 r_2 e^{i\theta_1} e^{i\theta_2} \;=\; r_1 r_2 e^{i(\theta_1 + \theta_2)}.$$

The addition formula (2.6) can be used to derive formulae for $\cos(n\theta)$ and $\sin(n\theta)$ in terms of $\cos\theta$ and $\sin\theta$. Write

$$\cos(n\theta) + i\sin(n\theta) \;=\; e^{in\theta} \;=\; (e^{i\theta})^n \;=\; (\cos\theta + i\sin\theta)^n,$$

expand the right-hand side, and equate real and imaginary parts. This yields expressions for $\cos(n\theta)$ and $\sin(n\theta)$ that are polynomials in $\cos\theta$ and $\sin\theta$. These identities are known as **de Moivre's formulae**. For instance, by equating $\cos(3\theta) + i\sin(3\theta)$ to

$$(\cos\theta + i\sin\theta)^3 \;=\; \cos^3\theta - 3\cos\theta\sin\theta + i(3\cos^2\theta\sin\theta - \sin^3\theta)$$

and taking real and imaginary parts, we obtain

$$\cos(3\theta) \;=\; \mathrm{Re}(\cos\theta + i\sin\theta)^3 \;=\; \cos^3\theta - 3\cos\theta\sin^2\theta,$$
$$\sin(3\theta) \;=\; \mathrm{Im}(\cos\theta + i\sin\theta)^3 \;=\; 3\cos^2\theta\sin\theta - \sin^3\theta.$$

A complex number z is an **nth root** of w if $z^n = w$. Thus the nth roots of w are precisely the zeros of the polynomial $z^n - w$ of degree n. Since this polynomial has degree n, w has at most n nth roots. If $w \neq 0$, then w has exactly n nth roots, and these are determined as follows. First express w in polar form,

$$w \;=\; \rho e^{i\varphi}.$$

The equation $z^n = w$ becomes

$$r^n e^{in\theta} \;=\; \rho e^{i\varphi}.$$

Thus $r^n = \rho$ and $n\theta = \varphi + 2\pi k$ for some integer k. This leads to the explicit solutions

$$r \;=\; \rho^{1/n},$$
$$\theta \;=\; \frac{\varphi}{n} + \frac{2\pi k}{n}, \qquad k = 0, 1, 2, \ldots, n-1,$$

where we take the usual positive root of ρ. Since these n roots are distinct, and there are at most n nth roots, this list includes all the nth roots of w. Other values of k do not give different roots, since any other integer k leads to a value of θ that is obtained from the above list by adding an integral multiple of 2π. Graphically, the roots are distributed in equal arcs on the circle centered at 0 of radius $|w|^{1/n}$.

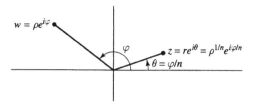

Example. To find and plot the square roots of $4i$, first express $4i$ in polar form $\rho e^{i\varphi}$. Here $\rho = |4i| = 4$ and $\varphi = \arg(4i) = \pi/2$. One root is given by $\sqrt{\rho}\, e^{i\varphi/2} = 2e^{i\pi/4}$. The other is $2e^{i(\pi/4+\pi)} = -2e^{i\pi/4}$. In Cartesian form, the roots are $\sqrt{4i} = \pm(\sqrt{2} + \sqrt{2}i)$.

Example. To find and plot the cube roots of $1 + i$, express $1 + i$ in polar form as $\sqrt{2}\, e^{i\pi/4}$. The polar form of the three cube roots is given by

$$2^{1/6}e^{i(\pi/12+2k\pi/3)}, \qquad k = 0, 1, 2.$$

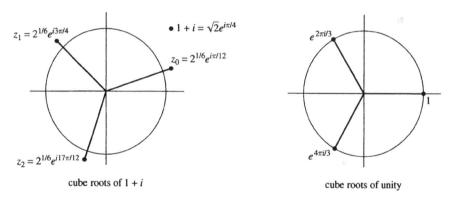

cube roots of $1 + i$ cube roots of unity

The nth roots of 1 are also called the **nth roots of unity**. They are given explicitly by

$$\omega_k = e^{2\pi i k/n}, \qquad 0 \le k \le n - 1.$$

Graphically, they are situated at equal intervals around the unit circle in the complex plane. Thus the two square roots of unity are $e^0 = 1$ and $e^{i\pi} = -1$.

The procedure for finding the nth roots of $w \ne 0$ can be rephrased in terms of the nth roots of unity. We express $w = \rho e^{i\varphi/n}$ in polar form as above. One root is given by $z_0 = \rho^{1/n}e^{i\varphi/n}$. The others are found by multiplying z_0 by the nth roots of unity:

$$z_k = z_0\omega_k = \rho^{1/n}e^{i\varphi/n}e^{2\pi i k/n}, \qquad 0 \le k \le n - 1.$$

Exercises for I.2

1. Express all values of the following expressions in both polar and cartesian coordinates, and plot them.

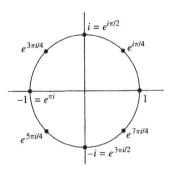

The eight eighth roots of unity

(a) $\sqrt{i}$ (c) $\sqrt[4]{-1}$ (e) $(-8)^{1/3}$ (g) $(1+i)^8$

(b) $\sqrt{i-1}$ (d) $\sqrt[4]{i}$ (f) $(3-4i)^{1/8}$ (h) $\left(\dfrac{1+i}{\sqrt{2}}\right)^{25}$

2. Sketch the following sets:
   (a) $|\arg z| < \pi/4$ (c) $|z| = \arg z$
   (b) $0 < \arg(z-1-i) < \pi/3$ (d) $\log|z| = -2\arg z$

3. For a fixed complex number b, sketch the curve $\{e^{i\theta} + be^{-i\theta} : 0 \le \theta \le 2\pi\}$. Differentiate between the cases $|b| < 1$, $|b| = 1$ and $|b| > 1$. *Hint.* First consider the case $b > 0$, and then reduce the general case to this case by a rotation.

4. For which n is i an nth root of unity?

5. For $n \ge 1$, show that
   (a) $1 + z + z^2 + \cdots + z^n = (1 - z^{n+1})/(1-z)$, $z \ne 1$,
   (b) $1 + \cos\theta + \cos 2\theta + \cdots + \cos n\theta = \dfrac{1}{2} + \dfrac{\sin\left(n+\frac{1}{2}\right)\theta}{2\sin\theta/2}$.

6. Fix $n \ge 1$. Show that the nth roots of unity $\omega_0, \ldots, \omega_{n-1}$ satisfy:
   (a) $(z - \omega_0)(z - \omega_1)\cdots(z - \omega_{n-1}) = z^n - 1$,
   (b) $\omega_0 + \cdots + \omega_{n-1} = 0$ if $n \ge 2$,
   (c) $\omega_0 \cdots \omega_{n-1} = (-1)^{n-1}$,
   (d) $\displaystyle\sum_{j=0}^{n-1} \omega_j^k = \begin{cases} 0, & 1 \le k \le n-1, \\ n, & k = n. \end{cases}$

7. Fix $R > 1$ and $n \ge 1$, $m \ge 0$. Show that

$$\left|\frac{z^m}{z^n + 1}\right| \le \frac{R^m}{R^n - 1}, \quad |z| = R.$$

Sketch the set where equality holds. *Hint.* See (1.1).

8. Show that $\cos 2\theta = \cos^2 \theta - \sin^2 \theta$ and $\sin 2\theta = 2\cos\theta\sin\theta$ using de Moivre's formulae. Find formulae for $\cos 4\theta$ and $\sin 4\theta$ in terms of $\cos\theta$ and $\sin\theta$.

3. Stereographic Projection

The **extended complex plane** is the complex plane together with the point at infinity. We denote the extended complex plane by $\mathbb{C}^*$, so that $\mathbb{C}^* = \mathbb{C}\cup\{\infty\}$. One way to visualize the extended complex plane is through stereographic projection. This is a function, or map, from the unit sphere in three-dimensional Euclidean space $\mathbb{R}^3$ to the extended complex plane, which is defined as follows. If $P = (X, Y, Z)$ is any point of the unit sphere other than the north pole $N = (0, 0, 1)$, we draw a straight line through N and P, and we define the **stereographic projection** of P to be the point $z = x + iy \sim (x, y, 0)$ where the straight line meets the coordinate plane $Z = 0$. The stereographic projection of the north pole N is defined to be ∞, the point at infinity.

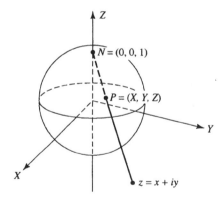

An explicit formula for the stereographic projection is derived as follows. We represent the line through P and N parametrically by $N + t(P - N)$, $-\infty < t < \infty$. The line meets the (x, y)-plane at a point $(x, y, 0)$ that satisfies

$$
\begin{aligned}
(x, y, 0) &= (0, 0, 1) + t[(X, Y, Z) - (0, 0, 1)] \\
&= (tX, tY, 1 + t(Z - 1))
\end{aligned}
$$

for some parameter value t. Equating the third components, we obtain $0 = 1 + t(Z - 1)$, which allows us to solve for the parameter value t,

$$
t = 1/(1 - Z).
$$

Equating the first two components and substituting this parameter value, we obtain equations for x and y in terms of X, Y, and Z,

$$\begin{cases} x &= tX &= X/(1-Z), \\ y &= tY &= Y/(1-Z). \end{cases}$$

To solve for X, Y, Z in terms of x and y, we use the defining equation $X^2 + Y^2 + Z^2 = 1$ of the sphere. Multiplying this equation by t^2 and substituting $tX = x$, $tY = y$, $tZ = t-1$, we obtain $x^2 + y^2 + t^2 - 2t + 1 = t^2$, which becomes

$$t = \frac{1}{2}(|z|^2 + 1).$$

This yields

$$\begin{cases} X &= 2x/(|z|^2 + 1), \\ Y &= 2y/(|z|^2 + 1), \\ Z &= 1 - 1/t &= (|z|^2 - 1)/(|z|^2 + 1). \end{cases}$$

The point (X, Y, Z) of the sphere is determined uniquely by the point $z = x + iy$ of the plane. Thus the stereographic projection provides a one-to-one correspondence between points P of the sphere, except the north pole N, and points $z = x + iy$ of the complex plane.

Lines of longitude on the sphere correspond to straight lines in the plane through 0, while lines of lattitude on the sphere correspond to circles centered at 0. As the radii of the circles tend to ∞, the lines of lattitude on the sphere tend to the north pole, so we are justified in making the north pole N correspond to the point at ∞.

Theorem. *Under the stereographic projection, circles on the sphere correspond to circles and straight lines in the plane.*

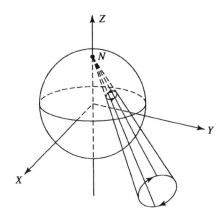

To see this, we will use the fact that the locus of points in the plane satisfying a quadratic equation of the form

$$(3.1) \qquad x^2 + y^2 + ax + by + c = 0$$

is either a circle, a point, or empty. This can be seen by completing the square and rewriting (3.1) as $(x + a/2)^2 + (y + b/2)^2 = (a^2 + b^2)/4 - c$. The three cases correspond respectively to whether $(a^2 + b^2)/4 - c$ is strictly positive, zero, or strictly negative.

We begin with a circle on the sphere, and we express it as the intersection of the sphere and a plane $AX + BY + CZ = D$. The stereographic projection of the circle then consists of points $z = x + iy$ that satisfy

$$A \frac{2x}{|z|^2 + 1} + B \frac{2y}{|z|^2 + 1} + C \frac{|z|^2 - 1}{|z|^2 + 1} = D.$$

We rewrite this as

$$(3.2) \qquad (C - D)(x^2 + y^2) + 2Ax + 2By - (C + D) = 0.$$

If $C = D$, the locus of (3.2) is a straight line. If $C \neq D$, then we divide by $C - D$, and the equation (3.2) has the form (3.1). Being the projection of a circle on the sphere, the locus cannot be a point or empty, so it must be a circle in the plane.

The argument is reversible. Every circle in the plane is the locus of solutions of an equation of the form

$$x^2 + y^2 + A'x + B'y + D' = 0.$$

Define A, B, C, D so that $2A = A'$, $2B = B'$, $C - D = 1$, $-(C + D) = D'$, and the corresponding set on the sphere is the intersection of the sphere with the plane $AX + BY + CZ = D$. The intersection cannot be empty or a point; hence it is a circle on the sphere. Similarly, every straight line in the plane is the locus of solutions of an equation of the form $A'x + B'y = D'$, which also determines a plane via $2A = A'$, $2B = B'$, $C = D = D'/2$, and this plane meets the sphere in a circle through the north pole.

Since straight lines in the plane correspond to circles on the sphere through the north pole, it is convenient to regard a straight line in the complex plane as a **circle through** ∞. With this convention the theorem asserts simply that stereographic projection maps circles on the sphere to circles in the extended complex plane.

Exercises for I.3

1. Sketch the image under the spherical projection of the following sets on the sphere: (a) the lower hemisphere $Z \leq 0$, (b) the polar cap $\frac{3}{4} \leq Z \leq 1$, (c) lines of lattitude $X = \sqrt{1 - Z^2} \cos \theta$, $Y = \sqrt{1 - Z^2} \sin \theta$, for Z fixed and $0 \leq \theta \leq 2\pi$, (d) lines of longitude $X = \sqrt{1 - Z^2} \cos \theta$, $Y = \sqrt{1 - Z^2} \sin \theta$, for θ fixed and $-1 \leq Z \leq 1$.

(e) the spherical cap $A \le X \le 1$, with center lying on the equator, for fixed A. Separate into cases, according to various ranges of A.

2. If the point P on the sphere corresponds to z under the stereographic projection, show that the antipodal point $-P$ on the sphere corresponds to $-1/\bar{z}$.

3. Show that as z traverses a small circle in the complex plane in the positive (counterclockwise) direction, the corresponding point P on the sphere traverses a small circle in the negative (clockwise) direction with respect to someone standing at the center of the circle and with body outside the sphere. (Thus the stereographic projection is orientation reversing, as a map from the sphere with orientation determined by the unit outer normal vector to the complex plane with the usual orientation.)

4. Show that a rotation of the sphere of $180°$ about the X-axis corresponds under stereographic projection to the inversion $z \mapsto 1/z$ of $\mathbb{C}$.

5. Suppose $(x, y, 0)$ is the spherical projection of (X, Y, Z). Show that the product of the distances from the north pole N to (X, Y, Z) and from N to $(x, y, 0)$ is 2. What is the situation when (X, Y, Z) lies on the equator of the sphere?

6. We define the **chordal distance** $d(z, w)$ between two points $z, w \in \mathbb{C}^*$ to be the length of the straight line segment joining the points P and Q on the unit sphere whose stereographic projections are z and w, respectively. (a) Show that the chordal distance is a metric, that is, it is symmetric, $d(z, w) = d(w, z)$; it satisfies the triangle inequality $d(z, w) \le d(z, \zeta) + d(\zeta, w)$; and $d(z, w) = 0$ if and only if $z = w$. (b) Show that the chordal distance from z to w is given by

$$d(z, w) \;=\; \frac{2|z - w|}{\sqrt{1 + |z|^2}\,\sqrt{1 + |w|^2}}\,, \qquad z, w \in \mathbb{C}.$$

(c) What is $d(z, \infty)$? *Remark.* The expression for $d(z, w)$ shows that infinitesimal arc length corresponding to the chordal metric is given by

$$d\sigma(z) \;=\; \frac{2\,ds}{1 + |z|^2}\,,$$

where $ds = |dz|$ is the usual Euclidean infinitesimal arc length. The infinitesimal arc length $d\sigma(z)$ determines another metric, the **spherical metric** $\sigma(z, w)$, on the extended complex plane. See Section IX.3.

7. Consider the sphere of radius $\frac{1}{2}$ in (X, Y, Z)-space, resting on the $(X, Y, 0)$-plane, with south pole at the origin $(0, 0, 0)$ and north pole at $(0, 0, 1)$. We define a stereographic projection of the sphere onto the complex plane as before, so that corresponding points (X, Y, Z) and $z \sim (x, y, 0)$ lie on the same line through the north pole. Find the equations for $z = x + iy$ in terms of X, Y, Z, and the equations for X, Y, Z in terms of z. What is the corresponding formula for the chordal distance? *Note.* In this case, the equation of the sphere is $X^2 + Y^2 + \left(Z - \frac{1}{2}\right)^2 = \frac{1}{4}$.

4. The Square and Square Root Functions

Real-valued functions of a real variable can be visualized by graphing them in the plane $\mathbb{R}^2$. The graph of a complex-valued function $f(z)$ of a complex variable z requires four (real) dimensions. Thus some techniques other than graphing in $\mathbb{R}^4$ must be developed for visualizing and understanding functions of a complex variable. One technique is to graph the modulus of the function $|f(z)|$ as a surface in three-dimensional space $\mathbb{R}^3$. Another is to graph separately the real and imaginary parts of $f(z)$ in $\mathbb{R}^3$.

We describe a different technique for gaining insight into the behavior of the function $f(z)$. We create two planes, a z-plane for the domain space and a w-plane for the range space. We then view $f(z)$ as a mapping from the z-plane to the w-plane, and we analyze how various geometric configurations in the z-plane are mapped by $w = f(z)$ to the w-plane. Which geometric configurations in the z-plane to consider depends very much on the specific function $f(z)$. To illustrate how this method works, we consider the simplest nontrivial function, the square function $w = z^2$.

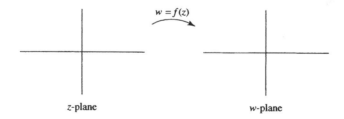

From the polar decomposition $w = z^2 = r^2 e^{2i\theta}$, we have

(4.1) $$|w| = |z|^2,$$

(4.2) $$\arg w = 2 \arg z.$$

Equation (4.1) shows that the circle $|z| = r_0$ in the z-plane is mapped to the circle $|w| = r_0^2$ in the w-plane. As $z = r_0 e^{i\theta}$ moves around the circle in the positive direction at constant angular velocity, the image $w = r_0^2 e^{2i\theta}$ moves around the image circle, in the same direction but at double the

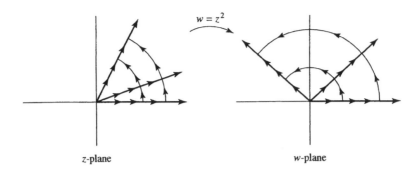

z-plane   w-plane

angular velocity. As z makes one complete loop, the image w makes two complete loops around the image circle.

Equation (4.2) shows that a ray $\{\arg z = \theta_0\}$ issuing from the origin in the z-plane is mapped to a ray in the w-plane of twice the angle. As z traverses the ray from the origin to ∞ at constant speed, the value w traverses the image ray from 0 to ∞, starting slowly and increasing its speed. The positive real axis in the z-plane, which is a ray with angle 0, is mapped to the positive real axis in the w-plane by the usual rule $x \mapsto x^2$. As z traverses the ray $\{\arg z = \pi/4\}$, the image w traverses the positive imaginary axis, and as z traverses the positive imaginary axis, the image w traverses the negative axis. As the rays in the z-plane sweep out the first quadrant, the image rays in the w-plane sweep out the upper half-plane, and as the rays in the z-plane sweep out the second quadrant, the image rays in the w-plane sweep out the lower half-plane. Eventually, we reach the ray along the negative real axis in the z-plane, which is mapped again to the positive real axis in the w-plane, and as we continue, the behavior is repeated in the lower half of the z-plane.

Now we turn to the problem of finding an inverse function for $w = z^2$. Every point $w \neq 0$ is hit by exactly two values of z, the two square roots $\pm\sqrt{w}$. In order to define an inverse function, we must restrict the domain in the z-plane so that values w are hit by only one z. There are many ways of doing this, and we proceed somewhat arbitrarily as follows.

Note that as rays sweep out the open right half of the z-plane, with the angle of the ray increasing from $-\pi/2$ to $\pi/2$, the image rays under $w = z^2$ sweep out the entire w-plane except for the negative axis, with the angle of the ray increasing from $-\pi$ to π. This leads us to draw a **slit**, or **branch cut**, in the w-plane along the negative axis from $-\infty$ to 0, and to define the inverse function on the **slit plane** $\mathbb{C}\backslash(-\infty, 0]$. Every value w in the slit plane is the image of exactly two z-values, one in the (open) right half-plane $\{\operatorname{Re} z > 0\}$, the other in the left half-plane $\{\operatorname{Re} z < 0\}$. Thus there are two possibilities for defining a (continuous) inverse function on the slit plane. We refer to each determination of the inverse function as a **branch** of the inverse. One branch $f_1(w)$ of the inverse function is defined by declaring that $f_1(w)$ is the value z such that $\operatorname{Re} z > 0$ and $z^2 = w$. Then $f_1(w)$ maps the slit plane $\mathbb{C}\backslash(-\infty, 0]$ onto the right half-plane $\{\operatorname{Re} z > 0\}$, and it forms an inverse for z^2 on that half-plane. To specify $f_1(w)$ explicitly, express $w = \rho e^{i\varphi}$ where φ lies in the range $-\pi < \varphi < \pi$, and then

$$f_1(w) = \sqrt{\rho}\, e^{i\varphi/2}, \qquad w = \rho e^{i\varphi}, \quad -\pi < \varphi < \pi.$$

The function $f_1(w)$ is called the **principal branch** of $\sqrt{w}$. It is expressed in terms of the principal branch of the argument function as

$$f_1(w) = |w|^{1/2} e^{i(\operatorname{Arg} w)/2}, \qquad w \in \mathbb{C}\backslash(-\infty, 0].$$

As w approaches a point $-r$ on the negative real axis $(-\infty, 0)$ from above, the values $f_1(w)$ approach the value $i\sqrt{r}$ on the positive imaginary axis. We express this by writing $f_1(-r + i0) = i\sqrt{r}$. Similarly, as w approaches $-r$ from below, the values $f_1(w)$ approach the value $-i\sqrt{r}$ on the negative imaginary axis, that is, $f_1(-r - i0) = -i\sqrt{r}$. The branch cut $(-\infty, 0]$ in the w-plane can be regarded as having two edges, and the function $f_1(w)$ extends continuously to each edge. The top edge, labeled "+" in the figure, is mapped to the positive imaginary axis by $f_1(w)$, and the bottom edge, labeled "−", is mapped to the negative imaginary axis by $f_1(w)$.

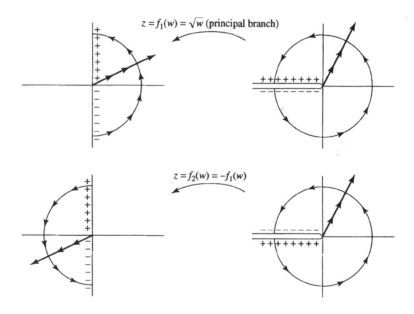

We use the other value of $\sqrt{w}$ to define a second branch $f_2(w)$ of the inverse function $\sqrt{w}$. For this we use a second copy of the w-plane, as in the figure. On this sheet the second branch of $\sqrt{w}$ is defined by $f_2(w) = -f_1(w)$. This branch maps the slit plane onto the left half-plane $\{\operatorname{Re} z < 0\}$. As w approaches a point $-r$ on the negative axis $(-\infty, 0)$ from above, the values $f_2(w)$ approach the value $-i\sqrt{r}$ on the negative imaginary axis, and as w approaches $-r$ from below, the values $f_2(w)$ approach the value $+i\sqrt{r}$ on the positive imaginary axis. Again we think of the slit as having two edges, though on this sheet the top edge is mapped to the negative imaginary axis and the bottom edge is mapped to the positive imaginary axis. Further, we have

$$f_1(-r+i0) = i\sqrt{r} = f_2(-r-i0), \quad f_1(-r-i0) = -i\sqrt{r} = f_2(-r+i0).$$

This leads us to the idea of constructing a surface to represent the inverse function by gluing together the edges where the functions $f_1(w)$ and $f_2(w)$ coincide. We glue the top edge of the branch cut on the sheet corresponding to $f_1(w)$ to the bottom edge of the branch cut on the sheet corresponding to $f_2(w)$, and similarly for the remaining two edges, to obtain a two-sheeted surface. Since the values of $f_1(w)$ and $f_2(w)$ coincide on the edges we have glued together, they determine a function $f(w)$ defined on the two-sheeted surface, with values in the z-plane that move continuously with w.

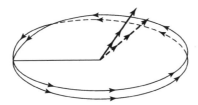

Since each sheet of the surface is a copy of the slit w-plane, we may think of the sheets as "lying over" the w-plane. Each $w \in \mathbb{C}\backslash\{0\}$ corresponds to exactly two points on the surface. The function $f(w)$ on the surface represents the multivalued function $\sqrt{w}$ in the sense that the values of $\sqrt{w}$ are precisely the values assumed by $f(w)$ at the points of the surface lying over w.

The surface we have constructed is called the **Riemann surface of** $\sqrt{w}$. The surface is essentially a sphere with two punctures corresponding to 0 and ∞. One way to see this is to note that the function $f(w)$ maps the surface one-to-one onto the z-plane punctured at 0. Another way to see this is to deform the surface by prying open each sheet at the slit, opening it to a hemisphere, and then joining the two hemispheres along the slit edges to form a sphere with two punctures corresponding to the endpoints 0 and ∞ of the slits.

Exercises for I.4

1. Sketch each curve in the z-plane, and sketch its image under $w = z^2$.
   (a) $|z - 1| = 1$ (c) $y = 1$ (e) $y^2 = x^2 - 1$, $x > 0$
   (b) $x = 1$ (d) $y = x + 1$ (f) $y = 1/x$, $x \neq 0$

2. Sketch the image of each curve in the preceding problem under the principal branch of $w = \sqrt{z}$, and also sketch, on the same grid but

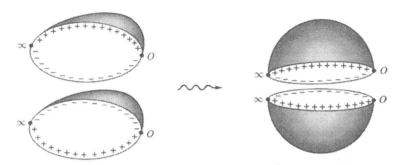

in a different color, the image of each curve under the other branch of $\sqrt{z}$.

3. (a) Give a brief description of the function $z \mapsto w = z^3$, considered as a mapping from the z-plane to the w-plane. (Describe what happens to w as z traverses a ray emanating from the origin, and as z traverses a circle centered at the origin.) (b) Make branch cuts and define explicitly three branches of the inverse mapping. (c) Describe the construction of the Riemann surface of $z^{1/3}$.

4. Describe how to construct the Riemann surfaces for the following functions: (a) $w = z^{1/4}$, (b) $w = \sqrt{z - i}$, (c) $w = (z-1)^{2/5}$. *Remark.* To describe the Riemann surface of a multivalued function, begin with one sheet for each branch of the function, make branch cuts so that the branches are defined continuously on each sheet, and identify each edge of a cut on one sheet to another appropriate edge so that the function values match up continuously.

5. The Exponential Function

We extend the definition of the exponential function to all complex numbers z by defining

$$e^z = e^x \cos y + i e^x \sin y, \qquad z = x + iy \in \mathbb{C}.$$

Since $e^{iy} = \cos y + i \sin y$, this is equivalent to

$$e^z = e^x e^{iy}, \qquad z = x + iy.$$

This identity is simply the polar representation of e^z,

(5.1) $$|e^z| = e^x,$$

(5.2) $$\arg e^z = y.$$

If z is real ($y = 0$), the definition of e^z agrees with the usual exponential function e^x. If z is imaginary ($x = 0$), the definition agrees with the definition of $e^{i\theta}$ given in Section 2.

A fundamental property of the exponential function is that it is periodic. The complex number λ is a **period** of the function $f(z)$ if $f(z + \lambda) = f(z)$ for all z for which $f(z)$ and $f(z + \lambda)$ are defined. The function $f(z)$ is **periodic** if it has a nonzero period. Since $\sin x$ and $\cos y$ are periodic functions with period 2π, the function e^z is periodic with period $2\pi i$:

$$e^{z+2\pi i} = e^z, \qquad z \in \mathbb{C}.$$

In fact, $2\pi i k$ is a period of e^z for any integer k.

Another fundamental property of the exponential function is the **addition formula**

(5.3) $$e^{z+w} = e^z e^w, \qquad z, w \in \mathbb{C}.$$

To check this, let $z = x + iy$ and $w = u + iv$. Then

$$e^{z+w} = e^{x+u} e^{i(y+v)} = e^x e^u e^{iy} e^{iv} = e^z e^w,$$

where we have used the addition formulae for e^x and $e^{i\theta}$.

From the addition formula (5.3) we have $e^z e^{-z} = e^0 = 1$. Consequently, the inverse of e^z is e^{-z},

$$1/e^z = e^{-z}, \qquad z \in \mathbb{C}.$$

To understand the exponential function better, we view $w = e^z$ as a mapping from the z-plane to the w-plane. If we restrict the exponential function to the real line $\mathbb{R}$, we obtain the usual exponential function $x \mapsto e^x$, $-\infty < x < \infty$, which maps the real line $\mathbb{R}$ to the positive real axis $(0, \infty)$. The equation (5.2) shows that an arbitrary horizontal line $x + iy_0$, $-\infty < x < \infty$, is mapped to the curve $e^x e^{iy_0}$, $-\infty < x < \infty$, which is a ray issuing from the origin at angle y_0. If we move the horizontal line up, the angle subtended by the ray increases, and the image ray is rotated in the positive (counterclockwise) direction. As we move the horizontal line upwards from the x-axis at $y_0 = 0$ to height $y_0 = \pi/2$, the image rays sweep out the first quadrant in the w-plane. The horizontal line at height $y_0 = \pi/2$ is mapped to the positive imaginary axis, the horizontal line of height $y_0 = \pi$ is mapped to the negative real axis, and when we reach the horizontal line of height $y_0 = 2\pi$, the image rays have swept out the full w-plane and returned to the positive real axis. The picture then repeats itself periodically. Each point in the w-plane, except $w = 0$, is hit infinitely often, by a sequence of z-values spaced at equal intervals of length 2π along a vertical line.

While the images of horizontal lines are rays issuing from the origin, the images of vertical lines are circles centered at the origin. The equation (5.1) shows that the image of the vertical line $x_0 + iy$, $-\infty < y < \infty$, is a circle in the w-plane of radius e^{x_0}. As z traverses the vertical line, the value w wraps infinitely often around the circle, completing one turn each time $y = \operatorname{Im} z$ increases by 2π.

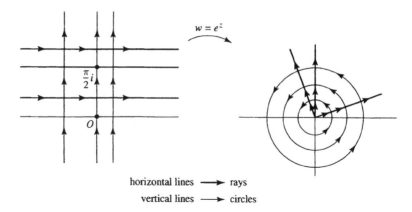

horizontal lines ⟶ rays
vertical lines ⟶ circles

Exercises for I.5

1. Calculate and plot e^z for the following points z:
   (a) 0 (c) $\pi(i-1)/3$ (e) $\pi i/m$, $m = 1, 2, 3, \ldots$
   (b) $\pi i + 1$ (d) $37\pi i$ (f) $m(i-1)$, $m = 1, 2, 3, \ldots$

2. Sketch each of the following figures and its image under the exponential map $w = e^z$. Indicate the images of horizontal and vertical lines in your sketch.
   (a) the vertical strip $0 < \operatorname{Re} z < 1$,
   (b) the horizontal strip $5\pi/3 < \operatorname{Im} z < 8\pi/3$,
   (c) the rectangle $0 < x < 1$, $0 < y < \pi/4$,
   (d) the disk $|z| \le \pi/2$,
   (e) the disk $|z| \le \pi$,
   (f) the disk $|z| \le 3\pi/2$.

3. Show that $e^{\bar{z}} = \overline{e^z}$.

4. Show that the only periods of e^z are the integral multiples of $2\pi i$, that is, if $e^{z+\lambda} = e^z$ for all z, then λ is an integer times $2\pi i$.

6. The Logarithm Function

For $z \ne 0$ we define $\log z$ to be the multivalued function

$$
\begin{aligned}
\log z &= \log |z| + i \arg z \\
&= \log |z| + i \operatorname{Arg} z + 2\pi i m, \qquad m = 0, \pm 1, \pm 2, \ldots .
\end{aligned}
$$

The values of $\log z$ are precisely the complex numbers w such that $e^w = z$. To see this, we plug in and compute. If $w = \log |z| + i \operatorname{Arg} z + 2\pi i m$, then

$$
e^w = e^{\log |z|} e^{i \operatorname{Arg} z} e^{2\pi i m} = |z| e^{i \operatorname{Arg} z} = z,
$$

where we have used the identities $e^{\log r} = r$ for $r > 0$ and $e^{2\pi i m} = 1$. On the other hand, suppose that $w = u + iv$ is an arbitrary complex number such that $e^w = z$. Then the polar representation of z is $z = re^{iv}$, where $r = |z| = e^u$. Thus $u = \log|z|$, and v is a value of $\arg z$, so that $v = \operatorname{Arg} z + 2\pi m$ for some integer m.

Recall that the principal value $\operatorname{Arg} z$ of $\arg z$ is the value θ satisfying $-\pi < \theta \le \pi$. We define the **principal value of** $\log z$ to be

$$(6.1) \qquad \operatorname{Log} z = \log|z| + i\operatorname{Arg} z, \qquad z \neq 0.$$

Thus $\operatorname{Log} z$ is a single-valued inverse for e^w, with values in the horizontal strip $-\pi < \operatorname{Im} w \le \pi$. Once we know the principal value of $\log z$, we obtain all values by

$$\log z = \operatorname{Log} z + 2\pi i m, \qquad m = 0, \pm 1, \pm 2, \dots .$$

Example. The values of $\log(1 + i)$ are given by

$$\begin{aligned} \log(1 + i) &= \log|1 + i| + i\arg(1 + i) \\ &= \log\sqrt{2} + i\pi/4 + 2\pi i m, \qquad m = 0, \pm 1, \pm 2, \dots . \end{aligned}$$

The principal value is

$$\operatorname{Log}(1 + i) = \log\sqrt{2} + i\pi/4.$$

The values form a vertical two-tailed sequence of equally spaced points.

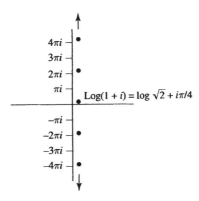

Now we regard $w = \operatorname{Log} z$ as a map from the slit z-plane $\mathbb{C}\backslash(-\infty, 0]$ to the w-plane. Since the exponential function maps horizontal lines to rays issuing from the origin, its inverse, the logarithm function, maps rays issuing from the origin to horizontal lines. In fact, formula (6.1) shows that the ray $\{\operatorname{Arg} z = \theta_0\}$ is mapped onto the horizontal line $\{\operatorname{Im} w = \theta_0\}$. As z traverses the ray from 0 to ∞, the image w traverses the entire horizontal line from left to right. As θ_0 increases between $-\pi$ and π, the rays sweep out the slit plane $\mathbb{C}\backslash(-\infty, 0]$, and the image lines fill out a horizontal strip

$\{-\pi < \operatorname{Im} w < \pi\}$ in the w-plane. Similarly, the formula (6.1) shows that the image of a punctured circle $\{|z| = r, -\pi < \arg z < \pi\}$ is the vertical interval $\{\operatorname{Re} w = \log|z|, -\pi < \operatorname{Im} z < \pi\}$, where the vertical line $\{\operatorname{Re} w = \log|z|\}$ meets the horizontal strip.

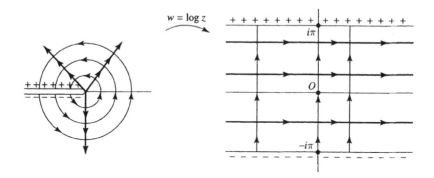

As with the inverse $\sqrt{z}$ of z^2, we can represent the multivalued function $\log z$ as a single-valued function on a Riemann surface spread over the z-plane, with one sheet for each branch of the function. The construction is as follows. This time we have infinitely many branches $f_m(z)$ of the logarithm function, defined for $z \neq 0$ and given explicitly by

$$f_m(z) = \operatorname{Log} z + 2\pi i m, \qquad -\infty < m < \infty.$$

For each branch, we take a copy of the complex plane and slit it along the negative real axis as before, to obtain a copy S_m of the slit plane $\mathbb{C}\backslash(-\infty, 0]$. We regard the function $f_m(z)$ as defined on the mth sheet S_m. Since the values of $f_m(z)$ at the top edge of the slit on S_m match the values of $f_{m+1}(z)$ at the bottom edge of the slit on S_{m+1}, we glue together these two edges. We do this for each m, and we obtain a surface resembling a spiral stairway leading infinitely far both up and down. The composite function $f(z)$ defined to be $f_m(z)$ on the mth sheet is then continuously defined on the surface. It represents the total function $\log z$, in the sense that the values of $\log z$ are precisely the values of $f(z)$ at the points of the surface that correspond to z.

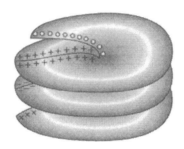

Exercises for I.6

1. Find and plot $\log z$ for the following complex numbers z. Specify the principal value. (a) 2, (b) i, (c) $1 + i$, (d) $(1 + i\sqrt{3})/2$.

2. Sketch the image under the map $w = \text{Log } z$ of each of the following figures.
   (a) the right half-plane $\text{Re } z > 0$,
   (b) the half-disk $|z| < 1$, $\text{Re } z > 0$,
   (c) the unit circle $|z| = 1$,
   (d) the slit annulus $\sqrt{e} < |z| < e^2$, $z \notin (-e^2, -\sqrt{e})$,
   (e) the horizontal line $y = e$,
   (f) the vertical line $x = e$.

3. Define explicitly a continuous branch of $\log z$ in the complex plane slit along the negative imaginary axis, $\mathbb{C}\backslash[0, -i\infty)$.

4. How would you make a branch cut to define a single-valued branch of the function $\log(z + i - 1)$? How about $\log(z - z_0)$?

7. Power Functions and Phase Factors

Let α be an arbitrary complex number. We define the power function z^α to be the multivalued function

$$z^\alpha = e^{\alpha \log z}, \qquad z \neq 0.$$

Thus the values of z^α are given by

$$z^\alpha = e^{\alpha[\log|z| + i \text{ Arg } z + 2\pi i m]}$$
$$= e^{\alpha \text{ Log } z} e^{2\pi i \alpha m}, \qquad m = 0, \pm 1, \pm 2, \ldots.$$

The various values of z^α are obtained from the principal value $e^{\alpha \text{ Log } z}$ by multiplying by the integral powers $(e^{2\pi i \alpha})^m$ of $e^{2\pi i \alpha}$. If α is itself an integer, then $e^{2\pi i \alpha} = 1$, and the function z^α is single-valued, the usual power function. If $\alpha = 1/n$ for some integer n, then the integral powers $e^{2\pi i m/n}$ of $e^{2\pi i/n}$ are exactly the nth roots of unity, and the values of $z^{1/n}$ are the n nth roots of z discussed earlier (Section 2).

Example. The values of i^i are given by

$$e^{i \log i} = e^{-\text{Arg } i - 2\pi m} = e^{-\pi/2} e^{-2\pi m}, \qquad m = 0, \pm 1, \pm 2, \ldots.$$

The values form a two-tailed sequence of positive real numbers, accumulating at 0 and at $+\infty$. Similarly, the values of i^{-i} are given by

$$e^{-i \log i} = e^{-\text{Arg}(-i) - 2\pi k} = e^{\pi/2} e^{-2\pi k}, \qquad k = 0, \pm 1, \pm 2, \ldots.$$

Danger! If we multiply the values of i^i by those of i^{-i}, we obtain infinitely many values $e^{2\pi n}$, $-\infty < n < \infty$. Thus

$$(i^i)(i^{-i}) \neq i^0 = 1,$$

and the usual algebraic rules do not apply to power functions when they are multivalued.

If α is not an integer, we cannot define z^α on the entire complex plane in such a way that the values move continuously with z. To define the function continuously, we must again make a branch cut. We could make the cut along the negative real axis, but this time let us make the cut along the positive real axis, from 0 to $+\infty$. We define a continuous branch of z^α on the slit plane $\mathbb{C}\backslash[0, \infty)$ explicitly by

$$w = r^\alpha e^{i\alpha\theta}, \qquad \text{for } z = re^{i\theta}, \ 0 < \theta < 2\pi.$$

At the top edge of the slit, corresponding to $\theta = 0$, we have the usual power function $r^\alpha = e^{\alpha \operatorname{Log} r}$. At the bottom edge of the slit, corresponding to $\theta = 2\pi$, we have the function $r^\alpha e^{2\pi i \alpha}$. If we fix r and let θ increase from 0 to 2π, $z = re^{i\theta}$ starts at the top edge of the slit and proceeds around a circle, ending at the bottom edge of the slit. As z describes this circle, the values $w = r^\alpha e^{i\theta\alpha}$ move continuously, starting from r^α at the top edge of the slit and ending at $r^\alpha e^{2\pi i \alpha}$ at the bottom edge. Thus the values of this branch of z^α on the bottom edge are $e^{2\pi i \alpha}$ times the values at the top edge. The multiplier $e^{2\pi i \alpha}$ is called the **phase factor** of z^α at $z = 0$.

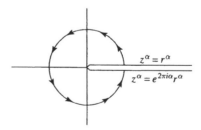

If we continue any other choice $w = r^\alpha e^{i\alpha(\theta+2\pi m)}$ of z^α around the same circle, the values of w move continuously from $r^\alpha e^{2\pi i \alpha m}$ at the top edge of the slit to $r^\alpha e^{i\alpha(2\pi+2\pi m)} = r^\alpha e^{2\pi i \alpha m} e^{2\pi i \alpha}$ at the bottom edge. Again the final w-value is the phase factor $e^{2\pi i \alpha}$ times the initial w-value.

The same analysis shows that the function $(z - z_0)^\alpha$ has a phase factor of $e^{2\pi i \alpha}$ at $z = z_0$, in the sense that if any branch of $w = (z - z_0)^\alpha$ is continued around a full circle centered at z_0 in the positive direction, the final w-value is $e^{2\pi i \alpha}$ times the initial w-value. This can be seen by making a change of variable $\zeta = z - z_0$. Further, this result does not change if we multiply $(z - z_0)^\alpha$ by any (single-valued) function. We state the result formally for emphasis.

Phase Change Lemma. *Let $g(z)$ be a (single-valued) function that is defined and continuous near z_0. For any continuously varying branch of $(z - z_0)^\alpha$ the function $f(z) = (z - z_0)^\alpha g(z)$ is multiplied by the phase factor $e^{2\pi i \alpha}$ when z traverses a complete circle about z_0 in the positive direction.*

Example. If α is an integer, the phase factor of z^α at 0 is $e^{2\pi i \alpha} = 1$, in accord with the fact that z^α is single-valued.

Example. The phase factor of $\sqrt{z - z_0}$ at z_0 is $e^{\pi i} = -1$. As z traverses a circle about z_0, the values of $f(z) = \sqrt{z - z_0}$ return to $-f(z)$. The phase factor of $1/\sqrt{z_0 - z} = i/\sqrt{z - z_0}$ at z_0 is also -1.

Example. The function $\sqrt{z(1 - z)}$ has two branch points, at 0 and at 1. At $z = 0$, each branch of $\sqrt{1 - z}$ is single-valued, so the phase factor of each branch of $\sqrt{z(1 - z)}$ at $z = 0$ is the same as that of $\sqrt{z}$, which is -1. Similarly, the phase factor of $\sqrt{z(1 - z)}$ at $z = 1$ is the same as that of $\sqrt{1 - z}$, which is -1. Now suppose we draw a branch cut from 0 to 1 and consider the branch $f(z)$ of $\sqrt{z(1 - z)}$ that is positive on the top edge of the slit. As z traverses a small circle around 0, the values of $f(z)$ return to $-f(z)$ on the bottom edge of the slit, corresponding to the phase factor -1 at $z = 0$. As z traverses the bottom edge of the slit and returns to the top edge around a small circle at $z = 1$, the values of $-f(z)$ are again multiplied by the phase factor -1. Thus the values of $f(z)$ return to the original positive value on the top edge of the slit when z traverses a dogbone path encircling both branch points. It follows that the branch $f(z)$ is a continuous single-valued function in the slit plane $\mathbb{C}\backslash[0,1]$. Now we may proceed, in analogy with $\sqrt{z}$ and $\log z$, to define a Riemann surface for the function $\sqrt{z(1 - z)}$ that captures both branches of the function. We require two sheets, since there are two choices of branches for the function $\sqrt{z(1 - z)}$. On each sheet we make the same cut, to form two copies of $\mathbb{C}\backslash[0,1]$. On one sheet we define $F(z)$ to be the branch $f(z)$ of $\sqrt{z(1 - z)}$ specified above, and on the other sheet we define $F(z)$ to be the other branch $-f(z)$ of $\sqrt{z(1 - z)}$. The sheets are then joined by identifying edges of the slits in such a way that $F(z)$ extends continuously to the surface. In this case, the top edge of the slit $[0, 1]$ on one sheet is identified to the bottom edge of the slit on the other sheet, and the remaining two edges are identified, to form the two-sheeted Riemann surface of $\sqrt{z(1 - z)}$.

In constructing the Riemann surface of a multivalued function, the number of sheets always coincides with the number of branches of the function. However, the branch cuts can be made in many ways, as long as there are

surface with closed path

enough branch cuts so that each branch of the function can be defined continuously in the slit plane. For instance, the branch cuts for the function $f(x) = \sqrt{z(1-z)}$ could as well be made from $-\infty$ to 0 along the negative real axis and from $+1$ to $+\infty$ along the positive real axis. The branch cuts could also be made along more complicated paths from 0 to 1.

Example. Consider $\sqrt{z - 1/z}$. We rewrite this as $\sqrt{z-1}\sqrt{z+1}/\sqrt{z}$. The function has three finite branch points, at 0 and ± 1. We must also consider ∞ as a branch point, since there is a phase change corresponding to a phase factor -1 as z traverses a very large circle centered at 0. Each branch point has phase factor -1, so any branch of the function returns to its original values when z traverse a path encircling two of the branch points. Thus it suffices to make two cuts, say $(-\infty, -1]$ and $[0, 1]$. Each branch of the function is continuous on $\mathbb{C} \backslash ((-\infty, -1] \cup [0, 1])$. Again top edges of slits on one sheet are identified to bottom edges of the others. The resulting surface is a torus (doughnut, or inner tube), with punctures corresponding to the branch points. What would happen if we were to make initially an additional branch cut along $[-1, 0]$, in addition to the two branch cuts above? The values of each branch at the top edge of the new cut would agree with the values of the same branch on the bottom edge. Consequently, we would identify the top and bottom edges of the slit $[-1, 0]$ on the same sheet, thereby effectively erasing the slits and arriving at the same doughnut surface.

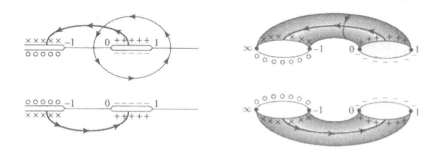

Exercises for I.7

1. Find all values and plot: (a) $(1+i)^i$, (b) $(-i)^{1+i}$, (c) $2^{-1/2}$, (d) $(1 + i\sqrt{3})^{(1-i)}$.

2. Compute and plot $\log\left[(1+i)^{2i}\right]$.

3. Sketch the image of the sector $\{0 < \arg z < \pi/6\}$ under the map $w = z^a$ for (a) $a = \frac{3}{2}$, (b) $a = i$, (c) $a = i + 2$. Use only the principal branch of z^a.

4. Show that $(zw)^a = z^a w^a$, where on the right we take all possible products.

5. Find i^{i^i}. Show that it does not coincide with $i^{i \cdot i} = i^{-1}$.

6. Determine the phase factors of the function $z^a(1 - z)^b$ at the branch points $z = 0$ and $z = 1$. What conditions on a and b guarantee that $z^a(1 - z)^b$ can be defined as a (continuous) single-valued function on $\mathbb{C}\backslash[0, 1]$?

7. Let $x_1 < x_2 < \cdots < x_n$ be n consecutive points on the real axis. Describe the Riemann surface of $\sqrt{(z - x_1) \cdots (z - x_n)}$. Show that for $n = 1$ and $n = 2$ the surface is topologically a sphere with certain punctures corresponding to the branch points and ∞. What is it when $n = 3$ or $n = 4$? Can you say anything for general n? (Any compact Riemann surface is topologically a sphere with handles. Thus a torus is topologically a sphere with one handle. For a given n, how many handles are there, and where do they come from?)

8. Show that $\sqrt{z^2 - 1/z}$ can be defined as a (single-valued) continuous function outside the unit disk, that is, for $|z| > 1$. Draw branch cuts so that the function can be defined continuously off the branch cuts. Describe the Riemann surface of the function.

9. Consider the branch of the function $\sqrt{z(z^3 - 1)(z + 1)^3}$ that is positive at $z = 2$. Draw branch cuts so that this branch of the function can be defined continuously off the branch cuts. Describe the Riemann surface of the function. To what value at $z = 2$ does this branch return if it is continued continuously once counterclockwise around the circle $\{|z| = 2\}$?

10. Consider the branch of the function $\sqrt{z(z^3 - 1)(z + 1)^3(z - 1)}$ that is positive at $z = 2$. Draw branch cuts so that this branch of the function can be defined continuously off the branch cuts. Describe the Riemann surface of the function. To what value at $z = 2$ does this branch return if it is continued continuously once counterclockwise around the circle $\{|z| = 2\}$?

11. Find the branch points of $\sqrt[3]{z^3 - 1}$ and describe the Riemann surface of the function.

8. Trigonometric and Hyperbolic Functions

If we solve the equations

$$
\begin{aligned}
e^{i\theta} &= \cos\theta + i\sin\theta, \\
e^{-i\theta} &= \cos\theta - i\sin\theta
\end{aligned}
$$

for $\cos\theta$ and $\sin\theta$, we obtain

$$
\cos\theta = \frac{e^{i\theta} + e^{-i\theta}}{2},
$$

$$
\sin\theta = \frac{e^{i\theta} - e^{-i\theta}}{2i}.
$$

This motivates us to extend the definition of $\cos z$ and $\sin z$ to complex numbers z by

$$
\cos z = \frac{e^{iz} + e^{-iz}}{2}, \qquad z \in \mathbb{C},
$$

$$
\sin z = \frac{e^{iz} - e^{-iz}}{2i}, \qquad z \in \mathbb{C}.
$$

This definition agrees with the usual definition when z is real. Evidently, $\cos z$ is an even function,

$$
\cos(-z) = \cos z, \qquad z \in \mathbb{C},
$$

while $\sin z$ is an odd function,

$$
\sin(-z) = -\sin z, \qquad z \in \mathbb{C}.
$$

As functions of a complex variable, $\cos z$ and $\sin z$ are periodic, with period 2π,

$$
\begin{aligned}
\cos(z + 2\pi) &= \cos z, \qquad z \in \mathbb{C}, \\
\sin(z + 2\pi) &= \sin z, \qquad z \in \mathbb{C}.
\end{aligned}
$$

After some algebraic manipulation, one checks (Exercise 1) that the addition formulae for $\cos z$ and $\sin z$ remain valid,

$$
\begin{aligned}
\cos(z + w) &= \cos z \cos w - \sin z \sin w, \qquad z, w \in \mathbb{C}, \\
\sin(z + w) &= \sin z \cos w + \cos z \sin w, \qquad z, w \in \mathbb{C}.
\end{aligned}
$$

If we substitute $w = -z$ in the addition formula for cosine, we obtain the familiar identity

$$
\cos^2 z + \sin^2 z = 1, \qquad z \in \mathbb{C}.
$$

We shall see, in fact, that any reasonable identity that holds for analytic functions of a real variable, such as $\cos x$ and $\sin x$, also holds when the functions are extended to be functions of a complex variable. This will be a

special case of the **principle of permanence of functional equations,**
proved in Chapter V.

The hyperbolic functions $\cosh x = (e^x + e^{-x})/2$ and $\sinh x = (e^x - e^{-x})/2$
are also extended to the complex plane in the obvious way, by

$$\cosh z = \frac{e^z + e^{-z}}{2}, \qquad z \in \mathbb{C},$$

$$\sinh z = \frac{e^z - e^{-z}}{2}, \qquad z \in \mathbb{C}.$$

Both $\cosh z$ and $\sinh z$ are periodic, with period $2\pi i$,

$$\cosh(z + 2\pi i) = \cosh z, \qquad z \in \mathbb{C},$$
$$\sinh(z + 2\pi i) = \sinh z, \qquad z \in \mathbb{C}.$$

Evidently, $\cosh z$ is an even function and $\sinh z$ is an odd function. There
are addition formulae for $\cosh z$ and $\sinh z$, derived easily from the addition
formulae for $\cos z$ and $\sin z$ (Exercise 1).

When viewed as functions of a complex variable, the trigonometric and
the hyperbolic functions exhibit a close relationship. They are obtained
from each other by rotating the domain space by $\pi/2$,

$$\cosh(iz) = \cos z, \qquad \cos(iz) = \cosh z,$$
$$\sinh(iz) = i \sin z, \qquad \sin(iz) = i \sinh z.$$

If we use these equations and the addition formula

$$\sin(x + iy) = \sin x \cos(iy) + \cos x \sin(iy),$$

we obtain the Cartesian representation for $\sin z$,

$$\sin z = \sin x \cosh y + i \cos x \sinh y, \qquad z = x + iy \in \mathbb{C}.$$

Thus

$$|\sin z|^2 = \sin^2 x \cosh^2 y + \cos^2 x \sinh^2 y.$$

Using $\cos^2 x + \sin^2 x = 1$ and $\cosh^2 y = 1 + \sinh^2 y$, we obtain

$$|\sin z|^2 = \sin^2 x + \sinh^2 y.$$

From this formula it is clear where the zeros of $\sin z$ are located; $\sin z = 0$
only when $\sin x = 0$ and $\sinh y = 0$, and this occurs only on the real axis
$y = 0$, at the usual zeros $0, \pm\pi, \pm 2\pi, \ldots$ of $\sin x$. Similarly, the only zeros
of $\cos z$ are the usual zeros of $\cos x$ on the real axis (Exercise 2).

Other trigonometric and hyperbolic functions are defined by the usual
formulae, such as

$$\tan z = \frac{\sin z}{\cos z}, \quad \tanh z = \frac{\sinh z}{\cosh z}, \qquad z \in \mathbb{C}.$$

Thus $\tan z$ and $\tanh z$ are odd functions, and $\tanh(iz) = i \tan z$.

The inverse trigonometric functions are multivalued functions, which can be expressed in terms of the logarithm function. Suppose $w = \sin^{-1} z$, that is,

$$\sin w = \frac{e^{iw} - e^{-iw}}{2i} = z.$$

Then $e^{2iw} - 2ize^{iw} - 1 = 0$. This is a quadratic equation in e^{iw}, which can be solved by the usual quadratic formula. The solutions are given by

$$e^{iw} = iz \pm \sqrt{1 - z^2}.$$

Taking logarithms we obtain

$$\sin^{-1} z = -i \log \left(iz \pm \sqrt{1 - z^2} \right).$$

This identity is to be understood as a set identity, in the sense that w satisfies $\sin w = z$ if and only if w is one of the values of $-i \log \left(iz \pm \sqrt{1 - z^2} \right)$. To obtain a genuine function, we must restrict the domain and specify the branch. One way to do this is to draw two branch cuts, from $-\infty$ to -1 and from $+1$ to $+\infty$ along the real axis, and to specify the branch of $\sqrt{1 - z^2}$ that is positive on the interval $(-1, 1)$. With this branch of $\sqrt{1 - z^2}$, we obtain a continuous branch $-i \operatorname{Log} \left(iz + \sqrt{1 - z^2} \right)$ of $\sin^{-1} z$.

Exercises for I.8

1. Establish the following addition formulae:
   (a) $\cos(z + w) = \cos z \cos w - \sin z \sin w$,
   (b) $\sin(z + w) = \sin z \cos w + \cos z \sin w$,
   (c) $\cosh(z + w) = \cosh z \cosh w + \sinh z \sinh w$,
   (d) $\sinh(z + w) = \sinh z \cosh w + \cosh z \sinh w$,

2. Show that $|\cos z|^2 = \cos^2 x + \sinh^2 y$, where $z = x + iy$. Find all zeros and periods of $\cos z$.

3. Find all zeros and periods of $\cosh z$ and $\sinh z$.

4. Show that

$$\tan^{-1} z = \frac{1}{2i} \log \left(\frac{1 + iz}{1 - iz} \right),$$

   where both sides of the identity are to be interpreted as subsets of the complex plane. In other words, show that $\tan w = z$ if and only if $2iw$ is one of the values of the logarithm featured on the right.

5. Let S denote the two slits along the imaginary axis in the complex plane, one running from i to $+i\infty$, the other from $-i$ to $-i\infty$. Show that $(1 + iz)/(1 - iz)$ lies on the negative real axis $(-\infty, 0]$ if and

only if $z \in S$. Show that the principal branch

$$\mathrm{Tan}^{-1}z \; = \; \frac{1}{2i}\,\mathrm{Log}\left(\frac{1+iz}{1-iz}\right)$$

maps the slit plane $\mathbb{C}\backslash S$ one-to-one onto the vertical strip $\{|\,\mathrm{Re}\,w| < \pi/2\}$.

6. Describe the Riemann surface for $\tan^{-1}z$.

7. Set $w = \cos z$ and $\zeta = e^{iz}$. Show that $\zeta = w \pm \sqrt{w^2 - 1}$. Show that

$$\cos^{-1}w \; = \; -i\log\left[w \pm \sqrt{w^2 - 1}\right],$$

where both sides of the identity are to be interpreted as subsets of the complex plane.

8. Show that the vertical strip $|\,\mathrm{Re}(w)| < \pi/2$ is mapped by the function $z(w) = \sin w$ one-to-one onto the complex z-plane with two slits $(-\infty, -1]$ and $[+1, +\infty)$ on the real axis. Show that the inverse function is the branch of $\sin^{-1}z = -i\,\mathrm{Log}\left(iz + \sqrt{1 - z^2}\right)$ obtained by taking the principal value of the square root. *Hint.* First show that the function $1 - z^2$ on the slit plane omits the negative real axis, so that the principal value of the square root is defined and continuous on the slit plane, with argument in the open interval between $-\pi/2$ and $\pi/2$.

II

Analytic Functions

In this chapter we take up the complex differential calculus. After reviewing some basic analysis in Section 1, we introduce complex derivatives and analytic functions in Section 2 and we show that the rules for complex differentiation are the same as the usual rules for differentiation. In Section 3 we characterize analytic functions in terms of the Cauchy-Riemann equations. In Sections 4 and 5 we give several applications of the Cauchy-Riemann equations, to inverses of analytic functions and to harmonic functions. In Section 6 we discuss conformality, which is a direct consequence of complex differentiability. We close in Section 7 with a discussion of fractional linear transformations, which form an important class of analytic functions.

1. Review of Basic Analysis

We begin by reviewing the background material in analysis that will (eventually) be called upon, and we say something about the language of formal mathematics. For the most part, we will not phrase our arguments completely formally, though any bilingual person will be able to translate easily to the language of formal mathematics in such a way that our development becomes completely rigorous.

Since the complex derivative is defined as a limit, we require some background material on limits and continuity. To be able to define and work with analytic functions, we also require some basic topological concepts, including open and closed sets, and domains. The confident reader may pass directly to the definitions of complex derivative and analytic function in the next section, and refer back to the material in this section only when needed.

We begin with the notion of a convergent sequence. For this we have two definitions.

Informal Definition. A sequence $\{s_n\}$ **converges to** s if the sequence eventually lies in any disk centered at s.

The language of formal mathematics serves to quantify this statement and make it precise. The "small disk" is traditionally given radius $\varepsilon > 0$. That "s_n lies in the disk" means that $|s_n - s| < \varepsilon$. That an event "eventually" occurs is translated to the statement that there is $N \geq 1$ such that the event occurs for $n \geq N$. Thus the translation of the definition of convergent sequence to the language of formal mathematics is as follows.

$$|s_n - s| < \varepsilon \text{ for } n \geq 4$$

Formal Definition. A sequence of complex numbers $\{s_n\}$ **converges to** s if for any $\varepsilon > 0$, there is an integer $N \geq 1$ such that $|s_n - s| < \varepsilon$ for all $n \geq N$.

If $\{s_n\}$ converges to s, we write $s_n \to s$, or $\lim s_n = s$. Some examples of convergent sequences that appear frequently are

(1.1)
$$\lim_{n \to \infty} \frac{1}{n^p} = 0, \qquad 0 < p < \infty,$$

(1.2)
$$\lim_{n \to \infty} |z|^n = 0, \qquad |z| < 1,$$

(1.3)
$$\lim_{n \to \infty} \sqrt[n]{n} = 1.$$

To prove (1.1) formally, we would for a given $\varepsilon > 0$ take N to be an integer satisfying $N > 1/\varepsilon^{1/p}$. Then for $n \geq N$ we have $n^p \geq N^p > 1/\varepsilon$, and $1/n^p < \varepsilon$. To prove (1.2) formally, we would take N to be an integer satisfying $N > (\log \varepsilon)/(\log |z|)$. To prove (1.3) formally, we resort to a trick. Let $t_n = \sqrt[n]{n} - 1$. We estimate t_n from the binomial expansion

$$n = (1 + t_n)^n = 1 + n t_n + \frac{n(n-1)}{2} t_n^2 + \cdots + t_n^n \geq \frac{n(n-1)}{2} t_n^2.$$

This yields $t_n^2 < 2/(n-1)$. Thus $|t_n| = |\sqrt[n]{n} - 1| < \varepsilon$ whenever $2/(n-1) < \varepsilon^2$, that is, for $n > 1 + 2/\varepsilon^2$. For the formal definition we can take N to be any integer satisfying $N > 1 + 2/\varepsilon^2$.

We give some definitions and state some theorems, without proofs, that we will be using.

A sequence of complex numbers $\{s_n\}$ is said to be **bounded** if there is some number $R > 0$ such that $|s_n| \leq R$ for all n. In other words, the sequence is bounded if it is contained in some disk.

Theorem. *A convergent sequence is bounded. Further, if $\{s_n\}$ and $\{t_n\}$ are sequences of complex numbers such that $s_n \to s$ and $t_n \to t$, then*
(a) $s_n + t_n \to s + t$,
(b) $s_n t_n \to st$,
(c) $s_n/t_n \to s/t$, *provided that $t \neq 0$.*

Thus the limit of a sum is the sum of the limits, the limit of a product is the product of the limits, and the limit of a quotient is the quotient of the limits, provided that the denominator is not 0.

Example. We can use these rules to evaluate the limit of a rational expression of the form

$$\lim_{n \to \infty} \frac{3n^2 + 2n - 1}{5n^2 - 4n + 8} = \frac{3}{5}.$$

As a preliminary trick, we divide numerator and denominator by the leading power, and rewrite the expression as

$$\frac{3 + (2/n) - 1/n^2}{5 - (4/n) + (8/n^2)}.$$

Since $1/n \to 0$ and $1/n^2 \to 0$, the sum and product statements show that the numerator converges to 3 and the denominator converges to 5. By the quotient statement, the quotient then converges to $\frac{3}{5}$.

The most useful criteria for convergence of sequences of real and complex numbers are gathered in the next several theorems. The first criterion is sometimes called the **in-between theorem**.

Theorem. *If $r_n \leq s_n \leq t_n$, and if $r_n \to L$ and $t_n \to L$, then $s_n \to L$.*

A sequence of real numbers $\{s_n\}$ is said to be **monotone increasing** if $s_{n+1} \geq s_n$ for all n, **monotone decreasing** if $s_{n+1} \leq s_n$ for all n, and **monotone** if it is either monotone increasing or decreasing. The following criterion is a version of the completeness axiom for the real numbers.

bounded monotone increasing sequence

Theorem. *A bounded monotone sequence of real numbers converges.*

A sequence $\{s_n\}$ of real numbers can behave rather wildly. It is still possible to assign an "upper limit" to $\{s_n\}$, denoted by $\limsup s_n$, which is the largest possible limit of a subsequence of $\{s_n\}$. Our working definition is that $\limsup s_n$ is the unique extended real number S, $-\infty \leq S \leq +\infty$, such that if $t > S$, then $s_n \geq t$ for only finitely many indices n, while if $t < S$, then $s_n > t$ for infinitely many indices n. It is easy to see that any such S is unique. The existence of such an S can be deduced from the preceding theorem. In fact, the existence is equivalent to the completeness axiom of the real numbers.

A "lower limit" of the sequence $\{s_n\}$, denoted by $\liminf s_n$, is defined similarly. It satisfies

$$\liminf s_n = -\limsup(-s_n).$$

The sequence $\{s_n\}$ converges if and only if its lim sup and lim inf are finite and equal.

Example. The sequence $\{(-1)^n\}_{n=0}^{\infty} = \{+1, -1, +1, -1, \ldots\}$ does not converge. Its upper and lower limits are

$$\limsup_{n\to\infty} (-1)^n = +1, \qquad \liminf_{n\to\infty} (-1)^n = -1.$$

For complex sequences, the following simple criterion is used very often.

Theorem. *A sequence $\{s_k\}$ of complex numbers converges if and only if the corresponding sequences of real and imaginary parts of the s_k's converge.*

We define a sequence of complex numbers $\{s_n\}$ to be a **Cauchy sequence** if the differences $s_n - s_m$ tend to 0 as n and m tend to ∞. In the language of formal mathematics, this means that for any $\varepsilon > 0$, there exists $N \geq 1$ such that $|s_n - s_m| < \varepsilon$ if $m, n \geq N$. The following theorem is an equivalent form of the completeness axiom. It is important because it provides a means of determining whether a sequence is convergent without producing explicitly the limit of the sequence.

Theorem. *A sequence of complex numbers converges if and only if it is a Cauchy sequence.*

We say that a complex-valued function $f(z)$ **has limit L as z tends to z_0** if the values $f(z)$ are near L whenever z is near z_0, $z \neq z_0$. The formal definition is that $f(z)$ has limit L as z tends to z_0 if for any $\varepsilon > 0$, there is $\delta > 0$ such that $|f(z) - L| < \varepsilon$ whenever z in the domain of $f(z)$ satisfies $0 < |z - z_0| < \delta$. In this case we write

$$\lim_{z \to z_0} f(z) = L,$$

or $f(z) \to L$ as $z \to z_0$. It is implicitly understood that there are points in the domain of $f(z)$ that are arbitrarily close to z_0 and different from z_0. The definition can be rephrased in terms of convergent sequences.

Lemma. *The complex-valued function $f(z)$ has limit L as $z \to z_0$ if and only if $f(z_n) \to L$ for any sequence $\{z_n\}$ in the domain of $f(z)$ such that $z_n \neq z_0$ and $z_n \to z_0$.*

From the theorem on limits, we obtain easily the following.

Theorem. *If a function has a limit at z_0, then the function is bounded near z_0. Further, if $f(z) \to L$ and $g(z) \to M$ as $z \to z_0$, then as $z \to z_0$ we have*

(a) $f(z) + g(z) \to L + M$,
(b) $f(z)g(z) \to LM$,
(c) $f(z)/g(z) \to L/M$, provided that $M \neq 0$.

We say that $f(z)$ is **continuous at** z_0 if $f(z) \to f(z_0)$ as $z \to z_0$. A **continuous function** is a function that is continuous at each point of its domain. The preceding theorem shows that sums and products of continuous functions are continuous, and so are quotients, provided that the denominator is not zero. Further, the composition of continuous functions is continuous.

Example. Any constant function is continuous. The coordinate function $f(z) = z$ is continuous. Thus any polynomial function $p(z) = a_n z^n + \cdots + a_1 z + a_0$ is continuous. Any rational function $p(z)/q(z)$ is continuous wherever the denominator $q(z)$ is not zero.

A useful strategy for showing that $f(z)$ is continuous at z_0 is to obtain an estimate of the form $|f(z) - f(z_0)| \leq C|z - z_0|$ for z near z_0. This guarantees that $|f(z) - f(z_0)| < \varepsilon$ whenever $|z - z_0| < \varepsilon/C$, so that we can take $\delta = \varepsilon/C$ in the formal definition of limit.

Example. The estimates

$$|\operatorname{Re}(z - z_0)| \leq |z - z_0|,$$
$$|\operatorname{Im}(z - z_0)| \leq |z - z_0|,$$
$$||z| - |z_0|| \leq |z - z_0|,$$

show respectively that the functions $\operatorname{Re}(z)$, $\operatorname{Im}(z)$, and $|z|$ are continuous.

A subset U of the complex plane is **open** if whenever $z \in U$, there is a disk centered at z that is contained in U.

open set: no boundary points   closed set: includes boundary

Any open disk $\{|z - z_0| < \rho\}$ is an open set. The closed disk $\{|z - z_0| \leq \rho\}$ is not an open set, since any open disk centered at a point on the boundary circle $\{|z - z_0| = \rho\}$ extends outside the closed disk.

In general, any set described by strict inequalities of continuous functions is open. For instance, the open upper half-plane is described by the strict inequality $\operatorname{Im}(z) > 0$, so that it is an open set. Other examples of open sets described by strict inequalities are the open sector $\{\theta_0 < \arg z < \theta_1\}$, the open horizontal strip $\{-1 < \operatorname{Im} z < +1\}$, the open annulus $\{r < |z| < s\}$, and the punctured disk $\{0 < |z| < r\}$.

A subset D of the complex plane is a **domain** if D is open and if any two points of D can be connected by a broken line segment in D. Open half-planes, open disks, open sectors, open annuli, and open punctured disks are all domains. An example of an open set that is not a domain is the union of the open upper and lower half-planes, $U = \mathbb{C}\backslash\mathbb{R}$. It is impossible to connect a point in the upper half-plane to a point in the lower half-plane by a broken line segment that does not cross the real line.

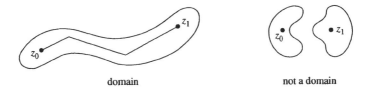

domain   not a domain

The most important property of domains for us is the following property, which actually characterizes domains (Exercise 18).

Theorem. *If $h(x,y)$ is a continuously differentiable function on a domain D such that $\nabla h = 0$ on D, then h is constant.*

This theorem is easy to justify. Since $\nabla h = 0$, the directional derivative of $h(x,y)$ in any direction is zero. Consequently, $h(x,y)$ is constant on any straight line segment contained in D, hence on any broken line segment. Since any two points of D can be joined by a broken line segment in D, $h(x,y)$ is constant on D.

A set is **convex** if whenever two points belong to the set, then the straight line segment joining the two points is contained in the set. An open or closed disk is convex, but a punctured disk is not convex.

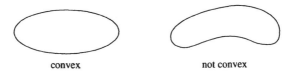

convex   not convex

A set is **star-shaped with respect to** z_0 if whenever a point belongs to the set, then the straight line segment joining z_0 to the point is contained in the set. In other words, a set is star-shaped with respect to z_0 if every point of the set is visible from z_0. Any convex set is star-shaped with respect to each of its points. The slit plane $\mathbb{C}\backslash(-\infty, 0]$ is star-shaped with respect to any point on the positive real axis. However, it is not convex, and it is not star-shaped with respect to any point not on the real axis.

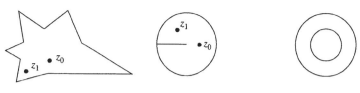

star-shaped with respect to z_0 but not z_1 not star-shaped

A **star-shaped domain** is a domain that is star-shaped with respect to one of its points. Thus $\mathbb{C}\backslash(-\infty,0]$ is a star-shaped domain. Any convex domain is a star-shaped domain. An open annulus is not star-shaped.

A subset E of the complex plane is **closed** if it contains the limit of every convergent sequence in E. The closed disk $\{|z - z_0| \leq r\}$ is a closed set, since if $|s_n - z_0| \leq r$ and $s_n \to s$, then $|s - z_0| \leq r$.

Sets of the form $\{f(z) \geq c\}$ or $\{f(z) \leq c\}$, where $f(z)$ is a continuous real-valued function, are closed. Thus for instance the closed upper half-plane, consisting of points z such that $\mathrm{Re}(z) \geq 0$, is a closed set.

The **boundary** of a set E consists of points z such that every disk centered at z contains both points in E and points not in E. Thus a set is closed if it contains its boundary, and a set is open if it does not include any of its boundary points. For example, the boundary of the closed disk $\{|z - z_0| \leq r\}$ is its boundary circle $\{|z - z_0| = r\}$, and the boundary of the open disk $\{|z - z_0| < r\}$ is also the boundary circle.

A subset of the complex plane that is closed and bounded is said to be **compact**. A closed disk $\{|z - z_0| \leq r\}$ is compact, as is a closed interval $[a, b]$ on the real line. We will use the following important property of compact sets in our discussion of the maximum principle for harmonic and analytic functions.

Theorem. *A continuous real-valued function on a compact set attains its maximum.*

Exercises for II.1

1. Establish the following:

   (a) $\displaystyle\lim_{n\to\infty} \frac{n}{n+1} = 1$

   (b) $\displaystyle\lim_{n\to\infty} \frac{n}{n^2+1} = 0$

   (c) $\displaystyle\lim_{n\to\infty} \frac{2n^p + 5n + 1}{n^p + 3n + 1} = 2, \qquad p > 1$

   (d) $\displaystyle\lim_{n\to\infty} \frac{z^n}{n!} = 0, \qquad z \in \mathbb{C}.$

2. For which values of z is the sequence $\{z^n\}_{n=1}^{\infty}$ bounded? For which values of z does the sequence converge to 0?

3. Show that $\{n^n z^n\}$ converges only for $z = 0$.

4. Show that $\displaystyle\lim_{N\to\infty} \frac{N!}{N^k(N-k)!} = 1, \qquad k \geq 0.$

5. Show that the sequence

$$b_n = 1 + \frac{1}{2} + \frac{1}{3} + \cdots + \frac{1}{n} - \log n, \qquad n \geq 1,$$

is decreasing, while the sequence $a_n = b_n - 1/n$ is increasing. Show that the sequences both converge to the same limit γ. Show that $\frac{1}{2} < \gamma < \frac{3}{5}$. *Remark.* The limit of the sequence is called **Euler's constant**. It is not known whether Euler's constant is a rational number or an irrational number.

6. For a complex number α, we define the binomial coefficient "α choose n" by

$$\binom{\alpha}{0} = 1, \qquad \binom{\alpha}{n} = \frac{\alpha(\alpha-1)\cdots(\alpha-n+1)}{n!}, \qquad n \geq 1.$$

Show the following.

   (a) The sequence $\binom{\alpha}{n}$ is bounded if and only if $\operatorname{Re}\alpha \geq -1$.

   (b) $\binom{\alpha}{n} \to 0$ if and only if $\operatorname{Re}\alpha > -1$.

   (c) If $\alpha \neq 0, 1, 2, \ldots$, then $\binom{\alpha}{n+1} \Big/ \binom{\alpha}{n} \to -1$.

   (d) If $\operatorname{Re}\alpha \leq -1$, $\alpha \neq -1$, then $\left|\binom{\alpha}{n+1}\right| > \left|\binom{\alpha}{n}\right|$ for all $n \geq 0$.

   (e) If $\operatorname{Re}\alpha > -1$ and α is not an integer, then $\left|\binom{\alpha}{n+1}\right| < \left|\binom{\alpha}{n}\right|$ for n large.

7. Define $x_0 = 0$, and define by induction $x_{n+1} = x_n^2 + \frac{1}{4}$ for $n \geq 0$. Show that $x_n \to \frac{1}{2}$. *Hint.* Show that the sequence is bounded and monotone, and that any limit satisfies $x = x^2 + \frac{1}{4}$.

8. Show that if $s_n \to s$, then $|s_n - s_{n-1}| \to 0$.

9. Plot each sequence and determine its $\liminf$ and $\limsup$.

   (a) $s_n = 1 + \dfrac{1}{n} + (-1)^n$ (c) $s_n = \sin(\pi n/4)$

   (b) $s_n = (-n)^n$ (d) $s_n = x^n$ ($x \in \mathbb{R}$ fixed)

10. At what points are the following functions continuous? Justify your answer. (a) z, (b) $z/|z|$, (c) $z^2/|z|$, (d) $z^2/|z|^3$.

11. At what points does the function $\operatorname{Arg} z$ have a limit? Where is $\operatorname{Arg} z$ continuous? Justify your answer.

12. Let $h(z)$ be the restriction of the function $\operatorname{Arg} z$ to the lower half-plane $\{\operatorname{Im} z < 0\}$. At what points does $h(z)$ have a limit? What is the limit?

13. For which complex values of α does the principal value of z^α have a limit as z tends to 0? Justify your answer.

14. Let $h(t)$ be a continuous complex-valued function on the unit interval $[0, 1]$, and consider

$$H(z) = \int_0^1 \frac{h(t)}{t - z} dt.$$

Where is $H(z)$ defined? Where is $H(z)$ continuous? Justify your answer. *Hint.* Use the fact that if $|f(t) - g(t)| < \varepsilon$ for $0 \le t \le 1$, then $\int_0^1 |f(t) - g(t)| dt < \varepsilon$.

15. Which of the following sets are open subsets of $\mathbb{C}$? Which are closed? Sketch the sets. (a) The punctured plane $\mathbb{C}\backslash\{0\}$, (b) the exterior of the open unit disk in the plane, $\{|z| \ge 1\}$, (c) the exterior of the closed unit disk in the plane, $\{|z| > 1\}$, (d) the plane with the open unit interval removed, $\mathbb{C}\backslash(0, 1)$, (e) the plane with the closed unit interval removed, $\mathbb{C}\backslash[0, 1]$, (f) the semidisk $\{|z| < 1, \, \text{Im}(z) \ge 0\}$, (g) the complex plane $\mathbb{C}$.

16. Show that the slit plane $\mathbb{C}\backslash(-\infty, 0]$ is star-shaped but not convex. Show that the slit plane $\mathbb{C}\backslash[-1, 1]$ is not star-shaped. Show that a punctured disk is not star-shaped.

17. Show that a set is convex if and only if it is star-shaped with respect to each of its points.

18. Show that the following are equivalent for an open subset U of the complex plane.
    (a) Any two points of U can be joined by a path consisting of straight line segments parallel to the coordinate axes.
    (b) Any continuously differentiable function $h(x, y)$ on U such that $\nabla h = 0$ is constant.
    (c) If V and W are disjoint open subsets of U such that $U = V \cup W$, then either $U = V$ or $U = W$. *Remark.* In the context of topological spaces, this latter property is taken as the definition of connectedness.

19. Give a proof of the fundamental theorem of algebra along the following lines. Show that if $p(z)$ is a nonconstant polynomial, then $|p(z)|$ attains its minimum at some point $z_0 \in \mathbb{C}$. Assume that the minimum is attained at $z_0 = 0$, and that $p(z) = 1 + az^m + \cdots$, where $m \ge 1$ and $a \ne 0$. Contradict the minimality by showing that $|P(\varepsilon e^{i\theta_0})| < 1$ for an appropriate choice of θ_0.

2. Analytic Functions

If the development in this section has a familiar ring, it should. The basic
definitions and rules for the complex derivative are exactly the same as
those for the usual derivative in elementary calculus. The only difference is
that multiplication and division are now performed with complex numbers
instead of real numbers.

A complex-valued function $f(z)$ is **differentiable** at z_0 if the difference
quotients

$$(2.1) \qquad \frac{f(z) - f(z_0)}{z - z_0}$$

have a limit as $z \to z_0$. The limit is denoted by $f'(z_0)$, or by $\frac{df}{dz}(z_0)$, and
we refer to it as the **complex derivative** of $f(z)$ at z_0. Thus

$$(2.2) \qquad \frac{df}{dz}(z_0) = f'(z_0) = \lim_{z \to z_0} \frac{f(z) - f(z_0)}{z - z_0}.$$

Example. A constant function $f(z) = c$ has derivative $f'(z_0) = 0$ at any
point z_0. In this case the difference quotients (2.1) are all zero, so that the
limit is also 0.

It is often useful to write the difference quotient (2.1) in the form

$$(2.3) \qquad \frac{f(z_0 + \Delta z) - f(z_0)}{\Delta z},$$

so that $z - z_0$ is replaced by Δz. The formula for the complex derivative
becomes

$$(2.4) \qquad f'(z_0) = \lim_{\Delta z \to 0} \frac{f(z_0 + \Delta z) - f(z_0)}{\Delta z}.$$

Occasionally we use z instead of z_0 in the expression (2.4).

Example. The power function $f(z) = z^m$ has derivative $f'(z) = mz^{m-1}$.
In this case the binomial expansion

$$(z + \Delta z)^m = z^m + mz^{m-1}\Delta z + \frac{m(m-1)}{2}z^{m-2}(\Delta z)^2 + \cdots + (\Delta z)^m$$

yields

$$\frac{f(z + \Delta z) - f(z)}{\Delta z} = mz^{m-1} + \frac{m(m-1)}{2}z^{m-2}\Delta z + \cdots + (\Delta z)^{m-1},$$

which has limit mz^{m-1} as $\Delta z \to 0$.

Example. The function $f(z) = \bar{z}$ is not differentiable at any point z. In
this case the difference quotient (2.3) becomes

$$\left(\overline{(z + \Delta z)} - \bar{z} \right)/\Delta z = \overline{\Delta z}/\Delta z.$$

If $\Delta z = \varepsilon$ is real, then this difference quotient is equal to 1, whereas if $\Delta z = i\varepsilon$ is imaginary, then the difference quotient is equal to -1. Thus the difference quotients do not have a limit as $\Delta z \to 0$.

The various properties of the complex derivative can be developed in exactly the same way as the properties of the usual derivative.

Theorem. *If $f(z)$ is differentiable at z_0, then $f(z)$ is continuous at z_0.*

This follows from the sum and product rules for limits. We write

$$f(z) \; = \; f(z_0) + \left(\frac{f(z) - f(z_0)}{z - z_0} \right)(z - z_0).$$

Since the difference quotient tends to $f'(z_0)$ as $z \to z_0$, and $z - z_0$ tends to 0 as $z \to z_0$, the product on the right tends to 0, and consequently, $f(z) \to f(z_0)$ as $z \to z_0$.

The complex derivative satisfies the usual rules for differentiating sums, products, and quotients. The rules are

(2.5) $\qquad (cf)'(z) \; = \; cf'(z),$

(2.6) $\qquad (f+g)'(z) \; = \; f'(z) + g'(z),$

(2.7) $\qquad (fg)'(z) \; = \; f(z)g'(z) + f'(z)g(z),$

(2.8) $\qquad (f/g)'(z) \; = \; \dfrac{g(z)f'(z) - f(z)g'(z)}{g(z)^2}, \qquad g(z) \neq 0.$

Here we are assuming that $f(z)$ and $g(z)$ are differentiable at z, and that c is any complex constant. The conclusion is that $cf(z)$, $f(z) + g(z)$, $f(z)g(z)$, and, provided that $g(z) \neq 0$, also $f(z)/g(z)$ are all differentiable at z and satisfy the rules (2.5) to (2.8) listed above. The proofs depend on the theorems for limits of sums, products, and quotients. For instance, to establish the product rule (2.7) we begin with the usual trick and rewrite the difference quotient $[(fg)(z + \Delta z) - (fg)(z)]/\Delta z$ for the product as

$$f(z + \Delta z)\frac{g(z + \Delta z) - g(z)}{\Delta z} \; + \; \frac{f(z + \Delta z) - f(z)}{\Delta z}g(z).$$

We now take a limit as $\Delta z \to 0$ and apply the rules for limits of sums and products, and we obtain (2.7).

The identities (2.5) and (2.6) express the fact that complex differentiation is a linear operation. Note that (2.5) is a consequence of the product rule (2.7) and the fact that the derivative of a constant function c is 0.

To establish the identity (2.8), it suffices to establish the simpler identity

(2.9) $\qquad (1/g)'(z) \; = \; -g'(z)/g(z)^2, \qquad g(z) \neq 0,$

and then to apply the product rule.

Example. Any polynomial

$$p(z) = a_n z^n + a_{n-1} z^{n-1} + \cdots + a_1 z + a_0$$

has a complex derivative, which is given by the usual formula

$$p'(z) = n a_n z^{n-1} + (n-1) a_{n-1} z^{n-2} + \cdots + a_1.$$

This follows from the linearity rules (2.5) and (2.6), since z^m is differentiable with derivative $m z^{m-1}$. Further, any rational function $p(z)/q(z)$ is differentiable at all points z except for the (finitely many) zeros of $q(z)$.

The chain rule is also valid for the complex derivative. We give a careful statement and proof.

Theorem (Chain Rule). *Suppose that $g(z)$ is differentiable at z_0, and suppose that $f(w)$ is differentiable at $w_0 = g(z_0)$. Then the composition $(f \circ g)(z) = f(g(z))$ is differentiable at z_0 and*

$$(2.10) \qquad (f \circ g)'(z_0) = f'(g(z_0)) g'(z_0).$$

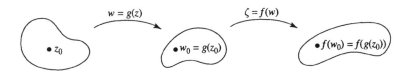

A useful mnemonic device for remembering the chain rule is

$$\frac{df}{dz} = \frac{df}{dw} \frac{dw}{dz},$$

where we have written $w = g(z)$. *Danger!* We regard f on the right-hand side as a function of w, and we regard f on the left-hand side as the function $f(g(z))$ of z. The mnemonic device can be justified by the proof, which involves multiplying and dividing by Δw. The proof goes as follows.

We consider two cases. For the first case, we assume that $g'(z_0) \neq 0$. Then $g(z) \neq g(z_0)$ for $0 < |z - z_0| < \varepsilon$, so we are justified in writing

$$(2.11) \qquad \frac{f(g(z)) - f(g(z_0))}{z - z_0} = \frac{f(g(z)) - f(g(z_0))}{g(z) - g(z_0)} \frac{g(z) - g(z_0)}{z - z_0}.$$

Since $g(z)$ is differentiable at z_0, it is continuous at z_0, that is, $g(z) \to g(z_0)$ as $z \to z_0$. Consequently,

$$\frac{f(g(z)) - f(g(z_0))}{g(z) - g(z_0)} \to f'(g(z_0))$$

as $z \to z_0$. Thus we can pass to the limit in (2.11), and we obtain (2.10).

For the second case, we assume that $g'(z_0) = 0$. Since $f(w)$ is differentiable at w_0, the difference quotients $(f(w) - f(w_0)/(w - w_0)$ are bounded near w_0, say

$$\left| \frac{f(w) - f(w_0)}{w - w_0} \right| \leq C$$

for some constant C and $0 < |w - w_0| < \varepsilon$. Hence $|f(g(z)) - f(g(z_0))| \leq C|g(z) - g(z_0)|$ for z near z_0, and consequently,

$$\left| \frac{f(g(z)) - f(g(z_0))}{z - z_0} \right| \leq C \left| \frac{g(z) - g(z_0)}{z - z_0} \right|.$$

Since the right-hand side tends to 0 as $z \to z_0$, we obtain $(f \circ g)'(z_0) = 0$. Thus both sides of (2.10) are 0, and in particular, the identity (2.10) holds.

Example. Suppose $f(w) = 1/w$, and $g(z) = z^2 - 1$. Then $f(g(z)) = 1/(z^2 - 1)$. Using $f'(w) = -1/w^2$ and $g'(z) = 2z$, we obtain from the chain rule

$$\frac{d}{dz} \frac{1}{z^2 - 1} = \left(-\frac{1}{w^2} \Big|_{w=z^2-1} \right)(2z) = -\frac{2z}{(z^2 - 1)^2}, \qquad z \neq \pm 1.$$

This is, of course, the same as the result we would have obtained by applying the quotient rule. More generally, the rule (2.9) follows from the chain rule and the formula for the derivative for $1/w$.

Now we turn to the definition of the class of functions that is the main subject of complex analysis. As usual, all our functions will be complex-valued functions defined on a subset of the complex plane.

Definition. A function $f(z)$ is **analytic on the open set** U if $f(z)$ is (complex) differentiable at each point of U and the complex derivative $f'(z)$ is continuous on U.

We have seen that any polynomial in z has a complex derivative at any point, and the complex derivative is a polynomial, hence continuous. Thus any polynomial in z is analytic on the entire complex plane. Rational functions are analytic wherever they are finite.

More generally, the rules established for complex derivatives show that sums and products of analytic functions are analytic. Quotients of analytic functions are analytic wherever the denominator does not vanish.

An example of a function that is not analytic is $f(z) = \overline{z}$, which does not have a complex derivative at any point.

The requirement that $f'(z)$ be continuous is a nuisance to verify. The student will be happy to learn that this condition is redundant. In Chapter IV we will prove Goursat's theorem, that if $f'(z)$ exists at each point of an open set U, then $f'(z)$ is automatically continuous on U. Meanwhile, the student who is willing to take this theorem on faith need not check the continuity of $f'(z)$, though in all cases we will treat, where $f'(z)$ can be shown to exist, it will also be apparent that $f'(z)$ is continuous.

Exercises for II.2

1. Find the derivatives of the following functions.
   (a) $z^2 - 1$ (c) $(z^2 - 1)^n$ (e) $1/(z^2 + 3)$ (g) $(az + b)/(cz + d)$
   (b) $z^n - 1$ (d) $1/(1 - z)$ (f) $z/(z^3 - 5)$ (h) $1/(cz + d)^2$

2. Show that
   $$1 + 2z + 3z^2 + \cdots + nz^{n-1} = \frac{1 - z^n}{(1 - z)^2} - \frac{nz^n}{1 - z}.$$

3. Show from the definition that the functions $x = \operatorname{Re} z$ and $y = \operatorname{Im} z$ are not complex differentiable at any point.

4. Suppose $f(z) = az^2 + bz\bar{z} + c\bar{z}^2$, where a, b, and c are fixed complex numbers. By differentiating $f(z)$ by hand, show that $f(z)$ is complex differentiable at z if and only if $bz + 2c\bar{z} = 0$. Where is $f(z)$ analytic?

5. Show that if f is analytic on D, then $g(z) = \overline{f(\bar{z})}$ is analytic on the reflected domain $D^* = \{\bar{z} : z \in D\}$, and $g'(z) = \overline{f'(\bar{z})}$.

6. Let $h(t)$ be a continuous complex-valued function on the unit interval $[0, 1]$, and define
   $$H(z) = \int_0^1 \frac{h(t)}{t - z}\, dt, \qquad z \in \mathbb{C}\backslash[0, 1].$$

   Show that $H(z)$ is analytic and compute its derivative. *Hint.* Differentiate by hand; that is, use the defining identity (2.4) of the complex derivative.

3. The Cauchy-Riemann Equations

Suppose $f = u + iv$ is analytic on a domain D. Fix a point $z \in D$. We will compute the complex derivative
$$f'(z) = \lim_{\Delta z \to 0} \frac{f(z + \Delta z) - f(z)}{\Delta z}$$

in two different ways, first by letting $z + \Delta z$ tend to z along the horizontal line through z (that is, $\Delta z = \Delta x$ real), then by letting $z + \Delta z$ tend to z along the vertical line through z (that is, $\Delta z = i\Delta y$ imaginary). This yields two expressions for $f'(z)$, which lead to the Cauchy-Riemann equations.

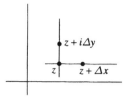

Expressing the difference quotient in terms of u and v and setting $\Delta z = \Delta x$, we obtain

$$\frac{f(z + \Delta x) - f(z)}{\Delta x} = \frac{u(x + \Delta x, y) + iv(x + \Delta x, y) - (u(x, y) + iv(x, y))}{\Delta x}$$

$$= \frac{u(x + \Delta x, y) - u(x, y)}{\Delta x} + i\frac{v(x + \Delta x, y) - v(x, y)}{\Delta x}.$$

Passing to the limit, we see that the x-derivatives of u and v exist, and

$$(3.1) \qquad f'(z) = \frac{\partial u}{\partial x}(x, y) + i\frac{\partial v}{\partial x}(x, y), \qquad z = x + iy.$$

Since $f'(z)$ is continuous, (3.1) shows that the x-derivatives of u and v are continuous.

Next we set $\Delta z = i\Delta y$, and we play the same game. The difference quotient becomes

$$\frac{f(z + i\Delta y) - f(z)}{i\Delta y} = \frac{u(x, y + \Delta y) + iv(x, y + \Delta y) - (u(x, y) + iv(x, y))}{i\Delta y}$$

$$= \frac{v(x, y + \Delta y) - v(x, y)}{\Delta y} - i\frac{u(x, y + \Delta y) - u(x, y)}{\Delta y}.$$

Passing to the limit as before, we see that the y-derivatives of u and v are continuous and satisfy

$$(3.2) \qquad f'(z) = \frac{\partial v}{\partial y}(x, y) - i\frac{\partial u}{\partial y}(x, y), \qquad z = x + iy.$$

Now we have two expressions, (3.1) and (3.2), for $f'(z)$. We equate their real and imaginary parts, and we obtain

$$(3.3) \qquad \frac{\partial u}{\partial x} = \frac{\partial v}{\partial y}, \qquad \frac{\partial u}{\partial y} = -\frac{\partial v}{\partial x}.$$

These equations are called the **Cauchy-Riemann equations** for u and v. We have proved half of the following theorem.

Theorem. Let $f = u + iv$ be defined on a domain D in the complex plane, where u and v are real-valued. Then $f(z)$ is analytic on D if and only if $u(x, y)$ and $v(x, y)$ have continuous first-order partial derivatives that satisfy the Cauchy-Riemann equations (3.3).

It remains to be shown that if the partial derivatives of u and v exist, are continuous, and satisfy the Cauchy-Riemann equations, then $f = u + iv$ is analytic. For this, we use Taylor's theorem. Fix $z \in D$. Taylor's theorem with remainder provides an approximation

$$u(x + \Delta x, y + \Delta y) = u(x, y) + \frac{\partial u}{\partial x}(x, y)\Delta x + \frac{\partial u}{\partial y}(x, y)\Delta y + R(\Delta x, \Delta y),$$

where $R(\Delta x, \Delta y)/|\Delta z|$ has limit 0 as Δz approaches 0. (The continuity of the first-order partial derivatives of u must be used to obtain the estimate for R.) Similarly,

$$v(x + \Delta x, y + \Delta y) = v(x, y) + \frac{\partial v}{\partial x}(x, y)\Delta x + \frac{\partial v}{\partial y}(x, y)\Delta y + S(\Delta x, \Delta y),$$

where $S(\Delta x, \Delta y)/|\Delta z| \to 0$ as $\Delta z \to 0$. Thus

$$f(z + \Delta z) = f(z) + \frac{\partial u}{\partial x}(x, y)\Delta x + \frac{\partial u}{\partial y}(x, y)\Delta y + R(\Delta z)$$
$$+ i\frac{\partial v}{\partial x}(x, y)\Delta x + i\frac{\partial v}{\partial y}(x, y)\Delta y + iS(\Delta z).$$

If we use the Cauchy-Riemann equations to replace the y-derivatives by x-derivatives, and we use $\Delta x + i\Delta y = \Delta z$, a minor miracle occurs. The identity becomes

$$f(z + \Delta z) = f(z) + \left(\frac{\partial u}{\partial x}(x, y) + i\frac{\partial v}{\partial x}(x, y)\right)\Delta z + R(\Delta z) + iS(\Delta z).$$

Thus

$$\frac{f(z + \Delta z) - f(z)}{\Delta z} = \frac{\partial u}{\partial x}(x, y) + i\frac{\partial v}{\partial x}(x, y) + \frac{R(\Delta z) + iS(\Delta z)}{\Delta z},$$

which tends to

$$\frac{\partial u}{\partial x}(x, y) + i\frac{\partial v}{\partial x}(x, y)$$

as Δz tends to 0. This shows that $f'(z)$ exists and is given by (3.1), so that $f'(z)$ is continuous, and thus $f(z)$ is analytic. Both directions of the theorem are now established.

Example. The functions $u(x, y) = x$ and $v(x, y) = y$, corresponding to $z = x + iy$, satisfy the Cauchy-Riemann equations, since

$$\frac{\partial u}{\partial x} = 1 = \frac{\partial v}{\partial y}, \qquad \frac{\partial u}{\partial y} = 0 = -\frac{\partial v}{\partial x}.$$

The functions $u(x, y) = x$ and $v(x, y) = -y$, corresponding to $\bar{z} = x - iy$, do not satisfy the Cauchy-Riemann equations, since

$$\frac{\partial u}{\partial x} = 1, \qquad \frac{\partial v}{\partial y} = -1.$$

We may use the Cauchy-Riemann equations to show that the function e^z is analytic and satisfies

$$\frac{d}{dz}e^z = e^z.$$

In this case, $u(x,y) = e^x \cos y$ and $v(x,y) = e^x \sin y$. We check that the Cauchy-Riemann equations hold:

$$\frac{\partial u}{\partial x} = \frac{\partial}{\partial x} e^x \cos y = e^x \cos y = \frac{\partial}{\partial y} e^x \sin y = \frac{\partial v}{\partial y},$$

$$\frac{\partial u}{\partial y} = \frac{\partial}{\partial y} e^x \cos y = -e^x \sin y = -\frac{\partial}{\partial x} e^x \sin y = -\frac{\partial v}{\partial x}.$$

Thus $f(z) = e^z$ is analytic, and (3.1) yields

$$f'(z) = \frac{\partial u}{\partial x} + i\frac{\partial v}{\partial x} = e^x \cos y + ie^x \sin y = e^z.$$

From the chain rule we deduce further that any complex exponential function of the form e^{az}, where a is a complex constant, is analytic and satisfies

$$\frac{d}{dz}e^{az} = ae^{az}.$$

Linear combinations, such as $\sin z$ and $\cos z$, of complex exponential functions are also analytic, and the usual formulae for the derivatives hold:

$$\frac{d}{dz}\sin z = \cos z,$$

$$\frac{d}{dz}\cos z = -\sin z,$$

$$\frac{d}{dz}\sinh z = \cosh z,$$

$$\frac{d}{dz}\cosh z = \sinh z.$$

To verify the formula for the derivative of $\sin z$, for instance, we compute

$$\frac{d}{dz}\sin z = \frac{d}{dz}\frac{e^{iz} - e^{-iz}}{2i} = \frac{ie^{iz} + ie^{-iz}}{2i} = \cos z.$$

Two important consequences of the Cauchy-Riemann equations and the equations for $f'(z)$ are as follows.

Theorem. If $f(z)$ is analytic on a domain D, and if $f'(z) = 0$ on D, then $f(z)$ is constant.

In this case, the equations (3.1) and (3.2) yield

$$\frac{\partial u}{\partial x} = \frac{\partial u}{\partial y} = \frac{\partial v}{\partial x} = \frac{\partial v}{\partial y} = 0.$$

Since D is a domain, the theorem in Section 1 shows that u and v are constant on D. Thus $f = u + iv$ is constant.

Theorem. *If $f(z)$ is analytic and real-valued on a domain D, then $f(z)$ is constant.*

In this case, $v = 0$ on D, and the Cauchy-Riemann equations become

$$\frac{\partial u}{\partial x} = 0, \qquad \frac{\partial u}{\partial y} = 0.$$

Since D is a domain, u is constant in D.

Exercises for II.3

1. Find the derivatives of the following functions.

   (a) $\tan z = \dfrac{\sin z}{\cos z}$ (b) $\tanh z = \dfrac{\sinh z}{\cosh z}$ (c) $\sec z = 1/\cos z$

2. Show that $u = \sin x \sinh y$ and $v = \cos x \cosh y$ satisfy the Cauchy-Riemann equations. Do you recognize the analytic function $f = u + iv$? (Determine its complex form.)

3. Show that if f and $\bar{f}$ are both analytic on a domain D, then f is constant.

4. Show that if f is analytic on a domain D, and if $|f|$ is constant, then f is constant. *Hint.* Write $\bar{f} = |f|^2/f$.

5. If $f = u + iv$ is analytic, then $|\nabla u| = |\nabla v| = |f'|$.

6. If $f = u + iv$ is analytic on D, then ∇v is obtained by rotating ∇u by 90°. In particular, ∇u and ∇v are orthogonal.

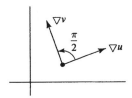

7. Sketch the vector fields ∇u and ∇v for the following functions $f = u + iv$. (a) iz, (b) z^2, (c) $1/z$.

8. Derive the polar form of the Cauchy-Riemann equations for u and v:

$$\frac{\partial u}{\partial r} = \frac{1}{r}\frac{\partial v}{\partial \theta}, \qquad \frac{\partial u}{\partial \theta} = -r\frac{\partial v}{\partial r}.$$

Check that for any integer m, the functions $u(re^{i\theta}) = r^m \cos(m\theta)$ and $v(re^{i\theta}) = r^m \sin(m\theta)$ satisfy the Cauchy-Riemann equations.

4. Inverse Mappings and the Jacobian

Let $f = u + iv$ be analytic on a domain D. We may regard D as a domain in the Euclidean plane $\mathbb{R}^2$ and f as a map from D to $\mathbb{R}^2$ with components $(u(x,y), v(x,y))$. The Jacobian matrix of this map is

$$J_f = \begin{pmatrix} \dfrac{\partial u}{\partial x} & \dfrac{\partial u}{\partial y} \\ \dfrac{\partial v}{\partial x} & \dfrac{\partial v}{\partial y} \end{pmatrix},$$

and the determinant of the Jacobian matrix is

$$\det J_f = \frac{\partial u}{\partial x}\frac{\partial v}{\partial y} - \frac{\partial u}{\partial y}\frac{\partial v}{\partial x}.$$

If we use the Cauchy-Riemann equations to replace the y-derivatives by x-derivatives, we obtain

$$\det J_f = \left(\frac{\partial u}{\partial x}\right)^2 + \left(\frac{\partial v}{\partial x}\right)^2 = \left|\frac{\partial u}{\partial x} + i\frac{\partial v}{\partial x}\right|^2.$$

By equation (3.2), this is equal to $|f'(z)|^2$. We have shown the following.

Theorem. *If $f(z)$ is analytic, then its Jacobian matrix J_f (as a map from $\mathbb{R}^2$ to $\mathbb{R}^2$) has determinant*

$$\det J_f(z) = |f'(z)|^2.$$

Now we can invoke the inverse function theorem from multivariable calculus, and this leads to the following.

Theorem. *Suppose $f(z)$ is analytic on a domain D, $z_0 \in D$, and $f'(z_0) \neq 0$. Then there is a (small) disk $U \subset D$ containing z_0 such that $f(z)$ is one-to-one on U, the image $V = f(U)$ of U is open, and the inverse function*

$$f^{-1} : V \longrightarrow U$$

is analytic and satisfies

(4.1) $$(f^{-1})'(f(z)) = 1/f'(z), \qquad z \in U.$$

All of the assertions of this theorem are consequences of the inverse function theorem, except for the assertions concerning the analyticity of f^{-1}. To check that f^{-1} is analytic, we write $g = f^{-1}$ on U and differentiate by hand. Fix $w, w_1 \in U$ with $w \neq w_1$, and set $z = g(w)$, $z_1 = g(w_1)$. Then $z \neq z_1$, $f(z) = w$, $f(z_1) = w_1$, and we have

$$\frac{g(w) - g(w_1)}{w - w_1} = \frac{z - z_1}{f(z) - f(z_1)} = 1 \bigg/ \left(\frac{f(z) - f(z_1)}{z - z_1} \right).$$

As w tends to w_1, z tends to z_1, and the right-hand side tends to $1/f'(z_1)$. Thus g is differentiable at w_1, and $g'(w_1) = 1/f'(z_1)$, which is the required identity (4.1) at z_1. Since $1/f'(z)$ is continuous, $(f^{-1})'$ is continuous, and thus f^{-1} is analytic.

If we write $w = g(z)$, the identity (4.1) becomes

$$\frac{dz}{dw} = \frac{1}{\dfrac{dw}{dz}},$$

which is the usual mnemonic device for remembering the derivative of the inverse function. The device is justified by the proof. *Danger!* Take care to evaluate the derivatives at the right points.

Once we know that f^{-1} is analytic, we can easily derive the formula (4.1) for the derivative from the chain rule. Since $f^{-1}(f(z)) = z$, the chain rule yields $(f^{-1})'(f(z))f'(z) = 1$, which is (4.1).

Example. The principal logarithm function $w = \operatorname{Log} z$ is a continuous inverse for $z = e^w$ for $-\pi < \arg w < \pi$. Since e^w is analytic and $(e^w)' \neq 0$, the preceding theorem applies, with z and w interchanged. From that theorem we conclude that $\operatorname{Log} z$ is analytic. If we use the chain rule to differentiate

$$z = e^{\operatorname{Log} z},$$

we obtain

$$1 = e^{\operatorname{Log} z} \frac{d}{dz}(\operatorname{Log} z) = z \frac{d}{dz}(\operatorname{Log} z).$$

Thus

(4.2)
$$\frac{d}{dz} \operatorname{Log} z = \frac{1}{z}.$$

Any other continuous branch of the logarithm differs from the principal branch by a constant, hence has the same derivative.

Example. Any continuous branch of $\sqrt{z}$ is analytic, and

(4.3)
$$\frac{d}{dz} \sqrt{z} = \frac{1}{2\sqrt{z}},$$

where we use the same branch of $\sqrt{z}$ on both sides of the identity. To see this, note first that no continuous branch of $\sqrt{z}$ can be defined on a domain containing 0, so that $z \neq 0$ in (4.3). Each branch $w = \sqrt{z}$ satisfies $w^2 = z$. Since $(w^2)' = 2w$ is not zero for $w \neq 0$, the continuous inverse branch $\sqrt{z}$ is analytic. Differentiating $w^2 = z$, we obtain

$$2w \frac{dw}{dz} = 1, \qquad \frac{dw}{dz} = \frac{1}{2w},$$

which is (4.3).

We will give in Section VIII.4 another proof of the existence and analyticity of the inverse of an analytic function, which does not depend on the inverse function theorem from calculus but rather on residue theory. That proof will provide an explicit integral representation formula for the inverse function.

Exercises for II.4

1. Sketch the gradient vector fields ∇u and ∇v for (a) $u + iv = e^z$, (b) $u + iv = \operatorname{Log} z$.

2. Let a be a complex number, $a \neq 0$, and let $f(z)$ be an analytic branch of z^a on $\mathbb{C}\backslash(-\infty, 0]$. Show that $f'(z) = af(z)/z$. (Thus $f'(z) = az^{a-1}$, where we pick the branch of z^{a-1} that corresponds to the original branch of z^a divided by z.)

3. Consider the branch of $f(z) = \sqrt{z(1-z)}$ on $\mathbb{C}\backslash[0,1]$ that has positive imaginary part at $z = 2$. What is $f'(z)$? Be sure to specify the branch of the expression for $f'(z)$.

4. Recall that the principal branch of the inverse tangent function was defined on the complex plane with two slits on the imaginary axis by

$$\operatorname{Tan}^{-1} z = \frac{1}{2i} \operatorname{Log}\left(\frac{1+iz}{1-iz}\right), \qquad z \notin (-i\infty, -i] \cup [i, i\infty).$$

Find the derivative of $\operatorname{Tan}^{-1} z$. Find the derivative of $\tan^{-1} z$ for any analytic branch of the function defined on a domain D.

5. Recall that $\cos^{-1}(z) = -i\log[z \pm \sqrt{z^2 - 1}]$. Suppose $g(z)$ is an analytic branch of $\cos^{-1}(z)$, defined on a domain D. Find $g'(z)$. Do different branches of $\cos^{-1}(z)$ have the same derivative?

6. Suppose $h(z)$ is an analytic branch of $\sin^{-1}(z)$, defined on a domain D. Find $h'(z)$. Do different branches of $\sin^{-1}(z)$ have the same derivative?

7. Let $f(z)$ be a bounded analytic function, defined on a bounded domain D in the complex plane, and suppose that $f(z)$ is one-to-

one. Show that the area of $f(D)$ is given by

$$\text{Area}\,(f(D)) \;=\; \iint_D |f'(z)|^2 dx\,dy.$$

8. Sketch the image of the circle $\{|z-1| \le 1\}$ under the map $w = z^2$.
    Compute the area of the image.

9. Compute

$$\iint_D |f'(z)|^2 dx\,dy,$$

for $f(z) = z^2$ and D the open unit disk $\{|z| < 1\}$. Interpret your
answer in terms of areas.

10. For smooth functions g and h defined on a bounded domain U, we
    define the **Dirichlet form** $D_U(g,h)$ by

$$D_U(g,h) \;=\; \iint_U \left[\frac{\partial g}{\partial x}\frac{\overline{\partial h}}{\partial x} + \frac{\partial g}{\partial y}\frac{\overline{\partial h}}{\partial y} \right] dx\,dy.$$

Show that if $z = f(\zeta)$ is a one-to-one analytic function from the
bounded domain V onto U, then

$$D_U(g,h) \;=\; D_V(g \circ f, h \circ f).$$

Remark. This shows that the Dirichlet form is a "conformal invariant."

5. Harmonic Functions

The equation

$$\frac{\partial^2 u}{\partial x_1^2} + \cdots + \frac{\partial^2 u}{\partial x_n^2} = 0$$

is called **Laplace's equation**. The operator

$$\Delta \;=\; \frac{\partial^2}{\partial x_1^2} + \cdots + \frac{\partial^2}{\partial x_n^2}$$

is called the **Laplacian**. In terms of this operator, Laplace's equation
becomes simply $\Delta u = 0$. Smooth functions $u(x_1, \ldots, x_n)$ that satisfy
Laplace's equation are called **harmonic functions**. Laplace's equation
is one of the most important partial differential equations of mathematical
physics. Some indication of the applications will be given in Chapter III.

We will be concerned with harmonic functions of two variables, that is,
solutions of

$$\Delta u = \frac{\partial^2 u}{\partial x^2} + \frac{\partial^2 u}{\partial y^2} = 0.$$

We say that a function $u(x, y)$ is **harmonic** if all its first- and second-order partial derivatives exist and are continuous and satisfy Laplace's equation. In the case of functions of two variables, there is an intimate connection between analytic functions and harmonic functions.

Theorem. If $f = u + iv$ is analytic, and the functions u and v have continuous second-order partial derivatives, then u and v are harmonic.

The second hypothesis of the theorem is redundant. We will show in Chapter IV that an analytic function has continuous partial derivatives of all orders.

The harmonicity of u and v is a simple consequence of the Cauchy-Riemann equations,

$$(5.1) \qquad \frac{\partial u}{\partial x} = \frac{\partial v}{\partial y},$$

$$(5.2) \qquad \frac{\partial u}{\partial y} = -\frac{\partial v}{\partial x}.$$

Using these, we obtain

$$\frac{\partial^2 u}{\partial x^2} = \frac{\partial}{\partial x}\frac{\partial v}{\partial y} = \frac{\partial}{\partial y}\frac{\partial v}{\partial x} = -\frac{\partial^2 u}{\partial y^2},$$

which shows that u is harmonic. The verification that v is harmonic is the same.

If u is harmonic on a domain D, and v is a harmonic function such that $u + iv$ is analytic, we say that v is a **harmonic conjugate** of u. The harmonic conjugate v is unique, up to adding a constant. Indeed, if v_0 is another harmonic conjugate for u, so that $u + iv_0$ is also analytic, then the difference $i(v - v_0)$ is analytic, and $v - v_0$ is a real-valued analytic function, hence constant on D.

Exercise. Show that $u(x, y) = xy$ is harmonic, and find a harmonic conjugate for u.
Solution. We have

$$\frac{\partial^2}{\partial x^2} xy = 0 = -\frac{\partial^2}{\partial y^2} xy,$$

so that xy is harmonic. To find a harmonic conjugate v, we solve the Cauchy-Riemann equations. From (5.1) we have

$$\frac{\partial u}{\partial x} = y = \frac{\partial v}{\partial y}.$$

Thus

$$v(x, y) = \frac{y^2}{2} + h(x),$$

where $h(x)$ depends only on x and not on y. Equation (5.2) becomes $x = -h'(x)$, which has solution $h(x) = -x^2/2 + C$. Thus

$$v(x,y) \;=\; \frac{y^2}{2} - \frac{x^2}{2} + C,$$

where C is a constant. The analytic function $f = u + iv$ is given by

$$f(z) \;=\; -i\frac{z^2}{2} + iC.$$

The method used above actually shows that any harmonic function on a rectangle with sides parallel to the axes has a harmonic conjugate on the rectangle. Indeed, let $u(x,y)$ be harmonic on such a rectangle D, and let (x_0, y_0) be any fixed point of D. If we integrate the first Cauchy-Riemann equation (5.1) along a vertical segment from y_0 to y, with x fixed, we obtain

$$v(x,y) \;=\; \int_{y_0}^{y} \frac{\partial u}{\partial x}(x,t)\,dt \;+\; h(x),$$

where $h(x)$ is the constant of integration with respect to y. Though $h(x)$ does not depend on y, it may depend on x. The second Cauchy-Riemann equation (5.2) then becomes

$$\frac{\partial u}{\partial y}(x,y) \;=\; -\frac{\partial}{\partial x}\int_{y_0}^{y} \frac{\partial u}{\partial x}(x,t)\,dt \;-\; h'(x).$$

If we differentiate under the integral sign (as we may) and use Laplace's equation, we obtain

$$\frac{\partial u}{\partial y}(x,y) \;=\; -\int_{y_0}^{y} \frac{\partial^2 u}{\partial x^2}(x,t)\,dt \;-\; h'(x) \;=\; \int_{y_0}^{y} \frac{\partial^2 u}{\partial y^2}(x,t)\,dt \;-\; h'(x)$$

$$=\; \frac{\partial u}{\partial y}(x,y) \;-\; \frac{\partial u}{\partial y}(x,y_0) \;-\; h'(x).$$

Thus we obtain

$$h'(x) \;=\; -\frac{\partial u}{\partial y}(x,y_0).$$

This has the solution

$$h(x) \;=\; -\int_{x_0}^{x} \frac{\partial u}{\partial y}(s,y_0)\,ds \;+\; C,$$

where C is a genuine constant. Thus we see that a harmonic conjugate $v(x,y)$ for $u(x,y)$ is given explicitly by

$$(5.3) \qquad v(x,y) \;=\; \int_{y_0}^{y} \frac{\partial u}{\partial x}(x,t)\,dt \;-\; \int_{x_0}^{x} \frac{\partial u}{\partial y}(s,y_0)\,ds \;+\; C.$$

The formula (5.3) is also valid if D is the entire complex plane, or if D is an open disk with center (x_0, y_0). Note that if we specify $v(x_0, y_0) = 0$, then $C = 0$, and the solution is unique.

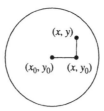

We summarize in the following theorem.

Theorem. *Let D be an open disk, or an open rectangle with sides parallel to the axes, and let $u(x, y)$ be a harmonic function on D. Then there is a harmonic function $v(x, y)$ on D such that $u + iv$ is analytic on D. The harmonic conjugate v is unique, up to adding a constant.*

We will see in Chapter III that this theorem holds in star-shaped domains. However, the theorem fails in annuli and in the punctured plane (Exercise 7). Roughly speaking, the theorem holds only in domains that have no "holes." Such domains are called "simply connected domains." They will be discussed in Chapter VIII.

Exercises for II.5

1. Show that the following functions are harmonic, and find harmonic conjugates:
   (a) $x^2 - y^2$ (c) $\sinh x \sin y$ (e) $\tan^{-1}(y/x)$, $x > 0$
   (b) $xy + 3x^2 y - y^3$ (d) $e^{x^2 - y^2} \cos(2xy)$ (f) $x/(x^2 + y^2)$

2. Show that if v is a harmonic conjugate for u, then $-u$ is a harmonic conjugate for v.

3. Define $u(z) = \operatorname{Im}(1/z^2)$ for $z \neq 0$, and set $u(0) = 0$.
   (a) Show that all partial derivatives of u with respect to x exist at all points of the plane $\mathbb{C}$, as do all partial derivative of u with respect to y.
   (b) Show that $\dfrac{\partial^2 u}{\partial x^2} + \dfrac{\partial^2 u}{\partial y^2} = 0$.
   (c) Show that u is *not* harmonic on $\mathbb{C}$.
   (d) Show that $\dfrac{\partial^2 u}{\partial x \partial y}$ does not exist at $(0,0)$.

4. Show that if $h(z)$ is a complex-valued harmonic function (solution of Laplace's equation) such that $zh(z)$ is also harmonic, then $h(z)$ is analytic.

5. Show that Laplace's equation in polar coordinates is

$$\frac{\partial^2 u}{\partial r^2} + \frac{1}{r} \frac{\partial u}{\partial r} + \frac{1}{r^2} \frac{\partial^2 u}{\partial \theta^2} = 0.$$

6. Show using Laplace's equation in polar coordinates that $\log|z|$ is harmonic on the punctured plane $\mathbb{C}\backslash\{0\}$.

7. Show that $\log|z|$ has no conjugate harmonic function on the punctured plane $\mathbb{C}\backslash\{0\}$, though it does have a conjugate harmonic function on the slit plane $\mathbb{C}\backslash(-\infty,0]$.

8. Show using Laplace's equation in polar coordinates that $u(re^{i\theta}) = \theta\log r$ is harmonic. Use the polar form of the Cauchy-Riemann equations (Exercise 3.8) to find a harmonic conjugate v for u. What is the analytic function $u + iv$?

6. Conformal Mappings

Let $\gamma(t) = x(t) + iy(t)$, $0 \leq t \leq 1$, be a smooth parameterized curve terminating at $z_0 = \gamma(0)$. We refer to

$$\gamma'(0) \;=\; \lim_{t\to 0}\frac{\gamma(t)-\gamma(0)}{t} \;=\; x'(0)+iy'(0)$$

as the **tangent vector** to the curve γ at z_0. It is the complex representation of the usual tangent vector. We define the **angle between two curves** at z_0 to be the angle between their tangent vectors at z_0.

Theorem. *If $\gamma(t)$, $0 \leq t \leq 1$, is a smooth parameterized curve terminating at $z_0 = \gamma(0)$, and $f(z)$ is analytic at z_0, then the tangent to the curve $f(\gamma(t))$ terminating at $f(z_0)$ is*

(6.1) $(f\circ\gamma)'(0) \;=\; f'(z_0)\gamma'(0).$

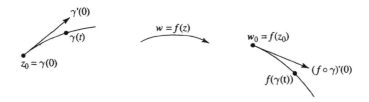

The proof is a close relative of the proof of the chain rule for the composition of analytic functions (Section 2). If $\gamma'(0) \neq 0$, then $\gamma(t) \neq \gamma(0)$ for t near 0, $t \neq 0$, so we may write

$$\frac{f(\gamma(t))-f(\gamma(0))}{t} \;=\; \frac{f(\gamma(t))-f(\gamma(0))}{\gamma(t)-\gamma(0)}\,\frac{\gamma(t)-\gamma(0)}{t}$$

and pass to the limit, to obtain the formula (6.1). If $\gamma'(0) = 0$, then proceeding as in Section 2, we obtain $(f\circ\gamma)'(0) = 0$, and again the formula holds.

We may think of the tangent vector as a vector in the plane with tail at z_0. Composing a parameterized curve with $f(z)$ then has the effect upon the tangent vector of multiplying it by $f'(z_0)$ (complex multiplication) and moving the tail to $w_0 = f(z_0)$. If the tangent vector is represented by $z - z_0$, then the tangent to the image curve is represented by $w - f(z_0) = f'(z_0)(z - z_0)$. As far as the tangent vector at z_0 is concerned, the effect of composing with $f(z)$ is the same as the effect of composing with the function $f(z_0) + f'(z_0)(z - z_0)$, which is the first-order Taylor approximation to $f(z)$ at z_0. The remainder term $R(z)$ in the Taylor approximation satisfies $R(z)/(z - z_0) \to 0$ as $z \to z_0$, so that $R(z)$ has no effect on tangent vectors.

A function is **conformal** if it preserves angles. More precisely, we say that a smooth complex-valued function $g(z)$ is **conformal at** z_0 if whenever γ_0 and γ_1 are two curves terminating at z_0 with nonzero tangents, then the curves $g \circ \gamma_0$ and $g \circ \gamma_1$ have nonzero tangents at $g(z_0)$ and the angle from $(g \circ \gamma_0)'(z_0)$ to $(g \circ \gamma_1)'(z_0)$ is the same as the angle from $\gamma_0'(z_0)$ to $\gamma_1'(z_0)$. A **conformal mapping** of one domain D onto another V is a continuously differentiable function that is conformal at each point of D and that maps D one-to-one onto V.

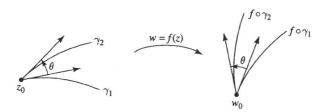

The translation $f(z) = z + b$ and the complex multiplication $g(z) = az$, where $a \neq 0$, evidently preserve angles, hence are conformal everywhere. They are conformal mappings of the complex plane onto itself. On the other hand, the function $a\bar{z}$ reverses angles and orientation, so it is not conformal. For $n > 1$, the function z^n multiplies angles at the origin by n, so it is not conformal at $z = 0$. The following theorem shows that z^n is conformal at any point z other than 0.

Theorem. *If $f(z)$ is analytic at z_0 and $f'(z_0) \neq 0$, then $f(z)$ is conformal at z_0.*

Let γ_0 and γ_1 be two curves terminating at z_0 with nonzero tangents. By the preceding theorem, the tangents to the curves $g \circ \gamma_0$ and $g \circ \gamma_1$ are obtained by multiplying the respective tangents to γ_0 and γ_1 by $f'(z_0)$. Thus the arguments of both tangents are increased by the same angle, namely the argument of $f'(z_0)$. Consequently, the angle between them is preserved.

There is a converse to this theorem, to the effect that conformal mappings are analytic. Though the result is elementary, we postpone it to Section IV.8. (But see Exercise 9.)

Example. The function $w = z^2$ maps the right half-plane $\{\operatorname{Re} z > 0\}$ conformally onto the slit plane $\mathbb{C} \backslash (-\infty, 0]$. For any fixed θ_0, $0 < \theta_0 \le \pi/2$, it maps the sector $\{|\arg z| < \theta_0\}$ conformally onto the sector $\{|\arg z| < 2\theta_0\}$ of twice the aperture.

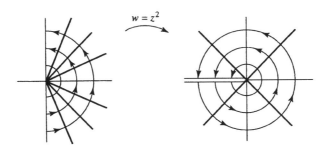

Example. Fix θ_0, $0 < \theta_0 \le \pi$. If $0 < a < \pi/\theta_0$, the function z^a maps the sector $\{|\arg z| < \theta_0\}$ conformally onto the sector $\{|\arg z| < a\theta_0\}$. In particular, the function $z^{\pi/2\theta_0}$ maps the sector $\{|\arg z| < \theta_0\}$ conformally onto the right half-plane.

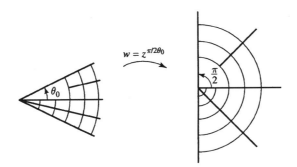

Example. The exponential function e^z is conformal at each point $z \in \mathbb{C}$, since its derivative does not vanish at z. Its image is the punctured plane $\mathbb{C} \backslash \{0\}$. However, it is not a conformal mapping of the plane onto the punctured plane, since it is not one-to-one. Its restriction to the horizontal strip $\{|\operatorname{Im} z| < \pi\}$ is a conformal mapping of the strip onto the slit plane $\mathbb{C} \backslash (-\infty, 0]$.

Example. The principal branch $\operatorname{Log} z$ of the logarithm is a conformal mapping of the slit plane $\mathbb{C} \backslash (-\infty, 0]$ onto the horizontal strip $\{|\operatorname{Im} w| < \pi\}$. See the figure in Section I.6.

Any conformal mapping carries orthogonal curves to orthogonal curves, and it carries orthogonal families of curves to orthogonal families of curves. In the case of the exponential function e^z, the orthogonal grid consisting of horizontal and vertical lines is mapped to an orthogonal grid consisting of rays emanating from the origin and circles centered at the origin.

Something similar happens for any nonconstant analytic function $f = u + iv$ on a domain D. Fix a point z_0 where $f'(z_0) \neq 0$, and consider the two curves $\{u(z) = u(z_0)\}$ and $\{v(z) = v(z_0)\}$, which meet at z_0. The function $f(z)$ is one-to-one near z_0, it maps the part of the level set $\{u(z) = u(z_0)\}$ near z_0 to a vertical line segment through $f(z_0)$, and it maps the part of the level set $\{v(z) = v(z_0)\}$ near z_0 to a horizontal line segment through $f(z_0)$. Since these line segments are orthogonal at $f(z_0)$, the level sets of u and v are orthogonal at z_0. Thus the two families of curves $\{u = \text{constant}\}$ and $\{v = \text{constant}\}$ are orthogonal except at points where $f'(z_0) = 0$.

Example. For $f(z) = z^2 = x^2 - y^2 + 2ixy$, the families of curves $u = \text{con-stant}$ and $v = \text{constant}$ form two families of hyperbolas that are orthogonal except at the origin.

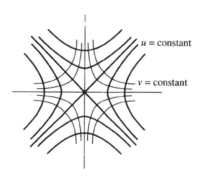

Exercises for II.6

1. Sketch the families of level curves of u and v for the following functions $f = u + iv$. (a) $f(z) = 1/z$, (b) $f(z) = 1/z^2$, (c) $f(z) = z^6$. Determine where $f(z)$ is conformal and where it is not conformal.

2. Sketch the families of level curves of u and v for $f(z) = \text{Log}\, z = u + iv$. Relate your sketch to a figure in Section I.6.

3. Sketch the families of level curves of u and v for the functions $f = u + iv$ given by (a) $f(z) = e^z$, (b) $f(z) = e^{\alpha z}$, where α is complex. Determine where $f(z)$ is conformal and where it is not conformal.

4. Find a conformal map of the horizontal strip $\{-A < \operatorname{Im} z < A\}$ onto the right half-plane $\{\operatorname{Re} w > 0\}$. *Hint.* Recall the discussion of the exponential function, or refer to the preceding problem.

5. Find a conformal map of the wedge $\{-B < \arg z < B\}$ onto the right half-plane $\{\operatorname{Re} w > 0\}$. Assume $0 < B < \pi$.

6. Determine where the function $f(z) = z+1/z$ is conformal and where it is not conformal. Show that for each w, there are at most two values z for which $f(z) = w$. Show that if $r > 1$, $f(z)$ maps the circle $\{|z| = r\}$ onto an ellipse, and that $f(z)$ maps the circle $\{|z| = 1/r\}$ onto the same ellipse. Show that $f(z)$ is one-to-one on the exterior domain $D = \{|z| > 1\}$. Determine the image of D under $f(z)$. Sketch the images under $f(z)$ of the circles $\{|z| = r\}$ for $r > 1$, and sketch also the images of the parts of the rays $\{\arg z = \beta\}$ lying in D.

7. For the function $f(z) = z + 1/z = u + iv$, sketch the families of level curves of u and v. Determine the images under $f(z)$ of the top half of the unit disk, the bottom half of the unit disk, the part of the upper half-plane outside the unit disk, and the part of the lower half-plane outside the unit disk. *Hint.* Start by locating the images of the curves where $u = 0$, where $v = 0$, and where $v = 1$. Note that the level curves are symmetric with respect to the real and imaginary axes, and they are invariant under the inversion $z \mapsto 1/z$ in the unit circle.

8. Consider $f(z) = z + e^{i\alpha}/z$, where $0 < \alpha < \pi$. Determine where $f(z)$ is conformal and where it is not conformal. Sketch the images under $f(z)$ of the unit circle $\{|z| = 1\}$ and the intervals $(-\infty, -1]$ and $[+1, +\infty)$ on the real axis. Show that $w = f(z)$ maps $\{|z| > 1\}$ conformally onto the complement of a slit in the w-plane. Sketch roughly the images of the segments of rays outside the unit circle $\{\arg z = \beta, |z| \geq 1\}$ under $f(z)$. At what angles do they meet the slit, and at what angles do they approach ∞?

9. Let $f = u+iv$ be a continuously differentiable complex-valued function on a domain D such that the Jacobian matrix of f does not vanish at any point of D. Show that if f maps orthogonal curves to orthogonal curves, then either f or $\bar{f}$ is analytic, with nonvanishing derivative.

7. Fractional Linear Transformations

A fractional linear transformation is a function of the form

$$(7.1) \qquad w = f(z) = \frac{az+b}{cz+d},$$

where a, b, c, d are complex constants satisfying $ad - bc \neq 0$. Fractional linear transformations are also called **Möbius transformations**. Since

$$f'(z) = \frac{ad-bc}{(cz+d)^2},$$

the condition $ad - bc \neq 0$ simply guarantees that $f(z)$ is not constant.

If we multiply each of the parameters a, b, c, d in (7.1) by the same nonzero constant, we obtain the same function. Thus different choices of the parameters may lead to the same fractional linear transformation.

Example. A function of the form $f(z) = az + b$, where $a \neq 0$, is called an **affine transformation**. These are the fractional linear transformations of the form (7.1) with $c = 0$. Special cases are the **translations** $z \mapsto z + b$ and the **dilations** $z \mapsto az$.

Example. The fractional linear transformation $f(z) = 1/z$ is called an **inversion**.

It is convenient to regard a fractional linear transformation as a map from the extended complex plane $\mathbb{C}^* = \mathbb{C} \cup \{\infty\}$ to itself. If $f(z)$ is affine, we define $f(\infty) = \infty$. Otherwise, $f(z)$ has the form (7.1) where $c \neq 0$, and we define $f(-d/c) = \infty$ and

$$f(\infty) = \lim_{z \to \infty} f(z) = \lim_{z \to \infty} \frac{a + b/z}{c + d/z} = \frac{a}{c}.$$

Thus translations and dilations map ∞ to ∞, while the inversion $z \mapsto 1/z$ interchanges 0 and ∞.

The inverse of a fractional linear transformation is again a fractional linear transformation. To see this, we solve (7.1) for z, to obtain

$$z = \frac{-dw + b}{cw - a}.$$

The condition on the coefficients is satisfied, since $(-d)(-a) - bc = ad - bc \neq 0$, or alternatively, since the function $z = z(w)$ is not constant. This shows that each fractional linear transformation is a one-to-one function from the extended complex plane onto itself.

The composition of two fractional linear transformations is again a fractional linear transformation. To see this, suppose $f(z) = (az + b)/cz + d)$ and $g(z) = (\alpha z + \beta)/(\gamma z + \delta)$, and compute

$$f(g(z)) = \frac{a((\alpha z + \beta)/(\gamma z + \delta)) + b}{c((\alpha z + \beta)/(\gamma z + \delta)) + d} = \frac{(a\alpha + b\gamma)z + a\beta + b\delta}{(c\alpha + d\gamma)z + c\beta + d\delta}.$$

Since the composition $f \circ g$ cannot be constant, the condition on the parameters is met, and $f \circ g$ is a fractional linear transformation.

Note that the composition corresponds to matrix multiplication,

$$\begin{pmatrix} a & b \\ c & d \end{pmatrix} \begin{pmatrix} \alpha & \beta \\ \gamma & \delta \end{pmatrix} = \begin{pmatrix} a\alpha + b\gamma & a\beta + b\delta \\ c\alpha + d\gamma & c\beta + d\delta \end{pmatrix}.$$

The condition $ad - bc \neq 0$ on the parameters is simply the condition that the matrix associated with the fractional linear transformation has nonzero determinant, that is, that the matrix is invertible.

The fact that matrix multiplication corresponds to composition can be reformulated in the language of group theory. If we assign to each 2×2 invertible matrix the corresponding fractional linear transformation, we obtain what is called a "group homomorphism," from the group of 2×2 invertible matrices with complex entries onto the group of fractional linear transformations with operation composition.

A fractional linear transformation depends on four complex parameters. One of these can be adjusted without changing the transformation, for instance by multiplying all the parameters by the same nonzero constant. That leaves three parameters to be specified. The next theorem shows that there are three independent complex parameters that describe fractional linear transformations uniquely, namely, the images of any three prescribed points.

Theorem. *Given any three distinct points z_0, z_1, z_2 in the extended complex plane, and given any three distinct values w_0, w_1, w_2 in the extended complex plane, there is a unique fractional linear transformation $w = w(z)$ such that $w(z_0) = w_0$, $w(z_1) = w_1$, and $w(z_2) = w_2$.*

To establish the existence assertion, it suffices to show that any three distinct points can be mapped by a fractional linear transformation to 0, 1, and ∞. Indeed, if f maps z_0, z_1, z_2 respectively to $0, 1, \infty$, and g maps w_0, w_1, w_2 respectively to $0, 1, \infty$, then the composition $g^{-1} \circ f$, of f followed by the inverse of g, maps z_0, z_1, z_2 to w_0, w_1, w_2. If now none of the points z_0, z_1, z_2 is ∞, a transformation mapping them to $0, 1, \infty$ is given explicitly by

$$(7.2) \qquad w = f(z) = \frac{z - z_0}{z - z_2} \frac{z_1 - z_2}{z_1 - z_0}.$$

If one of the z_j's is ∞, we define $f(z)$ by sending that z_j to ∞ in the above formula. For instance, if $z_0 = \infty$, we rewrite the right-hand side of (7.2) as

$$\frac{(z/z_0) - 1}{z - z_2} \frac{z_1 - z_2}{(z_1/z_0) - 1}$$

and take a limit as $z_0 \to \infty$, to obtain

$$w = f(z) = \frac{z_1 - z_2}{z - z_2}.$$

This maps ∞, z_1, z_2 to 0, 1, ∞. There are similar formulae for the cases $z_1 = \infty$ and $z_2 = \infty$.

For the uniqueness, suppose first that $f(z)$ is a fractional linear transformation that fixes 0, 1, and ∞. Since $f(\infty) = \infty$, $f(z) = az + b$ for some $a \neq 0$. From $f(0) = 0$ we obtain $b = 0$, and from $f(1) = 1$ we obtain $a = 1$. Thus $f(z) = z$ is the identity transformation.

Now suppose that $g(z)$ and $h(z)$ are both fractional linear transformations mapping the z_j's to the respective w_j's. Let $k(z)$ map the z_j's respectively to $0, 1, \infty$. Then $f = k \circ h^{-1} \circ g \circ k^{-1}$ maps 0 to 0, 1 to 1, and ∞ to ∞. Hence $f(z) = z$ is the identity, and $g = h \circ k^{-1} \circ f \circ k = h \circ k^{-1} \circ k = h$. This establishes the uniqueness assertion of the theorem.

Exercise. Find the fractional linear transformation mapping -1 to 0, ∞ to 1, and i to ∞.

Solution. We could use (7.2) and send z_1 to ∞. However, it is easier to proceed directly. Since $w(i) = \infty$, we place $z - i$ in the denominator, and since $w(-1) = 0$, we place $z + 1$ in the numerator, to obtain $w(z) = a(z + 1)/(z - i)$. Since $w(z) \to 1$ as $z \to \infty$, we obtain $a = 1$, and hence $w(z) = (z + 1)/(z - i)$.

Theorem. *Every fractional linear transformation is a composition of dilations, translations, and inversions.*

A fractional linear transformation mapping ∞ to ∞ has the form $w = az + b$ where $a \neq 0$. This is the composition of the translation $z \mapsto z + b/a$ and the dilation $z \mapsto az$:

$$z \mapsto z + b/a \mapsto a(z + b/a) = az + b.$$

If $w(\infty)$ is finite, then w has the form (7.1) where $c \neq 0$. In this case we may divide each of the parameter values by c and assume that $c = 1$. Then $w(z) = (az + b)/(z + d)$. Now we conjure up by magic the identity

$$w(z) = \frac{az + b}{z + d} = a + \frac{b - ad}{z + d}.$$

This expression allows us to represent $w(z)$ as

$$z \mapsto z + d \mapsto \frac{1}{z + d} \mapsto \frac{b - ad}{z + d} \mapsto a + \frac{b - ad}{z + d},$$

and consequently, $w(z)$ is a composition of a translation, an inversion, a dilation, and a translation.

Theorem. *A fractional linear transformation maps circles in the extended complex plane to circles.*

It suffices to establish the theorem for translations, dilations, and inversions, since every fractional linear transformation is a composition of these.

It is clear (or it should be) that translations and dilations map circles to circles. Thus it suffices to check that the inversion $w = 1/z$ maps circles to circles.

Consider first a circle that does not pass through ∞. It has an equation of the form $|z - a|^2 = r^2$, where a is its center and r its radius. The image of the circle under the inversion $w = 1/z$ consists of points satisfying $|1 - aw|^2 = r^2|w|^2$, that is,

$$0 = |1 - aw|^2 - r^2|w|^2 = (1 - aw)\left(\overline{1 - aw}\right) - r^2|w|^2$$
$$= \left(|a|^2 - r^2\right)|w|^2 - aw - \overline{aw} + 1.$$

Set $w = u + iv$ where u and v are real. The equation assumes the form

$$\left(|a|^2 - r^2\right)\left(u^2 + v^2\right) + Au + Bv + 1 = 0,$$

where A and B are real constants. If $r = |a|$, the equation represents a straight line in the plane, which is a circle through ∞. If $r \neq |a|$, this is a quadratic equation in u and v of the form met in Section I.3. It has more than one solution, so the solutions form a circle. (To obtain the center and radius, complete the square.)

Next consider a circle passing through ∞, that is, a straight line. It has an equation of the form $Cx + Dy = E$. A calculation similar to the one given above shows that the image is a circle if $E \neq 0$ and a straight line if $E = 0$ (Exercise 6).

Exercise. Find the equation of the fractional linear transformation mapping 0 to -1, i to 0, and ∞ to 1.
Solution. Since $i \to 0$, we can normalize a to be 1 and write the fractional linear transformation in the form $w = (z - i)/(cz + d)$. The condition $0 \mapsto -1$ yields $-i/d = -1$, and so $d = i$. Finally, the condition $\infty \mapsto 1$ yields $1/c = 1$, and so $c = 1$. Thus $w = (z - i)/(z + i)$.

Exercise. Determine the images of each of the following sets under the above fractional linear transformation: (a) the imaginary axis, (b) the right half-plane, (c) the real axis, (d) the upper half-plane, (e) the horizontal line through i. Sketch the images of horizontal lines and of vertical lines under the transformation.
Solution. We will solve this exercise without referring to the explicit formula for the transformation. We use two facts. First, to determine the image of a circle under a fractional linear transformation, it suffices to determine the images of three points on the circle. Since three points determine a circle, the image of the circle is then the circle passing through the three image points. Second, fractional linear transformations map orthogonal circles to orthogonal circles, since they are conformal.
(a) The three points 0, i, ∞ lie on the circle corresponding to the imaginary axis in the extended complex plane. The image of the imaginary axis is then the circle through the three image points $-1, 0, +1$, which is the

real line.

(b) The ordered triple $0, i, \infty$ is mapped to the ordered triple $-1, 0, 1$, so the image w moves forward on the real line as z moves upwards on the imaginary axis. Since orientations are preserved, the right half-plane is mapped to the domain on the right of the positively traversed real line, which is the lower half-plane.

(c) The real axis is mapped to a circle through $w(0) = -1$ and $w(\infty) = +1$. Since the real and imaginary axes are orthogonal, their images are orthogonal. Thus the image of the real line is a circle through ± 1 that is orthogonal to the real line. There is only one such circle, the unit circle. Consequently, the image of the real line is the unit circle $\{|w| = 1\}$.

(d) Since the image of the real line is the unit circle, the image of the upper half-plane does not cross the unit circle, and it must coincide either with the inside $\{|w| < 1\}$ or with the exterior domain $\{|w| > 1\}$ together with ∞. Since i is mapped to 0, which is inside the unit circle, the image of the upper half-plane is the inside, that is, it is the open unit disk $\{|w| < 1\}$.

(e) The image of the horizontal line through i is a circle passing through 0 and 1, and it lies inside the unit disk, so it must be the circle centered at $\frac{1}{2}$ of radius $\frac{1}{2}$.

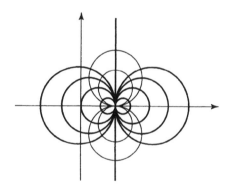

The image of any horizontal line is a circle through $w(\infty) = 1$, and it is orthogonal to the real line (the image of the imaginary axis). These images of the horizontal lines form a pencil of circles as sketched in the figure. The images of vertical lines are circles through $w(\infty) = 1$. Since the real axis is the image of the imaginary axis, these circles must be tangent to the real axis at 1. The images of the vertical lines are also sketched in the figure. Note that the images of the horizontal and vertical lines are orthogonal to each other.

Exercises for II.7

1. Compute explicitly the fractional linear transformations determined by the following correspondences of triples:

(a) $(1+i, 2, 0) \mapsto (0, \infty, i-1)$ (e) $(1, 2, \infty) \mapsto (0, 1, \infty)$
(b) $(0, 1, \infty) \mapsto (1, 1+i, 2)$ (f) $(0, \infty, i) \mapsto (0, 1, \infty)$
(c) $(\infty, 1+i, 2) \mapsto (0, 1, \infty)$ (g) $(0, 1, \infty) \mapsto (0, \infty, i)$
(d) $(-2, i, 2) \mapsto (1-2i, 0, 1+2i)$ (h) $(1, i, -1) \mapsto (1, 0, -1)$

2. Consider the fractional linear transformation in Exercise 1a above, which maps $1+i$ to 0, 2 to ∞, and 0 to $i-1$. Without referring to an explicit formula, determine the image of the circle $\{|z-1| = 1\}$, the image of the disk $\{|z-1| < 1\}$, and the image of the real axis.

3. Consider the fractional linear transformation that maps 1 to i, 0 to $1+i$, and -1 to 1. Determine the image of the unit circle $\{|z| = 1\}$, the image of the open unit disk $\{|z| < 1\}$, and the image of the imaginary axis. Illustrate with a sketch.

4. Consider the fractional linear transformation that maps -1 to $-i$, 1 to $2i$, and i to 0. Determine the image of the unit circle $\{|z| = 1\}$, the image of the open unit disk $\{|z| < 1\}$, and the image of the interval $[-1, +1]$ on the real axis. Illustrate with a sketch.

5. What is the image of the horizontal line through i under the fractional linear transformation that interchanges 0 and 1 and maps -1 to $1+i$? Illustrate with a sketch.

6. Show that the image of a straight line under the inversion $z \mapsto 1/z$ is a straight line or circle, depending on whether the line passes through the origin.

7. Show that the fractional linear transformation $f(z) = (az+b)/(cz+d)$ is the identity mapping z if and only if $b = c = 0$ and $a = d \neq 0$.

8. Show that any fractional linear transformation can be represented in the form $f(z) = (az + b)/(cz + d)$, where $ad - bc = 1$. Is this representation unique?

9. Show that the fractional linear transformations that are real on the real axis are precisely those that can be expressed in the form $(az + b)/(cz + d)$, where a, b, c, and d are real.

10. Suppose the fractional linear transformation $(az+b)/(cz+d)$ maps $\mathbb{R}$ to $\mathbb{R}$, and $ad - bc = 1$. Show that a, b, c, and d are real or they are all pure imaginary.

11. Two maps f and g are **conjugate** if there is h such that $g = h \circ f \circ h^{-1}$. Here the conjugating map h is assumed to be one-to-one, with appropriate domain and range. We can think of f and g as the "same" map, after the change of variable $w = h(z)$. A point z_0 is a **fixed point** of f if $f(z_0) = z_0$. Show the following. (a) If f

is conjugate to g, then g is conjugate to f. (b) If f_1 is conjugate to f_2 and f_2 to f_3, then f_1 is conjugate to f_3. (c) If f is conjugate to g, then $f \circ f$ is conjugate to $g \circ g$, and more generally, the m-fold composition $f \circ \cdots \circ f$ (m times) is conjugate to $g \circ \cdots \circ g$ (m times). (d) If f and g are conjugate, then the conjugating function h maps the fixed points of f to the fixed points of g. In particular, f and g have the same number of fixed points.

12. Classify the conjugacy classes of fractional linear transformations by establishing the following:

(a) A fractional linear transformation that is not the identity has either 1 or 2 fixed points, that is, points satisfying $f(z_0) = z_0$.

(b) If a fractional linear transformation $f(z)$ has two fixed points, then it is conjugate to the dilation $z \mapsto az$ with $a \neq 0$, $a \neq 1$, that is, there is a fractional linear transformation $h(z)$ such that $h(f(z)) = ah(z)$. Is a unique? *Hint.* Consider a fractional linear transformation that maps the fixed points to 0 and ∞.

(c) If a fractional linear transformation $f(z)$ has exactly one fixed point, then it is conjugate to the translation $\zeta \mapsto \zeta + 1$. In other words, there is a fractional linear transformation $h(z)$ such that $h(f(h^{-1}(\zeta))) = \zeta + 1$, or equivalently, such that $h(f(z)) = h(z) + 1$. *Hint.* Consider a fractional linear transformation that maps the fixed point to ∞.

III

Line Integrals and Harmonic Functions

In Sections 1 and 2 we review multivariable integral calculus in order to prepare for complex integration in the next chapter. The salient features are Green's theorem and independence of path for line integrals. In Section 3 we introduce harmonic functions, and in Sections 4 and 5 we discuss the mean value property and the maximum principle for harmonic functions. Sections 6 and 7 include various applications to physics. The student may proceed directly to complex integration in the next chapter after paging through the review of multivariable calculus in Sections 1 and 2 and reading about harmonic conjugates in Section 3.

1. Line Integrals and Green's Theorem

Line integrals play an important role in complex analysis. In this section and the next we review line integrals in the plane, without filling in all the details. We begin by saying something about paths and curves.

A **path** in the plane from A to B is a continuous function $t \mapsto \gamma(t)$ on some parameter interval $a \leq t \leq b$ such that $\gamma(a) = A$ and $\gamma(b) = B$. The path is **simple** if $\gamma(s) \neq \gamma(t)$ when $s \neq t$. The path is **closed** if it starts and ends at the same point, that is, $\gamma(a) = \gamma(b)$. A **simple closed path** is a closed path γ such that $\gamma(s) \neq \gamma(t)$ for $a \leq s < t < b$.

| simple path | path (not simple) | simple closed path |

If $\gamma(t)$, $a \leq t \leq b$, is a path from A to B, and if $\phi(s)$, $\alpha \leq s \leq \beta$, is a strictly increasing continuous function satisfying $\phi(\alpha) = a$ and $\phi(\beta) = b$, then the composition $\gamma(\phi(s))$, $\alpha \leq s \leq \beta$, is also a path from A to B. The composition $\gamma \circ \phi$ is a "reparametrization" of γ. For our purposes we can usually regard γ and any of its reparametrizations as being the same path. (Technically, we should consider equivalence classes of paths.) Note that reparametrization preserves the order of points of a path; that is, it preserves orientation.

The **trace** of the path γ is its image $\gamma([a, b])$, which is a subset of the plane. When it is clear from context, we will denote the trace of a path also by γ. It will not be until Chapter VIII that we need to be careful about distinguishing the path γ from its trace.

If one path ends where another begins, the two paths can be concatenated by following one and then the other, after suitable reparametrization.

A **smooth path** is a path that can be represented in the form $\gamma(t) = (x(t), y(t))$, $a \le t \le b$, where the functions $x(t)$ and $y(t)$ are smooth, that is, have as many derivatives as is necessary for whatever is being asserted to be true. A **piecewise smooth path** is a concatenation of smooth paths. By a **curve** we mean (usually) a smooth or piecewise smooth path.

Let γ be a path in the plane from A to B, and let $P(x, y)$ and $Q(x, y)$ be continuous complex-valued functions on γ. We consider successive points on the path, $A = (x_0, y_0)$, (x_1, y_1), $\ldots$, $B = (x_n, y_n)$, and we form the sum

$$(1.1) \qquad \sum P(x_j, y_j)(x_{j+1} - x_j) \; + \; \sum Q(x_j, y_j)(y_{j+1} - y_j).$$

If these sums have a limit as the distances between the successive points on γ tend to 0, we define the limit to be the **line integral of $P dx + Q dy$ along** γ, and we denote it by

$$(1.2) \qquad \int_\gamma P \, dx \; + \; Q \, dy.$$

Suppose the path $\gamma(t) = (x(t), y(t))$, $a \le t \le b$, is continuously differentiable, that is, the parameter functions $x(t)$ and $y(t)$ are continuously differentiable. Suppose the parameter values t_j satisfy $x(t_j) = x_j$, $y(t_j) = y_j$, where $a = t_0 < t_1 < \cdots < t_n = b$. By the mean value theorem, there are points t_j^* between t_j and t_{j+1} such that $x(t_{j+1}) - x(t_j) = x'(t_j^*)(t_{j+1} - t_j)$. If we substitute this into the first sum in (1.1), we obtain

$$\sum P(x(t_j), y(t_j))x'(t_j^*)(t_{j+1} - t_j),$$

which is a Riemann sum approximating the integral $\int_a^b P(x(t), y(t))x'(t)dt$. Similarly, the second sum in (1.1) is a Riemann sum approximating the integral $\int_a^b Q(x(t), y(t))y'(t)dt$. As the distances between the successive t_j's tend to 0, the sums in (1.1) converge to an ordinary garden-variety Riemann integral, and we obtain

$$(1.3) \quad \int_\gamma P \, dx \; + \; Q \, dy \; = \; \int_a^b P(x(t), y(t))\frac{dx}{dt} \, dt \; + \; \int_a^b Q(x(t), y(t))\frac{dy}{dt} \, dt.$$

Thus to evaluate a line integral over a smooth curve, we simply parametrize the curve by $t \mapsto (x(t), y(t))$, calculate the derivatives dx/dt and dy/dt of the components, and plug these into the definite integral in (1.3).   To evaluate the line integral over a path that is only piecewise smooth, we parametrize each smooth subpath, calculate the corresponding integrals by (1.3), and add them.

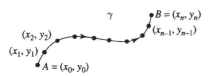

Note that the definition of the line integral over γ is independent of the parameterization of γ. The parameterization enters only in determining the ordering of the points on the curve γ. Thus different parameterizations give the same integral in (1.3). Also note that if we reverse the direction of γ, then the line integral is replaced by its negative.

Example. To evaluate $\int_\gamma xy\,dx$, where γ is the quarter-circle from $(1,0)$ to $(0,1)$ on the unit circle, we parametrize γ by

$$(x(\theta), y(\theta)) \;=\; (\cos\theta, \sin\theta), \qquad 0 \le \theta \le \pi/2,$$

and we substitute into (1.3). This gives

$$\int_\gamma xy\,dx \;=\; \int_0^{\pi/2} \cos\theta \sin\theta \, d(\cos\theta) \;=\; -\int_0^{\pi/2} \cos\theta \sin^2\theta \, d\theta$$

$$=\; -\left.\frac{\sin^3\theta}{3}\right|_0^{\pi/2} \;=\; -\frac{1}{3}.$$

Note that the sign is correct, since $xy \ge 0$ in the curve γ, while $dx < 0$ on the curve (since x decreases on the curve).

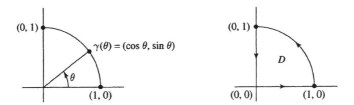

A domain D has **piecewise smooth boundary** if the boundary of D can be decomposed into a finite number of smooth curves meeting only at endpoints. By "smooth" we usually mean "continuously differentiable," though in applications the curves making up the boundary will usually be straight line segments or arcs of circles. We denote the boundary of D by ∂D. For purposes of integration, the **orientation of ∂D** is chosen so that D lies on the left of a curve in ∂D as we traverse the boundary curve in the positive direction, that is, as the parameter value increases.

Example. To evaluate $\int_{\partial D} xy\, dx$, where D is the quarter-disk in the first quadrant, we divide the integral into three pieces,

$$\int_{\partial D} xy\, dx \;=\; \int_{\gamma} xy\, dx \;+\; \int_{(1,0)}^{(0,0)} xy\, dx \;+\; \int_{(0,0)}^{(1,0)} xy\, dx,$$

where γ is the quarter-circle in the preceding example, and the other two paths are straight line segments. The integral along the horizontal interval on the x-axis is 0, because $xy = 0$ there. The integral along the vertical interval on the y-axis is 0, because $dx = 0$ there. (To see this, either parametrize the line segment explicitly, or go back to the definition (1.1) and observe that each of the x_j's is 0.) Using the result of the preceding example, we find that the value of the integral around ∂D is $-\frac{1}{3}$.

A very useful tool for evaluating line integrals is provided by Green's theorem, which converts a line integral around the boundary of a domain to an area integral over the domain.

Green's Theorem. *Let D be a bounded domain in the plane whose boundary ∂D consists of a finite number of disjoint piecewise smooth closed curves. Let P and Q be continuously differentiable functions on $D \cup \partial D$. Then*

$$(1.4) \qquad \int_{\partial D} P\, dx \;+\; Q\, dy \;=\; \iint_{D} \left(\frac{\partial Q}{\partial x} - \frac{\partial P}{\partial y} \right) dx\, dy.$$

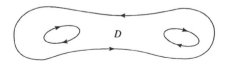

Example. We again evaluate $\int_{\partial D} xy\, dx$, where D is the quarter-disk in the first quadrant, this time using Green's theorem. In this case, $P(x,y) = xy$ and $Q(x,y) = 0$, so (1.4) becomes

$$\int_{\partial D} xy\, dx \;=\; -\iint_{D} x\, dx\, dy \;=\; -\iint r\cos\theta\, r\, dr\, d\theta$$

$$= \; -\int_{0}^{\pi/2} \cos\theta\, d\theta \int_{0}^{1} r^2 dr \;=\; -(1)(\frac{1}{3}),$$

as before.

Since Green's theorem is of fundamental importance, we provide a sketch of the ideas behind the derivation of the formula (1.4). One basic idea is to cut the domain into little curvilinear triangular pieces and treat each piece separately. Another is to reduce the double integral over a triangle to a line integral by applying the fundamental theorem of calculus in one variable with the other variable as a parameter. For convenience, we break the proof-sketch into three steps.

The first step is to establish the formula (1.4) for the triangle T with vertices at $(0,0)$, $(1,0)$, and $(0,1)$. We must establish the two identities

$$\int_{\partial T} P\,dx = -\iint_T \frac{\partial P}{\partial y}\,dx\,dy, \qquad \int_{\partial T} Q\,dy = \iint_T \frac{\partial Q}{\partial x}\,dx\,dy.$$

Consider just the first identity here. We represent the double integral as an iterated integral and use the fundamental theorem of calculus, to obtain

$$\iint_T \frac{\partial P}{\partial y}\,dx\,dy = \int_0^1 \left[\int_0^{1-x} \frac{\partial P}{\partial y}\,dy\right]dx = \int_0^1 P(x,1-x)\,dx - \int_0^1 P(x,0)\,dx.$$

The sum on the right we recognize as $-\int_{\partial T} P\,dx$, after we parametrize separately the three sides of ∂T. Indeed, the line integral of $P\,dx$ along the vertical edge of T is 0, since $dx = 0$ there; the line integral of $P\,dx$ along the bottom edge of T is $\int_0^1 P(x,0)dx$; and the line integral of $P\,dx$ back along the hypotenuse of T is $-\int_0^1 P(x,1-x)dx$, where we have used the parameterization $y = 1 - x$.

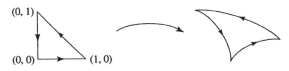

differentiable triangle

The second step of the proof is to establish the formula for any domain D that can be obtained from the triangle T by a change of variables. (See Exercise 7.)

The final step in the proof, for an arbitrary domain D, involves triangulating D, that is, cutting D into small triangular pieces, each of which can be obtained from the triangle T by a change of variables. Green's theorem is applied to each triangular piece, and the results are added. The sum of the area integrals over the triangular pieces is the area integral over D. The boundary integrals over the sides of the triangular pieces inside D cancel in pairs, since each curvilinear triangle side is traversed twice, once in each direction, and the opposing directions cancel. The boundary integrals over the curvilinear triangle sides in ∂D add up to the integral over ∂D.

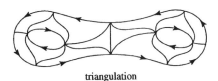

triangulation

Note that we will be using Green's theorem only for relatively simple domains, those whose boundaries consist of straight line segments and circular arcs, for which Green's theorem can be established relatively easily.

Exercises for III.1

1. Evaluate $\int_\gamma y^2 dx + x^2 dy$ along the following paths γ from $(0,0)$ to $(2,4)$: (a) the arc of the parabola $y = x^2$; (b) the horizontal interval from $(0,0)$ to $(2,0)$, followed by the vertical interval from $(2,0)$ to $(2,4)$; (c) the vertical interval from $(0,0)$ to $(0,4)$, followed by the horizontal interval from $(0,4)$ to $(2,4)$.

2. Evaluate $\int_\gamma xy\, dx$ both directly and using Green's theorem, where γ is the boundary of the square with vertices at $(0,0)$, $(1,0)$, $(1,1)$, and $(0,1)$.

3. Evaluate $\int_{\partial D} x^2 dy$ both directly and using Green's theorem, where D is the quarter-disk in the first quadrant bounded by the unit circle and the two coordinate axes.

4. Evaluate $\int_\gamma y\, dx$ both directly and using Green's theorem, where γ is the semicircle in the upper half-plane from R to $-R$.

5. Show that $\int_{\partial D} x\, dy$ is the area of D, while $\int_{\partial D} y\, dx$ is minus the area of D.

6. Show that if P and Q are continuous complex-valued functions on a curve γ, then

$$F(w) \;=\; \int_\gamma \frac{P dx}{z - w} + \int_\gamma \frac{Q dy}{z - w} \qquad (z = x + iy)$$

is analytic for $w \in \mathbb{C}\backslash\gamma$. Express $F'(w)$ as a line integral over γ.

7. Show that the formula in Green's theorem is invariant under coordinate changes, in the sense that if the theorem holds for a bounded domain U with piecewise smooth boundary, and if $F(x,y)$ is a smooth function that maps U one-to-one onto another such domain V and that maps the boundary of U one-to-one smoothly onto the boundary of V, then Green's theorem holds for V. *Hint.* First note the change of variable formulae for line and area integrals, given by

$$\int_{\partial V} P\, d\xi \;=\; \int_{\partial U} (P \circ F) \left(\frac{\partial \xi}{\partial x} dx \;+\; \frac{\partial \xi}{\partial y} dy \right),$$

$$\iint_V R\, d\xi\, d\eta \;=\; \iint_U (R \circ F) \det J_F\, dx\, dy,$$

where $F(x, y) = (\xi(x, y), \eta(x, y))$, and where J_F is the Jacobian matrix of F. Use these formulae, with $R = -\partial P / \partial \eta$. The summand $\int Q \, d\eta$ is treated similarly.

8. Prove Green's theorem for the rectangle defined by $x_0 < x < x_1$ and $y_0 < y < y_1$ (a) directly, and (b) using the result for triangles.

2. Independence of Path

In order to draw a useful analogy with single-variable calculus, we begin by reviewing the fundamental theorem of calculus. Recall that $F(t)$ is an **antiderivative** for $f(t)$ if its derivative is f, that is, $F' = f$.

Fundamental Theorem of Calculus.
Part I. If $F(t)$ is an antiderivative for the continuous function $f(t)$, then

$$\int_a^b f(t) \, dt = F(b) - F(a).$$

Part II. If $f(t)$ is a continuous function on $[a, b]$, then the indefinite integral

$$F(t) = \int_a^t f(s) \, ds, \qquad a \le t \le b,$$

is an antiderivative for $f(t)$. Further, each antiderivative for $f(t)$ differs from $F(t)$ by a constant.

If $h(x, y)$ is a continuously differentiable complex-valued function, we define the **differential** dh of h by

$$dh = \frac{\partial h}{\partial x} dx + \frac{\partial h}{\partial y} dy.$$

We say that a differential $P \, dx + Q \, dy$ is **exact** if $P \, dx + Q \, dy = dh$ for some function h. The function h plays the role of the antiderivative, and the following theorem is the analogue of Part I of the fundamental theorem of calculus. It provides a useful tool for evaluating line integrals.

Theorem (Part I). If γ is a piecewise smooth curve from A to B, and if $h(x, y)$ is continuously differentiable on γ, then

$$(2.1) \qquad\qquad \int_\gamma dh = h(B) - h(A).$$

To see this, let the curve be given by $t \mapsto (x(t), y(t))$, $a \le t \le b$. From (1.3) we have

$$\int_\gamma dh = \int_\gamma \frac{\partial h}{\partial x} dx + \frac{\partial h}{\partial y} dy = \int_a^b \frac{\partial h}{\partial x} \frac{dx}{dt} dt + \int_a^b \frac{\partial h}{\partial y} \frac{dy}{dt} dt.$$

By the chain rule and the fundamental theorem of calculus, this is

$$\int_a^b \frac{d}{dt} h(x(t), y(t))\, dt \;=\; h(x(t), y(t))\Big|_a^b \;=\; h(B) - h(A).$$

Example. To evaluate $\int_\gamma 2xy\, dx + (x^2 + 2y)\, dy$, where γ is the quarter-circle given by $\gamma(\theta) = (\cos\theta, \sin\theta)$, $0 \le \theta \le \pi/2$, we could proceed as in the preceding section and plug the parameterizing functions into (1.3). However, in this case it is easier to observe that $2xy\, dx + (x^2 + 2y)\, dy = dh$ for $h(x, y) = x^2 y + y^2$. Consequently,

$$\int_\gamma 2xy\, dx + (x^2 + 2y)\, dy \;=\; (x^2 y + y^2)\Big|_{(1,0)}^{(0,1)} \;=\; 1 - 0 \;=\; 1.$$

Unfortunately, not every differential $P\,dx + Q\,dy$ is exact. We aim to give some conditions that can be used to determine when a differential is exact.

Let P and Q be continuous complex-valued functions on a domain D. We say that the line integral $\int P\,dx + Q\,dy$ is **independent of path** in D if for any two points A and B of D, the integrals $\int_\gamma P\,dx + Q\,dy$ are the same for any path γ in D from A to B. This is tantamount to requiring $\int_\gamma P\,dx + Q\,dy = 0$ for any closed path γ in D. Indeed, if γ_1 and γ_2 are two paths in D from A to B, then we can form a closed path γ in D, starting and ending at A, by following γ_1 from A to B and then following γ_2 backwards from B to A. Since the reversal of direction along γ_2 changes the sign of the integral, we have $\int_\gamma = \int_{\gamma_1} - \int_{\gamma_2}$, so that $\int_\gamma = 0$ if and only if $\int_{\gamma_1} = \int_{\gamma_2}$.

Formula (2.1) shows that the integrals of exact differentials are independent of path. The converse is easily seen to be true also.

Lemma. *Let P and Q be continuous complex-valued functions on a domain D. Then $\int P\,dx + Q\,dy$ is independent of path in D if and only if $P\,dx + Q\,dy$ is exact, that is, there is a continuously differentiable function $h(x, y)$ such that $dh = P\,dx + Q\,dy$. Moreover, the function h is unique, up to adding a constant.*

Suppose that $\int P\,dx + Q\,dy$ is independent of path in D. Fix a point A in D, and define a function $h(x, y)$ on D by

$$h(B) \;=\; \int_A^B P\,dx + Q\,dy, \qquad B \in D,$$

where we may take any path in D from A to B. We compute the partial derivatives of $h(x, y)$ by choosing some special paths. Fix (x_0, y_0) in D, and fix a path γ from A to (x_0, y_0). For x near x_0, we evaluate $h(x, y_0)$ by following the path γ from A to (x_0, y_0) and then the straight line path $x(t) = t$, $y(t) = y_0$ from (x_0, y_0) to (x, y_0). This gives

$$h(x, y_0) \;=\; \int_\gamma P dx + Q dy \;+\; \int_{x_0}^x P(t, y_0)\, dt.$$

Since the first summand on the right is a constant, we obtain from the fundamental theorem of calculus that

$$\frac{\partial h}{\partial x}(x_0, y_0) \;=\; P(x_0, y_0).$$

Similarly, we obtain

$$\frac{\partial h}{\partial y}(x_0, y_0) \;=\; Q(x_0, y_0),$$

and consequently $dh = P dx + Q dy$. For the uniqueness, note that if h_1 is any other function such that $dh_1 = P dx + Q dy$, then $d(h - h_1) = 0$, that is,

$$\frac{\partial}{\partial x}(h - h_1) \;=\; 0 \;=\; \frac{\partial}{\partial y}(h - h_1).$$

Since D is a domain, $h - h_1$ is constant on D.

Let P and Q be continuously differentiable complex-valued functions on a domain D. We say that the differential $P dx + Q dy$ is **closed** on D if

$$(2.2) \qquad\qquad \frac{\partial P}{\partial y} \;=\; \frac{\partial Q}{\partial x}.$$

This is precisely the condition that the integrand in Green's theorem is zero. Thus Green's theorem implies that if $P dx + Q dy$ is closed on D, then $\int_{\partial U} P dx + Q dy = 0$ for any bounded domain U with piecewise smooth boundary such that U together with its boundary is contained in D.

Lemma. *Exact differentials are closed.*

Indeed, if $P dx + Q dy = dh$ is exact, then

$$\frac{\partial P}{\partial y} \;=\; \frac{\partial}{\partial y}\frac{\partial h}{\partial x} \;=\; \frac{\partial}{\partial x}\frac{\partial h}{\partial y} \;=\; \frac{\partial Q}{\partial x}.$$

Not every closed differential is exact. For certain domains, the so-called simply connected domains, any closed differential is exact; in fact, this statement characterizes simply connected domains. We content ourselves with the following theorem for star-shaped domains, which includes the simply connected domains of most interest to us. The theorem is the analogue of Part II of the fundamental theorem of calculus. It gives conditions on a smooth differential to have an antiderivative, that is, to be exact.

Theorem (Part II). *Let P and Q be continuously differentiable complex-valued functions on a domain D. Suppose*
(i) D is a star-shaped domain (as a disk or rectangle), and
(ii) the differential $P\,dx + Q\,dy$ is closed on D.
Then $P\,dx + Q\,dy$ is exact on D.

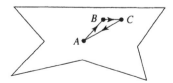

The proof is similar to the one given just above. Suppose that D is star-shaped with respect to the point $A \in D$. We define $h(B)$ at any point $B \in D$ by

$$h(B) \;=\; \int_A^B P\,dx \;+\; Q\,dy,$$

where the path of integration is the straight line segment from A to B. We claim that $dh = P\,dx + Q\,dy$. To see this, fix $B = (x_0, y_0)$, and let $C = (x, y_0)$ lie on the horizontal line through B and close enough to B so that the triangle with vertices A, B, C lies within D. We apply Green's theorem to the triangle, to obtain

$$\left(\int_A^B + \int_B^C + \int_C^A \right)(P\,dx \;+\; Q\,dy) \;=\; 0.$$

Thus

$$\int_A^C (P\,dx \;+\; Q\,dy) \;-\; \int_A^B (P\,dx \;+\; Q\,dy) \;=\; \int_B^C (P\,dx \;+\; Q\,dy),$$

or

$$h(x, y_0) - h(x_0, y_0) \;=\; \int_{x_0}^x P(t, y_0)\,dt.$$

From the fundamental theorem of calculus we obtain

$$\frac{\partial h}{\partial x}(x_0, y_0) \;=\; P(x_0, y_0).$$

Similarly,

$$\frac{\partial h}{\partial y}(x_0, y_0) \;=\; Q(x_0, y_0).$$

Consequently, $dh = P\,dx + Q\,dy$, and $P\,dx + Q\,dy$ is exact.

Example. Consider the differential

$$\frac{-y\,dx\; +\; x\,dy}{x^2 + y^2}, \qquad x + iy \in \mathbb{C}\backslash\{0\}.$$

A straightforward calculation (Exercise 2) reveals that the differential is closed on $\mathbb{C}\backslash\{0\}$. If we integrate the differential around the unit circle, using the parameterization $x = \cos\theta$, $y = \sin\theta$, we obtain

$$\oint_{|z|=1} \frac{-y\,dx\; +\; x\,dy}{x^2 + y^2} = \int_0^{2\pi} d\theta = 2\pi.$$

Thus the integral is not independent of path, and the differential is not exact on $\mathbb{C}\backslash\{0\}$. On the other hand, since $\mathbb{C}\backslash(-\infty, 0]$ is a star-shaped domain, the differential *is* exact on $\mathbb{C}\backslash(-\infty, 0]$. On this domain the differential coincides with $d(\operatorname{Arg} z)$.

Now suppose that $P\,dx + Q\,dy$ is a closed differential on a domain D. We fix points $A, B \in D$, and we consider paths γ in D from A to B. The integral $\int_\gamma P\,dx + Q\,dy$ may depend on the path γ. We claim, however, that if two paths γ_0 and γ are "sufficiently near" to each other, then $\int_{\gamma_0} P\,dx + Q\,dy = \int_\gamma P\,dx + Q\,dy$. By "sufficiently near" we mean that there are successive points $A = A_0, A_1, A_2, \ldots, A_n = B$ on γ_0 and $A = C_0, C_1, C_2, \ldots, C_n = B$ on γ such that the intervals on γ_0 from A_{k-1} to A_k and on γ from C_{k-1} to C_k are contained in the same disk Δ_k, which is contained in D. To see that the integrals are the same, we let γ_k be the path in D that follows γ from A to the point C_k, then follows a straight line segment in Δ_k from C_k to A_k, then follows γ_0 from A_k to B. Thus γ_k is obtained from γ_{k-1} by changing only the subpath in Δ_k from C_{k-1} to A_k, so that instead of following the straight line from C_{k-1} to A_{k-1} and then γ_0 from A_{k-1} to A_k, we follow γ from C_{k-1} to C_k and then the straight line from C_k to A_k. Since the integral of $P\,dx + Q\,dy$ is independent of path in the disk Δ_k, this change in γ_{k-1} to γ_k does not affect the integral,

$$\int_{\gamma_{k-1}} P\,dx + Q\,dy = \int_{\gamma_k} P\,dx + Q\,dy, \qquad 1 \le k \le n.$$

Since $\gamma_n = \gamma$, we obtain after n steps

$$\int_\gamma P\,dx + Q\,dy = \int_{\gamma_0} P\,dx + Q\,dy.$$

This identity holds not only if γ is near γ_0 but whenever γ can be obtained by deforming γ_0 continuously. We state the deformation theorem formally.

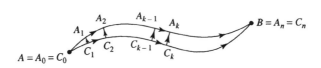

Theorem. *Let D be a domain, and let $\gamma_0(t)$ and $\gamma_1(t)$, $a \le t \le b$, be two paths in D from A to B. Suppose that γ_0 can be continuously deformed to γ_1, in the sense that for $0 \le s \le 1$ there are paths $\gamma_s(t)$, $a \le t \le b$, from A to B such that $\gamma_s(t)$ depends continuously on s and t for $0 \le s \le 1$, $a \le t \le b$. Then*

$$(2.3) \qquad \int_{\gamma_0} P\,dx + Q\,dy \;=\; \int_{\gamma_1} P\,dx + Q\,dy$$

for any closed differential $P\,dx + Q\,dy$ on D.

The idea of the proof is that any continuous deformation of γ_0 can be realized as a finite number of successive modifications by paths in D from A to B that are close to each other. The proof requires a compactness argument, but otherwise it is straightforward. The compactness argument allows us to find paths corresponding to $0 = s_0 < s_1 < \cdots < s_m = 1$, and successive t-values $a = t_0 < t_1 < \cdots < t_n = b$ such that the subpaths $\gamma_{s_{j-1}}(t)$ and $\gamma_{s_j}(t)$, $t_{k-1} \le t \le t_k$, lie in the same disk Δ_{jk} in D. This can be done on account of the continuity of $\gamma_s(t)$ in s and t. Then $\gamma_{s_{j-1}}$ is "sufficiently near" to γ_{s_j} for each j, and the rolling wave argument above shows that the integrals of $P\,dx + Q\,dy$ over the two paths are the same, this for $1 \le j \le n$, so that (2.3) holds.

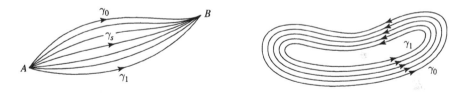

A slight variation of the argument establishes a deformation theorem for closed paths in D. These are paths in D that start and end at the same point. When we deform closed paths, we allow the starting point to move also.

Theorem. *Let D be a domain, and let $\gamma_0(t)$ and $\gamma_1(t)$, $a \le t \le b$, be two closed paths in D. Suppose that γ_0 can be continuously deformed to γ_1, in the sense that for $0 \le s \le 1$ there are closed paths $\gamma_s(t)$, $a \le t \le b$, such that $\gamma_s(t)$ depends continuously on s and t for $0 \le s \le 1$, $a \le t \le b$. Then*

$$\int_{\gamma_0} P\,dx + Q\,dy \;=\; \int_{\gamma_1} P\,dx + Q\,dy$$

for any closed differential $P\,dx + Q\,dy$ on D.

Summary. We have defined what it means for a differential $P\,dx + Q\,dy$ to be **exact**, to be **closed**, and to be **independent of path**. We have shown that

independent of path $\Longleftrightarrow$ exact $\Longrightarrow$ closed.

For star-shaped domains we have shown that

independent of path $\Longleftrightarrow$ exact $\Longleftrightarrow$ closed.

We have also shown that if $Pdx + Qdy$ is a closed differential, then a deformation in the path from A to B does not change the value of the integral of $Pdx + Qdy$ along the path.

Exercises for III.2

1. Determine whether each of the following line integrals is independent of path. If it is, find a function h such that $dh = Pdx + Qdy$. If it is not, find a closed path γ around which the integral is not zero. (a) $xdx + ydy$, (b) $x^2dx + y^5dy$, (c) $ydx + xdy$, (d) $ydx - xdy$.

2. Show that the differential

$$\frac{-ydx + xdy}{x^2 + y^2}, \qquad (x,y) \neq (0,0),$$

   is closed. Show that it is not independent of path on any annulus centered at 0.

3. Suppose that P and Q are smooth functions on the annulus $\{a < |z| < b\}$ that satisfy $\partial P/\partial y = \partial Q/\partial x$. Show directly using Green's theorem that $\oint_{|z|=r} P\,dx + Q\,dy$ is independent of the radius r, for $a < r < b$.

4. Let P and Q be smooth functions on D satisfying $\partial P/\partial y = \partial Q/\partial x$. Let γ_0 and γ_1 be two closed paths in D such that the straight line segment from $\gamma_0(t)$ to $\gamma_1(t)$ lies in D for every parameter value t. Then $\int_{\gamma_0} P\,dx + Q\,dy = \int_{\gamma_1} P\,dx + Q\,dy$. Use this to give another solution to the preceding exercise.

5. Let $\gamma_0(t)$ and $\gamma_1(t)$, $0 \le t \le 1$, be paths in the slit annulus $\{a < |z| < b\}\backslash(-b, -a)$ from A to B. Write down explicitly a family of paths $\gamma_s(t)$ from A to B in the slit annulus that deforms γ_0 continuously to γ_1. *Suggestion.* Deform separately the modulus and the principal value of the argument.

6. Show that any closed path $\gamma(t)$, $0 \le t \le 1$, in the annulus $\{a < |z| < b\}$ can be deformed continuously to the circular path $\sigma(t) = \gamma(0)e^{2\pi imt}$, $0 \le t \le 1$, for some integer m. *Hint.* Reduce to the case where $|\gamma(t)| \equiv |\gamma(0)|$ is constant. Then start by finding a subdivision $0 = t_0 < t_1 < \cdots < t_n = 1$ such that $\arg \gamma(t)$ has a continuous determination on each interval $t_{j-1} \le t \le t_j$.

7. Show that if 0 and ∞ lie in different connected components of the complement $\mathbb{C}^*\backslash D$ of D in the extended complex plane, then there is a closed path γ in D such that $\int_\gamma d\theta \neq 0$. *Hint.* The hypothesis means that there are $\delta > 0$ and a bounded subset E of $\mathbb{C}\backslash D$ such that $0 \in E$, and every point of E has distance at least 5δ from every point of $\mathbb{C}\backslash D$ not in E. Lay down a grid of squares in the plane with side length δ, and let F be the union of the closed squares in the grid that meet E or that border on a square meeting E. Show that ∂F is a finite union of closed paths in D, and that $\int_{\partial F} d\theta = 2\pi$.

3. Harmonic Conjugates

The basis for application of Green's theorem to harmonic functions is the following important observation.

Lemma. *If $u(x,y)$ is harmonic, then the differential*

$$(3.1) \qquad -\frac{\partial u}{\partial y}dx \;+\; \frac{\partial u}{\partial x}dy$$

is closed.

Indeed, for this differential the condition (2.1) for $P = -\partial u/\partial y$ and $Q = \partial u/\partial x$ becomes $-\partial^2 u/\partial y^2 = \partial^2 u/\partial x^2$, which is equivalent to Laplace's equation.

Now suppose that $u(x,y)$ is harmonic on a star-shaped domain D. If we apply the theorem in Section 2 to the differential given in (3.1), we obtain a smooth function $v(x,y)$ such that

$$(3.2) \qquad dv \;=\; -\frac{\partial u}{\partial y}dx \;+\; \frac{\partial u}{\partial x}dy.$$

The equation (3.2) is equivalent to

$$\frac{\partial v}{\partial x} \;=\; -\frac{\partial u}{\partial y}, \qquad \frac{\partial v}{\partial y} \;=\; \frac{\partial u}{\partial x},$$

which are the Cauchy-Riemann equations. Thus $u + iv$ is analytic, and we have established the following theorem.

Theorem. *Any harmonic function $u(x,y)$ on a star-shaped domain D (as a disk or rectangle) has a harmonic conjugate function $v(x,y)$ on D.*

By (3.2), the harmonic conjugate $v(x,y)$ is given explicitly up to an additive constant by

$$(3.3) \qquad v(B) \;=\; \int_A^B -\frac{\partial u}{\partial y}dx \;+\; \frac{\partial u}{\partial x}dy,$$

where A is fixed, and the integral is independent of path in D. If D is a disk, and we take the path from A to B to be a vertical interval followed by a horizontal interval, we obtain the formula (5.3) derived in Section II.5.

Example. To find a harmonic conjugate $v(z)$ for $u = \log|z|$ on the star-shaped domain $\mathbb{C}\backslash(-\infty, 0]$, we express u in the form

$$u(x, y) \;=\; \frac{1}{2}\log(x^2 + y^2),$$

and we compute

$$du \;=\; \frac{x}{x^2 + y^2}dx \;+\; \frac{y}{x^2 + y^2}dy.$$

Equation (3.2) becomes

$$dv \;=\; -\frac{y}{x^2 + y^2}dx \;+\; \frac{x}{x^2 + y^2}dy.$$

This leads to the identity

$$\operatorname{Arg} z \;=\; \int_1^z \frac{-y}{x^2 + y^2}dx \;+\; \frac{x}{x^2 + y^2}dy, \qquad z \notin (-\infty, 0],$$

since the principal branch $\operatorname{Arg} z$ is the unique harmonic conjugate of $\log|z|$ on $\mathbb{C}\backslash(-\infty, 0]$, normalized to vanish at $z = 1$.

Exercises for III.3

1. For each of the following harmonic functions u, find du, find dv, and find v, the conjugate harmonic function of u.
   (a) $u(x, y) = x - y$ (c) $u(x, y) = \sinh x \cos y$
   (b) $u(x, y) = x^3 - 3xy^2$ (d) $u(x, y) = \dfrac{y}{x^2 + y^2}$

2. Show that a complex-valued function $h(z)$ on a star-shaped domain D is harmonic if and only if $h(z) = f(z) + \overline{g(z)}$, where $f(z)$ and $g(z)$ are analytic on D.

3. Let $D = \{a < |z| < b\}\backslash(-b, -a)$, an annulus slit along the negative real axis. Show that any harmonic function on D has a harmonic conjugate on D. *Suggestion.* Fix c between a and b, and define $v(z)$ explicitly as a line integral along the path consisting of the straight line from c to $|z|$ followed by the circular arc from $|z|$ to z. Or map the slit annulus to a rectangle by $w = \operatorname{Log} z$.

4. Let $u(z)$ be harmonic on the annulus $\{a < |z| < b\}$. Show that there is a constant C such that $u(z) - C\log|z|$ has a harmonic conjugate on the annulus. Show that C is given by

$$C \;=\; \frac{r}{2\pi}\int_0^{2\pi} \frac{\partial u}{\partial r}\left(re^{i\theta}\right) d\theta,$$

where r is any fixed radius, $a < r < b$.

5. The flux of a function u across a curve γ is defined to be

$$\int_\gamma \frac{\partial u}{\partial n}\, ds \;=\; \int_\gamma \nabla u \cdot \mathbf{n}\, ds,$$

where $\mathbf{n}$ is the unit normal vector to γ and ds is arc length. Show that if a harmonic function u on a domain D has a conjugate harmonic function v on D, then the integral giving the flux is independent of path in D. Further, the flux across a path γ in D from A to B is $v(B) - v(A)$.

4. The Mean Value Property

Let $h(z)$ be a continuous real-valued function on a domain D. Let $z_0 \in D$, and suppose D contains the disk $\{|z - z_0| < \rho\}$. We define the **average value** of $h(z)$ on the circle $\{|z - z_0| = r\}$ to be

$$A(r) \;=\; \int_0^{2\pi} h\left(z_0 + re^{i\theta}\right) \frac{d\theta}{2\pi}, \qquad 0 < r < \rho.$$

Since $h(z)$ is continuous, the average value $A(r)$ varies continuously with the radius r. Since the values of $h(z)$ are all near $h(z_0)$ when z is near z_0, the averages $A(r)$ are also near $h(z_0)$ when r is small, and consequently $A(r)$ tends to $h(z_0)$ as r decreases to 0. This can be seen with complete rigor from the estimates

$$|A(r) - h(z_0)| \;=\; \left| \int_0^{2\pi} \left[h\left(z_0 + re^{i\theta}\right) - h(z_0) \right] \frac{d\theta}{2\pi} \right|$$

$$\leq \int_0^{2\pi} \left| h\left(z_0 + re^{i\theta}\right) - h(z_0) \right| \frac{d\theta}{2\pi},$$

where we have used the fact that $d\theta/2\pi$ is a probability measure, that is, it is positive and its integral is 1. The continuity of $h(z)$ at z_0 guarantees that the integrand in the right-hand side tends to 0 uniformly in θ, so that the integral tends to 0 as r tends to 0.

Theorem. *If $u(z)$ is a harmonic function on a domain D, and if the disk $\{|z - z_0| < \rho\}$ is contained in D, then*

(4.1) $$u(z_0) \;=\; \int_0^{2\pi} u\left(z_0 + re^{i\theta}\right) \frac{d\theta}{2\pi}, \qquad 0 < r < \rho.$$

In other words, the average value of a harmonic function on the boundary circle of any disk contained in D is its value at the center of the disk. To see this, we begin with the identity

$$0 = \oint_{|z-z_0|=r} -\frac{\partial u}{\partial y}dx + \frac{\partial u}{\partial x}dy,$$

which follows immediately from Green's theorem and the harmonicity of $u(z)$. (See the Lemma in Section 3.) We parametrize the circle by $x(\theta) = x_0 + r\cos\theta$ and $y(\theta) = y_0 + r\sin\theta$, and we obtain

$$(4.2)\quad 0 = r\int_0^{2\pi} \left[\frac{\partial u}{\partial x}\cos\theta + \frac{\partial u}{\partial y}\sin\theta\right]d\theta = r\int_0^{2\pi}\frac{\partial u}{\partial r}\left(z_0 + re^{i\theta}\right)d\theta.$$

Since $u(z)$ is smooth, we can interchange the order of integration and differentiation. We obtain after dividing by $2\pi r$ that

$$0 = \frac{\partial}{\partial r}\int_0^{2\pi} u\left(z_0 + re^{i\theta}\right)\frac{d\theta}{2\pi},\qquad 0 < r < \rho.$$

Thus

$$\int_0^{2\pi} u\left(z_0 + re^{i\theta}\right)\frac{d\theta}{2\pi}$$

is constant for $0 < r < \rho$. Since $u(z)$ is continuous at z_0, the average value tends to $u(z_0)$ as $r \to 0$, and the constant is $u(z_0)$. This establishes (4.1).

We say that a continuous function $h(z)$ on a domain D has the **mean value property** if for each point $z_0 \in D$, $h(z_0)$ is the average of its values over any small circle centered at z_0. In the language of formal mathematics, this means that for any $z_0 \in D$, there is $\varepsilon > 0$ such that

$$h(z_0) = \int_0^{2\pi} h\left(z_0 + re^{i\theta}\right)\frac{d\theta}{2\pi},\qquad 0 < r < \varepsilon.$$

Our theorem above has a simple restatement: Harmonic functions have the mean value property. We will show in Chapter X that the converse is true, that any continuous function on D with the mean value property is harmonic. This is rather remarkable, since the hypothesis requires only continuity but no differentiability of the function.

Exercises for III.4

1. Let $f(z)$ be a continuous function on a domain D. Show that if $f(z)$ has the mean value property with respect to circles, as defined above, then $f(z)$ has the mean value property with respect to disks, that is, if $z_0 \in D$ and D_0 is a disk centered at z_0 with area A and contained in D, then $f(z_0) = \dfrac{1}{A}\iint_{D_0} f(z)\,dx\,dy$.

2. Derive (4.2) from the polar form of the Cauchy-Riemann equations (Exercise II.3.8).

3. A function $f(t)$ on an interval $I = (a, b)$ has the **mean value prop-
    erty** if

$$f\left(\frac{s+t}{2}\right) = \frac{f(s) + f(t)}{2}, \qquad s, t \in I.$$

Show that any affine function $f(t) = At + B$ has the mean value
property. Show that any continuous function on I with the mean
value property is affine.

4. Formulate the mean value property for a function on a domain in $\mathbb{R}^3$,
    and show that any harmonic function has the mean value property.
    Hint. For $A \in \mathbb{R}^3$ and $r > 0$, let B_r be the ball of radius r centered
    at A, with volume element $d\tau$, and let ∂B_r be its boundary sphere,
    with area element $d\sigma$ and unit outward normal vector $\mathbf{n}$. Apply the
    Gauss divergence theorem

$$\iint_{\partial B_r} \mathbf{F} \cdot \mathbf{n} \, d\sigma = \iiint_{B_r} \nabla \cdot \mathbf{F} \, d\tau$$

to $\mathbf{F} = \nabla u$.

5. The Maximum Principle

The strict maximum principle asserts that if a real-valued harmonic function attains its maximum on a domain D, then it is constant.

Strict Maximum Principle (Real Version). *Let $u(z)$ be a real-valued
harmonic function on a domain D such that $u(z) \leq M$ for all $z \in D$. If
$u(z_0) = M$ for some $z_0 \in D$, then $u(z) = M$ for all $z \in D$.*

The idea of the proof is to use the mean value property to show that the
set of points for which $u(z) = M$ is open. Indeed, suppose $u(z_1) = M$, and
express the mean value equality (4.1) in the form

(5.1) $$0 = \int_0^{2\pi} \left[u(z_1) - u\left(z_1 + re^{i\theta}\right) \right] \frac{d\theta}{2\pi}, \qquad 0 < r < \rho.$$

Since the integrand is nonnegative (≥ 0) and continuous, the integral (5.1)
can be zero only if the integrand is zero. Thus $u(z_1 + re^{i\theta}) = u(z_1) = M$
for $0 \leq \theta \leq 2\pi$ and $0 < r < \rho$, and the set $\{u(z) = M\}$ contains a disk
centered at each of its points, hence is open. Now, the set $\{u(z) < M\}$ is
also open, since $u(z)$ is continuous. Since D is a domain, one of these two
sets is empty and the other coincides with all of D. (See Section II.1.) In
other words, either $u(z) < M$ for all $z \in D$, or $u(z) = M$ for all $z \in D$,
and this proves the theorem.

Recall that a complex-valued function is **harmonic** if its real and imaginary parts are harmonic, that is, if the function satisfies Laplace's equa-

tion. Thus any analytic function is harmonic. There is a strict maximum principle also for complex-valued harmonic functions. It asserts that if a complex-valued harmonic function attains its maximum modulus on a domain D, then it is constant.

Strict Maximum Principle (Complex Version). *Let h be a bounded complex-valued harmonic function on a domain D. If $|h(z)| \leq M$ for all $z \in D$, and $|h(z_0)| = M$ for some $z_0 \in D$, then $h(z)$ is constant on D.*

This can be derived easily from the real version of the strict maximum principle. We replace $h(z)$ by $\lambda h(z)$ for an appropriate unimodular constant λ, and we can assume that $h(z_0) = M$. Let $u(z) = \operatorname{Re} h(z)$. Then $u(z)$ is a harmonic function on D that attains its maximum at z_0. By the strict maximum principle for real-valued harmonic functions, $u(z) = M$ for all $z \in D$. Since $|h(z)| \leq M$ and $\operatorname{Re} h(z) = M$, we must have $\operatorname{Im} h(z) = 0$ for $z \in D$. Hence $h(z)$ is constant on D.

As a corollary of the strict maximum principle we also obtain the following version of the maximum principle. In words, it asserts that a complex-valued harmonic function on a bounded domain attains its maximum modulus on the boundary.

Maximum Principle. *Let $h(z)$ be a complex-valued harmonic function on a bounded domain D such that $h(z)$ extends continuously to the boundary ∂D of D. If $|h(z)| \leq M$ for all $z \in \partial D$, then $|h(z)| \leq M$ for all $z \in D$.*

The proof of the maximum principle hinges on the fact that a continuous function on a compact set attains its maximum modulus at some point of the set. (See Section II.1.) In this case the compact set is the union of the domain and its boundary, which is a closed bounded set. If the harmonic function attains its maximum modulus at some point of D, then it is constant. Thus in all cases it attains its maximum modulus on the boundary of D.

The maximum principle is useful, for instance, for demonstrating convergence of a sequence of harmonic functions. To show that a sequence of harmonic functions converges in a disk or rectangle, it suffices to obtain good estimates on the boundary of the disk or rectangle, since the boundary estimates automatically persist in the interior.

Exercises for III.5

1. Let D be a bounded domain, and let u be a real-valued harmonic function on D that extends continuously to the boundary ∂D. Show that if $a \leq u \leq b$ on ∂D, then $a \leq u \leq b$ on D.

2. Fix $n \geq 1$, $r > 0$, and $\lambda = \rho e^{i\varphi}$. What is the maximum modulus of $z^n + \lambda$ over the disk $\{|z| \leq r\}$? Where does $z^n + \lambda$ attain its maximum modulus over the disk?

3. Use the maximum principle to prove the fundamental theorem of algebra, that any polynomial $p(z)$ of degree $n \geq 1$ has a zero, by applying the maximum principle to $1/p(z)$ on a disk of large radius.

4. Let $f(z)$ be an analytic function on a domain D that has no zeros on D. (a) Show that if $|f(z)|$ attains its minimum on D, then $f(z)$ is constant. (b) Show that if D is bounded, and if $f(z)$ extends continuously to the boundary ∂D of D, then $|f(z)|$ attains its minimum on ∂D.

5. Let $f(z)$ be a bounded analytic function on the right half-plane. Suppose that $f(z)$ extends continuously to the imaginary axis and satisfies $|f(iy)| \leq M$ for all points iy on the imaginary axis. Show that $|f(z)| \leq M$ for all z in the right half-plane. *Hint.* For $\varepsilon > 0$ small, consider $(z + 1)^{-\varepsilon} f(z)$ on a large semidisk.

6. Let $f(z)$ be a bounded analytic function on the right half-plane. Suppose that $\limsup |f(z)| \leq M$ as z approaches any point of the imaginary axis. Show that $|f(z)| \leq M$ for all z in the half-plane. *Remark.* This is a technical improvement on the preceding exercise for students who can deal with a lim sup (see Section V.1).

7. Let $f(z)$ be a bounded analytic function on the open unit disk $\mathbb{D}$. Suppose there are a finite number of points on the boundary such that $f(z)$ extends continuously to the arcs of $\partial \mathbb{D}$ separating the points and satisfies $|f(e^{i\theta})| \leq M$ there. Show that $|f(z)| \leq M$ on $\mathbb{D}$. *Hint.* In the case that there is only one exceptional point $z = 1$, consider the function $(1 - z)^\varepsilon f(z)$.

8. Let $f(z)$ be a bounded analytic function on a horizontal strip in the complex plane. Suppose that $f(z)$ extends continuously to the boundary lines of the strip and satisfies $|f(z)| \leq M$ there. Show that $|f(z)| \leq M$ for all z in the strip. *Hint.* Find a conformal map of the strip onto $\mathbb{D}$ and apply Exercise 7.

9. Let D be an unbounded domain, $D \neq \mathbb{C}$, and let $u(z)$ be a harmonic function on D that extends continuously to the boundary ∂D. Suppose that $u(z)$ is bounded below on D, and that $u(z) \geq 0$ on ∂D. Show that $u(z) \geq 0$ on D. *Hint.* Suppose $0 \in \partial D$, and consider functions of the form $u(z) + \rho \log |z|$ on $D \cap \{|z| > \varepsilon\}$.

10. Let D be a bounded domain, and let $z_0 \in \partial D$. Let $u(z)$ be a harmonic function on D that extends continuously to each boundary point of D except possibly z_0. Suppose that $u(z)$ is bounded below

on D, and that $u(z) \geq 0$ for all $z \in \partial D$, $z \neq z_0$. Show that $u(z) \geq 0$ on D.

11. Let E be a bounded set of integer lattice points in the complex plane. A point $m + ni$ of E is an **interior point** of E if its four immediate neighbors $m \pm 1 + ni$, $m + ni \pm i$ belong to E. Otherwise, $m + ni$ is a **boundary point** of E. A function on E is **harmonic** if its value at any interior point of E is the average of its values at the four immediate neighbors. Show that a harmonic function on a bounded set of lattice points attains its maximum modulus on the boundary of the set.

6. Applications to Fluid Dynamics

We consider a fluid flow in a domain D in the plane. We think of the fluid as a collection of particles that move in the plane as time evolves. We associate with the particle at the point z its velocity vector $\mathbf{V}(z)$. The direction of $\mathbf{V}(z)$ is the direction the particle is moving, and the magnitude $|\mathbf{V}(z)|$ is its speed. We make the following assumptions on the flow.

1. The flow is independent of time, that is, the velocity vector field $\mathbf{V}(z)$ does not change with time.
2. There are no sources or sinks in D; that is, no fluid is created or destroyed in D.
3. The flow is incompressible; that is, the density of the fluid is the same at each point of D, and it does not change with time.
4. The flow is "irrotational"; that is, there is no circulation of fluid around small circles centered in D.

We will return shortly to explain this fourth condition. First we define "flux" across a curve and "circulation" around a closed curve.

Let γ be a curve in the plane. We denote the unit tangent vector to γ by $\mathbf{t}$ and the unit normal vector to γ by $\mathbf{n}$. If γ is parameterized by arc length s, $\gamma(s) = (x(s), y(s))$, then the unit tangent and normal vectors are given respectively by

$$\mathbf{t} = \left(\frac{dx}{ds}, \frac{dy}{ds} \right), \qquad \mathbf{n} = \left(\frac{dy}{ds}, -\frac{dx}{ds} \right).$$

The normal component of $\mathbf{V}$ to the curve γ is then $\mathbf{V} \cdot \mathbf{n}$. If the flow is parallel to γ, then $\mathbf{V} \cdot \mathbf{n} = 0$, and no fluid crosses γ. The maximum flow per unit length is obtained when γ is orthogonal to $\mathbf{V}$.

We define the **flux of the fluid flow** across γ to be the integral of the normal component of $\mathbf{V}$ with respect to arc length,

$$\text{flux across } \gamma = \int_\gamma \mathbf{V} \cdot \mathbf{n} \, ds.$$

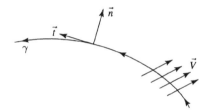

Except for a constant involving units and the density, the flux across γ can be regarded as the amount of fluid crossing γ. If we express $\mathbf{V} = (P, Q)$ in terms of its components P and Q, we obtain an expression for the flux across γ as a line integral:

$$(6.1) \qquad \int_\gamma \mathbf{V} \cdot \mathbf{n}\, ds = \int_\gamma P \frac{dy}{ds} ds - Q \frac{dx}{ds} ds = \int_\gamma P\, dy - Q\, dx.$$

If γ is a closed curve, we define the **circulation of the fluid flow** around γ to be the integral of the tangential component of $\mathbf{V}$ with respect to arc length:

$$\text{circulation around } \gamma = \int_\gamma \mathbf{V} \cdot \mathbf{t}\, ds.$$

This can also be expressed in terms of the components P and Q of $\mathbf{V}$ as a line integral,

$$(6.2) \qquad \int_\gamma \mathbf{V} \cdot \mathbf{t}\, ds = \int_\gamma P \frac{dx}{ds} ds + Q \frac{dy}{ds} ds = \int_\gamma P\, dx + Q\, dy.$$

Let $\gamma_\varepsilon = z_0 + \varepsilon e^{i\theta}$ be a small circle around a fixed point $z_0 \in D$. From (6.2) and Green's theorem, the amount of fluid circulating around γ_ε is given by

$$\oint_{\gamma_\varepsilon} \mathbf{V} \cdot \mathbf{t}\, ds = \iint_{|z - z_0| < \varepsilon} \left(\frac{\partial Q}{\partial x} - \frac{\partial P}{\partial y} \right) dx\, dy.$$

The irrotationality of the fluid flow (the fourth condition above) means simply that this integral is zero for all small $\varepsilon > 0$. This occurs if and only if the integrand is zero, that is,

$$(6.3) \qquad \frac{\partial P}{\partial y} = \frac{\partial Q}{\partial x}.$$

Thus the irrotationality of the flow on D means that (6.3) holds on D.

The mathematical formulation of the second and third conditions is that the net flow of fluid across the boundary of any small circle γ_ε centered at a point z_0 of D is zero. Using Green's theorem again, we obtain

$$0 = \oint_{\gamma_\varepsilon} \mathbf{V} \cdot \mathbf{n}\, ds = \int_{\gamma_\varepsilon} P\, dy - Q\, dx = \iint_{|z - z_0| < \varepsilon} \left(\frac{\partial P}{\partial x} + \frac{\partial Q}{\partial y} \right) dx\, dy.$$

We conclude as before that

(6.4)
$$\frac{\partial P}{\partial x} + \frac{\partial Q}{\partial y} = 0.$$

Thus the second and third conditions on the flow can be reinterpreted to assert that (6.4) holds on D.

The conditions (6.3) and (6.4) can be expressed directly in terms of the velocity $\mathbf{V}(z)$, as

$$\nabla \times \mathbf{V} = \mathbf{0}, \qquad \nabla \cdot \mathbf{V} = 0.$$

This is the form of the equations that is usually most recognizable to physical scientists. It generalizes to three dimensions.

Now, the condition (6.3) is just that the differential $P\,dx + Q\,dy$ be closed. On any disk in D, we can then find a smooth function ϕ such that $d\phi = P\,dx + Q\,dy$, that is,

$$P = \frac{\partial \phi}{\partial x}, \qquad Q = \frac{\partial \phi}{\partial y}.$$

In terms of the vector field $\mathbf{V}$, this is the same as $\nabla \phi = \mathbf{V}$. The function ϕ is called the **potential function** of $\mathbf{V}$. It is unique, up to adding a constant. In terms of ϕ, the condition (6.4) is that

$$\frac{\partial^2 \phi}{\partial x^2} + \frac{\partial^2 \phi}{\partial y^2} = 0,$$

that is, that the potential function ϕ is harmonic. On any disk in D, ϕ then has a conjugate harmonic function ψ, so that $f(z) = \phi(z) + i\psi(z)$ is analytic. The function $f(z)$ is called the **complex velocity potential** of the flow. It is also unique up to adding a constant. Note that while ϕ and ψ may be defined on any disk in D, neither ϕ nor ψ need be defined on all of D.

The velocity vector field $\mathbf{V}(z)$ is expressed in terms of the complex velocity potential $f(z) = \phi(z) + i\psi(z)$, in complex notation, by

$$\mathbf{V}(z) = \frac{\partial \phi}{\partial x} + i\frac{\partial \phi}{\partial y} = \frac{\partial \phi}{\partial x} - i\frac{\partial \psi}{\partial x} = \overline{f'(z)}.$$

Thus the speed of the fluid particles is given by $|\mathbf{V}(z)| = |f'(z)|$.

As we saw in Section II.6, the level curves of ϕ and of ψ are orthogonal to each other. Since the level curves of ϕ are also orthogonal to $\nabla \phi = \mathbf{V}$, the level curves of ψ are tangent to $\mathbf{V}$. Thus the fluid particles flow along the level curves of ψ.

We define the **streamlines** of the fluid flow to be the level curves of ψ, that is, the curves $\{\psi = c\}$, for c constant. The streamlines of the flow represent the paths of the fluid particles. The function $\psi(z)$ is called the **stream function** of the fluid flow.

The stream function ψ can be used to calculate the flux of the fluid flow across a curve γ. Indeed, the flux across γ is given with the help of the

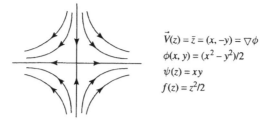

$$\vec{V}(z) = \bar{z} = (x, -y) = \nabla\phi$$
$$\phi(x, y) = (x^2 - y^2)/2$$
$$\psi(z) = xy$$
$$f(z) = z^2/2$$

Cauchy-Riemann equations by

$$\int_\gamma \mathbf{V} \cdot \mathbf{n}\,ds = \int_\gamma \frac{\partial\phi}{\partial x}dy - \frac{\partial\phi}{\partial y}dx = \int_\gamma \frac{\partial\psi}{\partial y}dy + \frac{\partial\psi}{\partial x}dx = \int_\gamma d\psi.$$

Thus the flux of the flow across a curve γ is equal to the increase of the stream function $\psi(z)$ along γ.

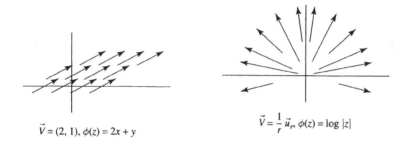

$$\vec{V} = (2, 1),\ \phi(z) = 2x + y \qquad\qquad \vec{V} = \frac{1}{r}\vec{u}_r,\ \phi(z) = \log|z|$$

Example. The simplest flow in the plane is a constant flow $\mathbf{V} = (\alpha, \beta)$, where α and β are constants. The velocity potential ϕ satisfying $\nabla\phi = \mathbf{V}$ is given, up to an additive constant, by $\phi(x, y) = \alpha x + \beta y$. The stream function ψ is the harmonic conjugate of ϕ, given up to an additive constant by $\psi(x, y) = \alpha y - \beta x$. The complex velocity potential of the flow is $f(z) = (\alpha - i\beta)z$.

Example. Consider the vector field on the punctured plane $\mathbb{C}\backslash\{0\}$ defined by

$$\mathbf{V} = \frac{1}{r}\mathbf{u}_r,$$

where $\mathbf{u}_r$ is the unit vector in the radial direction. We can express $\mathbf{V}(z)$ as $\nabla\phi(z)$ for $\phi(z) = \log|z|$. Since $\log|z|$ is harmonic, $\mathbf{V}(z)$ is the velocity vector field of a fluid flow. The stream function of the flow is $\psi(z) = \arg z$, which is defined only locally, and the complex velocity potential is $f(z) = \log z$, also defined only locally. The flux of the flow across a circle centered at the origin is calculated directly using $\mathbf{n} = \mathbf{u}_r$ and $ds = r\,d\theta$ by

$$\oint_{|z|=r} \mathbf{V} \cdot \mathbf{n}\,ds = \int \frac{1}{r}\mathbf{u}_r \cdot \mathbf{u}_r\,r\,d\theta = \int_0^{2\pi} d\theta = 2\pi.$$

This coincides with the increase of the stream function $\arg z$ around the circle. The origin is a source for the fluid flow. The speed of the fluid particles is given by $|\mathbf{V}(z)| = |f'(z)| = 1/|z| = 1/r$. The fluid particles emanate from the origin and follow along rays at continually diminishing speeds.

In addition to the four conditions above, there is one further important condition to be placed on a fluid flow velocity $\mathbf{V}(z)$, which is a boundary condition.

5. If no fluid is injected or extracted through a boundary curve, then the velocity vector field $\mathbf{V}(z)$ is parallel to the boundary along that curve.

Thus the boundary curve should be a streamline of the flow, and the stream function $\psi(z)$ should be constant on the curve.

source at 0

Example. We consider a fluid flow in the upper half-plane with a source or sink at 0. The fluid is injected or extracted from the upper half-plane at a constant rate at the origin. The stream function $\psi(z)$ should be a harmonic function in the upper half-plane that is constant on each of the boundary intervals $(-\infty, 0)$ and $(0, +\infty)$. Such a function is given by $\psi(z) = C \arg z$, where C is a constant. The corresponding complex velocity potential function is $f(z) = C \log z$, and the velocity vector field is $\mathbf{V}(z) = \overline{f'(z)} = 1/\overline{z} = z/|z|^2$. The flux of the fluid flow entering or exiting at the origin is the increase of $\psi(z)$ along a small semicircle in the upper half-plane centered at 0, which is πC. This determines the constant C. If $C > 0$, we have a source at 0, and if $C < 0$, we have a sink at 0.

One way to gain insight into a fluid flow is to map D conformally onto a domain for which the corresponding flow is simpler to understand. Flows are preserved by conformal maps, in the following sense. If $h : D \to U$ is a one-to-one analytic function (conformal map) from D onto a domain U, and if $f_0(w) = \phi_0(w) + i\psi_0(w)$ is a complex velocity potential for a flow on U, then the composition $f(z) = f_0(h(z))$ is analytic, hence the complex velocity potential for a flow on D.

One of the simplest flows to understand is the constant horizontal flow on the upper half-plane $\mathbb{H} = \{\operatorname{Im} z > 0\}$, for which no fluid enters or leaves across the bounding real line $\mathbb{R}$. A complex velocity potential for the flow is

$f_0(z) = z$. The streamlines of the flow are horizontal lines in $\mathbb{H}$, for which $\text{Im}(z)$ is constant. The boundary $\mathbb{R}$ of $\mathbb{H}$ is a streamline, corresponding to $\text{Im}(z) = 0$.

Suppose now γ is a curve in the plane extending to infinity in both directions, and suppose D is the domain lying on one side of γ. Suppose we wish to find a fluid flow in D such that no fluid enters or exits through γ. According to the boundary condition, γ should be a streamline of the flow. The problem can be solved by finding a conformal map $h : D \to \mathbb{H}$ that maps ∞ to ∞. In this case the complex velocity potential $f(z) = f_0(h(z))$ is simply $h(z)$ itself, and the vector velocity for the fluid flow is $\mathbf{V}(z) = \overline{h'(z)}$. The stream function is $\psi(z) = \text{Im}(h(z))$, and this is zero on γ. Thus γ is a streamline, and the boundary condition is satisfied. Note that any real multiple $Ch(z)$ is also a complex velocity potential for a fluid flow that satisfies the boundary conditions and has the same streamlines. It corresponds to multiplying the velocity vector field by C.

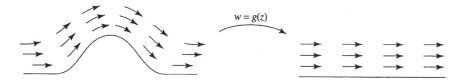

Example. Consider a fluid flow at a corner, represented by a sector $D = \{0 < \arg z < \alpha\}$, where $0 < \alpha \leq 2\pi$. We assume that no fluid passes through the boundary, so that the boundary of the sector should be a streamline. We map the sector conformally onto the upper half-plane by a power function

$$h(z) = z^{\pi/\alpha} = r^{\pi/\alpha}[\cos(\pi\theta/\alpha) + i\sin(\pi\theta/\alpha)].$$

We take $h(z)$ to be the complex velocity potential. Then the stream function is the imaginary part of $h(z)$, which is zero on the boundary of the sector. The streamlines are given in polar coordinates by

$$r = c(\sin(\pi\theta/\alpha))^{-\pi/\alpha}, \qquad 0 < \theta < \alpha,$$

where c is a positive constant. The velocity vector for this flow is given by $\mathbf{V}(z) = \overline{h'(z)}$, and the speed of the flow is

$$|\mathbf{V}(z)| = |h'(z)| = \frac{\pi}{\alpha}|z|^{(\pi/\alpha)-1}.$$

If $0 < \alpha < \pi$, the fluid particles traveling near the boundary slow down as they approach the corner, while if $\pi < \alpha \leq 2\pi$, the fluid particles traveling near the boundary speed up as they make the turn around the corner.

The successful analysis of a two-dimensional flow in D using the techniques of complex analysis often depends upon being able to find a conformal map of D onto an appropriate canonical domain, such as the upper

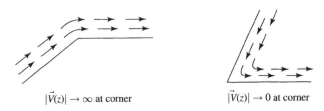

$|\vec{V}(z)| \to \infty$ at corner $|\vec{V}(z)| \to 0$ at corner

half-plane, where the solution is apparent. In Chapter XI we will focus on enlarging our stockpile of conformal maps.

Finally, note that the flows that can be modeled as two-dimensional flows form a rather narrow class of flows. The flow of air around an airplane wing profile can be analyzed using techniques of conformal mapping, but the flow of air around an entire airplane is a three-dimensional problem to which the techniques of complex analysis do not apply.

Exercises for III.6

1. Consider the fluid flow with constant velocity $\mathbf{V} = (2, 1)$. Find the velocity potential $\phi(z)$, the stream function $\psi(z)$, and the complex velocity potential $f(z)$ of the flow. Sketch the streamlines of the flow. Determine the flux of the flow across the interval $[0, 1]$ on the real axis and across the interval $[0, i]$ on the imaginary axis.

2. Fix real numbers α and β, and consider the vector field given in polar coordinates by

$$\mathbf{V}(r, \theta) \;=\; \frac{\alpha}{r}\mathbf{u_r} + \frac{\beta}{r}\mathbf{u_\theta},$$

   where $\mathbf{u_r}$ and $\mathbf{u_\theta}$ are the unit vectors in the r and θ directions, respectively. (a) Show that $\mathbf{V}(r, \theta)$ is the velocity vector field of a fluid flow, and find the velocity potential $\phi(z)$ of the flow. (b) Find the stream function $\psi(z)$ and the complex velocity potential $f(z)$ of the flow. (c) Determine the flux of the flow emanating from the origin. When is 0 a source and when is 0 a sink? (d) Sketch the streamlines of the flow in the case $\alpha = -1$ and $\beta = 1$.

3. Consider the fluid flow with velocity $\mathbf{V} = \nabla\phi$, where $\phi(r, \theta) = (\cos\theta)/r$. Show that the streamlines of the flow are circles and sketch them. Determine the flux of the flow emanating from the origin.

4. Consider the fluid flow with velocity $\mathbf{V} = \nabla\phi$, where

$$\phi(z) \;=\; \log\left|\frac{z-1}{z+1}\right|.$$

Show that the streamlines of the flow are arcs of circles and sketch them. Determine the flux of the flow emanating from each of the singularities at ± 1.

5. Consider the fluid flow in the horizontal strip $\{0 < \operatorname{Im} z < \pi\}$ with a sink at 0 and equal sources at $\pm\infty$. Find the stream function $\psi(z)$ and the velocity vector field $\mathbf{V}(z)$ of the flow. Sketch the streamlines of the flow. *Hint.* Map the strip to a half-plane by $\zeta = e^z$ and solve a Dirichlet problem with constant boundary values on the three intervals in the boundary separating sinks and sources.

6. For a fluid flow with velocity potential $\phi(z)$, we define the **conjugate flow** to be the flow whose velocity potential is the conjugate harmonic function $\psi(z)$ of $\phi(z)$. What is the stream function of the conjugate flow? What is the complex velocity potential of the conjugate flow?

7. Find the stream function and the complex velocity potential of the conjugate flow associated with the fluid flow with velocity vector $\mathbf{u_r}/r$. Sketch the streamlines of the conjugate flow. Do the particles near the origin travel faster or slower than particles on the unit circle?

8. Find the stream function of the conjugate flow of

$$\mathbf{V}(r, \theta) \;=\; \frac{1}{r}(-\mathbf{u_r} + \mathbf{u_\theta}).$$

Sketch the streamlines of both the flow and the conjugate flow on the same axes. (See Exercise 2d.)

7. Other Applications to Physics

There are two other physical phenomena that are completely analogous to fluid dynamics. They are steady-state heat flow and electrostatics. To draw the analogy, we address each topic area briefly, beginning with steady-state heat flow.

The normalized heat equation is $u_t = \Delta u$, where u_t denotes the derivative of u with respect to time. The steady-state (or time-independent) equation for the heat is obtained by setting $u_t = 0$. The heat equation reduces to Laplace's equation $\Delta u = 0$, and thus the steady-state heat distribution is a harmonic function. In connection with steady-state heat distribution, it is natural to consider boundary-value problems of the following types.

Dirichlet Problem. Given a prescribed function v on the boundary of D, interpreted as the distribution of heat on the boundary, find a harmonic function u on D such that $u = v$ on the boundary of D.

Neumann Problem. Given a prescribed function v on the boundary of D, interpreted as the rate of flow of heat through the boundary, find a harmonic function u on D whose normal derivative $\partial u / \partial n = \nabla u \cdot \mathbf{n}$ coincides with v on the boundary of D.

Under reasonable conditions on D and v, the Dirichlet problem has a unique solution u. For a steady-state solution of the Neumann problem, there can be no net flow of heat through the boundary, $\int_{\partial D} v \, ds = 0$. Subject to this condition, the Neumann problem also has a unique solution.

The vector field $\mathbf{Q} = \nabla u$ is called the **field of flow of thermal energy**. It satisfies

$$\nabla \times \mathbf{Q} = \nabla \times \nabla u = \mathbf{0},$$
$$\nabla \cdot \mathbf{Q} = \Delta u = 0.$$

The equation $\nabla \times \mathbf{Q} = \mathbf{0}$ means that the field $\mathbf{Q}$ is **irrotational**. The equation $\nabla \cdot \mathbf{Q} = 0$ means that there are no heat sources or sinks within D. The heat flux across an arbitrary curve γ is given by

$$\int_\gamma \mathbf{Q} \cdot \mathbf{n} \, ds,$$

where $\mathbf{n}$ is the unit normal to γ and ds is arc length.

Exercise. Find the steady-state heat distribution in a laminar plate represented by the unit disk $\{x^2 + y^2 < 1\}$ when the boundary is held at a constant temperature $+1$ on the edge of the top half $\{y > 0, \ x^2 + y^2 = 1\}$ of the disk, and at -1 on the edge of the bottom half $\{y < 0, \ x^2 + y^2 = 1\}$ of the disk.

Solution. We recall that the argument function $\mathrm{Arg}\, w$ has a similar behavior in the upper half-plane $\mathbb{H}$, in that it is constant on each of the intervals $(-\infty, 0)$ and $(0, +\infty)$ of the boundary of $\mathbb{H}$. We map the upper half-plane to the unit disk by a fractional linear transformation so that the interval $(-\infty, 0)$ corresponds to the lower semicircular edge of the disk and $(0, +\infty)$ to the upper semicircular edge. The transformation is given explicitly by

$$z = -\frac{w - i}{w + i}, \qquad w = i\frac{1 - z}{1 + z}.$$

Since $\mathrm{Log}\, w(z)$ is analytic on the disk, its imaginary part $\mathrm{Arg}\, w(z)$ is harmonic, and it attains the values 0 on the top edge of the disk and π on the bottom edge. Thus the solution is given by

$$u(z) = 1 - \frac{2}{\pi} \mathrm{Arg}\, w(z) = \frac{2}{\pi}[\mathrm{Arg}(1 + z) - \mathrm{Arg}(1 - z)].$$

This solution can be expressed as $u(z) = (2\pi)(\varphi + \psi)$, where φ and ψ are the angles represented in the figure.

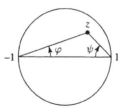

Now we turn to electrostatics. Here it is usual to start with the electric force field $\mathbf{E}$ on D, defined so that $\mathbf{E}(A)$ is the force exerted on a unit charge if it were placed at the point A. The work done on a unit charge in moving it along a path γ in D is then given by the integral $\int_\gamma \mathbf{E} \cdot \mathbf{t}\, ds$, where $\mathbf{t}$ is the unit tangent to the curve, and ds is arc length. The field $\mathbf{E}$ is assumed to be irrotational: $\nabla \times \mathbf{E} = \mathbf{0}$. This is equivalent, at least locally, to assuming that $\int_\gamma \mathbf{E} \cdot \mathbf{t}\, ds$ is independent of path. Thus $\mathbf{E} = \nabla \phi$ for some function ϕ, which is called the **electric field potential**. The work done on a unit charge in moving it along a path γ from A to B is then given by

$$\text{work done} = \int_\gamma \mathbf{E} \cdot \mathbf{t}\, ds = \phi(B) - \phi(A).$$

If there are no sources or sinks in D, then $\nabla \cdot \mathbf{E} = 0$, and this is equivalent to Laplace's equation $\Delta \phi = 0$. Thus the electric field potential ϕ is harmonic in D.

Example. By a "line charge" we mean a uniform distribution of charge along a line (infinite straight wire) in three-dimensional space. We assume that the line is perpendicular to the (x, y)-plane. The electric field is then independent of the z-direction, and it is sufficient to describe it in the (x, y)-plane. The electric field corresponding to a line charge through the origin is

$$\mathbf{E} = \frac{1}{r}\mathbf{u_r},$$

up to a constant multiple depending upon units, where $\mathbf{u_r}$ is the unit vector in the radial direction. The potential function associated with a line charge at 0 is $\phi = \log r$, since $\nabla \phi = \mathbf{E}$. Note that ϕ is harmonic; however, ϕ has no single-valued conjugate harmonic function. The equipotential lines for $\mathbf{E}$ are circles centered at the origin. Through each such circle $\{|z| = r\}$ there is a positive outward flux given by

$$\int_{|z|=r} \mathbf{E} \cdot \mathbf{n}\, ds = \int_{|z|=r} \mathbf{E} \cdot \mathbf{u_r}\, ds = \int_0^{2\pi} d\theta = 2\pi.$$

We say that the origin is a **source** for the electric field $\mathbf{E}$. The origin is a **sink** for the electric field $-\mathbf{E}$.

Exercise. Find the electric field $\mathbf{E}$ and the potential function ϕ in a circular ring whose inner edge is grounded ($\phi = 0$) and whose outer edge is conducting ($\phi = $ constant).
Solution. The problem is invariant under rotations, so we try a potential solution that is invariant under rotations, say $\phi = a \log r + b$. The constant values of the potential on the inner and outer circles determine the constants a and b. The electric field is given by $\mathbf{E} = \nabla \phi = (a/r)\mathbf{u_r}$, which is the electric field of a line charge.

As in the case of fluid flow, problems in electrostatics and in heat flow that are effectively two-dimensional and can be analyzed using complex analysis and conjugate harmonic functions are very special. A typical problem in three dimensions involves a vector field $\mathbf{F}$ that satisfies the conditions

$$\nabla \times \mathbf{F} = 0 \qquad (\mathbf{F} \text{ is irrotational}),$$
$$\nabla \cdot \mathbf{F} = 0 \qquad (\text{no sources or sinks in } D).$$

Again, the first condition is equivalent to $\int \mathbf{F} \cdot d\mathbf{x}$ being independent of path, and we say that $\mathbf{F}$ is a **conservative field**. In this case $\mathbf{F} = \nabla \phi$ for some potential function ϕ, at least locally. In terms of the potential function ϕ, the second condition becomes $\Delta \phi = 0$, and ϕ is harmonic. Thus harmonic functions play an important role, as do the Dirichlet and Neumann problems. However, in three-dimensional problems the harmonic functions cannot be analyzed as the real parts of analytic functions, and conformal mapping techniques are not available.

Exercises for III.7

1. Find the steady-state heat distribution $u(x, y)$ in a laminar plate corresponding to the half-disk $\{x^2 + y^2 < 1, y > 0\}$, where the semicircular top edge is held at temperature T_1 and the lower edge $(-1, 1)$ is held at temperature T_2. Find and sketch the isothermal curves for the heat distribution. *Hint.* Consider the steady-state heat distribution for the full unit disk with the top held at temperature T_1 and the bottom at temperature T_3, where $T_2 = (T_1 + T_3)/2$.

2. Find the potential function $\phi(x, y)$ for the electric field for a conducting laminar plate corresponding to the quarter-disk $\{x^2 + y^2 < 1, x > 0, y > 0\}$, where the two edges on the coordinate axes are grounded (that is, $\phi = 0$ on the edges), and the semicircular edge is held at constant potential V_1. Find and sketch the equipotential lines and the lines of force for the electric field. *Hint.* Use the conformal map $\zeta = z^2$ and the solution to the preceding exercise.

3. Find the potential function $\phi(x, y)$ for the electric field for a conducting laminar plate corresponding to the unit disk where the boundary

quarter-circles in each quadrant are held at constant voltages V_1, V_2, V_3, and V_4. *Hint.* Map the disk to the upper half-plane by $w = w(z)$ and consider potential functions of the form $\text{Arg}(w - a)$.

4. Find the steady-state heat distribution in a laminar plate corresponding to the vertical half-strip $\{|x| < \pi/2, y > 0\}$, where the vertical sides at $x = \pm\pi/2$ are held at temperature $T_0 = 0$ and the bottom edge $(-\pi/2, \pi/2)$ on the real axis is held at temperature $T_1 = 100$. Make a rough sketch of the isothermal curves and the lines of heat flow. *Hint.* Use $w = \sin z$ to map the strip to the upper half-plane, and make use of harmonic functions of the form $\text{Arg}(w - a)$.

5. Find the steady-state heat distribution in a laminar plate corresponding to the vertical half-strip $\{|x| < \pi/2, y > 0\}$, where the side $x = -\pi/2$ is held at constant temperature T_0, the side $x = \pi/2$ is held at constant temperature T_1, and the bottom edge $(-\pi/2, \pi/2)$ on the real axis is insulated; that is, no heat passes through the bottom edge, so the gradient ∇u of the solution $u(x, y)$ is parallel there to the x-axis. *Hint.* Try linear functions plus constants.

6. Find the steady-state heat distribution in a laminar plate corresponding to the upper half-plane $\{y > 0\}$, where the interval $(-1, 1)$ is insulated, the interval $(-\infty, -1)$ is held at temperature T_0, and the interval $(1, \infty)$ is held at temperature T_1. Make a rough sketch of the isothermal curves and the lines of heat flow. *Hint.* Use the solution to Exercise 5 and the conformal map from Exercise 4.

7. The gravitational field near the surface of the earth is approximately constant, of the form $\mathbf{F} = c\mathbf{k}$, where $\mathbf{k}$ is the unit vector in the z-direction in (x, y, z)-space and the surface of the earth is represented by the plane where $z = 0$ (the flat earth theory). Show that $\mathbf{F}$ is conservative, and find a potential function ϕ for $\mathbf{F}$.

8. Show that the inverse square force field $\mathbf{F} = \mathbf{u_r}/r^2$ on $\mathbb{R}^3$ is conservative. Find the potential function ϕ for $\mathbf{F}$, and show that ϕ is harmonic.

9. For $n \geq 3$, show that the function $1/r^{n-2}$ is harmonic on $\mathbb{R}^n \backslash \{0\}$. Find the vector field $\mathbf{F}$ that has this function as its potential.

IV

Complex Integration and Analyticity

In this chapter we take up the complex integral calculus. In Section 1 we introduce complex line integrals, and in Section 2 we develop the complex integral calculus, emphasizing the analogy with the usual one-variable integral calculus. In Section 3 we lay the cornerstone of the complex integral calculus, which is Cauchy's theorem. The version we prove is an immediate consequence of Green's theorem. In Section 4 we derive the Cauchy integral formula and use it to show that analytic functions have analytic derivatives. Each of the final four sections features a "named" theorem. In Section 5 we prove Liouville's theorem. In Section 6 we give a version of Morera's theorem that provides a useful criterion for determining whether a continuous function is analytic. Sections 7 and 8, on Goursat's theorem and the Pompeiu formula, can be omitted at first reading.

1. Complex Line Integrals

For complex analysis it is convenient to define $dz = dx + i\,dy$. According to this notation, if $h(z)$ is a complex-valued function on a curve γ, then

$$(1.1) \qquad \int_\gamma h(z)\,dz \;=\; \int_\gamma h(z)\,dx \;+\; i\int_\gamma h(z)\,dy.$$

Suppose γ is parameterized by $t \mapsto z(t) = x(t) + iy(t)$, $a \le t \le b$. The Riemann sum approximating $\int_\gamma h(z)(dx + i\,dy)$ corresponding to the subdivision $a = t_0 < t_1 < \cdots < t_n = b$ is given by

$$\sum h(z_j)(x_{j+1} - x_j) \;+\; i\sum h(z_j)(y_{j+1} - y_j),$$

where $z(t_j) = z_j = x_j + iy_j$. If we express these sums in terms of the z_j's, we obtain the Riemann sum approximation

$$(1.2) \qquad \int_\gamma h(z)\,dz \;\approx\; \sum h(z_j)(z_{j+1} - z_j).$$

102

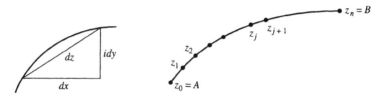

This expression justifies the notation dz.

Example. To compute $\int_0^{1+i} z^2\, dz$ along the straight line segment from 0 to $1+i$, we parametrize the line segment by $z(t) = t + ti$, $0 \le t \le 1$, so that $x(t) = t$ and $y(t) = t$. Then $dz = dx + i\, dy = (1+i)dt$, and we obtain

$$\int_0^{1+i} z^2\, dz \;=\; \int_0^1 [(1+i)t]^2(1+i)dt \;=\; (1+i)^3 \int_0^1 t^2 dt \;=\; \frac{(1+i)^3}{3}.$$

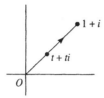 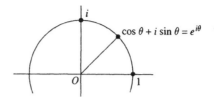

Example. We evaluate

$$\oint_{|z|=1} \frac{dz}{z},$$

where we integrate around the unit circle in the usual positive (counter-clockwise) direction. The unit circle is parameterized by $z(\theta) = e^{i\theta} = \cos\theta + i\sin\theta$, $0 \le \theta \le 2\pi$, so that $dx = -\sin\theta\, d\theta$, $dy = \cos\theta\, d\theta$, and

$$dz \;=\; dx + i\, dy \;=\; -\sin\theta\, d\theta + i\cos\theta\, d\theta \;=\; i(\cos\theta + i\sin\theta)d\theta.$$

Thus

$$\frac{dz}{z} \;=\; i\, d\theta,$$

and the integral becomes

$$\oint_{|z|=1} \frac{dz}{z} \;=\; i \int_0^{2\pi} d\theta \;=\; 2\pi i.$$

Note in this calculation that $dz = ie^{i\theta}d\theta$, which is what is obtained by applying the usual rule for differentiation of exponentials to $z(\theta) = e^{i\theta}$.

Example. For m an integer and $R > 0$, we show that

$$\oint_{|z-z_0|=R} (z - z_0)^m dz \;=\; \begin{cases} 0, & m \neq -1, \\ 2\pi i, & m = -1. \end{cases}$$

We parametrize the circle $|z - z_0| = R$, with the usual positive orientation, by $z(\theta) = z_0 + Re^{i\theta}$, and we calculate that $dz = iRe^{i\theta}\,d\theta$. Thus the integral becomes

$$\int_0^{2\pi} (Re^{i\theta})^m iRe^{i\theta}\,d\theta \;=\; iR^{m+1}\int_0^{2\pi} e^{i(m+1)\theta}\,d\theta.$$

The integral on the right is 0 unless $m = -1$, in which case it is 2π. This yields the formula.

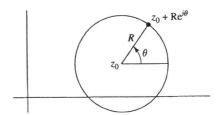

In complex analysis it is customary to denote the infinitesimal arc length ds by $|dz|$:

$$|dz| \;=\; ds \;=\; \sqrt{(dx)^2 + (dy)^2}.$$

This means that if a curve γ is parameterized by $z(t) = x(t) + iy(t)$, then

$$(1.3) \qquad \int_\gamma h(z)|dz| \;=\; \int_\gamma h(z)\,ds \;=\; \int_a^b h(z(t))\sqrt{\left(\frac{dx}{dt}\right)^2 + \left(\frac{dy}{dt}\right)^2}\,dt.$$

In particular, the length of γ is

$$L \;=\; \int_\gamma |dz| \;=\; \int_a^b \sqrt{\left(\frac{dx}{dt}\right)^2 + \left(\frac{dy}{dt}\right)^2}\,dt.$$

The notation can be justified by considering the sums approximating these integrals. For the subdivision of the parameter interval used earlier, the usual sum used in multivariable calculus to approximate $\int_\gamma h(x,y)\,ds$ is

$$\int_\gamma h(x,y)\,ds \;\approx\; \sum h(x_j, y_j)\sqrt{(x_{j+1} - x_j)^2 + (y_{j+1} - y_j)^2}.$$

In complex notation this becomes

$$(1.4) \qquad \int_\gamma h(z)\,|dz| \;\approx\; \sum h(z_j)|z_{j+1} - z_j|.$$

In particular, the sum approximating the length L of γ is given by

$$(1.5) \qquad L \;\approx\; \sum |z_{j+1} - z_j|.$$

Example. The parameterization $z(\theta) = z_0 + Re^{i\theta}$ of the circle $|z - z_0| = R$ can be used to derive the expression for the infinitesimal arc length for the circle in terms of the central angle θ. In this case, $x(\theta) = R\cos\theta$, so that $dx = -R\sin\theta\,d\theta$, and similarly $dy = R\cos\theta\,d\theta$. Thus $|dz| = ds = \sqrt{(dx)^2 + (dy)^2} = R\,d\theta$. The length of the circle is $\int_0^{2\pi} |dz| = 2\pi R$, as usual.

Theorem. *Suppose γ is a piecewise smooth curve. If $h(z)$ is a continuous function on γ, then*

(1.6)
$$\left| \int_\gamma h(z)\,dz \right| \le \int_\gamma |h(z)|\,|dz|.$$

Further, if γ has length L, and $|h(z)| \le M$ on γ, then

(1.7)
$$\left| \int_\gamma h(z)\,dz \right| \le ML.$$

The estimate (1.6) is the triangle inequality for integrals. It will be used frequently and without reference. The estimate (1.7) is called the ML-**estimate**. It may or may not be referred to. These estimates both follow from the corresponding estimates for the approximating sums. The triangle inequality (1.6) for integrals follows from the triangle inequality for the approximating Riemann sums,

$$\left| \sum h(z_j)(z_{j+1} - z_j) \right| \le \sum |h(z_j)||(z_{j+1} - z_j)|,$$

in view of (1.2) and (1.4). The ML-estimate (1.7) follows from the estimate

$$\left| \sum h(z_j)(z_{j+1} - z_j) \right| \le M \sum |z_{j+1} - z_j|,$$

in view of (1.2) and (1.5).

Example. If we apply the ML-estimate to the first example above, we obtain

$$\left| \int_0^{1+i} z^2 dz \right| \le 2\sqrt{2},$$

since $|z^2| \le |1+i|^2 = 2$ on the straight line segment from 0 to $1+i$, and the segment has length $L = \sqrt{2}$. Since the value of the integral has modulus $|1 + i|^3/3 = 2\sqrt{2}/3$, in this case the ML-estimate provides only a rough estimate. A better estimate is obtained by noting that $|dz| = \sqrt{2}dt$ on the line and applying (1.6):

$$\left| \int_0^{1+i} z^2 dz \right| \le \int_0^{1+i} |z^2|\,|dz| = \int_0^1 \left(\sqrt{2t}\right)^2 \sqrt{2}\,dt = \frac{2\sqrt{2}}{3}.$$

Since equality actually holds here, this estimate cannot be improved. We say that the estimate is a **sharp estimate**.

Example. We apply the ML-estimate to

$$\oint_{|z-z_0|=R} \frac{1}{z-z_0}\, dz \;=\; 2\pi i.$$

In this case the integrand has constant modulus $1/R$ on the circle of integration, so we take $M = 1/R$ and $L = 2\pi R$. The ML-estimate becomes

$$\left| \oint_{|z-z_0|=R} \frac{1}{z-z_0}\, dz \right| \;\le\; 2\pi.$$

Since equality actually holds here, the ML-estimate is sharp.

Exercises for IV.1

1. Let γ be the boundary of the triangle $\{0 < y < 1 - x,\ 0 < x < 1\}$, with the usual counterclockwise orientation. Evaluate the following integrals.

   (a) $\displaystyle\int_\gamma \operatorname{Re} z\, dz$ (b) $\displaystyle\int_\gamma \operatorname{Im} z\, dz$ (c) $\displaystyle\int_\gamma z\, dz$

2. Let γ be the unit circle $\{|z| = 1\}$, with the usual counterclockwise orientation. Evaluate the following integrals, for $m = 0, \pm 1, \pm 2, \dots$.

   (a) $\displaystyle\int_\gamma z^m\, dz$ (b) $\displaystyle\int_\gamma \bar{z}^m\, dz$ (c) $\displaystyle\int_\gamma z^m\, |dz|$

3. Let γ be the circle $\{|z| = R\}$, with the usual counterclockwise orientation. Evaluate the following integrals, for $m = 0, \pm 1, \pm 2, \dots$.

   (a) $\displaystyle\int_\gamma |z^m|\, dz$ (b) $\displaystyle\int_\gamma |z^m|\, |dz|$ (c) $\displaystyle\int_\gamma \bar{z}^m\, dz$

4. Show that if D is a bounded domain with smooth boundary, then

$$\int_{\partial D} \bar{z}\, dz \;=\; 2i\, \text{Area}(D).$$

5. Show that

$$\left| \oint_{|z-1|=1} \frac{e^z}{z+1}\, dz \right| \;\le\; 2\pi e^2.$$

6. Show that

$$\left| \oint_{|z|=R} \frac{\operatorname{Log} z}{z^2}\, dz \right| \;\le\; 2\sqrt{2}\,\pi\, \frac{\log R}{R}, \qquad R > e^\pi.$$

7. Show that there is a strict inequality

$$\left| \oint_{|z|=R} \frac{z^n}{z^m - 1}\, dz \right| \;<\; \frac{2\pi R^{n+1}}{R^m - 1}, \qquad R > 1,\ m \ge 1,\ n \ge 0.$$

8. Suppose the continuous function $f\left(e^{i\theta}\right)$ on the unit circle satisfies $|f(e^{i\theta})| \leq M$ and $|\int_{|z|=1} f(z)dz| = 2\pi M$. Show that $f(z) = c\bar{z}$ for some constant c with modulus $|c| = M$.

9. Suppose $h(z)$ is a continuous function on a curve γ. Show that

$$H(w) = \int_\gamma \frac{h(z)}{z - w} \, dz, \qquad w \in \mathbb{C}\backslash\gamma,$$

is analytic on the complement of γ, and find $H'(w)$.

2. Fundamental Theorem of Calculus for Analytic Functions

Let $f(z)$ be a continuous function on a domain D. A function $F(z)$ on D is a **(complex) primitive** for $f(z)$ if $F(z)$ is analytic and $F'(z) = f(z)$. The following theorem is the analogue of the first statement of the fundamental theorem of calculus.

Theorem (Part I). *If $f(z)$ is continuous on a domain D, and if $F(z)$ is a primitive for $f(z)$, then*

$$\int_A^B f(z) \, dz \; = \; F(B) \, - \, F(A),$$

where the integral can be taken over any path in D from A to B.

This formula follows from the corresponding formula for line integrals. In this case, we have

$$F'(z) \; = \; \frac{\partial F}{\partial x} \; = \; \frac{1}{i} \frac{\partial F}{\partial y},$$

so that

$$F(B) - F(A) \; = \; \int_A^B dF \; = \; \int_A^B \frac{\partial F}{\partial x} dx \, + \, \frac{\partial F}{\partial y} dy$$

$$= \; \int_A^B F'(z) \, (dx + i \, dy) \; = \; \int_A^B F'(z) \, dz.$$

This theorem provides a powerful tool for evaluating definite integrals. The problem of evaluating $\int_A^B f(z)dz$ is reduced to that of finding an analytic function $F(z)$ whose derivative is $f(z)$.

Example. To integrate z^2 from 1 to i, we observe that $z^3/3$ is a primitive for z^2, and then we proceed as we would to evaluate an ordinary garden-variety integral,

$$\int_0^{1+i} z^2 dz \; = \; \frac{z^3}{3}\bigg|_0^{1+i} \; = \; \frac{(1+i)^3}{3}.$$

This coincides with the result that was obtained in the preceding section
by parameterizing the straight line from 0 to $1 + i$. Note, though, that the
integral is independent of path.

Example. The function $1/z$ does not have an analytic primitive defined
on any domain containing the unit circle. To evaluate the integral of dz/z
around the unit circle we can still use the primitive $F(z) = \operatorname{Log} z$ in the
slit plane $\mathbb{C}\backslash(-\infty, 0]$, by taking the integral over a path counterclockwise
around the unit circle starting at a point $-1 - 0i$ just below the slit and
ending at a point $-1 + 0i$ just above the slit. Then

$$\int_{|z|=1} \frac{dz}{z} = \operatorname{Log} z \bigg|_{-1-0i}^{-1+0i} = i\pi - i(-\pi) = 2\pi i,$$

which coincides with the result obtained earlier by parameterizing the unit
circle.

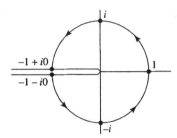

The analogue of the second statement in the fundamental theorem of
calculus is that every analytic function on a star-shaped domain has a
primitive, which can be defined as an indefinite complex integral.

Theorem (Part II). *Let D be a star-shaped domain, and let $f(z)$ be
analytic on D. Then $f(z)$ has a primitive on D, and the primitive is
unique up to adding a constant. A primitive for $f(z)$ is given explicitly by*

$$F(z) = \int_{z_0}^{z} f(\zeta)\, d\zeta, \qquad z \in D,$$

*where z_0 is any fixed point of D, and where the integral can be taken along
any path in D from z_0 to z.*

To see this, we write $f = u + iv$ as usual, and we consider the differential $u\, dx - v\, dy$. Since f is analytic, the Cauchy-Riemann equations yield
$\partial u/\partial y = -\partial v/\partial x$, and the differential is closed. By the fundamental theorem of calculus for line integrals (Section III.2), the differential is exact
on D, that is, there is a continuously differentiable function U on D such

that $dU = u\,dx - v\,dy$. In other words, U satisfies

$$\frac{\partial U}{\partial x} = u, \qquad \frac{\partial U}{\partial y} = -v.$$

These equations show that U is in fact twice continuously differentiable. Another application of the Cauchy-Riemann equations yields

$$\frac{\partial^2 U}{\partial x^2} + \frac{\partial^2 U}{\partial y^2} = \frac{\partial u}{\partial x} - \frac{\partial v}{\partial y} = 0.$$

Hence U is harmonic. Since D is star-shaped, there is a conjugate harmonic function V for U on D, so that $G = U + iV$ is analytic on D. Then

$$G' = \frac{\partial U}{\partial x} + i\frac{\partial V}{\partial x} = \frac{\partial U}{\partial x} - i\frac{\partial U}{\partial y} = u + iv = f,$$

and $G(z)$ is a primitive for $f(z)$. If $G_0(z)$ is another primitive for $f(z)$, then $G - G_0$ has derivative zero, and $G - G_0$ is constant on D. If we apply Part I above to the primitive $F(z) = G(z) - G(z_0)$ for $f(z)$, we obtain the formula in Part II.

Note in particular that integrals of analytic functions in star-shaped domains are independent of path. This is not true for arbitrary domains. The identity

$$\oint_{|z|=1} \frac{dz}{z} = 2\pi i \neq 0$$

shows that the differential dz/z does not have a primitive on the punctured plane $\mathbb{C}\backslash\{0\}$.

Exercises for IV.2

1. Evaluate the following integrals, for a path γ that travels from $-\pi i$ to πi in the right half-plane, and also for a path γ from $-\pi i$ to πi in the left half-plane.

   (a) $\displaystyle\int_\gamma z^4 dz$ (b) $\displaystyle\int_\gamma e^z dz$ (c) $\int_\gamma \cos z\, dz$ (d) $\int_\gamma \sinh z\, dz$

2. Using an appropriate primitive, evaluate $\int_\gamma 1/z\, dz$ for a path γ that travels from $-\pi i$ to πi in the right half-plane, and also for a path γ from $-\pi i$ to πi in the left half-plane. For each path give a precise definition of the primitive used to evaluate the integral.

3. Show that if $m \neq -1$, then z^m has a primitive on $\mathbb{C}\backslash\{0\}$.

4. Let $D = \mathbb{C}\backslash(-\infty, 1]$, and consider the branch of $\sqrt{z^2 - 1}$ on D that is positive on the interval $(1, \infty)$. (a) Show that $z + \sqrt{z^2 - 1}$ omits the negative real axis, that is, the range of the function on D does not include any values in the interval $(-\infty, 0]$ on the real axis. (b)

Show that $\text{Log}(z + \sqrt{z^2 - 1})$ is a primitive for $1/\sqrt{z^2 - 1}$ on D. (c) Evaluate

$$\int_\gamma \frac{dz}{\sqrt{z^2 - 1}},$$

where γ is the path from $-2i$ to $+2i$ in D counterclockwise around the circle $|z| = 2$. (d) Evaluate the integral above in the case γ is the entire circle $|z| = 2$, oriented counterclockwise. (Note that the primitive is discontinuous at $z = -2$.)

5. Show that an analytic function $f(z)$ has a primitive in D if and only if $\int_\gamma f(z)dz = 0$ for every closed path γ in D.

3. Cauchy's Theorem

We begin with a smooth complex-valued function $f(z) = u + iv$, and we express

$$f(z)\, dz = (u + iv)(dx + i\, dy) = (u + iv)dx + (-v + iu)dy.$$

The condition that $f(z)dz$ be a closed differential is

$$\frac{\partial}{\partial y}(u + iv) = \frac{\partial}{\partial x}(-v + iu).$$

Taking the real and imaginary parts, we see that this is equivalent to

$$\frac{\partial u}{\partial y} = -\frac{\partial v}{\partial x}, \qquad \frac{\partial v}{\partial y} = \frac{\partial u}{\partial x},$$

which are the Cauchy-Riemann equations for u and v. Thus we obtain the following theorem, which is the original form of Morera's theorem.

Theorem. *A continuously differentiable function $f(z)$ on D is analytic if and only if the differential $f(z)dz$ is closed.*

From Green's theorem (Section III.1.1) we obtain immediately the following far-reaching theorem.

Theorem (Cauchy's Theorem). *Let D be a bounded domain with piecewise smooth boundary. If $f(z)$ is an analytic function on D that extends smoothly to ∂D, then*

$$\int_{\partial D} f(z)\, dz = 0.$$

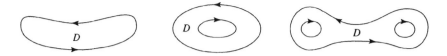

Example. If we apply Cauchy's theorem to a function $f(z)$ analytic on the annulus $D = \{r < |z| < R\}$, we obtain

$$0 = \int_{\partial D} f(z)\,dz = \oint_{|z|=R} f(z)\,dz - \oint_{|z|=r} f(z)\,dz.$$

Note the change in sign, due to the fact that the inside boundary circle is traversed in the negative (clockwise) direction according to its orientation as a boundary curve for D. Thus

$$\oint_{|z|=R} f(z)\,dz = \oint_{|z|=r} f(z)\,dz.$$

This is in accord with the discussion in Section III.2, since the differential $f(z)dz$ is closed, and since the outside circle of the annulus can be continuously deformed to the inside circle of the annulus through a family of closed curves (intermediate circles).

Exercises for IV.3

1. By integrating $e^{-z^2/2}$ around a rectangle with vertices $\pm R,\ it \pm R$, and sending R to ∞, show that

$$\frac{1}{\sqrt{2\pi}} \int_{-\infty}^{\infty} e^{-x^2/2} e^{-itx}\,dx = e^{-t^2/2}, \qquad \infty < t < \infty.$$

Use the known value of the integral for $t = 0$. *Remark.* This shows that $e^{-x^2/2}$ is an eigenfunction of the Fourier transform with eigenvalue 1. For more, see the next exercise.

2. We define the **Hermite polynomials** $H_n(x)$ and **Hermite orthogonal functions** $\phi_n(x)$ for $n \geq 0$ by

$$H_n(x) = (-1)^n e^{x^2} \frac{d^n}{dx^n}\left(e^{-x^2}\right), \qquad \phi_n(x) = e^{-x^2/2} H_n(x).$$

(a) Show that $H_n(x) = 2^n x^n + \cdots$ is a polynomial of degree n that is even when n is even and odd when n is odd.

(b) By integrating the function

$$e^{(z-it)^2/2} \frac{d^n}{dz^n}\left(e^{-z^2}\right)$$

around a rectangle with vertices $\pm R,\ it \pm R$ and sending R to ∞, show that

$$\frac{1}{\sqrt{2\pi}} \int_{-\infty}^{\infty} \phi_n(x) e^{-itx}\,dx = (-i)^n \phi_n(t), \qquad -\infty < t < \infty.$$

Hint. Use the identity from Exercise 1, and also justify and use the identity

$$\frac{d^n}{dx^n}e^{-(x+it)^2} = \frac{1}{i^n}\frac{d^n}{dt^n}e^{-(x+it)^2}.$$

(c) Show that $\phi_n'' - x^2\phi_n + (2n+1)\phi_n = 0$.

(d) Using $\int \phi_n''\phi_m\,dx = \int \phi_n\phi_m''\,dx$ and (c), show that

$$\int_{-\infty}^{\infty} \phi_n(x)\phi_m(x)\,dx = 0, \qquad n \neq m.$$

Remark. This shows that the ϕ_n's form an orthogonal system of eigenfunctions for the (normalized) Fourier transform operator $\mathcal{F}$ with eigenvalues ± 1 and $\pm i$. Thus $\mathcal{F}$ extends to a unitary operator on square-integrable functions. Further, $\mathcal{F}^4$ is the identity operator, and the inverse Fourier transform is given by $(\mathcal{F}^{-1}f)(x) = (\mathcal{F}f)(-x)$.

3. Let $f(z) = c_0 + c_1 z + \cdots + c_n z^n$ be a polynomial.

   (a) If the c_k's are real, show that

$$\int_{-1}^{1} f(x)^2\,dx \leq \pi \int_0^{2\pi} |f(e^{i\theta})|^2\,\frac{d\theta}{2\pi} = \pi \sum_{k=0}^{n} c_k^2.$$

Hint. For the first inequality, apply Cauchy's theorem to the function $f(z)^2$ separately on the top half and the bottom half of the unit disk.

   (b) If the c_k's are complex, show that

$$\int_{-1}^{1} |f(x)|^2\,dx \leq \pi \int_0^{2\pi} |f(e^{i\theta})|^2\,\frac{d\theta}{2\pi} = \pi \sum_{k=0}^{n} |c_k|^2.$$

   (c) Establish the following variant of **Hilbert's inequality**, that

$$\left| \sum_{j,k=0}^{n} \frac{c_j c_k}{j+k+1} \right| \leq \pi \sum_{k=0}^{n} |c_k|^2,$$

with strict inequality unless the complex numbers $c_0, \ldots, c_n$ are all zero. *Hint.* Start by evaluating $\int_0^1 f(x)^2\,dx$.

4. Prove that a polynomial in z without zeros is constant (the fundamental theorem of algebra) using Cauchy's theorem, along the following lines. If $P(z)$ is a polynomial that is not a constant, write $P(z) = P(0) + zQ(z)$, divide by $zP(z)$, and integrate around a large circle. This will lead to a contradiction if $P(z)$ has no zeros.

5. Suppose that D is a bounded domain with piecewise smooth boundary, and that $f(z)$ is analytic on $D \cup \partial D$. Show that

$$\sup_{z \in \partial D} |\bar{z} - f(z)| \geq 2 \frac{\text{Area}(D)}{\text{Length}(\partial D)}.$$

Show that this estimate is sharp, and that in fact there exist D and $f(z)$ for which equality holds. *Hint.* Consider $\int_{\partial D} [\bar{z} - f(z)] dz$, and use Exercise 4 in Section 1.

6. Suppose $f(z)$ is continuous in the closed disk $\{|z| \leq R\}$ and analytic on the open disk $\{|z| < R\}$. Show that $\oint_{|z|=R} f(z) dz = 0$. *Hint.* Approximate $f(z)$ uniformly by $f_r(z) = f(rz)$.

4. The Cauchy Integral Formula

Integral representation formulae are powerful tools for studying functions. One application of an integral representation is to estimate the size of the function being represented. Another is to obtain formulae for derivatives, by differentiating under the integral sign. The prototype of the integral representation is provided by the Cauchy integral formula, representing an analytic function. The integral representation will allow us to show that all the derivatives of an analytic function are analytic. It will also allow us to obtain power series expansions for analytic functions.

Theorem (Cauchy Integral Formula). *Let D be a bounded domain with piecewise smooth boundary. If $f(z)$ is analytic on D, and $f(z)$ extends smoothly to the boundary of D, then*

$$(4.1) \qquad f(z) = \frac{1}{2\pi i} \int_{\partial D} \frac{f(w)}{w - z} dw, \qquad z \in D.$$

To establish the formula, fix a point $z \in D$, let $\varepsilon > 0$ be small, and consider the domain $D_\varepsilon = D \backslash \{|w - z| \leq \varepsilon\}$ obtained from D by punching out a disk centered at z of radius ε. The boundary ∂D_ε is the union of ∂D and the circle $\{|w - z| = \varepsilon\}$, oriented clockwise. Since $f(w)/(w - z)$ is analytic for $w \in D_\varepsilon$, Cauchy's theorem yields

$$\int_{\partial D_\varepsilon} \frac{f(w)}{w - z} dw = 0.$$

Reversing the orientation of the circle to counterclockwise produces a sign change, and we obtain

$$\int_{|w-z|=\varepsilon} \frac{f(w)}{w - z} dw = \int_{\partial D} \frac{f(w)}{w - z} dw.$$

Writing $w = z + \varepsilon e^{i\theta}$, $dw = i\varepsilon e^{i\theta} d\theta$, and dividing by $2\pi i$, we obtain

$$\int_0^{2\pi} f\left(z + \varepsilon e^{i\theta}\right) \frac{d\theta}{2\pi} = \frac{1}{2\pi i} \int_{\partial D} \frac{f(w)}{w - z} dw.$$

By the mean value property of harmonic functions, the integral on the left-hand side coincides with $f(z)$, and the formula is established.

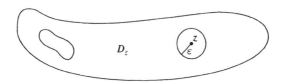

We remark that without the mean value property at our disposal, we could complete the argument by observing that the integral on the left is the average value of $f(w)$ on the circle centered at z of radius ε. Since $f(w)$ is continuous at z, these averages tend to $f(z)$ as $\varepsilon \to 0$. This can be justified with complete rigor by writing

$$\int_0^{2\pi} f\left(z + \varepsilon e^{i\theta}\right) \frac{d\theta}{2\pi} = f(z) + \int_0^{2\pi} \left[f\left(z + \varepsilon e^{i\theta}\right) - f(z)\right] \frac{d\theta}{2\pi},$$

and showing by a direct estimate that the latter integral tends to 0 as $\varepsilon \to 0$.

If we differentiate under the integral sign and use

$$\frac{d^m}{dz^m} \frac{1}{w - z} = \frac{m!}{(w - z)^{m+1}},$$

we obtain integral formulae for the derivatives $f^{(m)}(z)$ of $f(z)$.

Theorem. *Let D be a bounded domain with piecewise smooth boundary. If $f(z)$ is an analytic function on D that extends smoothly to the boundary of D, then $f(z)$ has complex derivatives of all orders on D, which are given by*

$$(4.2) \qquad f^{(m)}(z) = \frac{m!}{2\pi i} \int_{\partial D} \frac{f(w)}{(w - z)^{m+1}} dw, \qquad z \in D, \ m \geq 0.$$

The case $m = 0$ of (4.2) is the Cauchy integral formula (4.1). Though it is not really necessary, we give separately the argument for the case $m = 1$. This special case already includes all the ideas for the proof of the general case.

Using (4.1) and the identity

$$\frac{1}{w - (z + \Delta z)} - \frac{1}{w - z} = \frac{\Delta z}{(w - (z + \Delta z))(w - z)},$$

we express the difference quotient approximating $f'(z)$ as

$$\frac{f(z+\Delta z) - f(z)}{\Delta z} = \frac{1}{\Delta z}\left[\frac{1}{2\pi i}\int_{\partial D}\frac{f(w)}{w-(z+\Delta z)}dw - \frac{1}{2\pi i}\int_{\partial D}\frac{f(w)}{w-z}dw\right]$$

$$= \frac{1}{2\pi i}\int_{\partial D} f(w)\frac{1}{(w-(z+\Delta z))(w-z)}dw.$$

As Δz tends to 0, the integrand converges to $f(w)/(w-z)^2$, uniformly for $w \in \partial D$. Hence the integrals converge, and we obtain in the limit that

$$f'(z) = \frac{1}{2\pi i}\int_{\partial D}\frac{f(w)}{(w-z)^2}dw, \qquad z \in D,$$

which is the case $m = 1$ of (4.2).

The general case of (4.2) is proved by induction on m. We assume that $f^{(m-2)}(z)$ is complex differentiable and the formula holds for $f^{(m-1)}(z)$. We must show that $f^{(m-1)}(z)$ is complex differentiable and the formula holds for $f^{(m)}(z)$. Using the binomial expansion

$$(w-z-\Delta z)^m = (w-z)^m - m(w-z)^{m-1}\Delta z + \frac{m(m-1)}{2}(w-z)^{m-2}(\Delta z)^2 + \cdots$$

and simplifying, we obtain

$$\frac{1}{(w-(z+\Delta z))^m} - \frac{1}{(w-z)^m} = \frac{(w-z)^m - (w-z-\Delta z)^m}{(w-z)^m(w-z-\Delta z)^m}$$

$$= \frac{m\Delta z}{(w-z)(w-z-\Delta z)^m} - \frac{m(m-1)(\Delta z)^2}{2(w-z)^2(w-z-\Delta z)^m} + \cdots,$$

where the dots indicate terms with powers of Δz up to $(\Delta z)^m$. The integral formula for $f^{(m-1)}(z)$ then yields the expression

$$\frac{(m-1)!}{2\pi i}\int_{\partial D} f(w)\left[\frac{m}{(w-z)(w-z-\Delta z)^m} + \Delta z(\cdots)\right]dw$$

for the difference quotient $\left(f^{(m-1)}(z+\Delta z) - f^{(m-1)}(z)\right)/\Delta z$. Again the integrand converges as $\Delta z \to 0$, uniformly for $w \in \partial D$, and we can pass to the limit to conclude that $f^{(m)}(z)$ exists and is given by (4.2).

Since each of the successive complex derivatives of $f(z)$ is complex differentiable, each is continuous, and thus each is analytic. The hypothesis that the domain has smooth boundary is irrelevant for determining analyticity, as we can restrict the function to an appropriate small disk. Thus we have proved the following.

Corollary. If $f(z)$ is analytic on a domain D, then $f(z)$ is infinitely differentiable, and the successive complex derivatives $f'(z)$, $f''(z)$, ..., are all analytic on D.

Example. The Cauchy integral formula for z^2 yields

$$\oint_{|z|=2} \frac{z^2}{z-1}\,dz = 2\pi i z^2\Big|_{z=1} = 2\pi i.$$

Example. The Cauchy integral formula for the derivative of $z^2 \sin z$ yields

$$\oint_{|z|=2\pi} \frac{z^2 \sin z}{(z-\pi)^3}\,dz = \frac{2\pi i}{2!}\frac{d^2}{dz^2}\left(z^2 \sin z\right)\Big|_{z=\pi} = \pi i(4\pi\cos\pi) = -4\pi^2 i.$$

Example. Consider the integral

$$\oint_{|z|=2} \frac{e^z}{z^2(z-1)}\,dz,$$

which does not have the form given in the Cauchy integral formula. By applying Cauchy's theorem to the domain D_ε obtained by excising two small disks centered at 0 and 1 from the disk $\{|z| < 2\}$, we can express the integral above as the sum of two integrals, each of which can be evaluated by the Cauchy formula:

$$\oint_{|z|=2} \frac{e^z}{z^2(z-1)}\,dz = \oint_{|z|=\varepsilon} \frac{e^z}{z^2(z-1)}\,dz + \oint_{|z-1|=\varepsilon} \frac{e^z}{z^2(z-1)}\,dz$$

$$= 2\pi i\,\frac{d}{dz}\frac{e^z}{(z-1)}\Big|_{z=0} + 2\pi i\,\frac{e^z}{z^2}\Big|_{z=1}$$

$$= -2\pi i - 2\pi i + 2\pi i e = 2\pi i(e-2).$$

These integrals can also be handled by residue theory, as we shall soon see.

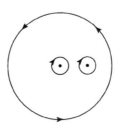

Exercises for IV.4

1. Evaluate the following integrals, using the Cauchy integral formula:

(a) $\displaystyle\oint_{|z|=2} \frac{z^n}{z-1}\,dz, \quad n \geq 0$

(b) $\displaystyle\oint_{|z|=1} \frac{z^n}{z-2}\,dz, \quad n \geq 0$

(c) $\displaystyle\oint_{|z|=1} \frac{\sin z}{z}\,dz$

(e) $\displaystyle\oint_{|z|=1} \frac{e^z}{z^m}\,dz, \quad -\infty < m < \infty$

(f) $\displaystyle\int_{|z-1-i|=5/4} \frac{\mathrm{Log}\,z}{(z-1)^2}\,dz$

(g) $\displaystyle\oint_{|z|=1} \frac{dz}{z^2(z^2-4)e^z}$

(d) $\displaystyle\oint_{|z|=1} \frac{\cosh z}{z^3}\,dz$

(h) $\displaystyle\oint_{|z-1|=2} \frac{dz}{z(z^2-4)e^z}$

2. Show that a harmonic function is C^∞, that is, a harmonic function has partial derivatives of all orders.

3. Use the Cauchy integral formula to derive the mean value property of harmonic functions, that

$$u(z_0) = \int_0^{2\pi} u\left(z_0 + \rho e^{i\theta}\right) \frac{d\theta}{2\pi}, \qquad z_0 \in D,$$

whenever $u(z)$ is harmonic in a domain D and the closed disk $|z - z_0| \le \rho$ is contained in D.

4. Let D be a bounded domain with smooth boundary ∂D, and let $z_0 \in D$. Using the Cauchy integral formula, show that there is a constant C such that

$$|f(z_0)| \le C \,\sup\{|f(z)| : z \in \partial D\}$$

for any function $f(z)$ analytic on $D \cup \partial D$. By applying this estimate to $f(z)^n$, taking nth roots, and letting $n \to \infty$, show that the estimate holds with $C = 1$. *Remark.* This provides an alternative proof of the maximum principle for analytic functions.

5. Liouville's Theorem

Suppose that $f(z)$ is analytic on the closed disk $\{|z - z_0| \le \rho\}$, that is, it is analytic on some domain containing the closed disk. By the Cauchy integral formula for $f^{(m)}(z)$,

$$f^{(m)}(z_0) = \frac{m!}{2\pi i} \int_{|z-z_0|=\rho} \frac{f(z)}{(z-z_0)^{m+1}}\,dz.$$

We parametrize the boundary circle by $z = z_0 + \rho e^{i\theta}$, $dz = i\rho e^{i\theta} d\theta$. Then

$$\frac{1}{2\pi i} \frac{f(z)}{(z-z_0)^{m+1}}\,dz = \frac{f\left(z_0 + \rho e^{i\theta}\right)}{\rho^m e^{im\theta}} \frac{d\theta}{2\pi},$$

and we obtain

$$f^{(m)}(z_0) = \frac{m!}{\rho^m} \int_0^{2\pi} f\left(z_0 + \rho e^{i\theta}\right) e^{-im\theta} \frac{d\theta}{2\pi}.$$

The obvious estimate

$$\left|f^{(m)}(z_0)\right| \le \frac{m!}{\rho^m} \int_0^{2\pi} \left|f\left(z_0 + \rho e^{i\theta}\right)\right| \frac{d\theta}{2\pi}$$

now leads to the following version of the Cauchy estimates.

Theorem (Cauchy Estimates). *Suppose $f(z)$ is analytic for $|z-z_0| \leq \rho$. If $|f(z)| \leq M$ for $|z - z_0| = \rho$, then*

$$\left| f^{(m)}(z_0) \right| \leq \frac{m!}{\rho^m} M, \qquad m \geq 0.$$

Note that this estimate scales correctly with respect to M, in the sense that if we multiply $f(z)$ by a positive constant, the estimate is multiplied by the same constant. It also scales correctly with respect to ρ, in the sense that if we dilate the disk by a factor $c > 0$, then both the mth derivative of the dilated function and the factor $1/\rho^m$ are multiplied by $1/c^m$. The estimate is invariant under translations. In effect it would have sufficed to check the estimate in the special case of analytic functions on the unit disk that are bounded in modulus by one on the unit circle. In this special case the estimate asserts that all derivatives of the function are bounded in modulus by one at the origin.

As an application of the Cauchy estimates, we prove the following.

Theorem (Liouville's Theorem). *Let $f(z)$ be an analytic function on the complex plane. If $f(z)$ is bounded, then $f(z)$ is constant.*

Indeed, suppose $|f(z)| \leq M$ for all $z \in \mathbb{C}$. We can apply the Cauchy estimate to a disk centered at any z_0, of any radius ρ, to obtain

$$|f'(z_0)| \leq \frac{M}{\rho}.$$

Letting ρ tend to $+\infty$, we obtain $f'(z_0) = 0$. Since this is true for all z_0, $f(z)$ is constant.

We define an **entire function** to be a function that is analytic on the entire complex plane. The polynomials $a_n z^n + \cdots + a_1 z + a_0$ are entire functions. The transcendental functions e^z, $\cos z$, $\sin z$, $\cosh z$, $\sinh z$ are also entire. Any linear combination of entire functions is entire, and any product of entire functions is entire. Examples of functions that are not entire are $1/z$, $\log z$, and $\sqrt{z}$.

In terms of entire functions, Liouville's theorem has a succinct statement: *A bounded entire function is constant.*

As a test of the strength of Liouville's theorem, we apply it to give yet another proof (one of hundreds of proofs) of the fundamental theorem of algebra, that every polynomial in z of degree $n \geq 1$ has a zero. (See Section I.1.) The proof is by contradiction. Suppose $p(z) = z^n + a_{n-1} z^{n-1} + \cdots + a_0$ is a polynomial with no complex root. Then $1/p(z)$ is an entire function. Since

$$\frac{p(z)}{z^n} = 1 + \frac{a_{n-1}}{z} + \cdots + \frac{a_0}{z^n}$$

tends to 1 as $z \to \infty$, $p(z) \to \infty$ and $1/p(z) \to 0$ as $z \to \infty$. Consequently, $1/p(z)$ is bounded. By Liouville's theorem, $1/p(z)$ is constant. Since the constant cannot be 0, we have a contradiction. Our supposition is false, and $p(z)$ must have a zero.

Exercises for IV.5

1. Show that if u is a harmonic function on $\mathbb{C}$ that is bounded above, then u is constant. *Hint.* Express u as the real part of an analytic function, and exponentiate.

2. Show that if $f(z)$ is an entire function, and there is a nonempty disk such that $f(z)$ does not attain any values in the disk, then $f(z)$ is constant.

3. A function $f(z)$ on the complex plane is **doubly periodic** if there are two periods ω_0 and ω_1 of $f(z)$ that do not lie on the same line through the origin (that is, ω_0 and ω_1 are linearly independent over the reals, and $f(z + \omega_0) = f(z + \omega_1) = f(z)$ for all complex numbers z). Prove that the only entire functions that are doubly periodic are the constants.

4. Suppose that $f(z)$ is an entire function such that $f(z)/z^n$ is bounded for $|z| \geq R$. Show that $f(z)$ is a polynomial of degree at most n. What can be said if $f(z)/z^n$ is bounded on the entire complex plane?

5. Show that if $\mathbf{V}(z)$ is the velocity vector field for a fluid flow in the entire complex plane, and if the speed $|\mathbf{V}(z)|$ is bounded, then $\mathbf{V}(z)$ is a constant flow.

6. Morera's Theorem

What Morera did was to observe that the differential $f(z)dz$ is closed if and only if $f(z)$ is analytic. The following more precise variant of this observation is often referred to as "Morera's theorem." It has a number of useful applications.

Theorem (Morera's Theorem). *Let $f(z)$ be a continuous function on a domain D. If $\int_{\partial R} f(z)dz = 0$ for every closed rectangle R contained in D with sides parallel to the coordinate axes, then $f(z)$ is analytic on D.*

The power of Morera's theorem resides in the fact that no hypothesis is made concerning the smoothness of $f(z)$. Only continuity is required.

To prove the theorem, we can assume that D is a disk with center z_0. Define

$$F(z) = \int_{z_0}^{z} f(\zeta)\,d\zeta, \qquad z \in D,$$

where the path of integration runs along a horizontal line and then a vertical line. We could as well define $F(z)$ using the path starting from z_0 along a vertical line followed by a horizontal line. The hypothesis guarantees that these two paths yield the same integral, as the difference is the integral of $f(z)\,dz$ over the boundary of a rectangle. Now we differentiate $F(z)$ by hand. We have

$$F(z + \Delta z) - F(z) = \int_{z}^{z+\Delta z} f(\zeta)\,d\zeta,$$

where the path of integration is the path from z to $z + \Delta z$ that follows a horizontal line and then a vertical line, as in the figure. Here we have again used the fact that the integral along the boundary of a rectangle is zero, as indicated in the figure. To deal with the integral on the right, we use the trick of adding and subtracting $f(z)$ from the integrand. Since z is fixed, the value $f(z)$ is constant for the integration, and we obtain

$$F(z + \Delta z) - F(z) = f(z) \int_{z}^{z+\Delta z} d\zeta + \int_{z}^{z+\Delta z} (f(\zeta) - f(z))\,d\zeta$$

$$= f(z)\Delta z + \int_{z}^{z+\Delta z} (f(\zeta) - f(z))\,d\zeta.$$

Now, the length of the contour from z to $z + \Delta z$ is at most $2|\Delta z|$. If we divide by Δz and use the ML-estimate on the last integral, we obtain

$$\left| \frac{F(z + \Delta z) - F(z)}{\Delta z} - f(z) \right| \leq 2M_\varepsilon, \qquad |\Delta z| < \varepsilon,$$

where M_ε is the maximum of $|f(\zeta) - f(z)|$ over all ζ satisfying $|\zeta - z| \leq \varepsilon$. Since $f(\zeta)$ is continuous at z, M_ε tends to 0 as $\varepsilon \to 0$. Consequently, $F(z)$ is complex differentiable, with complex derivative $F'(z) = f(z)$. Since $f(z)$ is continuous, $F(z)$ is analytic, and since $f(z)$ is the derivative of an analytic function, $f(z)$ is also analytic.

There is a metatheorem to the effect that *if the integrand depends analytically on a parameter, then the integral depends analytically on the parameter.* This sort of theorem is easy to prove using Morera's theorem and switching the orders of integration. To illustrate this idea, we prove the following typical theorem on analyticity of integrals.

Theorem. *Suppose that $h(t, z)$ is a continuous complex-valued function, defined for $a \leq t \leq b$ and $z \in D$. If for each fixed t, $h(t, z)$ is an analytic function of $z \in D$, then*

$$H(z) = \int_a^b h(t, z)\, dt, \qquad z \in D,$$

is analytic on D.

To see this, note first that $H(z)$ is continuous on D. Indeed, if $z_n \to z$, then $h(t, z_n) \to h(t, z)$ uniformly for $a \leq t \leq b$, so $H(z_n) \to H(z)$. Let R be a closed rectangle in D. By Cauchy's theorem, we have

$$\int_{\partial R} h(t, z)\, dz = 0.$$

Consequently,

$$\int_a^b \int_{\partial R} h(t, z)\, dz\, dt = 0.$$

Parameterization of the sides of ∂R converts the inside integral to a sum of four garden-variety integrals of a continuous function, and we can interchange the order of integration. This yields

$$0 = \int_{\partial R} \int_a^b h(t, z)\, dt\, dz = \int_{\partial R} H(z)\, dz.$$

The hypotheses of Morera's theorem are met, and $H(z)$ is analytic.

As another typical application, we prove the following useful result.

Theorem. *Suppose that $f(z)$ is a continuous function on a domain D that is analytic on $D \backslash \mathbb{R}$, that is, on the part of D not lying on the real axis. Then $f(z)$ is analytic on D.*

Let R be a closed rectangle contained in D, with sides parallel to the coordinate axes. To prove the theorem, it suffices to show that $\int_{\partial R} f(z) dz = 0$. There are three cases to consider. If the closed rectangle R does not meet the real axis $\mathbb{R}$, then $f(z)$ is analytic on R, so the integral is zero by Cauchy's theorem. For the second case, suppose that R has one edge that lies on the real axis, say the lower edge of R is an interval $[a, b]$ on the real axis. For $\varepsilon > 0$ small, let R_ε be the closed rectangle in the upper half-plane consisting of $z \in R$ such that $\operatorname{Im} z \geq \varepsilon$. By Cauchy's theorem, $\int_{\partial R_\varepsilon} f(z) dz = 0$. We claim that

(6.1) $$\int_{\partial R_\varepsilon} f(z) dz \to \int_{\partial R} f(z) dz$$

as $\varepsilon \to 0$. Indeed the integral along the bottom edge of R_ε has the form

$$\int_a^b f(t + i\varepsilon)\, dt,$$

and since $f(t + i\varepsilon)$ converges to $f(t)$ as $\varepsilon \to 0$, uniformly for $a \le t \le b$, the integrals converge to the integral $\int_a^b f(t)dt$ of $f(z)$ along the bottom edge of R. The top edge of R_ε coincides with the top edge of R, and the vertical sides of R_ε differ from the vertical sides of R only by vertical intervals of length ε, whose contribution to the integral tends to 0 with ε. Thus (6.1) holds, and we conclude in this case that $\int_{\partial R} f(z)dz = 0$. Finally, if the top edge of R is in the upper half-plane and the bottom edge of R is in the lower half-plane, we define R_+ to be the part of R in the closed upper half-plane and R_- to be the part of R in the closed lower half-plane. Then the integrals of $f(z)$ around ∂R_+ and ∂R_- are both zero, by case two. Thus

$$\int_{\partial R} f(z)dz \;=\; \int_{\partial R_+} f(z)dz \;+\; \int_{\partial R_-} f(z)dz \;=\; 0,$$

and the analyticity follows from Morera's theorem.

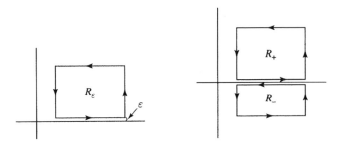

Exercises for IV.6

1. Let L be a line in the complex plane. Suppose $f(z)$ is a continuous complex-valued function on a domain D that is analytic on $D\backslash L$. Show that $f(z)$ is analytic on D.

2. Let $h(t)$ be a continuous function on the interval $[a, b]$. Show that the Fourier transform

$$H(z) = \int_a^b h(t)e^{-itz}dt$$

is an entire function that satisfies

$$|H(z)| \le Ce^{A|y|}, \quad z = x + iy \in \mathbb{C},$$

for some constants $A, C > 0$. *Remark.* An entire function satisfying such a growth restriction is called an **entire function of finite type**.

3. Let $h(t)$ be a continuous function on a subinterval $[a, b]$ of $[0, \infty)$. Show that the Fourier transform $H(z)$, defined as above, is bounded in the lower half-plane.

4. Let γ be a smooth curve in the plane $\mathbb{R}^2$, let D be a domain in the complex plane, and let $P(x, y, \zeta)$ and $Q(x, y, \zeta)$ be continuous complex-valued functions defined for (x, y) on γ and $\zeta \in D$. Suppose that the functions depend analytically on ζ for each fixed (x, y) on γ. Show that

$$G(\zeta) = \int_\gamma P(x, y, \zeta)\, dx + Q(x, y, \zeta)\, dy$$

is analytic on D.

7. Goursat's Theorem

We have defined $f(z)$ to be analytic on D if the complex derivative $f'(z)$ exists at each point of D and further, $f'(z)$ is a continuous function of z. Goursat's theorem asserts that the requirement that $f'(z)$ be continuous is redundant.

Theorem (Goursat's Theorem). *If $f(z)$ is a complex-valued function on a domain D such that*

$$f'(z_0) = \lim_{z \to z_0} \frac{f(z) - f(z_0)}{z - z_0}$$

exists at each point z_0 of D, then $f(z)$ is analytic on D.

Goursat's theorem is as useless as it is aesthetically pleasing. In applications it has always turned out that if one can show the existence of a complex derivative at each point, then one can see with little more effort that the complex derivative is continuous. Nevertheless, the idea of the proof has proved to be very useful in other contexts.

The proof is based on Morera's theorem. Let R be a closed rectangle in D. We subdivide R into four equal subrectangles. Since the integral of $f(z)$ around ∂R is the sum of the integrals of $f(z)$ around the four subrectangles, there is at least one of the subrectangles, call it R_1, for which

$$\left| \int_{\partial R_1} f(z)\, dz \right| \geq \frac{1}{4} \left| \int_{\partial R} f(z)\, dz \right|.$$

Now subdivide R_1 into four equal subrectangles and repeat the procedure. This yields a nested sequence of rectangles $\{R_n\}$ such that

$$\left| \int_{\partial R_n} f(z)\, dz \right| \geq \frac{1}{4} \left| \int_{\partial R_{n-1}} f(z)\, dz \right| \geq \cdots \geq \frac{1}{4^n} \left| \int_{\partial R} f(z)\, dz \right|.$$

Since the R_n's are decreasing and have diameters tending to 0, the R_n's converge to some point $z_0 \in D$. Since $f(z)$ is differentiable at z_0, we have an estimate of the form

$$\left| \frac{f(z) - f(z_0)}{z - z_0} - f'(z_0) \right| \leq \varepsilon_n , \qquad z \in R_n ,$$

where $\varepsilon_n \to 0$ as $n \to \infty$. Let L be the length of ∂R. Then the length of ∂R_n is $L/2^n$. For z belonging to R_n we have the estimate

$$|f(z) - f(z_0) - f'(z_0)(z - z_0)| \leq \varepsilon_n |z - z_0| \leq 2\varepsilon_n L/2^n .$$

From the ML-estimate and Cauchy's theorem, we obtain

$$\left| \int_{\partial R_n} f(z)\, dz \right| = \left| \int_{\partial R_n} [f(z) - f(z_0) - f'(z_0)(z - z_0)]\, dz \right|$$

$$\leq (2\varepsilon_n L/2^n) \cdot (L/2^n) = 2L^2 \varepsilon_n / 4^n .$$

Hence

$$\left| \int_{\partial R} f(z)\, dz \right| \leq 4^n \left| \int_{\partial R_n} f(z)\, dz \right| \leq 2L^2 \varepsilon_n .$$

Since $\varepsilon_n \to 0$ as $n \to \infty$, we must have

$$\int_{\partial R} f(z)\, dz = 0 .$$

By Morera's theorem, $f(z)$ is analytic.

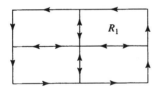
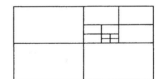

Exercises for IV.7

1. Find an application for Goursat's theorem in which it is not patently clear by other means that the function in question is analytic.

8. Complex Notation and Pompeiu's Formula

Many results in complex analysis can be expressed very simply in terms of the first-order differential operators

$$\frac{\partial}{\partial z} = \frac{1}{2}\left[\frac{\partial}{\partial x} - i\frac{\partial}{\partial y} \right], \qquad\qquad \frac{\partial}{\partial \bar{z}} = \frac{1}{2}\left[\frac{\partial}{\partial x} + i\frac{\partial}{\partial y} \right].$$

We may think of $\dfrac{\partial f}{\partial z}$ as an average of the derivatives of $f(z)$ in the x and the iy directions,

$$\frac{\partial f}{\partial z} = \frac{1}{2}\left[\frac{\partial f}{\partial x} + \frac{\partial f}{\partial(iy)}\right].$$

When we derived the Cauchy-Riemann equations in Chapter II, we found two expressions for the derivative of an analytic function $f(z)$,

$$f'(z) = \frac{\partial f}{\partial x} \quad \text{and} \quad f'(z) = -i\frac{\partial f}{\partial y} = \frac{\partial f}{\partial(iy)}.$$

If we take the average of these two expressions for $f'(z)$, we obtain

(8.1)
$$f'(z) = \frac{\partial f}{\partial z},$$

again provided that $f(z)$ is analytic.

To understand the operator $\partial/\partial\bar{z}$, we write $f = u + iv$ and compute

$$\frac{\partial f}{\partial\bar{z}} = \frac{1}{2}\left[\frac{\partial u}{\partial x} - \frac{\partial v}{\partial y}\right] + \frac{i}{2}\left[\frac{\partial u}{\partial y} + \frac{\partial v}{\partial x}\right].$$

If we equate the real and imaginary parts of the right-hand side to zero, we obtain the Cauchy-Riemann equations for u and v. Thus the equation

(8.2)
$$\frac{\partial f}{\partial\bar{z}} = 0$$

is equivalent to the Cauchy-Riemann equations for u and v. Equation (8.2) is referred to as the **complex form of the Cauchy-Riemann equations**. We summarize our observations.

Theorem. Let $f(z)$ be a continuously differentiable function on a domain D. Then $f(z)$ is analytic if and only if $f(z)$ satisfies the complex form (8.2) of the Cauchy-Riemann equations. If $f(z)$ is analytic, then the derivative of $f(z)$ is given by (8.1).

We list some rules for operating with $\partial/\partial z$ and $\partial/\partial\bar{z}$. Since these are both first-order differential operators with constant coefficients, they are linear,

$$\frac{\partial}{\partial z}(af + bg) = a\frac{\partial f}{\partial z} + b\frac{\partial g}{\partial z}, \qquad \frac{\partial}{\partial\bar{z}}(af + bg) = a\frac{\partial f}{\partial\bar{z}} + b\frac{\partial g}{\partial\bar{z}},$$

and they satisfy the Leibniz rule,

$$\frac{\partial}{\partial z}(fg) = f\frac{\partial g}{\partial z} + g\frac{\partial f}{\partial z}, \qquad \frac{\partial}{\partial\bar{z}}(fg) = f\frac{\partial g}{\partial\bar{z}} + g\frac{\partial f}{\partial\bar{z}}.$$

The z-derivative and the $\bar{z}$-derivative are related to each other by

$$\frac{\partial f}{\partial\bar{z}} = \overline{\frac{\partial\bar{f}}{\partial z}}, \qquad \frac{\partial\bar{f}}{\partial\bar{z}} = \overline{\frac{\partial f}{\partial z}}.$$

These formulae can be verified by writing $f = u + iv$, $\bar{f} = u - iv$, and carrying out the calculations.

The Taylor series expansion of a smooth function $f(z)$ at z_0, through only the linear terms, is given by

$$f(z) = f(z_0) + \frac{\partial f}{\partial z}(z_0)(z - z_0) + \frac{\partial f}{\partial \bar{z}}(z_0)\left(\overline{z - z_0}\right) + \mathcal{O}\left(|z - z_0|^2\right),$$

where the "big oh" term is a remainder term bounded by $C|z - z_0|^2$. The complex Taylor expansion can be derived from the usual Taylor expansion by substituting the definitions and calculating. It can also be derived by observing that it suffices to check the formula for the three functions 1, $z - z_0$, and $\overline{z - z_0}$.

As an application of Taylor's formula, we derive the complex form of the formula for the tangent vector to a curve. Let $f(z)$ be a smooth function, and let $\gamma(t)$ be a smooth curve terminating at $\gamma(0) = z_0$. From the linear Taylor series approximation we have

$$f(\gamma(t)) - f(\gamma(0)) = \frac{\partial f}{\partial z}(z_0)(\gamma(t) - z_0) + \frac{\partial f}{\partial \bar{z}}(z_0)\left(\overline{\gamma(t) - z_0}\right) + \mathcal{O}\left(|\gamma(t) - z_0|^2\right).$$

Dividing by t and taking the limit as $t \to 0$, we obtain

(8.3)
$$(f \circ \gamma)'(0) = \frac{\partial f}{\partial z}(z_0)\gamma'(0) + \frac{\partial f}{\partial \bar{z}}(z_0)\overline{\gamma'(0)}.$$

We use this formula to show that conformal maps are analytic.

Theorem. *Let $f(z)$ be a continuously differentiable function on a domain D. Suppose that the gradient of $f(z)$ does not vanish at any point of D, and that $f(z)$ is conformal. Then $f(z)$ is analytic on D, and $f'(z) \neq 0$ on D.*

To prove this, fix a point $z_0 \in D$, and consider the straight line $\gamma(t) = z_0 + te^{i\theta}$, $0 \leq t \leq \varepsilon$, terminating at z_0 with tangent $e^{i\theta}$. By (8.3), the image curve $f \circ \gamma$ has tangent at $f(z_0)$ given by

$$(f \circ \gamma)'(0) = \frac{\partial f}{\partial z}(z_0)e^{i\theta} + \frac{\partial f}{\partial \bar{z}}(z_0)e^{-i\theta}.$$

The condition on the gradient guarantees that this is not identically zero. In order for $f(z)$ to preserve angles at z_0, the difference in the arguments of $(f \circ \gamma)'(0)$ and $\gamma'(0)$ must be constant, independent of θ. Hence the argument of

$$\frac{(f \circ \gamma)'(0)}{\gamma'(0)} = \frac{\partial f}{\partial z}(z_0) + \frac{\partial f}{\partial \bar{z}}(z_0)e^{-2i\theta}$$

must be independent of θ. However, this occurs only when $\partial f/\partial \bar{z} = 0$. Hence $f(z)$ is analytic on D. The gradient condition implies that $f'(z) = \partial f/\partial z \neq 0$, and the theorem is established. Note that this theorem could as well be proved using x and y derivatives rather than z and $\bar{z}$ derivatives. However, the complex notation makes the proof more transparent.

For the case of complex line integrals, Green's theorem yields the following formula, which can be regarded as an extension of Cauchy's theorem to arbitrary smooth functions.

Theorem. *If D is a bounded domain in the complex plane with piecewise smooth boundary, and if $g(z)$ is a smooth function on $D \cup \partial D$, then*

$$(8.4) \qquad \int_{\partial D} g(z)\, dz \;=\; 2i \iint_D \frac{\partial g}{\partial \bar{z}}\, dx\, dy.$$

To see this we replace dz by $dx + i\, dy$ and apply Green's formula:

$$\int_{\partial D} g\, dx + \int_{\partial D} ig\, dy \;=\; \iint_D \left(i\frac{\partial g}{\partial x} - \frac{\partial g}{\partial y} \right) dx\, dy \;=\; 2i \iint_D \frac{\partial g}{\partial \bar{z}}\, dx\, dy.$$

Note that if $g(z)$ is analytic on D, then $\partial g / \partial \bar{z} = 0$ on D, so the integral over D vanishes, and we obtain Cauchy's theorem.

Cauchy's integral formula can also be extended to apply to arbitrary smooth functions.

Theorem (Pompeiu's Formula). *Suppose D is a bounded domain with piecewise smooth boundary. If $g(z)$ is a smooth complex-valued function on $D \cup \partial D$, then*

$$(8.5) \quad g(w) \;=\; \frac{1}{2\pi i} \int_{\partial D} \frac{g(z)}{z - w}\, dz \;-\; \frac{1}{\pi} \iint_D \frac{\partial g}{\partial \bar{z}} \frac{1}{z - w}\, dx\, dy, \qquad w \in D.$$

Pompeiu's formula is established by the same argument as was used in Section 4 to establish Cauchy's integral formula, except that now the correction term appears in the calculation. Let D_ε be the domain obtained from D by punching out a disk centered at w of radius ε. We apply the complex version (8.4) of Green's theorem to the function $g(z)/(z - w)$. For this, note that

$$\frac{\partial}{\partial \bar{z}} \left(\frac{g(z)}{z - w} \right) = \frac{\partial g}{\partial \bar{z}} \frac{1}{z - w} + g \frac{\partial}{\partial \bar{z}} \left(\frac{1}{z - w} \right) = \frac{\partial g}{\partial \bar{z}} \frac{1}{z - w}, \qquad z \in D_\varepsilon,$$

so that by (8.4),

$$(8.6) \qquad \int_{\partial D_\varepsilon} \frac{g(z)}{z - w}\, dz \;=\; 2i \iint_{D_\varepsilon} \frac{\partial g}{\partial \bar{z}} \frac{1}{z - w}\, dx\, dy.$$

The singularity of $1/(z - w)$ at $z = w$ is absolutely integrable:

$$\iint_{|z-w| \le 1} \frac{1}{|z - w|}\, dx\, dy \;=\; \int_0^{2\pi} \int_0^1 \frac{1}{r} \cdot r\, dr\, d\theta \;=\; 2\pi \;<\; \infty.$$

Hence the area integral in (8.6) over D_ε tends to the (improper) area integral over D as $\varepsilon \to 0$. The boundary integral in (8.6) has the form

$$\int_{\partial D_\varepsilon} \frac{g(z)}{z-w} dz = \int_{\partial D} \frac{g(z)}{z-w} dz - \int_{|z-w|=\varepsilon} \frac{g(z)}{z-w} dz.$$

If we parametrize the circle $|z - w| = \varepsilon$, we obtain for the integral on the right

$$\int_{|z-w|=\varepsilon} \frac{g(z)}{z-w} dz = i \int_0^{2\pi} g(w + \varepsilon e^{i\theta})\, d\theta.$$

This tends to $2\pi i g(w)$ as $\varepsilon \to 0$, since $g(z)$ is continuous at w. Thus if we let $\varepsilon \to 0$ in (8.6), we obtain

$$\int_{\partial D} \frac{g(z)}{z-w} dz - 2\pi i g(w) = 2i \iint_D \frac{\partial g}{\partial \bar z} \frac{1}{z-w} dx\, dy,$$

which is equivalent to (8.5).

The formula (8.5) is also known as the **Cauchy-Green formula**, since it is proved using Green's theorem. The formula can be regarded as Cauchy's integral formula with a correction term added to account for the fact that $g(z)$ may not be analytic. If $g(z)$ is analytic on D, it reduces to the Cauchy integral formula (4.1) for $g(z)$.

Exercises for IV.8

1. Show from the definition that

$$\frac{\partial}{\partial z} z = 1, \qquad \frac{\partial}{\partial \bar z} z = 0, \qquad \frac{\partial}{\partial z} \bar z = 0, \qquad \frac{\partial}{\partial \bar z} \bar z = 1.$$

2. Compute $\frac{\partial}{\partial \bar z}(az^2 + bz\bar z + c\bar z^2)$. Use the result to determine where $az^2 + bz\bar z + c\bar z^2$ is complex-differentiable and where it is analytic. (See Problem II.2.3.)

3. Show that the Jacobian of a smooth function f is given by

$$\det J_f = \left| \frac{\partial f}{\partial z} \right|^2 - \left| \frac{\partial f}{\partial \bar z} \right|^2.$$

4. Show that

$$\frac{\partial^2}{\partial x^2} + \frac{\partial^2}{\partial y^2} = 4\frac{\partial^2}{\partial z \partial \bar z}.$$

Deduce the following, for a smooth complex-valued function h.
(a) h is harmonic if and only if $\partial^2 h / \partial z \partial \bar z = 0$.
(b) h is harmonic if and only if $\partial h / \partial z$ is analytic.
(c) h is harmonic if and only if $\partial h / \partial \bar z$ is conjugate-analytic.

(d) If h is harmonic, then any mth order partial derivative of h is a linear combination of $\partial^m h/\partial z^m$ and $\partial^m h/\partial \bar{z}^m$.

5. With $d\bar{z} = dx - i\,dy$, show for a smooth function $f(z)$ that
$$df = \frac{\partial f}{\partial z}\,dz + \frac{\partial f}{\partial \bar{z}}\,d\bar{z}.$$

6. Show that if D is a domain with smooth boundary, and if $f(z)$ and $g(z)$ are analytic on $D \cup \partial D$, then
$$\int_{\partial D} f(z)\overline{g(z)}\,dz = 2i \iint_D f(z)\overline{g'(z)}\,dx\,dy.$$
Compare this formula with Exercise 1.4.

7. Show that the Taylor series expansion at $z_0 = 0$ of a smooth function $f(z)$, through the quadratic terms, is given by
$$f(z) = f(0) + \frac{\partial f}{\partial z}(0)z + \frac{\partial f}{\partial \bar{z}}(0)\bar{z}$$
$$+ \frac{1}{2}\left[\frac{\partial^2 f}{\partial z^2}(0)z^2 + 2\frac{\partial^2 f}{\partial z\partial \bar{z}}(0)|z|^2 + \frac{\partial^2 f}{\partial \bar{z}^2}(0)\bar{z}^2\right] + \mathcal{O}\left(|z|^3\right).$$

8. Establish the following version of the chain rule for smooth complex-valued functions $w = w(z)$ and $h = h(w)$:
$$\frac{\partial}{\partial \bar{z}}(h \circ w) = \frac{\partial h}{\partial w}\frac{\partial w}{\partial \bar{z}} + \frac{\partial h}{\partial \bar{w}}\frac{\partial \bar{w}}{\partial \bar{z}},$$
$$\frac{\partial}{\partial z}(h \circ w) = \frac{\partial h}{\partial w}\frac{\partial w}{\partial z} + \frac{\partial h}{\partial \bar{w}}\frac{\partial \bar{w}}{\partial z}.$$

9. Show with the aid of the preceding exercise that if both $h(w)$ and $w(z)$ are analytic, then $(h \circ w)(z)$ is analytic, and $(h \circ w)'(z) = h'(w(z))w'(z)$.

10. Let $g(z)$ be a continuously differentiable function on the complex plane that is zero outside of some compact set. Show that
$$g(w) = -\frac{1}{\pi}\iint_{\mathbb{C}} \frac{\partial g}{\partial \bar{z}}\frac{1}{z-w}\,dx\,dy, \qquad w \in \mathbb{C}.$$
Remark. If we integrate this formally by parts, we obtain
$$g(w) = \frac{1}{\pi}\iint_{\mathbb{C}} g(z)\frac{\partial}{\partial \bar{z}}\left(\frac{1}{z-w}\right)\,dx\,dy.$$
Thus the "distribution derivative" of $1/(\pi(z-w))$ with respect to z is the point mass at w ("Dirac delta-function"), in the sense that it is equal to 0 away from w, and it is infinite at w in such a way that its integral (total mass) is equal to 1.

V

Power Series

In this chapter we show that the analytic functions are exactly the functions that can be expanded in a convergent power series about any point. Since power series can be treated very much as polynomials, this provides a powerful tool for dealing with analytic functions. In Sections 1 and 2 we review infinite series and series of functions. Sections 3 through 6 contain the basic material on power series. In Section 7 we use power series to show that the zeros of an analytic function are isolated. This leads to the uniqueness principle for analytic functions. Section 8 contains a formal definition of analytic continuation, which can be omitted at first reading.

1. Infinite Series

In this section we review some basic material on infinite series of complex numbers. In the next section we provide some background material on sequences and series of functions. The reader may wish to skip to Section 3 and refer to the background sections only when necessary.

A series $\sum_{k=0}^{\infty} a_k$ of complex numbers is said to **converge to** S if the sequence of partial sums $\{S_k\}$, defined by $S_k = a_0 + \cdots + a_k$, converges to S. For notation, the sum S of the series is denoted by $\sum_{k=0}^{\infty} a_k$, or simply by $\sum a_k$.

Any statement concerning series can be reinterpreted as a statement about sequences, by phrasing the statement in terms of the sequence of partial sums of the series. For instance, we know that if $s_n \to s$ and $t_n \to t$, then $s_n + t_n \to s + t$. If we apply this statement to the partial sums of series, we conclude that if $\sum a_k = A$ and $\sum b_k = B$, then $\sum (a_k + b_k) = A + B$. Similarly, since $s_n \to s$ implies $cs_n \to cs$, we deduce that if $\sum a_k = A$, then $\sum ca_k = cA$. Thus taking limits of sequences is a linear operation, and this implies that summing series is also a linear operation.

If the r_k's are positive real numbers, then the partial sums $S_k = r_0 + \cdots + r_k$ form a monotone increasing sequence. Since a monotone sequence of real numbers converges if and only if it is bounded, we see that a series of positive numbers converges if and only if the partial sums are bounded. If $0 \leq a_k \leq r_k$, and if the partial sums of $\sum r_k$ are bounded, then the partial sums of $\sum a_k$ are bounded by the same bound. This observation leads to the following convergence test for series with positive terms.

Theorem (Comparison Test). *If $0 \leq a_k \leq r_k$, and if $\sum r_k$ converges, then $\sum a_k$ converges, and $\sum a_k \leq \sum r_k$.*

The terms of a series can be recovered from the partial sums, by $a_k = S_k - S_{k-1}$. Suppose the series $\sum a_k$ converges to S. Then S_k converges to S as $k \to \infty$, and also S_{k-1} converges to S as $k \to \infty$. Hence $a_k \to 0$ as $k \to \infty$. This simple necessary condition provides a useful screening test for convergence.

Theorem. *If $\sum a_k$ converges, then $a_k \to 0$ as $k \to \infty$.*

Example. The most important series for us will be the geometric series $\sum_{k=0}^{\infty} z^k$. The kth partial sum of the geometric series is given by

$$S_k = 1 + z + z^2 + \cdots + z^{k-1} + z^k = \frac{1 - z^{k+1}}{1 - z}, \qquad z \neq 1.$$

To see this, we multiply and divide the sum by $1 - z$, and we note that the numerator telescopes:

$$(1 - z)S_k = 1 - z + z - z^2 + \cdots + z^{k-1} - z^k + z^k - z^{k+1} = 1 - z^{k+1}$$

If $|z| < 1$, then $z^{k+1} \to 0$ and $S_k \to 1/(1 - z)$ as $k \to \infty$. Hence

$$\sum_{k=0}^{\infty} z^k = \frac{1}{1 - z}, \qquad |z| < 1.$$

On the other hand, if $|z| \geq 1$, then the kth term z^k does not converge to 0, so that the series does not converge.

The series $\sum a_k$ is said to **converge absolutely** if $\sum |a_k|$ converges. Thus for a series of positive terms, convergence and absolute convergence are the same. There are convergent series that are not absolutely convergent. (See Exercises 4 and 5.) However, every absolutely convergent series is convergent.

Theorem. *If $\sum a_k$ converges absolutely, then $\sum a_k$ converges, and*

$$(1.1) \qquad \left| \sum_{k=0}^{\infty} a_k \right| \leq \sum_{k=0}^{\infty} |a_k|.$$

The proof is easy, modulo a little trick, which is to express $\operatorname{Re} a_k$ as a difference $\operatorname{Re} a_k = (\operatorname{Re} a_k + |a_k|) - |a_k|$. Since $|\operatorname{Re} a_k| \leq |a_k|$, we have

$0 \leq \operatorname{Re} a_k + |a_k| \leq 2|a_k|$. Hence $\sum(\operatorname{Re} a_k + |a_k|)$ is a series of non-negative real numbers. Since its partial sums are bounded, by $2\sum|a_k|$, the series converges. Now $\sum \operatorname{Re} a_k$ is the difference of two convergent series, $\sum(\operatorname{Re} a_k + |a_k|)$ and $\sum|a_k|$, and so $\sum \operatorname{Re} a_k$ converges. Similarly, $\sum \operatorname{Im} a_k$ converges, and consequently $\sum a_k$ converges. For each N we have $|\sum_{k=0}^{N} a_k| \leq \sum_{k=0}^{N} |a_k|$. Letting $N \to \infty$, we obtain the estimate (1.1).

Example. The geometric series converges absolutely when $|z| < 1$, and

$$\left| \frac{1}{1-z} \right| = \left| \sum_{k=0}^{\infty} z^k \right| \leq \sum_{k=0}^{\infty} |z|^k = \frac{1}{1-|z|}, \qquad |z| < 1.$$

This leads also to a useful estimate for the difference between the partial sums of the geometric series and the full sum. From

$$(1.2) \qquad \frac{1}{1-z} - \sum_{k=0}^{n} z^k = \sum_{k=n+1}^{\infty} z^k = z^{n+1} \sum_{j=0}^{\infty} z^j = \frac{z^{n+1}}{1-z}$$

we obtain

$$\left| \frac{1}{1-z} - \sum_{k=0}^{n} z^k \right| \leq \frac{|z|^{n+1}}{1-|z|}, \qquad |z| < 1.$$

Exercises for V.1

1. (Harmonic Series) Show that

$$\sum_{k=1}^{n} \frac{1}{k} \geq \log n.$$

Deduce that the series $\sum \frac{1}{k}$ does not converge. *Hint.* Use the estimate

$$\frac{1}{k} \geq \int_{k}^{k+1} \frac{1}{x} \, dx.$$

2. Show that if $p < 1$, then the series $\sum_{k=1}^{\infty} 1/k^p$ diverges. *Hint.* Use Exercise 1 and the comparison test.

3. Show that if $p > 1$, then the series $\sum_{k=1}^{\infty} 1/k^p$ converges to S, where

$$\left| S - \sum_{k=1}^{n} \frac{1}{k^p} \right| < \frac{1}{(p-1)n^{p-1}}.$$

Hint. Use the estimate $\dfrac{1}{k^p} < \displaystyle\int_{k-1}^{k} \frac{dx}{x^p}$.

4. Show that the series

$$\sum_{k=1}^{\infty} \frac{(-1)^{k+1}}{k} = 1 - \frac{1}{2} + \frac{1}{3} - \frac{1}{4} + \cdots$$

converges. *Hint.* Show that the partial sums of the series satisfy $S_2 < S_4 < S_6 < \cdots < S_5 < S_3 < S_1$.

5. Show that the series

$$1 + \frac{1}{3} - \frac{1}{2} + \frac{1}{5} + \frac{1}{7} - \frac{1}{4} + \frac{1}{9} + \frac{1}{11} - \frac{1}{6} + \cdots$$

converges to $3S/2$, where S is the sum of the series in Exercise 4. (It turns out that $S = \log 2$.) *Hint.* Organize the terms in the series in Exercise 4 in groups of four, and relate it to the groups of three in the above series.

6. Show that $\sum \frac{1}{k \log k}$ diverges while $\sum \frac{1}{k(\log k)^2}$ converges.

7. Show that the series $\sum a_k$ converges if and only if $\sum_{k=m}^{k=n} a_k$ tends to 0 as $m, n \to \infty$. *Remark.* This is the **Cauchy criterion for series**.

2. Sequences and Series of Functions

Let $\{f_j\}$ be a sequence of complex-valued functions defined on some set E. We say that the sequence $\{f_j\}$ **converges pointwise** on E if for each point $x \in E$ the sequence of complex numbers $\{f_j(x)\}$ converges. The limit $f(x)$ of $\{f_j(x)\}$ is then a complex-valued function on E.

Example. The sequence of functions $f_j(x) = x^j$, $0 \le x \le 1$, converges pointwise on the unit interval to the function $f(x)$ defined on the unit interval by $f(x) = 0$ for $0 \le x < 1$ and $f(1) = 1$. Note that the pointwise limit of a sequence of continuous functions need not be continuous.

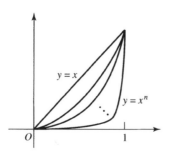

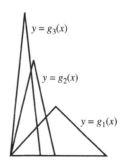

Example. We define a sequence of "tent functions" g_j on the unit interval by $g_j(x) = j^2 x$ for $0 \leq x \leq 1/j$, $g_j(x) = 2j - j^2 x$ for $1/j \leq x \leq 2/j$, and $g_j(x) = 0$ for $2/j \leq x \leq 1$. The height of the jth tent is j, and the width of the base is $2/j$, so that the area under the tent is $\int_0^1 g_j(x)\,dx = 1$. On the other hand, the sequence of functions $g_j(x)$ converges pointwise to 0 on the unit interval, and the integral of the pointwise limit is 0.

To guarantee that the limit of the integrals of a sequence of functions is the integral of the limit, we must require that the functions converge in some stronger sense than pointwise. Toward this goal we introduce the notion of uniform convergence. We say that the sequence $\{f_j\}$ of functions on E **converges uniformly** to f on E if $|f_j(x) - f(x)| \leq \varepsilon_j$ for all $x \in E$, where $\varepsilon_j \to 0$ as $j \to \infty$. We may regard ε_j as a worst-case estimator for the difference $f_j(x) - f(x)$, and usually we take ε_j to be the supremum (maximum) of $|f_j(x) - f(x)|$ over $x \in E$,

$$\varepsilon_j \;=\; \sup_{x \in E} |f_j(x) - f(x)|.$$

Note that if $\{f_j\}$ converges uniformly to f on E, then it converges pointwise to f on E.

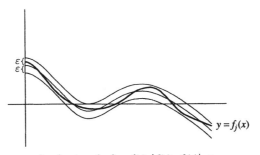

ε-tube about graph of $y = f(x)$, $|f(x) - f_j(x)| < \varepsilon$

In the two examples above, the sequences do not converge uniformly. In the first example the worst-case estimator is $\varepsilon_j = \max_{0 \leq x \leq 1} x^j = 1$, which does not converge to 0. In the example of the sequence of tent functions, the worst-case estimator is worse. It is the height of the tent, which is $\varepsilon_j = j \to +\infty$.

It turns out that with uniform instead of pointwise convergence, the two theorems we desire are valid: (1) a uniform limit of continuous functions is continuous, and (2) an integral of a uniform limit is the limit of the integrals. We state the theorems more precisely.

Theorem. *Let $\{f_j\}$ be a sequence of complex-valued functions defined on a subset E of the complex plane. If each f_j is continuous on E, and if $\{f_j\}$ converges uniformly to f on E, then f is continuous on E.*

Theorem. *Let γ be a piecewise smooth curve in the complex plane. If $\{f_j\}$ is a sequence of continuous complex-valued functions on γ, and if $\{f_j\}$ converges uniformly to f on γ, then $\int_\gamma f_j(z)dz$ converges to $\int_\gamma f(z)dz$.*

The first theorem above on the continuity of a uniform limit has a standard formal proof, which we omit. The second theorem above on the limit of the integrals is an easy consequence of the ML-estimate. Indeed, suppose $\{f_j\}$ converges uniformly to f on γ. Let ε_j be the worst-case estimator for $f_j - f$ on γ, so that $|f_j - f| \le \varepsilon_j$ on γ, and let L be the length of γ. Then the ML-estimate gives

$$\left| \int_\gamma f_j(z)dz - \int_\gamma f(z)dz \right| \le \varepsilon_j L,$$

and this tends to 0, since the f_j's converge uniformly to f. Hence $\int f_j dz$ tends to $\int f\, dz$.

Now we turn to series of functions. Let $\sum g_j(x)$ be a series of complex-valued functions defined on a set E. The partial sums of the series are the functions

$$S_n(x) = \sum_{k=0}^{n} g_j(x) = g_0(x) + g_1(x) + \cdots + g_n(x).$$

We say that the series **converges pointwise** on E if the sequence of partial sums converges pointwise on E, and the series **converges uniformly** on E if the sequence of partial sums converges uniformly on E. The following criterion for uniform convergence of a series of functions is extremely useful. In fact, it is the only test for uniform convergence of series that we will ever need.

Theorem (Weierstrass M-Test). *Suppose $M_k \ge 0$ and $\sum M_k$ converges. If $g_k(x)$ are complex-valued functions on a set E such that $|g_k(x)| \le M_k$ for all $x \in E$, then $\sum g_k(x)$ converges uniformly on E.*

The proof is straightforward. For each fixed x, the estimate for $g_k(x)$ shows that the series $\sum g_k(x)$ is absolutely convergent, and $\sum |g_k(x)| \le \sum M_k$. By the theorem in Section 1, the series $\sum g_k(x)$ converges to some complex number $g(x)$, and by (1.1), $|g(x)| \le \sum |g_k(x)| \le \sum M_k$. The same estimate, applied to the tail of the series, shows that

$$\left| g(x) - S_n(x) \right| = \left| \sum_{k=n+1}^{\infty} g_k(x) \right| \le \sum_{k=n+1}^{\infty} M_k.$$

If we set $\varepsilon_n = \sum_{k=n+1}^{\infty} M_k$, then $\varepsilon_n \to 0$ as $n \to \infty$, and the estimate shows that the partial sums $S_n(x)$ converge uniformly on E to $g(x)$.

Example. For the geometric series

$$\sum_{k=0}^{\infty} z^k = \frac{1}{1-z}, \qquad |z| < 1,$$

we have from (1.2) that

$$\left| \frac{1}{1-z} - S_n(z) \right| = \left| \frac{z^{n+1}}{1-z} \right|,$$

which tends to $+\infty$ as $z \to 1$. Hence the partial sums do not converge uniformly for $|z| < 1$. However, suppose we fix a radius $r < 1$. Define $M_k = r^k$. Then $\sum M_k$ converges, and $|z^k| \le M_k$ for $|z| \le r$. By the Weierstrass M-test, $\sum z^k$ converges uniformly for $|z| \le r$. Thus the geometric series converges uniformly on each disk $\{|z| \le r\}$, for each $r < 1$, but it does not converge uniformly on the disk $\{|z| < 1\}$.

Now we return to analytic functions. We begin by proving that a uniform limit of analytic functions is analytic.

Theorem. If $\{f_k(z)\}$ is a sequence of analytic functions on a domain D that converges uniformly to $f(z)$ on D, then $f(z)$ is analytic on D.

This can be proved using the Cauchy integral formula. An easier way to see it is to apply Morera's theorem. Since analytic functions are continuous, and the limit of a uniformly convergent sequence of continuous functions is continuous, $f(z)$ is continuous. Let E be a closed rectangle contained in D. By Cauchy's theorem, $\int_{\partial E} f_k(z) dz = 0$ for each k. From the theorem above we obtain in the limit that $\int_{\partial E} f(z) dz = 0$. By Morera's theorem, $f(z)$ is analytic.

Theorem. Suppose that $f_k(z)$ is analytic for $|z - z_0| \le R$, and suppose that the sequence $\{f_k(z)\}$ converges uniformly to $f(z)$ for $|z - z_0| \le R$. Then for each $r < R$ and for each $m \ge 1$, the sequence of mth derivatives $\left\{ f_k^{(m)}(z) \right\}$ converges uniformly to $f^{(m)}(z)$ for $|z - z_0| \le r$.

To prove this, suppose $\varepsilon_k \to 0$ are such that $|f_k(z) - f(z)| < \varepsilon_k$ for $|z - z_0| < R$. Fix s such that $r < s < R$. The Cauchy integral formula for the mth derivative of $f_k(z) - f(z)$ on the disk $|z - z_0| \le s$ yields

$$f_k^{(m)}(z) - f^{(m)}(z) = \frac{m!}{2\pi i} \int_{|z-z_0|=s} \frac{f_k(\zeta) - f(\zeta)}{(\zeta - z)^{m+1}} d\zeta, \qquad |z - z_0| \le r.$$

If $|\zeta - z_0| = s$ and $|z - z_0| \le r$, then $|\zeta - z| \ge s - r$, and so

$$\left| \frac{f_k(\zeta) - f(\zeta)}{(\zeta - z)^{m+1}} \right| \le \frac{\varepsilon_k}{(s-r)^{m+1}}.$$

From the ML-estimate, we obtain

$$\left| f_k^{(m)}(z) - f^{(m)}(z) \right| \le \frac{m!}{2\pi} \frac{\varepsilon_k}{(s-r)^{m+1}} \, 2\pi s \;=\; \rho_k, \qquad |z - z_0| \le r.$$

Since $\rho_k \to 0$ as $k \to \infty$, we obtain uniform convergence of the mth derivatives for $|z - z_0| \le r$.

We say that a sequence $\{f_k(z)\}$ of analytic functions on a domain D **converges normally** to the analytic function $f(z)$ on D if it converges uniformly to $f(z)$ on each closed disk contained in D. It is easy to see that this occurs if and only if $\{f_k(z)\}$ converges to $f(z)$ uniformly on each bounded subset E of D at a strictly positive distance from the boundary of D. (See Exercises 4 and 5.) Since any closed disk contained in D can be dilated to a larger disk contained in D, we can apply the preceding theorem, and we obtain the following.

Theorem. *Suppose that $\{f_k(z)\}$ is a sequence of analytic functions on a domain D that converges normally on D to the analytic function $f(z)$. Then for each $m \ge 1$, the sequence of mth derivatives $\left\{ f_k^{(m)}(z) \right\}$ converges normally to $f^{(m)}(z)$ on D.*

Exercises for V.2

1. Show that $f_k(x) = x^k/(k + x^{2k})$ converges uniformly to 0 on $[0, \infty)$. *Hint.* Determine the worst-case estimator ε_k by calculus.

2. Show that $g_k(x) = x^k/(1 + x^k)$ converges pointwise on $[0, \infty)$ but not uniformly. What is the limit function? On which subsets of $[0, \infty)$ does the sequence converge uniformly?

3. Show that $f_k(z) = z^k/k$ converges uniformly for $|z| < 1$. Show that $f_k'(z)$ does not converge uniformly for $|z| < 1$. What *can* be said about the uniform convergence of $f_k'(z)$?

4. Show that $\displaystyle\sum \frac{1}{k^2} \frac{x^k}{1 + x^{2k}}$ converges uniformly for $-\infty < x < +\infty$.

5. For which real numbers x does $\displaystyle\sum \frac{1}{k} \frac{x^k}{1 + x^{2k}}$ converge?

6. Show that for each $\varepsilon > 0$, the series $\displaystyle\sum \frac{1}{k} \frac{x^k}{1 + x^{2k}}$ converges uniformly for $x \ge 1 + \varepsilon$.

7. Let a_n be a bounded sequence of complex numbers. Show that for each $\varepsilon > 0$, the series $\sum_{n=1}^{\infty} a_n n^{-z}$ converges uniformly for $\operatorname{Re} z \ge 1 + \varepsilon$. Here we choose the principal branch of n^{-z}.

8. Show that $\sum \dfrac{z^k}{k^2}$ converges uniformly for $|z| < 1$.

9. Show that $\sum \dfrac{z^k}{k}$ does not converge uniformly for $|z| < 1$.

10. Show that if a sequence of functions $\{f_k(x)\}$ converges uniformly on E_j for $1 \le j \le n$, then the sequence converges uniformly on the union $E = E_1 \cup E_2 \cup \cdots \cup E_n$.

11. Suppose that E is a bounded subset of a domain $D \subset \mathbb{C}$ at a positive distance from the boundary of D, that is, there is $\delta > 0$ such that $|z - w| \ge \delta$ for all $z \in E$ and $w \in \mathbb{C} \backslash D$. Show that E can be covered by a finite number of closed disks contained in D. *Hint.* Consider all closed disks with centers at points $(m + ni)\delta/10$ and radius $\delta/10$ that meet E.

12. Let $f(z)$ be analytic on a domain D, and suppose $|f(z)| \le M$ for all $z \in D$. Show that for each $\delta > 0$ and $m \ge 1$, $|f^{(m)}(z)| \le m!M/\delta^m$ for all $z \in D$ whose distance from ∂D is at least δ. Use this to show that if $\{f_k(z)\}$ is a sequence of analytic functions on D that converges uniformly to $f(z)$ on D, then for each m the derivatives $f_k^{(m)}(z)$ converge uniformly to $f^{(m)}(z)$ on each subset of D at a positive distance from ∂D.

3. Power Series

A **power series** (centered at z_0) is a series of the form $\sum_{k=0}^{\infty} a_k(z - z_0)^k$. By making a change of variable $w = z - z_0$, we can always reduce to the case of power series centered at $z = 0$. The main result on convergence of power series is the following.

Theorem. *Let $\sum a_k z^k$ be a power series. Then there is R, $0 \le R \le +\infty$, such that $\sum a_k z^k$ converges absolutely if $|z| < R$, and $\sum a_k z^k$ does not converge if $|z| > R$. For each fixed r satisfying $r < R$, the series $\sum a_k z^k$ converges uniformly for $|z| \le r$.*

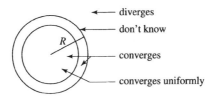

We call R the **radius of convergence** of the series $\sum a_k z^k$. The radius of convergence depends only on the tail of the series. If we alter a finite number of coefficients of the series, the radius of convergence remains the same.

For the general case of a power series $\sum a_k(z - z_0)^k$, the domain of convergence is a disk $|z - z_0| < R$. The series diverges if $|z - z_0| > R$, and anything can happen when $|z - z_0| = R$.

For the proof of the theorem, note first that if the sequence $|a_k|r^k$ is bounded for some value $r = r_0$, then it is bounded for all r satisfying $0 \leq r < r_0$. We define R, $0 \leq R \leq +\infty$, to be the supremum of the r's such that $|a_k|r^k$ is bounded. Thus $|a_k|r^k$ is bounded if $r < R$, while if $r > R$, then there is a sequence of terms with $|a_{k_j}|r^{k_j} \to +\infty$. In the borderline case $r = R$, anything can happen. The sequence $|a_k|R^k$ might be bounded and it might not.

If $|z| > R$, then the terms $a_k z^k$ do not tend to 0, so that the series does not converge. On the other hand, suppose $r < R$. Choose s such that $r < s < R$. Then the sequence $|a_k|s^k$ is bounded, say $|a_k|s^k \leq C$ for $k \geq 0$. If $|z| \leq r$, then

$$|a_k z^k| \leq |a_k|r^k = |a_k|s^k\left(\frac{r}{s}\right)^k \leq C\left(\frac{r}{s}\right)^k.$$

Set $M_k = C(r/s)^k$. Since $\sum M_k$ converges, the Weierstrass M-test applies, and the series $\sum a_k z^k$ converges uniformly for $|z| \leq r$, and also absolutely for each z. This proves the theorem.

Example. The geometric series $\sum z^k$ has radius of convergence $R = 1$. The series does not converge on the boundary circle $|z| = 1$, since the terms do not tend to 0.

Example. The power series $\sum z^k/k^2$ converges uniformly for $|z| \leq 1$. This follows from the Weierstrass M-test, with majorants $M_k = 1/k^2$. On the other hand, if $r > 1$, then $r^k/k^2 \to \infty$ as $k \to \infty$. Thus the series does not converge for $|z| > 1$, and the radius of convergence of the series is $R = 1$.

Example. The series

$$(3.1) \qquad \sum_{k=0}^{\infty} \frac{(-1)^k}{2^k} z^{2k} = 1 - \frac{z^2}{2} + \frac{z^4}{2^2} - \frac{z^6}{2^3} + \cdots$$

becomes a geometric series if we set $w = -z^2/2$,

$$\sum_{k=0}^{\infty} \frac{(-1)^k}{2^k} z^{2k} = \sum_{k=0}^{\infty} w^k.$$

The series converges precisely when $|w| < 1$, that is, when $|z^2| < 2$. The radius of convergence is thus $R = \sqrt{2}$. The series converges to $1/(1-w) = 2/(2 - z^2)$.

Example. The series $\sum k^k z^k$ has radius of convergence $R = 0$. It converges only for $z = 0$, since $k^k r^k \to +\infty$ for all $r > 0$.

Example. The series $\sum k^{-k} z^k$ has radius of convergence $R = +\infty$. It converges for all z.

The partial sums of a power series are polynomials in z, and in particular they are analytic functions. From the convergence theorem of Section 2, we obtain the following.

Theorem. *Suppose* $\sum a_k z^k$ *is a power series with radius of convergence* $R > 0$. *Then the function*

$$f(z) = \sum_{k=0}^{\infty} a_k z^k, \qquad |z| < R,$$

is analytic. The derivatives of $f(z)$ *are obtained by differentiating the series term by term,*

$$f'(z) = \sum_{k=1}^{\infty} k a_k z^{k-1}, \qquad f''(z) = \sum_{k=2}^{\infty} k(k-1) a_k z^{k-2}, \qquad |z| < R,$$

and similarly for the higher-order derivatives. The coefficients of the series are given by

(3.2) $$a_k = \frac{1}{k!} f^{(k)}(0), \qquad k \geq 0.$$

The formula for the coefficient a_k is obtained by differentiating k times the series for $f(z)$ and plugging in $z = 0$.

Example. By differentiating the representation of $1/(1-z)$ as a geometric series, we obtain a power series representation of $1/(1-z)^2$,

$$\frac{1}{(1-z)^2} = \sum_{k=1}^{\infty} k z^{k-1} = \sum_{m=0}^{\infty} (m+1) z^m, \qquad |z| < 1.$$

On account of the uniform convergence of power series on subdisks of radius strictly smaller than R, a power series can be integrated term by term. Thus if $\sum a_k z^k$ has radius of convergence R, then

$$\int_0^z \sum \left(a_k \zeta^k \right) d\zeta = \sum a_k \int_0^z \zeta^k \, d\zeta = \sum \frac{a_k}{k+1} z^{k+1}, \qquad |z| < R.$$

Example. If we integrate the geometric series term by term, we obtain

$$-\operatorname{Log}(1-z) = \int_0^z \frac{d\zeta}{1-\zeta} = \int_0^z \sum_{k=0}^{\infty} \zeta^k \, d\zeta$$

$$= \sum_{k=0}^{\infty} \frac{z^{k+1}}{k+1} = z + \frac{z^2}{2} + \frac{z^3}{3} + \frac{z^4}{4} + \cdots, \qquad |z| < 1.$$

Making the substitution $w = 1 - z$, we obtain a series expansion for $\operatorname{Log} w$ centered at $w = 1$,

$$\operatorname{Log} w = \sum_{k=1}^{\infty} \frac{(-1)^{k+1}}{k} (w-1)^k$$

$$= (w-1) - \frac{(w-1)^2}{2} + \frac{(w-1)^3}{3} - \frac{(w-1)^4}{4} + \cdots, \qquad |w-1| < 1.$$

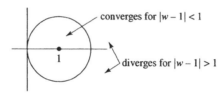

converges for $|w-1| < 1$

diverges for $|w-1| > 1$

Now we turn to two formulae for determining the radius of convergence of a power series from its coefficients. The first of these is based on the **ratio test**. It is especially convenient for determining the radius of convergence of many series that arise as solutions of linear differential equations.

Theorem. If $|a_k/a_{k+1}|$ has a limit as $k \to \infty$, either finite or $+\infty$, then the limit is the radius of convergence R of $\sum a_k z^k$,

$$R = \lim_{k \to \infty} \left| \frac{a_k}{a_{k+1}} \right|.$$

To see this, let $L = \lim |a_k/a_{k+1}|$. If $r < L$, then $|a_k/a_{k+1}| > r$ eventually, say for $k \geq N$. Then $|a_k| > r|a_{k+1}|$ for $k \geq N$, and

$$|a_N|r^N \geq |a_{N+1}|r^{N+1} \geq |a_{N+2}|r^{N+2} \geq \cdots.$$

Hence the sequence $|a_k|r^k$ is bounded. From the definition of R we have $r \leq R$, and since $r < L$ is arbitrary, we also have $L \leq R$.

Suppose next that $s > L$. Then $|a_k/a_{k+1}| < s$ eventually, say for $k \geq N$. Then $|a_k| < s|a_{k+1}|$ for $k \geq N$, and

$$|a_N|s^N \leq |a_{N+1}|s^{N+1} \leq |a_{N+2}|s^{N+2} \leq \cdots.$$

Hence the terms $a_k z^k$ do not converge to 0 for $|z| \geq s$, so that the series does not converge, and $s \geq R$. Since $s > L$ is arbitrary, we also have $L \geq R$. We conclude that $L = R$.

Example. For the series $\sum k z^k$, the ratio test gives

$$\left| \frac{a_k}{a_{k+1}} \right| = \frac{k}{k+1} \to 1.$$

Hence the radius of convergence is $R = 1$.

Example. For the series $\sum \dfrac{z^k}{k!}$, the ratio test gives

$$\left| \frac{a_k}{a_{k+1}} \right| = \frac{(k+1)!}{k!} = k+1 \to +\infty.$$

Hence the radius of convergence is $R = +\infty$.

The second formula is based on the **root test**.

Theorem. If $\sqrt[k]{|a_k|}$ has a limit as $k \to \infty$, either finite or $+\infty$, then the radius of convergence of $\sum a_k z^k$ is given by

$$(3.3) \qquad\qquad R = \frac{1}{\lim \sqrt[k]{|a_k|}}.$$

If $r > 1/\lim \sqrt[k]{|a_k|}$, then $\sqrt[k]{|a_k|}\, r > 1$ eventually, so that $|a_k| r^k > 1$ eventually, the terms of the series $\sum a_k z^k$ do not converge to 0 for $|z| = r$, and $r \geq R$. On the other hand, if $r < 1/\lim \sqrt[k]{|a_k|}$, then $\sqrt[k]{|a_k|}\, r < 1$ eventually, so that $|a_k| r^k < 1$ eventually, the sequence $|a_k| r^k < 1$ is bounded, and from the definition of R we have $r \leq R$. It follows that (3.3) holds.

Example. For the series $\sum k z^k$, the root test gives

$$R = 1/\lim \sqrt[k]{k} = 1.$$

There is a more general form of the formula (3.3), called the **Cauchy-Hadamard formula**, that gives the radius of convergence for *any* power series in terms of a lim sup. Recall (Section II.1) that the lim sup of a sequence $\{s_n\}$ is characterized as the number S, $-\infty \leq S \leq +\infty$, with the property that if $t > S$, then only finitely many terms of the sequence satisfy $s_n > t$, while if $t < S$, then infinitely many terms of the sequence satisfy $s_n > t$. If the sequence s_n has a limit, then the lim sup of the sequence coincides with the limit. However, every sequence has a lim sup. The Cauchy-Hadamard formula is obtained simply by replacing the limit in (3.3) by a lim sup,

$$(3.4) \qquad\qquad R = \frac{1}{\limsup \sqrt[k]{|a_k|}}.$$

The proof is identical to the proof given above, except that the characterizing property of the lim sup is used.

Example. We return to the series (3.1) treated earlier. This can be expressed as a power series $\sum a_k z^k$, where $a_k = (-1)^{k/2}/2^{k/2}$ if k is even, and $a_k = 0$ if k is odd. Thus $\sqrt[k]{|a_k|}$ is $1/\sqrt{2}$ if k is even and 0 if k is odd. The lim sup of this sequence is $1/\sqrt{2}$. By the Cauchy-Hadamard formula, the radius of convergence of the series is $\sqrt{2}$, as before.

Exercises for V.3

1. Find the radius of convergence of the following power series:

   (a) $\displaystyle\sum_{k=0}^{\infty} 2^k z^k$ (d) $\displaystyle\sum_{k=0}^{\infty} \frac{3^k z^k}{4^k + 5^k}$ (g) $\displaystyle\sum_{k=1}^{\infty} \frac{k^k}{1 + 2^k k^k} z^k$

   (b) $\displaystyle\sum_{k=0}^{\infty} \frac{k}{6^k} z^k$ (e) $\displaystyle\sum_{k=1}^{\infty} \frac{2^k z^{2k}}{k^2 + k}$ (h) $\displaystyle\sum_{k=3}^{\infty} (\log k)^{k/2} z^k$

   (c) $\displaystyle\sum_{k=1}^{\infty} k^2 z^k$ (f) $\displaystyle\sum_{k=1}^{\infty} \frac{z^{2k}}{4^k k^k}$ (i) $\displaystyle\sum_{k=1}^{\infty} \frac{k! z^k}{k^k}$

2. Determine for which z the following series converge.

   (a) $\displaystyle\sum_{k=1}^{\infty} (z-1)^k$ (c) $\displaystyle\sum_{m=0}^{\infty} 2^m (z-2)^m$ (e) $\displaystyle\sum_{n=1}^{\infty} n^n (z-3)^n$

   (b) $\displaystyle\sum_{k=10}^{\infty} \frac{(z-i)^k}{k!}$ (d) $\displaystyle\sum_{m=1}^{\infty} \frac{(z+i)^m}{m^2}$ (f) $\displaystyle\sum_{n=3}^{\infty} \frac{2^n}{n^2} (z-2-i)^n$

3. Find the radius of convergence of the following series.

   (a) $\displaystyle\sum_{n=0}^{\infty} z^{3^n} = z + z^3 + z^9 + z^{27} + z^{81} + \cdots,$

   (b) $\displaystyle\sum_{p \text{ prime}} z^p = z^2 + z^3 + z^5 + +z^7 + z^{11} + \cdots.$

4. Show that the function defined by $f(z) = \sum z^{n!}$ is analytic on the open unit disk $\{|z| < 1\}$. Show that $|f(r\lambda)| \to +\infty$ as $r \to 1$ whenever λ is a root of unity. *Remark.* Thus $f(z)$ does not extend analytically to any larger open set than the open unit disk.

5. What functions are represented by the following power series?

   (a) $\displaystyle\sum_{k=1}^{\infty} k z^k$,   (b) $\sum_{k=1}^{\infty} k^2 z^k.$

6. Show the series $\sum a_k z^k$, the differentiated series $\sum k a_k z^{k-1}$, and the integrated series $\sum \dfrac{a_k}{k+1} z^{k+1}$ all have the same radius of convergence.

7. Consider the series

$$\sum_{k=0}^{\infty} \left(2 + (-1)^k\right)^k z^k.$$

Use the Cauchy-Hadamard formula to find the radius of convergence of the series. What happens when the ratio test is applied? Evaluate explicitly the sum of the series.

8. Write out a proof of the Cauchy-Hadamard formula (3.4).

4. Power Series Expansion of an Analytic Function

We have seen that power series expansions $\sum a_k(z-z_0)^k$ are analytic inside the disk of convergence $\{|z - z_0| < R\}$. It is an important and far-reaching fact that conversely, any function analytic on a disk can be expanded in a power series that converges on the disk.

Theorem. *Suppose that $f(z)$ is analytic for $|z - z_0| < \rho$. Then $f(z)$ is represented by the power series*

$$(4.1) \qquad f(z) \; = \; \sum_{k=0}^{\infty} a_k(z - z_0)^k, \qquad |z - z_0| < \rho,$$

where

$$(4.2) \qquad\qquad a_k \; = \; \frac{f^{(k)}(z_0)}{k!}, \qquad k \geq 0,$$

and where the power series has radius of convergence $R \geq \rho$. For any fixed r, $0 < r < \rho$, we have

$$(4.3) \qquad a_k \; = \; \frac{1}{2\pi i} \oint_{|\zeta - z_0| = r} \frac{f(\zeta)}{(\zeta - z_0)^{k+1}} \, d\zeta, \qquad k \geq 0.$$

Further, if $|f(z)| \leq M$ for $|z - z_0| = r$, then

$$(4.4) \qquad\qquad |a_k| \; \leq \; \frac{M}{r^k}, \qquad k \geq 0.$$

The proof amounts to expanding the integrand in the Cauchy integral formula as a geometric series and using the uniform convergence of the series to integrate term by term. We assume for simplicity that $z_0 = 0$.

Fix z such that $|z| < r$. For $|\zeta| = r$ we have

$$\frac{1}{\zeta - z} = \frac{1}{\zeta} \frac{1}{1 - z/\zeta} = \frac{1}{\zeta} \sum_{k=0}^{\infty} \left(\frac{z}{\zeta}\right)^k = \sum_{k=0}^{\infty} \frac{z^k}{\zeta^{k+1}},$$

where the series converges uniformly for $|\zeta| = r$. Hence

$$f(z) = \frac{1}{2\pi i} \int_{|\zeta|=r} \frac{f(\zeta)}{\zeta - z} d\zeta = \frac{1}{2\pi i} \int_{|\zeta|=r} \left(\sum f(\zeta) \frac{z^k}{\zeta^{k+1}}\right) d\zeta$$

$$= \frac{1}{2\pi i} \sum \left(\int_{|\zeta|=r} \frac{f(\zeta)}{\zeta^{k+1}} d\zeta\right) z^k = \sum a_k z^k,$$

where a_k is given by (4.3) with $z_0 = 0$. Thus $f(z)$ is represented by a power series, which converges whenever $|z| < r$. Since $r < \rho$ is arbitrary, the radius of convergence of the power series satisfies $R \geq \rho$. The formula (4.2) for a_k coincides with formula (3.2), which was obtained by differentiating the power series k times and substituting $z = z_0$. The estimates (4.4) are another version of the Cauchy estimates already derived in Section IV.4. Recall that they are obtained by applying the ML-estimate to the integral in (4.3).

Example. The exponential function e^z has power series $\sum a_k z^k$, where

$$a_k = \frac{1}{k!} \frac{d^k}{dz^k} e^z \bigg|_{z=0} = \frac{1}{k!} e^z \bigg|_{z=0} = \frac{1}{k!}.$$

Hence

$$e^z = \sum_{k=0}^{\infty} \frac{z^k}{k!} = 1 + z + \frac{z^2}{2!} + \frac{z^3}{3!} + \cdots.$$

Since e^z is entire, the radius of convergence of the power series is $R = +\infty$. Similarly, the entire functions $\sin z$ and $\cos z$ have power series expansions with infinite radius of convergence, given by

$$\sin z = \sum_{k=0}^{\infty} \frac{(-1)^k z^{2k+1}}{(2k+1)!} = z - \frac{z^3}{3!} + \frac{z^5}{5!} - \cdots,$$

$$\cos z = \sum_{k=0}^{\infty} \frac{(-1)^k z^{2k}}{(2k)!} = 1 - \frac{z^2}{2!} + \frac{z^4}{4!} - \cdots.$$

We state for emphasis two results that are immediate consequences of the theorem. The first is that an analytic function on a disk is completely determined by its value and the values of its derivatives at the center of the disk. This is because the power series representing $f(z)$ is determined by the derivatives of $f(z)$ at the center z_0 via the formula (4.2) for the power series coefficients.

Corollary. *Suppose that $f(z)$ and $g(z)$ are analytic for $|z - z_0| < r$. If $f^{(k)}(z_0) = g^{(k)}(z_0)$ for $k \geq 0$, then $f(z) = g(z)$ for $|z - z_0| < r$.*

The second corollary provides another method for determining the radius of convergence of a power series. The slogan form of the method is: *The radius of convergence is the distance to the nearest singularity.* The formal statement is as follows.

Corollary. *Suppose that $f(z)$ is analytic at z_0, with power series expansion $f(z) = \sum a_k(z - z_0)^k$ centered at z_0. Then the radius of convergence of the power series is the largest number R such that $f(z)$ extends to be analytic on the disk $\{|z - z_0| < R\}$.*

Note that the analytic extension of $f(z)$ to this largest disk is unique, by the first corollary.

Example. The function $1/(1 + x^2)$ is a beautiful function of the real variable x, which is expandable as a real power series about any point on the real axis. Yet the power series about $x = 0$, given by

$$\frac{1}{1 + x^2} = 1 - x^2 + x^4 - x^6 + \cdots ,$$

has radius of convergence only 1. The reason is that considered as a function $1/(1 + z^2)$ of a complex variable z, the function has singularities at the points $\pm i$. Thus the radius of convergence is necessarily $R = 1$, which is the distance from 0 to the nearest singularities at $\pm i$. This example teaches us that to understand real-valued functions of a real variable, we must look into the complex plane.

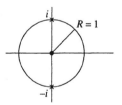

singularities at $\pm i$

Example. Consider the power series expansion of the function $f(z) = (z^3 - 1)/(z^2 - 1)$ about $z = 2$, $f(z) = \sum a_k(z - 2)^k$. The function $f(z)$ is analytic in the entire complex plane except for apparent singularities at $z = \pm 1$. However, the singularity at $z = +1$ is illusory. If we eliminate the common factor of $z - 1$ from the numerator and the denominator, we obtain $f(z) = (z^2 + z + 1)/(z + 1)$, which is analytic except at $z = -1$. The apparent singularity at $z = +1$ is called a "removable singularity." We

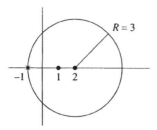

singularity at −1, analytic at +1

will return to these in the next chapter. Since $|f(z)| \to +\infty$ as $z \to -1$, the function $f(z)$ cannot be extended analytically to $z = -1$, and the singularity at $z = -1$ is genuine. The largest disk centered at $z = 2$ to which $f(z)$ extends analytically is then the disk $\{|z-2| < 3\}$. Consequently, the radius of convergence of the power series is $R = 3$.

Example. The radius of convergence of the power series $\sum a_k(z - 5)^k$ of the function $(\mathrm{Log}\, z)/(z - 1)$ about $z = 5$ is $R = 5$. While the function has an apparent singularity at $z = 1$, in fact it extends analytically to that point. Its power series representation about $z = 1$ can be obtained by dividing the power series representation of $\mathrm{Log}\, z$ by $z - 1$,

$$\frac{\mathrm{Log}\, z}{z - 1} = 1 - \frac{1}{2}(z - 1) + \frac{1}{3}(z - 1)^2 - \frac{1}{4}(z - 1)^3 + \cdots .$$

Since the function tends to $+\infty$ as z tends to 0 from along the positive real axis, the function cannot be extended analytically to $z = 0$. Thus $z = 0$ is a genuine singularity of the function, of a type called a "branch point," and the radius of the largest disk centered at $z = 5$ to which $(\mathrm{Log}\, z)/(z - 1)$ extends analytically is $R = 5$.

Exercises for V.4

1. Find the radius of convergence of the power series for the following functions, expanded about the indicated point.

   (a) $\dfrac{1}{z - 1}$,　　about $z = i$,

   (b) $\dfrac{1}{\cos z}$,　　about $z = 0$,

   (c) $\dfrac{1}{\cosh z}$,　　about $z = 0$,

   (d) $\mathrm{Log}\, z$,　　about $z = 1 + 2i$,

   (e) $z^{3/2}$,　　about $z = 3$,

   (f) $\dfrac{z - i}{z^3 - z}$,　　about $z = 2i$.

2. Show that the radius of convergence of the power series expansion of $(z^2 - 1)/(z^3 - 1)$ about $z = 2$ is $\sqrt{7}$.

3. Find the power series expansion of $\mathrm{Log}\, z$ about the point $z = i - 2$. Show that the radius of convergence of the series is $R = \sqrt{5}$. Explain why this does not contradict the discontinuity of $\mathrm{Log}\, z$ at $z = -2$.

4. Suppose $f(z)$ is analytic at $z = 0$ and satisfies $f(z) = z + f(z)^2$. What is the radius of convergence of the power series expansion of $f(z)$ about $z = 0$?

5. Deduce the identity $e^{iz} = \cos z + i \sin z$ from the power series expansions.

6. Find the power series expansions of $\cosh z$ and $\sinh z$ about $z = 0$. What are the radii of convergence of the series?

7. Find the power series expansion of the principal branch $\mathrm{Tan}^{-1}(z)$ of the inverse tangent function about $z = 0$. What is the radius of convergence of the series? *Hint.* Find it by integrating its derivative (a geometric series) term by term.

8. Expand $\mathrm{Log}(1 + iz)$ and $\mathrm{Log}(1 - iz)$ in power series about $z = 0$. By comparing power series expansions (see the preceding exercise), establish the identity

$$\mathrm{Tan}^{-1}z = \frac{1}{2i}\,\mathrm{Log}\left(\frac{1 + iz}{1 - iz}\right).$$

(See Exercise 5 in Section I.8.)

9. Let a be real, and consider the branch of z^a that is real and positive on $(0, \infty)$. Expand z^a in a power series about $z = 1$. What is the radius of convergence of the series? Write down the series explicitly.

10. Recall that for a complex number α, the binomial coefficient "α choose n" is defined by

$$\binom{\alpha}{0} = 1, \quad \text{and} \quad \binom{\alpha}{n} = \frac{\alpha(\alpha - 1)\cdots(\alpha - n + 1)}{n!}, \quad n \geq 1.$$

Find the radius of convergence of the binomial series

$$\sum_{n=0}^{\infty} \binom{\alpha}{n} z^n.$$

Show that the binomial series represents the principal branch of the function $(1 + z)^\alpha$. For which α does the binomial series reduce to a polynomial?

11. For fixed $n \geq 0$, define the function $J_n(z)$ by the power series

$$J_n(z) = \sum_{k=0}^{\infty} \frac{(-1)^k \, z^{n+2k}}{k!(n + k)! \, 2^{n+2k}}.$$

Show that $J_n(z)$ is an entire function. Show that $w = J_n(z)$ satisfies the differential equation

$$w'' + \frac{1}{z}w' + \left(1 - \frac{n^2}{z^2}\right)w = 0.$$

Remark. This is **Bessel's differential equation**, and $J_n(z)$ is **Bessel's function** of order n.

12. Suppose that the analytic function $f(z)$ has power series expansion $\sum a_n z^n$. Show that if $f(z)$ is an even function, then $a_n = 0$ for n odd. Show that if $f(z)$ is an odd function, then $a_n = 0$ for n even.

13. Prove the following version of L'Hospital's rule. If $f(z)$ and $g(z)$ are analytic, $f(z_0) = g(z_0) = 0$, and $g(z)$ is not identically zero, then

$$\lim_{z \to z_0} \frac{f(z)}{g(z)} = \lim_{z \to z_0} \frac{f'(z)}{g'(z)},$$

in the sense that either both limits are finite and equal, or both limits are infinite.

14. Let f be a continuous function on the unit circle $T = \{|z| = 1\}$. Show that f can be approximated uniformly on T by a sequence of polynomials in z if and only if f has an extension F that is continuous on the closed disk $\{|z| \leq 1\}$ and analytic on the interior $\{|z| < 1\}$. *Hint.* To approximate such an F, consider dilates $F_r(z) = F(rz)$.

5. Power Series Expansion at Infinity

We say that the function $f(z)$ is **analytic at** $z = \infty$ if the function $g(w) = f(1/w)$ is analytic at $w = 0$. Thus we make a change of variable $w = 1/z$, $z = 1/w$, and we study the behavior of $f(z)$ at $z = \infty$ by studying the behavior of $g(w)$ at $w = 0$.

If $f(z)$ is analytic at ∞, then $g(w) = f(1/w)$ has a power series expansion centered at $w = 0$,

$$g(w) = \sum_{k=0}^{\infty} b_k w^k = b_0 + b_1 w + b_2 w^2 + b_3 w^3 + \cdots, \qquad |w| < \rho.$$

Thus $f(z)$ is represented by a convergent series expansion in descending powers of z,

$$(5.1) \qquad f(z) = \sum_{k=0}^{\infty} \frac{b_k}{z^k} = b_0 + \frac{b_1}{z} + \frac{b_2}{z^2} + \frac{b_3}{z^3} + \cdots, \qquad |z| > \frac{1}{\rho}.$$

This series converges absolutely for $|z| > 1/\rho$, and for any $r > 1/\rho$ it converges uniformly for $|z| \geq r$.

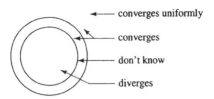

A formula for the coefficients can be obtained by multiplying the series by z^m and integrating term by term around the circle $|z| = r$. We have

$$\int_{|z|=r} f(z)z^m \, dz = \int_{|z|=r} \left(\sum b_k z^{-k} \right) z^m \, dz$$

$$= \sum b_k \int_{|z|=r} z^{m-k} \, dz = 2\pi i b_{m+1}.$$

Thus the coefficient b_k of $1/z^k$ is given by

$$b_k = \frac{1}{2\pi i} \int_{|z|=r} f(z)z^{k-1} \, dz, \qquad k \geq 0.$$

This formula should not be memorized. It will be superseded by the formula for the coefficients of a Laurent expansion, to be derived in Chapter VI.

Example. If $n \geq 0$, the function $f(z) = 1/z^n$ is analytic at ∞, since $g(w) = w^n$ is analytic at $w = 0$.

Example. Consider the function $f(z) = 1/(z^2 + 1)$. In this case,

$$g(w) = f(1/w) = \frac{1}{(1/w)^2 + 1} = \frac{w^2}{1 + w^2}.$$

Since $g(w)$ is analytic at $w = 0$, $f(z)$ is analytic at ∞. The power series for $g(w)$ is obtained by expanding $1/(1 + w^2)$ in a geometric series, to obtain

$$g(w) = w^2 \sum_{k=0}^{\infty} (-1)^k w^{2k} = w^2 - w^4 + w^6 - w^8 + \cdots, \qquad |w| < 1.$$

Thus

$$f(z) = \sum_{k=1}^{\infty} \frac{(-1)^{k+1}}{z^{2k}} = \frac{1}{z^2} - \frac{1}{z^4} + \frac{1}{z^6} - \frac{1}{z^8} + \cdots, \qquad |z| > 1.$$

This expansion can also be obtained by expressing

$$\frac{1}{1 + z^2} = \frac{1}{z^2} \frac{1}{1 + 1/z^2}$$

and expanding $1/(1 + 1/z^2)$ in a geometric series.

Exercises for V.5

1. Expand the following functions in power series about ∞:

   (a) $\dfrac{1}{z^2+1}$ (b) $\dfrac{z^2}{z^3-1}$ (c) e^{1/z^2} (d) $z\sinh(1/z)$

2. Suppose $f(z)$ is analytic at ∞, with series expansion (5.1). With the notation $f(\infty)=b_0$ and $f'(\infty)=b_1$, show that
$$f'(\infty) = \lim_{z\to\infty} z[f(z)-f(\infty)].$$

3. Suppose $f(z)$ is analytic at ∞, with series expansion (5.1). Let $\sigma \geq 0$ be the smallest number such that $f(z)$ extends to be analytic for $|z| > \sigma$. Show that the series (5.1) converges absolutely for $|z| > \sigma$ and diverges for $|z| < \sigma$.

4. Let E be a bounded subset of the complex plane $\mathbb{C}$ over which area integrals can be defined, and set
$$f(w) = \iint_E \frac{dx\,dy}{w-z}, \qquad w \in \mathbb{C}\backslash E,$$
where $z = x + iy$. Show that $f(w)$ is analytic at ∞, and find a formula for the coefficients of the power series of $f(w)$ at ∞ in descending powers of w. *Hint.* Use a geometric series expansion.

5. Determine explicitly the function $f(w)$ defined in Exercise 4, in the case that $E = \{|w| \leq 1\}$ is the unit disk. *Hint.* There are two formulae for $f(w)$, one valid for $|w| \geq 1$ and the other for $|w| \leq 1$. Be sure they agree for $|w| = 1$.

6. Manipulation of Power Series

Power series are easy to work with. For all practical purposes, power series can be treated like polynomials. We have already seen that power series can be differentiated term by term and that they can be integrated term by term. Power series can also be added and multiplied, just like polynomials.

Suppose, for instance, that $f(z)$ and $g(z)$ are analytic at 0, with power series representations
$$f(z) = \sum_{k=0}^{\infty} a_k z^k, \qquad g(z) = \sum_{k=0}^{\infty} b_k z^k.$$

Then the power series of the sum $f(z) + g(z)$ is obtained by simply adding coefficients,
$$f(z) + g(z) = \sum_{k=0}^{\infty} (a_k + b_k)z^k.$$

If c is a complex constant, the power series of $cf(z)$ is obtained by multiplying coefficients by c,

$$cf(z) \ = \ \sum_{k=0}^{\infty} c a_k z^k.$$

The formula for the coefficients of the power series of the product $f(z)g(z)$ is more complicated, though it is the same as for products of polynomials. If

$$f(z)g(z) \ = \ \sum_{k=0}^{\infty} c_k z^k,$$

then the coefficients c_k are given by

$$(6.1) \qquad c_k \ = \ a_k b_0 + a_{k-1} b_1 + \cdots + a_1 b_{k-1} + a_0 b_k, \qquad k \ge 0.$$

This can be justified as follows. The partial sums $f_n(z) = \sum_{j=0}^{n} a_j z^j$ and $g_n(z) = \sum_{k=0}^{n} a_k z^k$ are sequences of polynomials that converge uniformly to $f(z)$ and $g(z)$, respectively, in some disk centered at 0 as $n \to \infty$. Consequently, $f_n(z)g_n(z)$ converges uniformly to $f(z)g(z)$ on the disk. It follows (Section 2) that the derivatives of $f_n(z)g_n(z)$ converge to the corresponding derivatives of $f(z)g(z)$, and by formula (4.2), the coefficient of z^k in the polynomial $f_n(z)g_n(z)$ converges to the power series coefficient of z^k for $f(z)g(z)$. Now the coefficients of $f_n(z)g_n(z)$ are obtained by multiplying the polynomials and gathering terms,

$$f_n(z)g_n(z) \ = \ \left(\sum_{j=0}^{n} a_j z^j \right) \left(\sum_{j=0}^{n} a_j z^j \right) \ = \ a_0 b_0 + (a_1 b_0 + a_0 b_1)z + \cdots$$
$$\cdots + (a_n b_0 + a_{n-1} b_1 + \cdots + a_0 b_n)z^n + \mathcal{O}(z^{n+1}),$$

where here the notation $\mathcal{O}(z^m)$ is used for terms involving powers z^k for $k \ge m$. (This is consistent with our earlier use of the "big-oh" notation. See Exercise VI.2.6.) For $k \le n$ the coefficient of z^k in this polynomial is exactly c_k given by (6.1). Passing to the limit as $n \to \infty$, we find that $f(z)g(z)$ also has power series coefficient of z^k equal to c_k.

The power series of a quotient $f(z)/g(z)$ can also be effectively computed. It suffices to compute the power series of $1/g(z)$. For this, we assume that $g(z)$ is analytic at $z = 0$, and we suppose for simplicity that $g(0) = 1$. The power series expansion of $g(z)$ then has the form

$$g(z) \ = \ 1 + \sum_{k=1}^{\infty} b_k z^k \ = \ 1 + b_1 z + b_2 z^2 + \cdots .$$

If z is near 0, the sum $\sum_{k=1}^{\infty} b_k z^k$ is small, and we can expand $1/g(z)$ in a geometric series

$$\frac{1}{g(z)} = \frac{1}{1 + \sum_{k=1}^{\infty} b_k z^k}$$

$$= 1 - \left(\sum_{k=1}^{\infty} b_k z^k\right) + \left(\sum_{k=1}^{\infty} b_k z^k\right)^2 - \left(\sum_{k=1}^{\infty} b_k z^k\right)^3 + \cdots.$$

The terms involving z^m occur only in the first $m+1$ summands. We can compute the coefficients of the power series expansion of $1/g(z)$ by collecting the coefficients of z^m in each of the first $m+1$ summands and discarding the remaining summands. The procedure is justified by the uniform convergence of the geometric series for z near 0.

Example. To find the coefficients of z^m for $m \leq 5$ in the power series expansion of $\tan z = \sin z / \cos z$ about $z = 0$, we calculate as follows, again using the notation $\mathcal{O}(z^m)$ for terms involving powers z^k for $k \geq m$:

$$\frac{1}{\cos z} = \frac{1}{1 - (z^2/2!) + (z^4/4!) + \mathcal{O}(z^6)}$$

$$= 1 + \left(\frac{z^2}{2!} - \frac{z^4}{4!} + \mathcal{O}(z^6)\right) + \left(\frac{z^2}{2!} - \frac{z^4}{4!} + \mathcal{O}(z^6)\right)^2 + \mathcal{O}(z^6)$$

$$= 1 + \frac{1}{2}z^2 + \frac{5}{24}z^4 + \mathcal{O}(z^6),$$

so that

$$\frac{\sin z}{\cos z} = \left(z - \frac{z^3}{3!} + \frac{z^5}{5!} + \mathcal{O}(z^7)\right)\left(1 + \frac{1}{2}z^2 + \frac{5}{24}z^4 + \mathcal{O}(z^6)\right)$$

$$= z + \frac{1}{3}z^3 + \frac{2}{15}z^5 + \mathcal{O}(z^7).$$

The end result can be checked by differentiating $\tan z$ five times. Note that $\tan z$ is an odd function, so that only odd terms appear in the power series.

Exercises for V.6

1. Calculate the terms through order seven of the power series expansion about $z = 0$ of the function $1/\cos z$.

2. Calculate the terms through order five of the power series expansion about $z = 0$ of the function $z/\sin z$.

3. Show that

$$\frac{e^z}{1+z} = 1 + \frac{1}{2}z^2 - \frac{1}{3}z^3 + \frac{3}{8}z^4 - \frac{11}{30}z^5 + \cdots.$$

Show that the general term of the power series is given by

$$a_n = (-1)^n \left[\frac{1}{2!} - \frac{1}{3!} + \cdots + \frac{(-1)^n}{n!} \right], \qquad n \geq 2.$$

What is the radius of convergence of the series?

4. Define the **Bernoulli numbers** B_n by

$$\frac{z}{2} \cot(z/2) = 1 - B_1 \frac{z^2}{2!} - B_2 \frac{z^4}{4!} - B_3 \frac{z^6}{6!} - \cdots.$$

Explain why there are no odd terms in this series. What is the radius of convergence of the series? Find the first three Bernoulli numbers.

5. Define the **Euler numbers** E_n by

$$\frac{1}{\cosh z} = \sum_{n=0}^{\infty} \frac{E_n}{n!} z^n.$$

What is the radius of convergence of the series? Show that $E_n = 0$ for n odd. Find the first four nonzero Euler numbers.

6. Show that the coefficients of a power series "depend continuously" on the function they represent, in the following sense. If $\{f_m(z)\}$ is a sequence of analytic functions that converges uniformly to $f(z)$ for $|z| < \rho$, and

$$f_m(z) = \sum_{k=0}^{\infty} a_{k,m} z^k, \qquad\qquad f(z) = \sum_{k=0}^{\infty} a_k z^k,$$

then for each $k \geq 0$, we have $a_{k,m} \to a_k$ as $m \to \infty$.

7. The Zeros of an Analytic Function

Let $f(z)$ be analytic at z_0, and suppose that $f(z_0) = 0$ but $f(z)$ is not identically zero. We say that $f(z)$ has a **zero of order** N at z_0 if $f(z_0) = f'(z_0) = \cdots = f^{(N-1)}(z_0) = 0$, while $f^{(N)}(z_0) \neq 0$. In view of the formula (4.2) for the power series coefficients, this occurs if and only if the power series expansion of $f(z)$ has the form

$$f(z) = a_N (z - z_0)^N + a_{N+1}(z - z_0)^{N+1} + \cdots,$$

where $a_N \neq 0$. We can factor out the term $(z - z_0)^N$ from the power series and write

(7.1) $$f(z) = (z - z_0)^N h(z),$$

where $h(z)$ is analytic at z_0 and $h(z_0) = a_N \neq 0$. Conversely, if there is a factorization (7.1) where $h(z)$ is analytic at z_0 and $h(z_0) \neq 0$, then the leading term in the power series for $f(z)$ is $h(z_0)(z - z_0)^N$, and $f(z)$ has a zero of order N at z_0.

A zero of order one is called a **simple zero**, and a zero of order two is called a **double zero**.

Example. The zeros of $\sin z$ are at the points $n\pi$, $-\infty < n < +\infty$, and the derivative $\cos z$ of $\sin z$ is ± 1 at each of these points. Hence all zeros of $\sin z$ are simple zeros.

Example. The monomial $(z - z_0)^n$ has a zero of order n at z_0 and no other zeros.

Example. From the power series expansion

$$\sin z = -(z - \pi) + \frac{1}{3!}(z - \pi)^3 - \frac{1}{5!}(z - \pi)^5 + \cdots,$$

we see that the function $\sin z + z - \pi$ has a triple zero at $z = \pi$. The function $\dfrac{\sin z}{z - \pi} + 1$, defined to be 0 at $z = \pi$, is entire and has a double zero at $z = \pi$, since the leading term of its power series expansion is $(z - \pi)^2/3!$.

A useful rule for determining orders of zeros is that the order of a zero of a product $f(z)g(z)$ is the sum of the orders of the corresponding zeros of the factors $f(z)$ and $g(z)$. Indeed, if $f(z) = a_n(z - z_0)^n + \cdots$ has a zero of order n at z_0 and $g(z) = b_m(z - z_0)^m + \cdots$ has a zero of order m at z_0, then $f(z)g(z) = a_n b_m(z - z_0)^{n+m} + \cdots$ has a zero of order $m + n$ at z.

If $f(z)$ is analytic at ∞ and $f(\infty) = 0$, we define the order of the zero of $f(z)$ at $z = \infty$ in the usual way, by making the change of variable $w = 1/z$. We say that $f(z)$ has a **zero at $z = \infty$ of order N** if $g(w) = f(1/w)$ has a zero at $w = 0$ of order N. In this case, $g(w) = b_N w^N + b_{N+1} w^{N+1} + \cdots$, where $b_N \neq 0$. Thus $f(z)$ has the series representation

$$f(z) = \frac{b_N}{z^N} + \frac{b_{N+1}}{z^{N+1}} + \cdots, \qquad |z| > R,$$

where $b_N \neq 0$.

Example. The function $1/(1+z^2)$ has a double zero at ∞. Its power series expansion, derived in Section 5, is $1/z^2 - 1/z^4 + \cdots$, which has leading term $1/z^2$.

Example. The rational monomial $1/(z - z_0)^n$ has a zero of order n at ∞.

We say that a point $z_0 \in E$ is an **isolated point** of the set E if there is $\rho > 0$ such that $|z - z_0| \geq \rho$ for all points $z \in E$ other than z_0. In other words, z_0 is an isolated point of E if z_0 is at a positive distance from $E \backslash \{z_0\}$. If E is a set such that each point of E is an isolated point of E, we say that the "points of E are isolated."

Example. If $E = [-1, 0] \cup \{1/n : n \geq 1\}$, then each of the points $1/n$ is an isolated point of E, while no point of the interval $[-1, 0]$ is an isolated point of E.

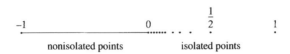

$$
\begin{array}{cccc}
-1 & 0 & \frac{1}{2} & 1
\end{array}
$$

nonisolated points   isolated points

Theorem. *If D is a domain, and $f(z)$ is an analytic function on D that is not identically zero, then the zeros of $f(z)$ are isolated.*

The proof of this theorem breaks into two parts, an observation about the local behavior of an analytic function near a zero, and a connectedness argument that depends on D being a domain. We begin with the connectedness argument, which is deceptively subtle.

Let U be the set of all $z \in D$ such that $f^{(m)}(z) = 0$ for all $m \geq 0$. If $z_0 \in U$, then the power series expansion $f(z) = \sum a_k (z - z_0)^k$ has $a_k = f^{(k)}(z_0)/k! = 0$ for all $k \geq 0$. Hence $f(z) = 0$ for z belonging to a disk centered at z_0. The points of this disk all belong to U. This shows that U is an open set. On the other hand, if $z_0 \in D\backslash U$, then $f^{(k)}(z_0) \neq 0$ for some k. Therefore, $f^{(k)}(z) \neq 0$ for z in some disk centered at z_0, and this disk is contained in $D\backslash U$, so $D\backslash U$ is also open. Since D is connected, either $U = D$ or U is empty. If $U = D$, then $f(z) = 0$ for all $z \in D$, contrary to our hypothesis. Hence U is empty. Thus we conclude from the connectedness argument that each zero of $f(z)$ has finite order.

The closing argument is easier. If z_0 is a zero of $f(z)$, say of order N, we can factor $f(z) = (z - z_0)^N h(z)$, where $h(z)$ is analytic at z_0 and $h(z_0) \neq 0$. Then for $\rho > 0$ sufficiently small, we have $h(z) \neq 0$ for $|z - z_0| < \rho$, and consequently $|f(z)| \neq 0$ for $0 < |z - z_0| < \rho$. Thus z_0 has distance at least ρ from any other zero of $f(z)$, and the zeros of $f(z)$ are isolated.

By applying the preceding theorem to $f(z) - g(z)$, we obtain immediately the following important result, which is also referred to as the **identity principle**.

Theorem (Uniqueness Principle). *If $f(z)$ and $g(z)$ are analytic on a domain D, and if $f(z) = g(z)$ for z belonging to a set that has a nonisolated point, then $f(z) = g(z)$ for all $z \in D$.*

Example. Once we know that $\sin z$ and $\cos z$ are entire functions that satisfy $\sin^2 x + \cos^2 x = 1$ for all real numbers x, then necessarily $\sin^2 z + \cos^2 z = 1$ for all complex numbers z. This follows from the uniqueness principle, applied to $f(z) = \sin^2 z + \cos^2 z$ and $g(z) = 1$.

The uniqueness principle has a natural extension to functions of two complex variables, which is sometimes referred to as the **principle of permanence of functional equations.**

Theorem. *Let D be a domain, and let E be a subset of D that has a nonisolated point. Let $F(z, w)$ be a function defined for $z, w \in D$ such that $F(z, w)$ is analytic in z for each fixed $w \in D$ and analytic in w for each fixed $z \in D$. If $F(z, w) = 0$ whenever z and w both belong to E, then $F(z, w) = 0$ for all $z, w \in D$.*

This follows from two applications of the uniqueness principle, one for each variable. First fix $z_0 \in E$. Then $F(z_0, w)$ is analytic for $w \in D$ and vanishes for $w \in E$. By the uniqueness principle, $F(z_0, w) = 0$ for all $w \in D$. Now fix $w \in D$. We have shown that $F(z_0, w) = 0$ for all $z_0 \in E$. By the uniqueness principle, $F(z, w) = 0$ for all $z \in D$.

Example. As a typical application, we derive the addition formula for the exponential function, assuming that $e^{s+t} = e^s e^t$ for s and t real. The function $F(z, w) = e^{z+w} - e^z e^w$ is an entire function of each variable for fixed values of the other variable, and it vanishes when both the variables are real. By the permanence principle, it then vanishes for all values of z and w. Thus $e^{z+w} = e^z e^w$.

Exercises for V.7

1. Find the zeros and orders of zeros of the following functions.

   (a) $\dfrac{z^2 + 1}{z^2 - 1}$ (d) $\cos z - 1$ (g) $e^z - 1$

   (b) $\dfrac{1}{z} + \dfrac{1}{z^5}$ (e) $\dfrac{\cos z - 1}{z}$ (h) $\sinh^2 z + \cosh^2 z$

   (c) $z^2 \sin z$ (f) $\dfrac{\cos z - 1}{z^2}$ (i) $\dfrac{\operatorname{Log} z}{z}$ (principal value)

2. Determine which of the functions in the preceding exercise are analytic at ∞, and determine the orders of any zeros at ∞.

3. Show that the zeros of $\sin z$ and $\tan z$ are all simple.

4. Show that $\cos(z + w) = \cos z \cos w - \sin z \sin w$, assuming the corresponding identity for z and w real.

5. Show that
$$\int_{-\infty}^{\infty} e^{-zt^2 + 2wt}\, dt = \sqrt{\frac{\pi}{z}}\, e^{w^2/z}, \qquad z, w \in \mathbb{C}, \ \operatorname{Re} z > 0,$$

   where we take the principal branch of the square root. Compare the result to Exercise IV.3.1. *Hint.* Show that the integral is analytic

in z and w, and evaluate it for $z = x > 0$ and w real by making
a change of variable and using the known value $\sqrt{\pi}$ for $z = 1$ and
$w = 0$.

6. Suppose $f(z)$ is analytic on a domain D and $z_0 \in D$. Show that if
   $f^{(m)}(z_0) = 0$ for $m \geq 1$, then $f(z)$ is constant on D.

7. Show that if $u(x, y)$ is a harmonic function on a domain D such that
   all the partial derivatives of $u(x, y)$ vanish at the same point of D,
   then $u(x, y)$ is constant on D.

8. With the convention that the function that is identically zero has a
   zero of infinite order at each point, show that if $f(z)$ and $g(z)$ have
   zeros of order n and m respectively at z_0, then $f(z) + g(z)$ has a
   zero of order $k \geq \min(n, m)$. Show that strict inequality can occur
   here, but that equality holds whenever $m \neq n$.

9. Show that if the analytic function $f(z)$ has a zero of order N at z_0,
   then $f(z) = g(z)^N$ for some function $g(z)$ analytic near z_0 and
   satisfying $g'(z_0) \neq 0$.

10. Show that if $f(z)$ is a continuous function on a domain D such that
    $f(z)^N$ is analytic on D for some integer N, then $f(z)$ is analytic
    on D.

11. Show that if $f(z)$ is a nonconstant analytic function on a domain D,
    then the image under $f(z)$ of any open set is open. *Remark.* This
    is the **open mapping theorem** for analytic functions. The proof
    is easy when $f'(z) \neq 0$, since the Jacobian of $f(z)$ coincides with
    $|f'(z)|^2$. Use Exercise 9 to deal with the points where $f'(z)$ is zero.

12. Show that the open mapping theorem for analytic functions implies
    the maximum principle for analytic functions.

13. Let $f_n(z)$ be a sequence of analytic functions on a domain D such
    that $f_n(D) \subset D$, and suppose that $f_n(z)$ converges normally to $f(z)$
    on D. Show that either $f(D) \subset D$, or else $f(D)$ consists of a single
    point in ∂D.

14. A set E is **discrete** if every point of E is isolated. Show that a closed
    discrete subset of a domain D either is finite or can be arranged in
    a sequence $\{z_k\}$ that accumulates only on $\{\infty\} \cup \partial D$.

8. Analytic Continuation

In Chapter I we analyzed the branches of the functions $\sqrt{z}$ and $\log z$ by
following the values of the functions along curves in the complex plane. By

this method we saw that these functions cannot be extended continuously to $\mathbb{C}\backslash\{0\}$. We constructed Riemann surfaces to which the functions *do* extend continuously, by defining branches of the functions on separate sheets and pasting these sheets together.

In this section we develop more formally the idea of tracking an analytic function along a path. The power series expansion of an analytic function about a point contains complete information about the function near the point. Rather than track only the values of the analytic function, we will track the power series expansions of the function about points of the path as we move along the path. We begin by observing that the radius of convergence of a power series depends continuously on the center of the expansion.

Lemma. *Suppose D is a disk, $f(z)$ is analytic on D, and $R(z_1)$ is the radius of convergence of the power series expansion of $f(z)$ about a point $z_1 \in D$. Then*

$$(8.1) \qquad |R(z_1) - R(z_2)| \le |z_1 - z_2|, \qquad z_1, z_2 \in D.$$

We use the characterization of $R(z_1)$ as the radius of the largest disk centered at z_1 to which $f(z)$ extends analytically. Thus $f(z)$ does not extend analytically to any disk containing $\{|z - z_1| \le R(z_1)\}$, and consequently $R(z_2) \le R(z_1) + |z_2 - z_1|$. Interchanging z_1 and z_2, we obtain also $R(z_1) \le R(z_2) + |z_2 - z_1|$. These two inequalities yield (8.1).

Now we start with a power series $\sum a_n(z-z_0)^n$ that represents a function $f(z)$ near z_0. We are interested in the behavior of $f(z)$ only near z_0, and we say that the power series represents the "germ" of $f(z)$ at z_0. Let $\gamma(t)$, $a \le t \le b$, be a path starting at $z_0 = \gamma(a)$. We say that $f(z)$ is **analytically continuable along** γ if for each t there is a convergent power series

$$(8.2) \qquad f_t(z) = \sum_{n=0}^{\infty} a_n(t)(z - \gamma(t))^n, \qquad |z - \gamma(t)| < r(t),$$

such that $f_a(z)$ is the power series representing $f(z)$ at z_0, and such that when s is near t, then $f_s(z) = f_t(z)$ for z in the intersection of the disks of convergence. By the uniqueness principle, the series $f_t(z)$ determines uniquely each of the series $f_s(z)$ for s near t. It follows that the series $f_b(z)$ is uniquely determined by $f_a(z)$. (Otherwise, we could define t_0 to be the infimum of the parameter values t such that $f_t(z)$ is *not* uniquely determined by $f_a(z)$, and we would soon have a contradiction to the local uniqueness assertion at t_0.) We refer to $f_b(z)$ as the **analytic continuation** of $f(z)$ along γ, where we regard $f_b(z)$ either as a power series or as an analytic function defined near $\gamma(b)$. Since the coefficients $a_n(t)$ in (8.2) are given by $a_n(s) = f_t^{(m)}(\gamma(s))/m!$ for s near t, the coefficients depend continuously on the parameter t. The preceding lemma shows that the

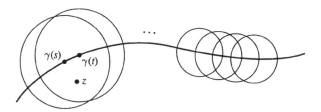

radius of convergence of the power series (8.2) also depends continuously on the parameter t. We summarize and give some examples.

Theorem. *Suppose $f(z)$ can be continued analytically along the path $\gamma(t)$, $a \leq t \leq b$. Then the analytic continuation is unique. Further, for each $n \geq 0$ the coefficient $a_n(t)$ of the series (8.2) depends continuously on t, and the radius of convergence of the series (8.2) depends continuously on t.*

Example. Suppose $f(z)$ is analytic in a domain D. Then $f(z)$ has an analytic continuation along any path in D. Simply define $f_t(z)$ to be the power series expansion of $f(z)$ about $\gamma(t)$.

Example. Suppose $f(z)$ is the principal branch of $\sqrt{z}$, with series expansion

$$f(z) \;=\; 1 + \frac{1}{2}(z-1) - \frac{1}{8}(z-1)^2 + \cdots$$

about $z = 1$. Let $\gamma(t) = e^{it}$, $0 \leq t \leq 2\pi$, be the closed path around the unit circle starting at $\gamma(0) = 1$. Then $f(z)$ has an analytic continuation along γ, and the power series $f_t(z)$ in (8.2) is given explicitly by

$$f_t(z) \;=\; e^{it/2} + \frac{e^{-it/2}}{2}(z - e^{it}) - \frac{e^{-3it/2}}{8}(z - e^{it})^2 + \cdots.$$

Thus the analytic continuation of $f(z)$ around the circle is

$$f_{2\pi}(z) \;=\; -1 - \frac{1}{2}(z-1) + \frac{1}{8}(z-1)^2 + \cdots,$$

which is just the other branch of $\sqrt{z}$.

Now suppose $f(z)$ is analytic at z_0, and suppose that $\gamma(t)$, $a \leq t \leq b$, is a path from $z_0 = \gamma(a)$ to $z_1 = \gamma(b)$ along which $f(z)$ has an analytic continuation $f_t(z)$. The radius of convergence $R(t)$ of the power series (8.2) varies continuously with t. Hence there is $\delta > 0$ such that $R(t) \geq \delta$ for all t, $a \leq t \leq b$.

Lemma. *Let f, γ, and δ be as above. If $\sigma(t)$, $a \leq t \leq b$, is another path from z_0 to z_1 such that $|\sigma(t) - \gamma(t)| < \delta$ for $a \leq t \leq b$, then there is an analytic continuation $g_t(z)$ of $f_t(z)$ along σ, and the terminal series $g_b(z)$ centered at $\sigma(b) = z_1$ coincides with $f_b(z)$.*

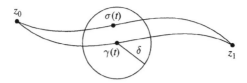

We can define $g_t(z)$ to be the power series expansion of $f_t(z)$ about $\sigma(t)$. This is possible, since $\sigma(t)$ lies within the disk of convergence of $f_t(z)$. In fact, $\sigma(s)$ lies inside the disk of convergence of $f_t(z)$ for s near t, and since $f_s(z) = f_t(z)$ for s near t, we see that $g_s(z) = g_t(z)$ for s near t, and so $g_t(z)$ does represent an analytic continuation. By its definition, $g_b(z)$ coincides with $f_b(z)$ at $z_1 = \sigma(b) = \gamma(b)$.

Just as in Section III.2, we can deduce from this "local" deformation result a global deformation theorem.

Theorem (Monodromy Theorem). *Let $f(z)$ be analytic at z_0. Let $\gamma_0(t)$ and $\gamma_1(t)$, $a \le t \le b$, be two paths from z_0 to z_1 along which $f(z)$ can be continued analytically. Suppose $\gamma_0(t)$ can be deformed continuously to $\gamma_1(t)$ by paths $\gamma_s(t)$, $0 \le s \le 1$, from z_0 to z_1 such that $f(z)$ can be continued analytically along each path γ_s. Then the analytic continuations of $f(z)$ along γ_0 and along γ_1 coincide at z_1.*

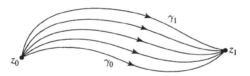

By a continuous deformation, we mean that the function $(s, t) \mapsto \gamma_s(t)$ is continuous for $0 \le s \le 1$ and $a \le t \le b$. If $f_{s,t}(z)$ denotes the analytic continuation of $f(z)$ along γ_s, the power series $f_{s,t}(z)$ and $f_{s',t'}(z)$ determine the same function near $\gamma_s(t)$ for s' near s and t' near t. Thus the radius of convergence $R_{s,t}$ of $f_{s,t}$ varies continuously with s and t, and there is $\delta > 0$ such that $R_{s,t} \ge \delta$ for all s and t. We can choose, then, parameter values $0 = s_0 < s_1 < \cdots < s_n = 1$ such that $|\gamma_{s_j}(t) - \gamma_{s_{j-1}}(t)| < \delta$ for $a \le t \le b$. By the preceding lemma, the analytic continuation along $\gamma_{s_{j-1}}$ leads to the same power series at z_1 as that along γ_{s_j}. If we apply this now successively to $\gamma_0 = \gamma_{s_0}, \gamma_{s_1}, \gamma_{s_2}, \ldots$, we conclude after n steps that analytic continuation along γ_0 leads to the same power series as analytic continuation along γ_1. This proves the monodromy theorem.

Exercise. Suppose that the principal branch $w = f(z)$ of the algebraic function $(z^4 - 1)^{1/3}$ is continued analytically from $z = 2$ around the figure-eight path indicated below. What is the analytic continuation of the function at the end of the path?

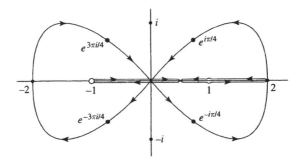

Solution. We deform the figure-eight path to a path for which it is possible to follow the analytic continuation more easily. By the monodromy theorem, the analytic continuation is the same if we follow a path along the real axis except for indentations at ± 1, as in the figure. For the various segments of this path, we track the behavior of w as follows:

path from $+2$ to $+1$ $\rightsquigarrow$ $\arg w = 0$,
semicircle around $+1$ $\rightsquigarrow$ phase change of $e^{i\pi/3}$,
path from $+1$ to -1 $\rightsquigarrow$ $\arg w = e^{i\pi/3}$,
circle around -1 $\rightsquigarrow$ phase change of $e^{-2\pi i/3}$,
path from -1 to $+1$ $\rightsquigarrow$ $\arg w = e^{-i\pi/3}$,
semicircle around $+1$ $\rightsquigarrow$ phase change of $e^{i\pi/3}$,
path from $+1$ to $+2$ $\rightsquigarrow$ $\arg w = 0$.

Thus at the end of the path we return to the same branch $w = f(z)$ that we started with. This can be checked by factoring $z^4 - 1$ and tracking separately each factor.

Exercises for V.8

1. Suppose that the principal branch of $\sqrt{z^2 - 1}$ is continued analytically from $z = 2$ around the figure-eight path indicated above. What is the analytic continuation of the function at the end of the path? Answer the same question for the functions $(z^3 - 1)^{1/3}$ and $(z^6 - 1)^{1/3}$.

2. Show that $f(z) = \operatorname{Log} z = (z - 1) - \frac{1}{2}(z - 1)^2 + \cdots$ has an analytic continuation around the unit circle $\gamma(t) = e^{it}$, $0 \le t \le 2\pi$. Determine explicitly the power series f_t for each t. How is $f_{2\pi}$ related to f_0?

3. Show that each branch of $\sqrt{z}$ can be continued analytically along any path γ in $\mathbb{C}\backslash\{0\}$, and show that the radius of convergence of the power series $f_t(z)$ representing the continuation is $|\gamma(t)|$. Show that $\sqrt{z}$ cannot be continued analytically along any path containing 0.

4. Let $f(z)$ be analytic on a domain D, fix $z_0 \in D$, and let $f(z) = \sum a_n(z-z_0)^n$ be the expansion of $f(z)$ about z_0. Let

$$F(z) = \int_{z_0}^{z} f(\zeta)\, d\zeta = \sum_{n=0}^{\infty} \frac{a_n}{n+1} (z-z_0)^{n+1}$$

be the indefinite integral of $f(z)$ for z near z_0. Show that $F(z)$ can be continued analytically along any path in D starting at z_0. What happens in the case $D = \mathbb{C}\backslash\{0\}$, $z_0 = 1$, and $f(z) = 1/z$? What happens in the case that D is star-shaped?

5. Show that the function defined by

$$f(z) = \sum z^{2^n} = z + z^2 + z^4 + z^8 + \cdots$$

is analytic on the open unit disk $\{|z| < 1\}$, and that it cannot be extended analytically to any larger open set. *Hint.* Observe that $f(z) = z + f(z^2)$, and that $f(r) \to +\infty$ as $r \to 1$.

6. Suppose $f(z) = \sum a_n z^n$, where $a_n = 0$ except for n in a sequence n_k that satisfies $n_{k+1}/n_k \geq 1 + \delta$ for some $\delta > 0$. Suppose further that the series has radius of convergence $R = 1$. Show that $f(z)$ does not extend analytically to any point of the unit circle. *Remark.* Such a sequence with large gaps between successive nonzero terms is called a **lacunary sequence**. This result is the **Hadamard gap theorem**. There is a slick proof. If $f(z)$ extends analytically across $z = 1$, consider $g(w) = f(w^m(1+w)/2)$, where m is a large integer. Show that the power series for $g(w)$ has radius of convergence $r > 1$, and that this implies that the power series of $f(z)$ converges for $|z - 1| < \varepsilon$.

7. Suppose $f(z) = \sum a_n z^n$, where the series has radius of convergence $R < \infty$. Show that there is an angle α such that $f(z)$ does not have an analytic continuation along the path $\gamma(t) = te^{i\alpha}$, $0 \leq t \leq R$. Determine the radius of convergence of the power series expansion of $f(z)$ about $te^{i\alpha}$.

8. Let $f(z)$ be analytic at z_0, and let $\gamma(t)$, $a \leq t \leq b$, be a path such that $\gamma(a) = z_0$. If $f(z)$ cannot be continued analytically along γ, show that there is a parameter value t_1 such that there is an analytic continuation $f_t(z)$ for $a \leq t < t_1$, and the radius of convergence of the power series $f_t(z)$ tends to 0 as $t \to t_1$.

9. Let $P(z, w)$ be a polynomial in z and w, of degree n in w. Suppose that $f(z)$ is analytic at z_0 and satisfies $P(z, f(z)) = 0$. Show that if $f_t(z)$ is any analytic continuation of $f(z)$ along any path starting at z_0, then $P(z, f_t(z)) = 0$ for all t. *Remark.* An analytic function $f(z)$ that satisfies a polynomial equation $P(z, f(z)) = 0$ is called an

algebraic function. For instance, the branches of $\sqrt[n]{z}$ are algebraic functions, since they satisfy $z - w^n = 0$.

10. Let D be the punctured disk $\{0 < |z| < \varepsilon\}$, suppose $f(z)$ is analytic at $z_0 \in D$, and $e^{w_0} = z_0$. Show that $f(z)$ has an analytic continuation along any path in D starting at z_0 if and only if there is an analytic function $g(w)$ in the half-plane $\{\operatorname{Re} w < \log \varepsilon\}$ such that $f(e^w) = g(w)$ for w near w_0. *Remark.* If $f(z)$ does not extend analytically to D but has an analytic continuation along any path in D, we say that $f(z)$ has a **branch point** at $z = 0$. For the proof, use the fact that any path in D starting at z_0 is the composition of a unique path in the half-plane starting at w_0 and the exponential function e^w.

VI

Laurent Series and Isolated Singularities

In Section 1 we derive the Laurent decomposition of a function that is analytic on an annulus, and in Section 2 we use the Laurent decomposition on a punctured disk to study isolated singularities of analytic functions. We classify these as removable singularities, essential singularities, or poles, and we characterize each type of singularity. In Section 3 we define isolated singularities at ∞, and in Section 4 we derive the partial fractions decomposition of a rational function. In Sections 5 and 6 we use the Laurent decomposition to study periodic functions and we relate Laurent series to Fourier series. Sections 5 and 6 can be omitted at first reading.

1. The Laurent Decomposition

The Laurent decomposition splits a function analytic in an annulus as the sum of a function analytic inside the annulus and a function analytic outside the annulus.

Theorem (Laurent Decomposition). *Suppose $0 \le \rho < \sigma \le +\infty$, and suppose $f(z)$ is analytic for $\rho < |z - z_0| < \sigma$. Then $f(z)$ can be decomposed as a sum*

$$(1.1) \qquad\qquad f(z) \;=\; f_0(z) \;+\; f_1(z),$$

where $f_0(z)$ is analytic for $|z - z_0| < \sigma$, and $f_1(z)$ is analytic for $|z - z_0| > \rho$ and at ∞. If we normalize the decomposition so that $f_1(\infty) = 0$, then the decomposition is unique.

If $f(z)$ is already analytic for $|z - z_0| < \sigma$, the Laurent decomposition becomes the trivial decomposition $f(z) = f_0(z)$, with $f_1(z) = 0$. If $f(z)$ is already analytic for $|z - z_0| > \rho$ and vanishes at ∞, the Laurent decomposition is the trivial decomposition $f(z) = f_1(z)$, with $f_0(z) = 0$.

The uniqueness of the decomposition follows from Liouville's theorem by the following argument. Suppose that $f(z) = g_0(z) + g_1(z)$ is another decomposition with the properties of the theorem. Then

$$(1.2) \qquad g_0(z) - f_0(z) \ = \ f_1(z) - g_1(z), \qquad \rho < |z - z_0| < \sigma.$$

Define $h(z)$ to be equal to $g_0(z) - f_0(z)$ in the disk $\{|z - z_0| < \sigma\}$, and equal to $f_1(z) - g_1(z)$ in the exterior domain $\{|z - z_0| > \rho\}$. These domains overlap in the annulus $\{\rho < |z - z_0| < \sigma\}$, and the identity (1.2) shows that the two definitions agree in the overlap. Thus $h(z)$ is defined for all $z \in \mathbb{C}$. Evidently, $h(z)$ is an entire function, and $h(z)$ tends to 0 as $z \to \infty$. By Liouville's theorem, $h(z)$ is identically zero. Consequently $g_0(z) = f_0(z)$ and $g_1(z) = f_1(z)$, and there is at most one such decomposition.

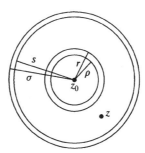

To find such a decomposition, we apply the Cauchy integral representation theorem on an annulus, as follows. Choose r and s such that $\rho < r < s < \sigma$. The Cauchy integral formula for an annulus yields

$$f(z) \ = \ \frac{1}{2\pi i} \oint_{|\zeta - z_0| = s} \frac{f(\zeta)}{\zeta - z} \, d\zeta \ - \ \frac{1}{2\pi i} \oint_{|\zeta - z_0| = r} \frac{f(\zeta)}{\zeta - z} \, d\zeta,$$

which is valid for $r < |z - z_0| < s$. The function

$$f_0(z) \ = \ \frac{1}{2\pi i} \oint_{|\zeta - z_0| = s} \frac{f(\zeta)}{\zeta - z} \, d\zeta, \qquad |z - z_0| < s,$$

is analytic for $|z - z_0| < s$, and the function

$$f_1(z) \ = \ -\frac{1}{2\pi i} \oint_{|\zeta - z_0| = r} \frac{f(\zeta)}{\zeta - z} \, d\zeta, \qquad |z - z_0| > r,$$

is analytic for $|z - z_0| > r$ and tends to 0 as $z \to \infty$. Thus we obtain the decomposition $f(z) = f_0(z) + f_1(z)$ for $r < |z - z_0| < s$. Technically, this decomposition depends on r and s. However, the uniqueness assertion already established shows that the decomposition is independent of r and s, so that $f_0(z)$ and $f_1(z)$ are defined for $\rho < |z - z_0| < \sigma$ and have the properties of the theorem.

Example. The function $f(z) = 1/(z-1)(z-2)$ has three Laurent decompositions centered at 0. One represents the function in the punctured disk $\{0 < |z| < 1\}$, one in the annulus $\{1 < |z| < 2\}$, and one in the exterior domain $\{2 < |z| < \infty\}$. Since the function is already analytic in the disk $\{|z| < 1\}$, the decomposition in the punctured disk is given by $f_0(z) = f(z)$ and $f_1(z) = 0$. Since the function is analytic at ∞ and vanishes there, its Laurent decomposition with respect to the exterior domain $\{2 < |z| < \infty\}$ is given by $f(z) = f_1(z)$ and $f_0(z) = 0$. The only nontrivial Laurent decomposition is with respect to the annulus $\{1 < |z| < 2\}$. To see what it is, we consider the partial fractions decomposition

$$\frac{1}{(z-1)(z-2)} = \frac{1}{z-2} - \frac{1}{z-1}.$$

The summand $f_0(z) = 1/(z-2)$ is analytic for $|z| < 2$, and the summand $f_1(z) = -1/(z-1)$ is analytic for $|z| > 1$ and it vanishes at ∞. Thus this partial fractions decomposition coincides with the Laurent decomposition with respect to the annulus.

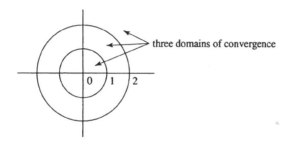

three domains of convergence

Example. The function $1/\sin z$ has a Laurent decomposition with respect to each annulus $\{n\pi < |z| < (n+1)\pi\}$, for $n = 0, 1, 2, \ldots$. We will see how to obtain the Laurent decompositions in the next section.

Suppose now $f(z) = f_0(z) + f_1(z)$ is the Laurent decomposition for a function analytic for $\rho < |z - z_0| < \sigma$. We can express $f_0(z)$ as a power series in $z - z_0$,

$$f_0(z) = \sum_{k=0}^{\infty} a_k(z - z_0)^k, \qquad |z - z_0| < \sigma,$$

where the series converges absolutely, and for any $s < \sigma$ it converges uniformly for $|z - z_0| \leq s$. Further, we can also express $f_1(z)$ as a series of negative powers of $z - z_0$, with zero constant term, since $f_1(z)$ tends to 0 at ∞,

$$f_1(z) = \sum_{k=-\infty}^{-1} a_k(z - z_0)^k, \qquad |z - z_0| > \rho.$$

This series converges absolutely, and for any $r > \rho$ it converges uniformly for $|z - z_0| \geq r$. If we add the two series, we obtain a two-tailed expansion for $f(z)$,

$$(1.3) \qquad f(z) = \sum_{k=-\infty}^{\infty} a_k (z - z_0)^k, \qquad \rho < |z - z_0| < \sigma,$$

that converges absolutely, and that converges uniformly for $r \leq |z - z_0| \leq s$. The series (1.3) is called the **Laurent series expansion** of $f(z)$ with respect to the annulus $\rho < |z - z_0| < \sigma$.

To obtain a formula for the coefficients in the expansion, we divide $f(z)$ by $(z - z_0)^{n+1}$ in (1.3) and integrate around the circle $\{|z - z_0| = r\}$. Since the series converges uniformly on the circle, we can interchange the summation and integration. The result is

$$\oint_{|z-z_0|=r} \frac{1}{(z-z_0)^{n+1}} f(z)\, dz = \oint \frac{1}{(z-z_0)^{n+1}} \sum_{k=-\infty}^{\infty} a_k (z - z_0)^k \, dz$$

$$= \sum_{k=-\infty}^{\infty} a_k \oint_{|z-z_0|=r} (z - z_0)^{k-n-1} \, dz.$$

The integral of $(z - z_0)^m$ is $2\pi i$ if $m = -1$, otherwise zero, so all the terms in the series disappear except one, and the series reduces to $2\pi i a_n$. Thus

$$(1.4) \qquad a_n = \frac{1}{2\pi i} \oint_{|z-z_0|=r} \frac{f(z)}{(z-z_0)^{n+1}} \, dz, \qquad -\infty < n < \infty.$$

Note that this formula for a_n coincides with the usual formula for the power series coefficients in the case that $f(z)$ is analytic for $|z - z_0| < \sigma$. In the case that $f(z)$ is analytic at ∞ the formula for a_n agrees with the formula given in Section V.5.

We summarize our results in the following theorem.

Theorem (Laurent Series Expansion). *Suppose $0 \leq \rho < \sigma \leq \infty$, and suppose $f(z)$ is analytic for $\rho < |z - z_0| < \sigma$. Then $f(z)$ has a Laurent expansion (1.3) that converges absolutely at each point of the annulus, and that converges uniformly on each subannulus $r \leq |z - z_0| \leq s$, where $\rho < r < s < \sigma$. The coefficients are uniquely determined by $f(z)$, and they are given by (1.4) for any fixed r, $\rho < r < \sigma$.*

Example. To expand the function $f(z) = 1/(z - 1)(z - 2)$ in a Laurent series centered at $z = 0$ and converging in the annulus $\{1 < |z| < 2\}$, we expand each of the partial fractions in a geometric series,

$$\frac{1}{z-2} = -\frac{1}{2}\frac{1}{1-z/2} = -\frac{1}{2}\left(1+\frac{z}{2}+\frac{z^2}{4}+\cdots\right),$$

$$\frac{1}{z-1} = \frac{1}{z}\frac{1}{1-1/z} = \frac{1}{z}+\frac{1}{z^2}+\frac{1}{z^3}+\cdots.$$

This leads to the Laurent series representation

$$f(z) = \sum_{k=-\infty}^{\infty} a_k z^k, \qquad 1 < |z| < 2,$$

where $a_k = -1$ if $k < 0$, and $a_k = -1/2^{k+1}$ if $k \geq 0$.

Example. The function $f(z) = 1/(z-1)(z-2)$ can also be expanded in a Laurent series centered at $z = 1$, convergent in the punctured disk $\{0 < |z-1| < 1\}$. Again we rely upon a geometric series,

$$\frac{1}{z-2} = -\frac{1}{1-(z-1)} = -\sum_{k=0}^{\infty}(z-1)^k, \qquad |z-1| < 1,$$

to obtain

$$\frac{1}{(z-1)(z-2)} = -\frac{1}{z-1} - 1 - (z-1) - (z-1)^2 - (z-1)^3 - \cdots$$

$$= -\sum_{k=-1}^{\infty}(z-1)^k, \qquad 0 < |z-1| < 1.$$

The tail of the series (1.3) with the positive powers of $z - z_0$ converges on the largest open disk centered at z_0 to which $f_0(z)$ extends to be analytic, while the tail of the series with the negative powers of $z - z_0$ converges on the largest exterior domain of the form $\{|z - z_0| > \tau\}$ to which $f_1(z)$ extends analytically. Thus the largest open domain on which the full Laurent series (1.3) converges is the largest open annular set centered at z_0 containing the annulus $\{\rho < |z - z_0| < \sigma\}$ to which $f(z)$ extends analytically. This annular set might extend to z_0 or to ∞, to be a punctured disk or a full disk, or a punctured complex plane or the full complex plane.

Exercise. Consider the Laurent series for $f(z) = (z^2 - \pi^2)/\sin z$ that is centered at 0 and that converges for $|z| = 1$. What is the largest open set on which the series converges?

Solution. Since $\sin z$ has a simple zero at π, the function $(\sin z)/(z - \pi)$ extends to be analytic and nonzero at $z = \pi$. Hence $(z^2 - \pi^2)/\sin z$ extends to be analytic at $z = \pi$. Similarly, it extends to be analytic at $z = -\pi$. (We say that the singularities at $z = \pm\pi$ are "removable.") The function tends to ∞ at $z = 0$ and at $z = \pm 2\pi$. Thus the largest annular domain containing the circle $\{|z| = 1\}$ to which the function extends analytically is the punctured disk $\{0 < |z| < 2\pi\}$. This is then the largest open set on which the series converges.

Exercises for VI.1

1. Find all possible Laurent expansions centered at 0 of the following functions:

   (a) $\dfrac{1}{z^2 - z}$ (b) $\dfrac{z-1}{z+1}$ (c) $\dfrac{1}{(z^2-1)(z^2-4)}$

2. For each of the functions in Exercise 1, find the Laurent expansion centered at $z = -1$ that converges at $z = \frac{1}{2}$. Determine the largest open set on which each series converges.

3. Recall the power series for the Bessel function $J_n(z)$, $n \geq 0$, given in Exercise V.4.11, and define $J_{-n}(z) = (-1)^n J_n(z)$. For fixed $w \in C$, establish the Laurent series expansion

$$\exp\left[\frac{w}{2}(z - 1/z)\right] = \sum_{n=-\infty}^{\infty} J_n(w)z^n, \qquad 0 < |z| < \infty.$$

From the coefficient formula (1.4), deduce that

$$J_n(z) = \frac{1}{2\pi}\int_0^{2\pi} e^{i(z\sin\theta - n\theta)}, \qquad z \in \mathbb{C}.$$

Remark. This Laurent expansion is called the **Schlömilch formula**.

4. Suppose that $f(z) = f_0(z) + f_1(z)$ is the Laurent decomposition of an analytic function $f(z)$ on the annulus $\{A < |z| < B\}$. Show that if $f(z)$ is an even function, then $f_0(z)$ and $f_1(z)$ are even functions, and the Laurent series expansion of $f(z)$ has only even powers of z. Show that if $f(z)$ is an odd function, then $f_0(z)$ and $f_1(z)$ are odd functions, and the Laurent series expansion of $f(z)$ has only odd powers of z.

5. Suppose $f(z)$ is analytic on the punctured plane $D = \mathbb{C}\backslash\{0\}$. Show that there is a constant c such that $f(z) - c/z$ has a primitive in D. Give a formula for c in terms of an integral of $f(z)$.

6. Fix an annulus $D = \{a < |z| < b\}$, and let $f(z)$ be a continuous function on its boundary ∂D. Show that $f(z)$ can be approximated uniformly on ∂D by polynomials in z and $1/z$ if and only if $f(z)$ has a continuous extension to the closed annulus $D \cup \partial D$ that is analytic on D.

7. Show that a harmonic function u on an annulus $\{A < |z| < B\}$ has a unique expansion

$$u(re^{i\theta}) = \sum_{n=-\infty}^{\infty} a_n r^n \cos(n\theta) + \sum_{n\neq 0} b_n r^n \sin(n\theta) + c \, \log r,$$

which is uniformly convergent on each circle in the annulus. Show that for each r, $A < r < B$, the coefficients a_n, b_n, and c satisfy

$$a_n r^n + a_{-n} r^{-n} = \frac{1}{\pi} \int_{-\pi}^{\pi} u\left(re^{i\theta}\right) \cos(n\theta) d\theta, \qquad n \neq 0,$$

$$b_n r^n - b_{-n} r^{-n} = \frac{1}{\pi} \int_{-\pi}^{\pi} u\left(re^{i\theta}\right) \sin(n\theta) d\theta, \qquad n \neq 0,$$

$$a_0 + c \log r = \frac{1}{2\pi} \int_{-\pi}^{\pi} u\left(re^{i\theta}\right) d\theta.$$

Hint. Use a decomposition of the form $u = \operatorname{Re} f + c \log |z|$, where f is analytic on the annulus. (See Exercise III.3.4.)

2. Isolated Singularities of an Analytic Function

A point z_0 is an **isolated singularity** of $f(z)$ if $f(z)$ is analytic in some punctured disk $\{0 < |z - z_0| < r\}$ centered at z_0. For example, the function $1/z$ has an isolated singularity at $z = 0$, while $1/\sin z$ has isolated singularities at each of the points $z = 0, \pm\pi, \pm 2\pi, \ldots$. The functions $\sqrt{z}$ and $\log z$ do not have isolated singularities at $z = 0$; they cannot be defined even continuously on any punctured disk centered at 0.

isolated   not isolated

Suppose that $f(z)$ has an isolated singularity at z_0. Then $f(z)$ has a Laurent series expansion

$$(2.1) \qquad f(z) = \sum_{k=-\infty}^{\infty} a_k (z - z_0)^k, \qquad 0 < |z - z_0| < r.$$

We classify the isolated singularity at z_0 as one of three types according to whether no negative powers of $z - z_0$ appear in the expansion, or at least one but only finitely many negative powers appear, or infinitely many negative powers appear. These are three mutually exclusive cases that cover all possibilities. We discuss each of these cases in turn, and we prove one theorem for each case.

The isolated singularity of $f(z)$ at z_0 is defined to be a **removable singularity** if $a_k = 0$ for all $k < 0$. In this case the Laurent series (2.1) becomes a power series

$$f(z) = \sum_{k=0}^{\infty} a_k(z - z_0)^k, \qquad 0 < |z - z_0| < r.$$

If we define $f(z_0) = a_0$, the function $f(z)$ becomes analytic on the entire disk $\{|z - z_0| < r\}$.

Example. The function $(\sin z)/z$ has an isolated singularity at $z = 0$, where it is not defined. From the power series expansion of $\sin z$ we obtain the Laurent series expansion

$$\frac{\sin z}{z} = 1 - \frac{z^2}{3!} + \frac{z^4}{5!} - \cdots,$$

valid for all z. We may extend $(\sin z)/z$ to be an entire function by defining $(\sin z)/z$ to be 1 at $z = 0$.

If $f(z)$ has a removable singularity at z_0, then $f(z)$ is bounded near z_0. There is a converse statement, which provides a useful criterion for determining whether a singularity is removable.

Theorem (Riemann's Theorem on Removable Singularities). *Let z_0 be an isolated singularity of $f(z)$. If $f(z)$ is bounded near z_0, then $f(z)$ has a removable singularity at z_0.*

To see this, we expand $f(z)$ in a Laurent series (2.1) and use the formula (1.4) for the coefficients given in the preceding section. Suppose $|f(z)| \leq M$ for z near z_0, and let $r > 0$ be small. Using the ML-estimate to estimate the integral in (1.4), we obtain

$$|a_n| \leq \frac{1}{2\pi} \frac{M}{r^{n+1}} (2\pi r) = \frac{M}{r^n}.$$

If $n < 0$, the right-hand side tends to 0 as $r \to 0$. We conclude that $a_n = 0$ for $n < 0$, and consequently the singularity at z_0 is removable.

The isolated singularity of $f(z)$ at z_0 is defined to be a **pole** if there is $N > 0$ such that $a_{-N} \neq 0$ but $a_k = 0$ for all $k < -N$. The integer N is the **order** of the pole. In this case the Laurent series (2.1) becomes

$$f(z) = \sum_{k=-N}^{\infty} a_k(z-z_0)^k = \frac{a_{-N}}{(z - z_0)^N} + \cdots + \frac{a_{-1}}{z - z_0} + a_0 + a_1(z-z_0) + \cdots.$$

The sum of the negative powers,

$$P(z) = \sum_{k=-N}^{-1} a_k(z - z_0)^k = \frac{a_{-N}}{(z - z_0)^N} + \cdots + \frac{a_{-1}}{z - z_0},$$

is called the **principal part** of $f(z)$ at the pole z_0. The principal part $P(z)$ coincides with the summand $f_1(z)$ in the Laurent decomposition $f(z) = f_0(z) + f_1(z)$ given in the preceding section. The bad behavior of $f(z)$ at z_0 is incorporated into $P(z)$, in the sense that $f(z) - P(z)$ is analytic at z_0.

A pole of order one is called a **simple pole**, and a pole of order two is called a **double pole**. Thus $1/z$ has a simple pole at $z = 0$, and $1/(z-i)^2$ has a double pole at $z = i$.

Theorem. *Let z_0 be an isolated singularity of $f(z)$. Then z_0 is a pole of $f(z)$ of order N if and only if $f(z) = g(z)/(z - z_0)^N$, where $g(z)$ is analytic at z_0 and $g(z_0) \neq 0$.*

The proof is straightforward. Suppose that $f(z)$ has a pole of order N at z_0, and that $f(z)$ has the above Laurent series. Then the power series $a_{-N} + a_{-N+1}(z - z_0) + a_{-N+2}(z - z_0)^2 + \cdots$ converges to a function $g(z)$ that is analytic at z_0 and satisfies $f(z) = g(z)/(z - z_0)^N$, and further, $g(z_0) = a_{-N} \neq 0$. Conversely, if $g(z)$ is analytic at z_0 and satisfies $g(z_0) \neq 0$, then $f(z) = g(z)/(z - z_0)^N$ has Laurent series with leading term $g(z_0)/(z - z_0)^N$, so that $f(z)$ has a pole of order N at z_0.

Theorem. *Let z_0 be an isolated singularity of $f(z)$. Then z_0 is a pole of $f(z)$ of order N if and only if $1/f(z)$ is analytic at z_0 and has a zero of order N.*

Again the proof is easy. Suppose $f(z)$ has a pole of order N at z_0. Let $g(z) = (z - z_0)^N f(z)$ be as above. Since $g(z_0) \neq 0$, the function $h(z) = 1/g(z)$ is analytic at z_0 and satisfies $h(z_0) \neq 0$. Thus $1/f(z) = (z - z_0)^N h(z)$ has a zero of order N at z_0. This argument is also reversible. If $1/f(z)$ has a zero of order N at z_0, then $1/f(z) = (z - z_0)^N h(z)$ for some analytic function $h(z)$ satisfying $h(z_0) \neq 0$, and then $g(z) = 1/h(z)$ is analytic and nonzero at z_0, so $f(z) = g(z)/(z - z_0)^N$ has a pole of order N at z_0.

Example. The function $1/\sin z$ has poles at each of the zeros of $\sin z$. Since the zeros of $\sin z$ are simple, they are simple poles for $1/\sin z$.

Exercise. Consider the Laurent series expansion for $1/\sin z$ that converges on the circle $\{|z| = 4\}$. Find the coefficients a_0, a_{-1}, a_{-2}, and a_{-3} of 1, $1/z$, $1/z^2$, and $1/z^3$, respectively. Determine the largest open set on which the series converges.
Solution. The only zeros of $\sin z$ are at the integral multiples of π. These are then the only singularities of $1/\sin z$, and they are all simple poles. The largest open annular set containing the circle and to which $1/\sin z$ extends analytically is then the annulus $\{\pi < |z| < 2\pi\}$. This annulus is then the largest open set on which the Laurent series converges. From the

expansion $\sin z = z + \mathcal{O}(z^3)$ near $z = 0$, we see that

$$\frac{1}{\sin z} = \frac{1}{z} + \text{analytic}$$

near $z = 0$, and $1/\sin z - 1/z$ is analytic at $z = 0$. Similarly, from the expansion $\sin z = -(z - \pi) + \mathcal{O}((z - \pi)^3)$ at $z = \pi$, we see that

$$\frac{1}{\sin z} = -\frac{1}{z - \pi} + \text{analytic}$$

near $z = \pi$, and $1/\sin z + 1/(z - \pi)$ is analytic at $z = \pi$. By the same token, $1/\sin z + 1/(z + \pi)$ is analytic at $z = -\pi$. We conclude that if

$$f_1(z) = \frac{1}{z} - \frac{1}{z + \pi} - \frac{1}{z - \pi},$$

then $f_0(z) = 1/\sin z - f_1(z)$ is analytic for $|z| < 2\pi$. Thus $1/\sin z = f_0(z) + f_1(z)$ is the Laurent decomposition of $1/\sin z$. To obtain the negative powers of the Laurent expansion, we expand $f_1(z)$ in a series of descending powers of z, using the geometric series expansion. This is done most easily by combining summands,

$$f_1(z) = \frac{1}{z} - \frac{2z}{z^2 - \pi^2} = \frac{1}{z} - \frac{2}{z}\sum_{k=0}^{\infty}\frac{\pi^{2k}}{z^{2k}} = -\frac{1}{z} - \sum_{k=1}^{\infty}\frac{2\pi^{2k}}{z^{2k+1}}.$$

Note that all the even powers of z disappear, as they must, since $1/\sin z$ is an odd function. We read off the coefficients $a_{-1} = -1$ and $a_{-3} = -2\pi^2$.

We say that a function $f(z)$ is **meromorphic** on a domain D if $f(z)$ is analytic on D except possibly at isolated singularities, each of which is a pole. Sums and products of meromorphic functions are meromorphic. Quotients of meromorphic functions are meromorphic, provided that the denominator is not identically zero. Note also that if there are infinitely many poles of $f(z)$ in D, then we can arrange them in a sequence that accumulates only at the boundary of D. Otherwise, there would be a point of accumulation in D of the poles of $f(z)$, and this point would not be an isolated singularity of $f(z)$.

Example. The function $1/\sin z$ is meromorphic on the entire complex plane. As another example, let $R(z)$ be any rational function. We can express $R(z)$ as a quotient of polynomials in the form

$$R(z) = c\frac{(z - \zeta_1)^{m_1} \cdots (z - \zeta_k)^{m_k}}{(z - z_1)^{n_1} \cdots (z - z_l)^{n_l}},$$

where the ζ_j's and z_j's are all distinct. Evidently, $R(z)$ has a zero of order m_j at each ζ_j and a pole of order n_j at each z_j. Thus $R(z)$ is meromorphic on the entire complex plane.

We add one more useful characterization of poles.

Theorem. *Let z_0 be an isolated singularity of $f(z)$. Then z_0 is a pole if and only if $|f(z)| \to \infty$ as $z \to z_0$.*

One direction of the theorem is trivial. If $f(z)$ has a pole of order N at z_0, then $g(z) = (z - z_0)^N f(z)$ is analytic and nonzero at z_0, so that

$$|f(z)| = |z - z_0|^{-N}|g(z)| \to \infty$$

as $z \to z_0$. For the converse, we use Riemann's theorem on removable singularities. Suppose $|f(z)| \to \infty$ as $z \to z_0$. Then $f(z) \neq 0$ for z near z_0, so that $h(z) = 1/f(z)$ is analytic in some punctured neighborhood of z_0. Further, $h(z) \to 0$ as $z \to z_0$. By Riemann's theorem, $h(z)$ extends to be analytic at z_0, and moreover, $h(z_0) = 0$. If N is the order of the zero of $h(z)$ at z_0, then $f(z) = 1/h(z)$ has a pole of order N at z_0.

The isolated singularity of $f(z)$ at z_0 is defined to be an **essential singularity** if $a_k \neq 0$ for infinitely many $k < 0$. Thus an isolated singularity that is neither removable nor a pole is declared to be essential.

Example. The Laurent expansion of $e^{1/z}$ at $z = 0$ is given by

$$e^{1/z} = 1 + \frac{1}{z} + \frac{1}{2!}\frac{1}{z^2} + \frac{1}{3!}\frac{1}{z^3} + \cdots, \qquad z \neq 0.$$

Since infinitely many negative powers of z appear in the expansion, the isolated singularity at $z = 0$ is essential. That the singularity is essential can also be seen from the behavior of $e^{1/z}$ as $z \to 0$. Since $e^{1/x} \to +\infty$ as $x > 0$ tends to 0, the singularity is not removable. And since $e^{1/(iy)}$ has unit modulus, the modulus of $e^{1/z}$ does not tend to $+\infty$ as z tends to 0 along the imaginary axis, and the singularity is not a pole. The only remaining possibility is that the singularity is essential.

At an essential singularity, the values of $f(z)$ cluster towards the entire complex plane. That is the content of the following theorem.

Theorem (Casorati-Weierstrass Theorem). *Suppose z_0 is an essential isolated singularity of $f(z)$. Then for every complex number w_0, there is a sequence $z_n \to z_0$ such that $f(z_n) \to w_0$.*

We argue the contrapositive. Suppose that there is some complex number w_0 that is not a limit of values of $f(z)$ as above. Then there is some small $\varepsilon > 0$ such that $|f(z) - w_0| > \varepsilon$ for all z near z_0. Hence $h(z) = 1/(f(z) - w_0)$ is bounded near z_0. By Riemann's theorem, $h(z)$ has a removable singularity at z_0. Hence $h(z) = (z - z_0)^N g(z)$ for some $N \geq 0$ and some analytic function $g(z)$ satisfying $g(z_0) \neq 0$. Thus $f(z) - w_0 = 1/h(z) = (z-z_0)^{-N}(1/g(z))$, where $1/g(z)$ is analytic at z_0. If $N = 0$, $f(z)$ extends to be analytic at z_0, while if $N > 0$, $f(z)$ has a pole of order N at z_0. This establishes the theorem.

Later we will prove Picard's theorem, that if $f(z)$ is an analytic function with an essential isolated singularity at z_0, then for all complex numbers w_0 with possibly one exception, there is a sequence $z_n \to z_0$ such that $f(z_n) = w_0$. The function $f(z) = e^{1/z}$, which omits the value $w = 0$, shows that we must allow for the exceptional point.

Exercises for VI.2

1. Find the isolated singularities of the following functions, and determine whether they are removable, essential, or poles. Determine the order of any pole, and find the principal part at each pole.

   (a) $z/(z^2-1)^2$ (d) $\tan z = \dfrac{\sin z}{\cos z}$ (g) $\operatorname{Log}\left(1 - \dfrac{1}{z}\right)$

   (b) $\dfrac{ze^z}{z^2-1}$ (e) $z^2 \sin\left(\dfrac{1}{z}\right)$ (h) $\dfrac{\operatorname{Log} z}{(z-1)^3}$

   (c) $\dfrac{e^{2z}-1}{z}$ (f) $\dfrac{\cos z}{z^2 - \pi^2/4}$ (i) $e^{1/(z^2+1)}$

2. Find the radius of convergence of the power series for the following functions, expanded about the indicated point.

   (a) $\dfrac{z-1}{z^4-1}$, about $z = 3+i$,   (c) $\dfrac{z}{\sin z}$, about $z = \pi i$,

   (b) $\dfrac{\cos z}{z^2 - \pi^2/4}$, about $z = 0$,   (d) $\dfrac{z^2}{\sin^3 z}$, about $z = \pi i$.

3. Consider the function $f(z) = \tan z$ in the annulus $\{3 < |z| < 4\}$. Let $f(z) = f_0(z) + f_1(z)$ be the Laurent decomposition of $f(z)$, so that $f_0(z)$ is analytic for $|z| < 4$, and $f_1(z)$ is analytic for $|z| > 3$ and vanishes at ∞. (a) Obtain an explicit expression for $f_1(z)$. (b) Write down the series expansion for $f_1(z)$, and determine the largest domain on which it converges. (c) Obtain the coefficients a_0, a_1, and a_2 of the power series expansion of $f_0(z)$. (d) What is the radius of convergence of the power series expansion for $f_0(z)$?

4. Suppose $f(z)$ is meromorphic on the disk $\{|z| < s\}$, with only a finite number of poles in the disk. Show that the Laurent decomposition of $f(z)$ with respect to the annulus $\{s - \varepsilon < |z| < s\}$ has the form $f(z) = f_0(z) + f_1(z)$, where $f_1(z)$ is the sum of the principal parts of $f(z)$ at its poles.

5. By estimating the coefficients of the Laurent series, prove that if z_0 is an isolated singularity of f, and if $(z-z_0)f(z) \to 0$ as $z \to z_0$, then z_0 is removable. Give a second proof based on Morera's theorem.

6. Show that if $f(z)$ is continuous on a domain D, and if $f(z)^8$ is analytic on D, then $f(z)$ is analytic on D.

7. Show that if z_0 is an isolated singularity of $f(z)$, and if $(z-z_0)^N f(z)$ is bounded near z_0, then z_0 is either removable or a pole of order at most N.

8. A meromorphic function f at z_0 is said to have **order** N at z_0 if $f(z) = (z - z_0)^N g(z)$ for some analytic function g at z_0 such that $g(z_0) \neq 0$. The order of the function 0 is defined to be $+\infty$. Show that
   (a) $\operatorname{order}(fg, z_0) = \operatorname{order}(f, z_0) + \operatorname{order}(g, z_0)$,
   (b) $\operatorname{order}(1/f, z_0) = -\operatorname{order}(f, z_0)$,
   (c) $\operatorname{order}(f + g, z_0) \geq \min\{\operatorname{order}(f, z_0), \operatorname{order}(g, z_0)\}$.
   Show that strict inequality can occur in (c), but that equality holds in (c) whenever f and g have different orders at z_0.

9. Recall that "$f(z) = \mathcal{O}(h(z))$ as $z \to z_0$" means that there is a constant C such that $|f(z)| \leq C|h(z)|$ for z near z_0. Show that if z_0 is an isolated singularity of an analytic function $f(z)$, and if $f(z) = \mathcal{O}((z-z_0)^m)$ as $z \to z_0$, then the Laurent coefficients of $f(z)$ are 0 for $k < m$, that is, the Laurent series of $f(z)$ has the form

$$f(z) = a_m(z - z_0)^m + a_{m+1}(z - z_0)^{m+1} + \cdots.$$

Remark. This shows that the use of the notation $\mathcal{O}(z^m)$ in Section V.6 is consistent.

10. Show that if $f(z)$ and $g(z)$ are analytic functions that both have the same order $N \geq 0$ at z_0, then

$$\lim_{z \to z_0} \frac{f(z)}{g(z)} = \frac{f^{(N)}(z_0)}{g^{(N)}(z_0)}.$$

11. Suppose $f(z) = \sum a_k z^k$ is analytic for $|z| < R$, and suppose that $f(z)$ extends to be meromorphic for $|z| < R + \varepsilon$, with only one pole z_0 on the circle $|z| = R$. Show that $a_k/a_{k+1} \to z_0$ as $k \to \infty$.

12. Show that if z_0 is an isolated singularity of $f(z)$ that is not removable, then z_0 is an essential singularity for $e^{f(z)}$.

13. Let S be a sequence converging to a point $z_0 \in \mathbb{C}$, and let $f(z)$ be analytic on some disk centered at z_0 except possibly at the points of S and at z_0. Show that either $f(z)$ extends to be meromorphic on some neighborhood of z_0, or else for any complex number L there is a sequence $\{w_j\}$ such that $w_j \to z_0$ and $f(w_j) \to L$.

14. Suppose $u(re^{i\theta})$ is harmonic on the punctured disk $\{0 < r < 1\}$, with Laurent series as in Exercise 7 of Section 1. Suppose $\alpha > 0$ is such that $r^\alpha u(re^{i\theta}) \to 0$ as $r \to 0$. Show that $a_n = 0 = b_n$ for $n \leq -\alpha$.

15. Suppose $u(z)$ is harmonic on the punctured disk $\{0 < |z| < \rho\}$. Show that if

$$\frac{u(z)}{\log(1/|z|)} \to 0$$

as $z \to 0$, then $u(z)$ extends to be harmonic at 0. What can you say if you know only that $|u(z)| \le C \log(1/|z|)$ for some fixed constant C and $0 < |z| < \rho$?

3. Isolated Singularity at Infinity

We say that $f(z)$ has an **isolated singularity at** ∞ if $f(z)$ is analytic outside some bounded set, that is, if there is $R > 0$ such that $f(z)$ is analytic for $|z| > R$. Thus $f(z)$ has an isolated singularity at ∞ if and only if $g(w) = f(1/w)$ has an isolated singularity at $w = 0$. We classify the isolated singularity of $f(z)$ at ∞ according to the isolated singularity of $g(w)$ at $w = 0$. Suppose that $f(z)$ has a Laurent series expansion

$$f(z) = \sum_{k=-\infty}^{\infty} b_k z^k, \qquad |z| > R.$$

The singularity of $f(z)$ at ∞ is **removable** if $b_k = 0$ for all $k > 0$, in which case $f(z)$ is analytic at ∞. The singularity of $f(z)$ at ∞ is **essential** if $b_k \ne 0$ for infinitely many $k > 0$. For fixed $N \ge 1$, $f(z)$ has a **pole** of order N at ∞ if $b_N \ne 0$ while $b_k = 0$ for $k > N$.

Suppose $f(z)$ has a pole of order N at ∞. The Laurent series expansion of $f(z)$ becomes

$$f(z) = b_N z^N + b_{N-1} z^{N-1} + \cdots + b_1 z + b_0 + \frac{b_{-1}}{z} + \cdots, \qquad |z| > R,$$

where $b_N \ne 0$. We define the **principal part of** $f(z)$ **at** ∞ to be the polynomial

$$P(z) = b_N z^N + b_{N-1} z^{N-1} + \cdots + b_1 z + b_0.$$

The inclusion of the constant b_0 in the principal part is a matter of convenience. It guarantees that $f(z) - P(z)$ is not only analytic at ∞ but also vanishes there.

Example. Any polynomial of degree $N \ge 1$ has a pole of order N at ∞. The principal part of the polynomial coincides with the polynomial itself.

Example. The function $e^z = 1 + z + z^2/2! + \cdots$ has an essential singularity at ∞.

Exercises for VI.3

1. Consider the functions in Exercise 1 of Section 2 above. Determine which have isolated singularities at ∞, and classify them.

2. Suppose that $f(z)$ is an entire function that is not a polynomial. What kind of singularity can $f(z)$ have at ∞?

3. Show that if $f(z)$ is a nonconstant entire function, then $e^{f(z)}$ has an essential singularity at $z = \infty$.

4. Show that each branch of the following functions is meromorphic at ∞, and obtain the series expansion for each branch at ∞.

   (a) $(z^2 - 1)^{5/2}$ (b) $\sqrt[3]{(z^3 - 1)}$ (c) $\sqrt{z^2 - \dfrac{1}{z}}$

4. Partial Fractions Decomposition

Proceeding in analogy with our earlier definition, we say that a function $f(z)$ is **meromorphic** on a domain D in the extended complex plane $\mathbb{C}^*$ if $f(z)$ is analytic on D except possibly at isolated singularities, each of which is a pole. Again, sums and products of meromorphic functions are meromorphic. Quotients of meromorphic functions are meromorphic, provided that the denominator is not identically zero.

Any rational function is meromorphic on the extended complex plane $\mathbb{C}^*$, including at ∞. We aim to establish the converse.

Theorem. *A meromorphic function on the extended complex plane $\mathbb{C}^*$ is rational.*

To see this, note first that a meromorphic function $f(z)$ on the extended complex plane can have only a finite number of poles. Otherwise, they would accumulate at a point that would not be an isolated singularity of $f(z)$. If $f(z)$ is analytic at ∞, we define $P_\infty(z)$ to be the constant function $f(\infty)$. Otherwise, $f(z)$ has a pole at ∞ and we define $P_\infty(z)$ to be the principal part of $f(z)$ at ∞. In any event, $P_\infty(z)$ is a polynomial, and $f(z) - P_\infty(z) \to 0$ as $z \to \infty$. Let $z_1, \ldots, z_m$ be the poles of $f(z)$ in the finite complex plane $\mathbb{C}$, and let $P_k(z)$ be the principal part of $f(z)$ at z_k. It has the form

$$P_k(z) = \frac{\alpha_1}{z - z_k} + \frac{\alpha_2}{(z - z_k)^2} + \cdots + \frac{\alpha_n}{(z - z_k)^n},$$

and in particular, $P_k(z)$ is analytic at ∞ and vanishes there. Consider the function

$$g(z) = f(z) - P_\infty(z) - \sum_{j=1}^{m} P_j(z).$$

Since $f(z) - P_k(z)$ is analytic at z_k, and each $P_j(z)$ is analytic at z_k for $j \neq k$, $g(z)$ is analytic at each z_k. Hence $g(z)$ is an entire function, and further, $g(z) \to 0$ as $z \to \infty$. By Liouville's theorem, $g(z)$ is identically zero. Thus

$$(4.1) \qquad f(z) \;=\; P_\infty(z) + \sum_{j=1}^{m} P_j(z),$$

which shows in particular that $f(z)$ is a rational function.

The decomposition (4.1) is called the **partial fractions decomposition** of the rational function $f(z)$. As a byproduct of the proof we obtain the following.

Theorem. *Every rational function has a partial fractions decomposition, expressing it as the sum of a polynomial in z and its principal parts at each of its poles in the finite complex plane.*

Suppose that $p(z)$ and $q(z)$ are polynomials. If the degree of $q(z)$ is strictly less than the degree of $p(z)$, then $p(z)/q(z) \to 0$ as $z \to \infty$. Thus $f(z) = p(z)/q(z)$ is analytic at ∞ and vanishes there, and the principal part $P_\infty(z)$ is zero. Formula (4.1) expresses $f(z)$ as the sum of the principal parts at each of its finite poles.

Example. The function $1/(z^2 - 1)$ is analytic at ∞ and vanishes there, and it has poles at ± 1. The partial fractions decomposition is

$$\frac{1}{z^2 - 1} \;=\; \frac{1}{2} \frac{1}{z - 1} - \frac{1}{2} \frac{1}{z + 1},$$

and these summands are the principal parts at $+1$ and -1, respectively.

For arbitrary polynomials $p(z)$ and $q(z)$, we can use the division algorithm to find the principal part $P_\infty(z)$ of $p(z)/q(z)$ at ∞. The division algorithm is a procedure that produces in a finite number of steps polynomials $P_\infty(z)$ and $r(z)$ that satisfy

$$(4.2) \qquad p(z) = P_\infty(z)q(z) + r(z), \qquad \deg r(z) < \deg q(z).$$

It proceeds as follows. We assume that $q(z)$ is a monic polynomial of degree m, so that $q(z) = z^m + \cdots$. We start with a polynomial $p(z)$ of degree $n \geq m$, say $p(z) = c_0 z^n + \cdots$. For the first step we kill the top coefficient of $p(z)$ by defining $p_1(z) = p(z) - c_0 z^{n-m} q(z)$. Thus $p_1(z)$ has degree $n_1 < n$, say $p_1(z) = c_1 z^{n_1} q(z) + \cdots$. Then we repeat the first step and define $p_2(z) = p_1(z) - c_1 z^{n_1 - m} q(z)$, which has degree $n_2 < n_1$. We proceed in this fashion until we reach a polynomial $p_k(z) = p_{k-1}(z) - c_{k-1} z^{n_{k-1} - m} q(z)$ such that the degree of $p_k(z)$ is less than m. This occurs after at most

$n - m + 1$ steps, and then we have

$$p(z) = c_0 z^{n-m} q(z) + p_1(z) = c_0 z^{n-m} q(z) + c_1 z^{n_1-m} q(z) + p_2(z) = \cdots$$
$$= c_0 z^{n-m} q(z) + c_1 z^{n_1-m} q(z) + \cdots + c_{k-1} z^{n_{k-1}-m} q(z) + p_k(z).$$

We set $r(z) = p_k(z)$, and we let $P_\infty(z)$ be the sum of the terms multiplying $q(z)$, and we obtain the decomposition (4.2).

Now the function $p(z)/q(z) - P_\infty(z) = r(z)/q(z)$ is analytic at ∞ and tends to 0 there. Thus the polynomial $P_\infty(z)$ coincides with the principal part of $p(z)/q(z)$ at ∞. Our recipe for obtaining the partial fractions decomposition of an arbitrary rational function $p(z)/q(z)$ is then first to obtain the polynomials $P_\infty(z)$ and $r(z)$ that satisfy (4.2) from the division algorithm, and then to find the principal parts of $r(z)/q(z)$ at each of the zeros of $q(z)$.

Example. To obtain the partial fractions decomposition of $z^3/(z^2 + 1)$, first express z^3 in the form given by the division algorithm (4.2), which is

$$z^3 = z(z^2 + 1) - z,$$

corresponding to $P_\infty(z) = z$ and $r(z) = -z$. Thus

$$\frac{z^3}{z^2 + 1} = z - \frac{z}{z^2 + 1}.$$

Then observe that the poles at $\pm i$ are simple poles, so that

$$\frac{z}{z^2 + 1} = \frac{\alpha}{z - i} + \frac{\beta}{z + i}$$

for some constants α and β. We put this expression over a common denominator and solve for α and β. This leads to the partial fractions decomposition

$$\frac{z^3}{z^2 + 1} = z - \frac{1}{2}\frac{1}{z - i} - \frac{1}{2}\frac{1}{z + i}.$$

Exercises for VI.4

1. Find the partial fractions decompositions of the following functions.

(a) $\dfrac{1}{z^2 - z}$ (c) $\dfrac{1}{(z + 1)(z^2 + 2z + 2)}$ (e) $\dfrac{z - 1}{z + 1}$

(b) $\dfrac{z^2 + 1}{z(z^2 - 1)}$ (d) $\dfrac{1}{(z^2 + 1)^2}$ (f) $\dfrac{z^2 - 4z + 3}{z^2 - z - 6}$

2. Use the division algorithm to obtain the partial fractions decomposition of the following functions.

(a) $\dfrac{z^3 + 1}{z^2 + 1}$ (b) $\dfrac{z^9 + 1}{z^6 - 1}$ (c) $\dfrac{z^6}{(z^2 + 1)(z - 1)^2}$

3. Let V be the complex vector space of functions that are analytic on the extended complex plane except possibly at the points 0 and i, where they have poles of order at most two. What is the dimension of V? Write down explicitly a vector space basis for V.

5. Periodic Functions

A complex number ω is a **period** of a function $f(z)$ if $f(z+\omega) = f(z)$ wherever defined. The function $f(z)$ is **periodic** if it has a period $\omega \neq 0$.

The exponential function e^z is periodic with periods 0, $\pm 2\pi i$, $\pm 4\pi i$, ... The exponential function e^{iz} is periodic with periods 0, $\pm 2\pi$, $\pm 4\pi$, ... Sums of exponential functions with the same periods are periodic. Thus $\cos z = (e^{iz} + e^{-iz})/2$ and $\sin z = (e^{iz} - e^{-iz})/2i$ are periodic. One of our goals in this section is to show that any periodic analytic function in a half-plane or strip can be represented as a sum of exponential functions.

If $\omega \neq 0$ is a period of $f(z)$, the function $g(z) = f(\omega z)$ satisfies $g(z+1) = f(\omega z + \omega) = f(\omega z) = g(z)$, so $g(z)$ has period 1. Thus we can always make a change of variable to arrange that one of the periods of a given periodic function is $\omega = 1$. We focus on functions that are analytic on a horizontal strip and that are periodic with period 1, that is, that satisfy $f(z+1) = f(z)$. This includes the exponentials $e^{2\pi i k z}$ for k an integer, and any linear combination of these exponentials.

Theorem. *If $f(z)$ is analytic on the horizontal strip $\{\alpha < \operatorname{Im}(z) < \beta\}$, and $f(z)$ is periodic with period 1, then $f(z)$ can be expanded in an absolutely convergent series of exponentials*

$$f(z) = \sum_{k=-\infty}^{\infty} a_k e^{2\pi i k z}, \qquad \alpha < \operatorname{Im}(z) < \beta.$$

The series converges uniformly on any smaller strip $\{\alpha_0 \leq \operatorname{Im}(z) \leq \beta_0\}$, where $\alpha < \alpha_0 < \beta_0 < \beta$.

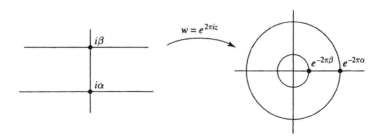

To see this, we make an exponential change of variable

$$w = e^{2\pi i z}, \qquad z = -\frac{i}{2\pi} \log |w| + \frac{\arg w}{2\pi},$$

and we set $g(w) = f(z)$ with z as above. Since $f(z)$ is periodic with period 1, the value $g(w)$ does not depend on the choice of the argument of w. Thus $g(w)$ is well-defined, and $g(w)$ is analytic for $e^{-2\pi\beta} < |w| < e^{-2\pi\alpha}$. We expand $g(w)$ as a Laurent series $\sum a_k w^k$ in this annulus, and this yields the exponential series for $f(z)$.

Theorem. *Suppose $f(z)$ is analytic on the half-plane $\{\mathrm{Im}(z) > \alpha\}$, and $f(z)$ is periodic with period 1. If $f(z)$ is bounded as $\mathrm{Im}(z) \to +\infty$, then $f(z)$ can be expanded in an absolutely convergent series of exponentials*

$$f(z) = \sum_{k=0}^{\infty} a_k e^{2\pi i k z}, \qquad \mathrm{Im}(z) > \alpha.$$

The series converges uniformly on any smaller half-plane $\{\mathrm{Im}(z) \geq \alpha_0\}$, where $\alpha_0 > \alpha$.

In this case the change of variable $w = e^{2\pi i z}$ converts $f(z)$ to an analytic function $g(w)$ on the punctured disk $0 < |w| < e^{-2\pi\alpha}$. The hypothesis on $f(z)$ implies that $g(w)$ is bounded as $w \to 0$. By Riemann's theorem on removable singularities, $g(w)$ extends to be analytic at 0. Hence $g(w)$ has a power series expansion

$$g(w) = \sum_{k=0}^{\infty} a_k w^k, \qquad |w| < e^{-2\pi\alpha},$$

and this yields the exponential series for $f(z)$.

Example. The meromorphic function $1/\sin(2\pi z)$ is analytic in the upper half-plane and has period 1. From

$$|\sin(2\pi z)|^2 = \sin^2(2\pi x) + \sinh^2(2\pi y), \qquad z = x + iy,$$

we see that $1/\sin(2\pi z) \to 0$ as $y = \mathrm{Im}(z) \to +\infty$. By the preceding theorem, $1/\sin(2\pi z)$ can be expanded in a series of exponentials $e^{2\pi i k z}$ that converges absolutely in the upper half-plane. The expansion can be obtained directly from a geometric series,

$$\frac{1}{\sin(2\pi z)} = \frac{-2i e^{2\pi i z}}{1 - e^{4\pi i z}} = -2i \left[e^{2\pi i z} + e^{6\pi i z} + e^{10\pi i z} + \cdots \right].$$

Now we change our point of view. We fix a function $f(z)$, say $f(z)$ is a meromorphic function on the complex plane, and we study the set of periods of $f(z)$. If $f(z)$ is constant, then every complex number is a period of $f(z)$, and there is not much to say. So we assume that $f(z)$ is nonconstant.

If ω_1 and ω_2 are periods of $f(z)$, then so are $m\omega_1 + n\omega_2$ for all integers m and n. (The periods form an "additive subgroup" of the complex numbers.)

Let z_0 be any point at which $f(z)$ is analytic. Since the zeros of a nonconstant analytic function are isolated, $f(z_0 + \zeta) - f(z_0) \neq 0$ on some punctured disk $\{0 < |\zeta| < \rho\}$. Since $f(z_0 + \omega) - f(z_0) = 0$ for any period ω of $f(z)$, there can be no periods of $f(z)$ in the punctured disk. Thus any nonzero period ω of $f(z)$ satisfies $|\omega| \geq \rho$.

If ω and ω_2 are two different periods of $f(z)$, then $\omega_1 - \omega_2$ is a nonzero period, so $|\omega_1 - \omega_2| \geq \rho$. Since any bounded subset of the complex plane can be covered by a finite number of disks of radius $\rho/2$, and each of these contains at most one period, we conclude that any bounded subset of the complex plane contains only finitely many periods. (The periods form a "discrete subgroup" of the complex numbers.)

Let L be a straight line through 0 that contains a nonzero period, and let ω_1 be a nonzero period on L that is closest to 0. Then there cannot be a period between two consecutive integral multiples $k\omega_1$ and $(k+1)\omega_1$ of ω_1, or by subtracting we would obtain a period on L closer to 0 than ω_1. Thus the periods on L are precisely the integral multiples of ω_1.

It may occur that all the periods of $f(z)$ lie on the same straight line through 0. Otherwise, from among the periods not on the line L, choose a period ω_2 that is closest to the line segment $[0, \omega_1]$. We claim that all periods of $f(z)$ have the form $m\omega_1 + n\omega_2$ for integers m and n. We argue as follows.

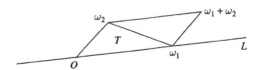

Let P be the closed parallelogram with vertices 0, ω_1, ω_2, and $\omega_1 + \omega_2$. Thus P consists of precisely the complex numbers of the form $s\omega_1 + t\omega_2$ where $0 \leq s, t \leq 1$. Every complex number can be expressed in the form $m\omega_1 + n\omega_2 + z$, where m, n are integers and $z \in P$. Geometrically, this means that the parallelograms $P + m\omega_1 + n\omega_2$ fill out the complex plane. We cut P into two triangles, the triangle with vertices $0, \omega_1, \omega_2$, which we denote by T, and the triangle $\omega_1 + \omega_2 - T$ with vertices ω_1, ω_2, and $\omega_1 + \omega_2$. The only periods in T are the vertices, since any other point of T either lies on L closer to 0 than ω_1 or lies off L closer to $[0, \omega_1]$ than ω_2. It follows then that the only periods in P are the four vertices, since if ω is a period in P, then either ω or $\omega_1 + \omega_2 - \omega$ is a period belonging to T, hence a vertex of T. Since any period can be expressed in the form $m\omega_1 + n\omega_2 + \omega$, where m, n are integers and ω is a period in P, in fact every period is an integral combination of ω_1 and ω_2, as asserted.

We summarize our results as follows.

Theorem. *Suppose that $f(z)$ is a nonconstant meromorphic function on the complex plane that is periodic. Either there is a period ω_1 for $f(z)$ such that the periods of $f(z)$ are the integral multiples $m\omega_1$, $-\infty < m < \infty$, or there are two periods ω_1 and ω_2 for $f(z)$ that do not lie on the same line through the origin such that the periods of $f(z)$ are the integral combinations $m\omega_1 + n\omega_2$, $-\infty < m, n < \infty$.*

In the case that the periods of $f(z)$ all lie on the same straight line through the origin, we say that $f(z)$ is **simply periodic**. Otherwise, we say that $f(z)$ is **doubly periodic**. The entire functions e^z and $\sin z$ are simply periodic. It is possible to construct (as in Exercise 7) meromorphic functions on the complex plane that are doubly periodic. However, the only entire functions that are doubly periodic are the trivial ones.

Theorem. *An entire function that is doubly periodic is constant.*

Indeed, if the entire function $f(z)$ is doubly periodic, and if $|f(z)| \le M$ on the parallelogram P constructed above, then by periodicity $|f(z)| \le M$ on each translate $m\omega_1 + n\omega_2 + P$ of P. Since these translates fill out the complex plane, $f(z)$ is a bounded entire function. By Liouville's theorem, $f(z)$ is constant.

Exercises for VI.5

1. Show that if $f(z)$ and $g(z)$ have period ω, then so do $f(z) + g(z)$ and $f(z)g(z)$.

2. Expand $1/\cos(2\pi z)$ in a series of powers of $e^{2\pi i z}$ that converges in the upper half-plane. Determine where the series converges absolutely and where it converges uniformly.

3. Expand $\tan z$ in a series of powers of exponentials e^{ikz}, $-\infty < k < \infty$, that converges in the upper half-plane. Also find an expansion of $\tan z$ as an exponential series that converges in the lower half-plane.

4. Let $f(z)$ be an analytic function in the upper half-plane that is periodic, with real period $2\pi\lambda > 0$. Suppose that there are $A, C > 0$ such that $|f(x + iy)| \le Ce^{Ay}$ for $y > 0$. Show that

$$ f(z) \;=\; \sum_{n \ge -A\lambda} a_n e^{inz/\lambda}, $$

where the series converges uniformly in each half-plane $\{y \ge \varepsilon\}$, for fixed $\varepsilon > 0$.

5. Suppose that ± 1 are periods of a nonzero doubly periodic function $f(z)$, and suppose that there are no periods ω of $f(z)$ satisfying $0 < |\omega| < 1$. How many periods of $f(z)$ lie on the unit circle?

Describe the possibilities, and sketch the set of periods for each possibility.

6. We say that ω_1 and ω_2 **generate** the periods of a doubly periodic function if the periods of the function are precisely the complex numbers of the form $m\omega_1 + n\omega_2$ where m and n are integers. Show that if ω_1 and ω_2 generate the periods of a doubly periodic function $f(z)$, and if λ_1 and λ_2 are complex numbers, then λ_1 and λ_2 generate the periods of $f(z)$ if and only if there is a 2×2 matrix A with integer entries and with determinant ± 1 such that $A(\omega_1, \omega_2) = (\lambda_1, \lambda_2)$.

7. Let ω_1 and ω_2 be two complex numbers that do not lie on the same line through 0. Let $k \geq 3$. Show that the series

$$\sum_{m,n=-\infty}^{\infty} \frac{1}{(z - (m\omega_1 + n\omega_2))^k}$$

converges uniformly on any bounded subset of the complex plane to a doubly periodic meromorphic function $f(z)$, whose periods are generated by ω_1 and ω_2. *Strategy.* Show that the number of periods in any annulus $\{N \leq |z| \leq N+1\}$ is bounded by CN for some constant C.

6. Fourier Series

A **complex Fourier series** is a two-tailed series of the form

$$(6.1) \quad \sum_{k=-\infty}^{\infty} c_k e^{ik\theta} = \cdots + c_{-2}e^{-2i\theta} + c_{-1}e^{-i\theta} + c_0 + c_1 e^{i\theta} + c_2 e^{2i\theta} + \cdots .$$

Laurent expansions are intimately related to Fourier series. If the Laurent series

$$f(z) = \sum_{k=-\infty}^{\infty} a_k z^k$$

converges uniformly on the circle $\{|z| = r\}$, then

$$f(re^{i\theta}) = \sum_{k=-\infty}^{\infty} a_k r^k e^{ik\theta}$$

is the Fourier series expansion of $f(re^{i\theta})$, regarded as a function of θ. The Fourier coefficients of the expansion are the coefficients $c_k = a_k r^k$.

Suppose the series (6.1) converges uniformly to a function $f(e^{i\theta})$,

$$f(e^{i\theta}) = \sum_{j=-\infty}^{\infty} c_j e^{ij\theta}.$$

We can capture the coefficients of the series by multiplying by the exponential function $e^{-ik\theta}$ and integrating with respect to the probability measure $d\theta/2\pi$. The orthogonality relations for the exponential functions,

(6.2)
$$\int_{-\pi}^{\pi} e^{ij\theta} e^{-ik\theta} \frac{d\theta}{2\pi} = \begin{cases} 1, & j = k, \\ 0, & j \neq k, \end{cases}$$

then yield

$$\int_{-\pi}^{\pi} f\left(re^{i\theta}\right) e^{-ik\theta} \frac{d\theta}{2\pi} = \sum_{j=-\infty}^{\infty} c_j \int_{-\pi}^{\pi} e^{ij\theta} e^{-ik\theta} \frac{d\theta}{2\pi} = c_k.$$

This leads us to define the **Fourier coefficients** of any piecewise continuous function (or any integrable function) $f\left(e^{i\theta}\right)$ to be

(6.3)
$$c_k = \int_{-\pi}^{\pi} f\left(e^{i\theta}\right) e^{-ik\theta} \frac{d\theta}{2\pi}, \qquad -\infty < k < \infty,$$

and we associate to $f\left(e^{i\theta}\right)$ the Fourier series

(6.4)
$$f\left(e^{i\theta}\right) \sim \sum_{k=-\infty}^{\infty} c_k e^{ik\theta},$$

where the c_k's are defined by (6.3). We call $\sum c_k e^{ik\theta}$ the **Fourier series** of $f\left(e^{i\theta}\right)$. However, we face now two big problems. Does the Fourier series of $f\left(e^{i\theta}\right)$ converge? And if so, to what?

Example. Define $f\left(e^{i\theta}\right)$ to be -1 for $-\pi < \theta < 0$, and $+1$ for $0 < \theta < \pi$. The Fourier coefficient c_0 is the average value of $f\left(e^{i\theta}\right)$, which is 0. If $k \neq 0$, then

$$c_k = -\int_{-\pi}^{0} e^{-ik\theta} \frac{d\theta}{2\pi} + \int_{0}^{\pi} e^{-ik\theta} \frac{d\theta}{2\pi} = \frac{e^{-ik\theta}}{2\pi i k} \bigg|_{-\pi}^{0} - \frac{e^{-ik\theta}}{2\pi i k} \bigg|_{0}^{\pi}$$

$$= \frac{1}{2\pi i k} \left[1 - (-1)^k - (-1)^k + 1\right] = \frac{1}{\pi i k} \left[1 - (-1)^k\right].$$

This is 0 if k is even and $2/\pi i k$ if k is odd. Thus the complex Fourier series of $f\left(e^{i\theta}\right)$ is

$$f\left(e^{i\theta}\right) \sim \frac{2}{\pi i} \sum_{k \text{ odd}} \frac{1}{k} e^{ik\theta} = \cdots + \left(\frac{-2}{3\pi i}\right) e^{-3i\theta} + \left(\frac{-2}{\pi i}\right) e^{-i\theta}$$

$$+ \left(\frac{2}{\pi i}\right) e^{i\theta} + \left(\frac{2}{3\pi i}\right) e^{3i\theta} + \cdots.$$

If we combine the terms for $\pm k$, we obtain sine functions, and the series becomes

(6.5)
$$f\left(e^{i\theta}\right) \sim \frac{4}{\pi} \left(\sin\theta + \frac{1}{3}\sin 3\theta + \frac{1}{5}\sin 5\theta + \cdots\right).$$

The terms of the series are all zero at $\theta = 0$ and at $\theta = \pm\pi$. We will soon see that the series converges to $f\left(e^{i\theta}\right)$ for $0 < |\theta| < \pi$.

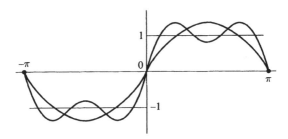

Theorem. *If $f\left(e^{i\theta}\right)$ is piecewise continuous (or more generally, square-integrable), with Fourier series $f\left(e^{i\theta}\right) \sim \sum c_k e^{ik\theta}$, then for $m, n \geq 0$ we have*

$$(6.6) \quad \sum_{k=-m}^{n} |c_k|^2 + \int_{-\pi}^{\pi} \left| f\left(e^{i\theta}\right) - \sum_{k=-m}^{n} c_k e^{ik\theta} \right|^2 \frac{d\theta}{2\pi} = \int_{-\pi}^{\pi} \left| f\left(e^{i\theta}\right) \right|^2 \frac{d\theta}{2\pi}.$$

This is established by writing

$$\left| f\left(e^{i\theta}\right) - \sum_{k=-m}^{n} c_k e^{ik\theta} \right|^2 = \left(f\left(e^{i\theta}\right) - \sum_{j=-m}^{n} c_j e^{ij\theta} \right) \left(\overline{f\left(e^{i\theta}\right)} - \sum_{k=-m}^{n} \overline{c_k} e^{-ik\theta} \right),$$

multiplying out the product on the right, and integrating both sides from $-\pi$ to π. The integrals featuring the products of exponentials for $j \neq k$ drop out, on account of the orthogonality relations (6.2). What remains on the right after we integrate is

$$\int_{-\pi}^{\pi} \left| f\left(e^{i\theta}\right) \right|^2 \frac{d\theta}{2\pi} - \sum \overline{c_k} \int_{-\pi}^{\pi} f\left(e^{i\theta}\right) e^{-ik\theta} \frac{d\theta}{2\pi}$$

$$- \sum c_j \int_{-\pi}^{\pi} \overline{f\left(e^{i\theta}\right)} e^{ij\theta} \frac{d\theta}{2\pi} + \sum |c_j|^2.$$

In view of formula (6.3), we recognize the second and third integrals appearing here as c_k and $\overline{c_j}$ respectively. Thus the sum is equal to

$$\int_{-\pi}^{\pi} \left| f\left(e^{i\theta}\right) \right|^2 \frac{d\theta}{2\pi} - \sum |c_k|^2 - \sum |c_j|^2 + \sum |c_j|^2.$$

If we cancel the two sums over j and move $\sum |c_k|^2$ to the other side, we obtain (6.6).

The identity (6.6) shows that the partial sums of the series $\sum |c_k|^2$ are bounded. Consequently, the series $\sum |c_k|^2$ converges, and from (6.6) we obtain the following estimate.

Theorem (Bessel's Inequality). *If $f\left(e^{i\theta}\right)$ is piecewise continuous (or more generally, square-integrable), with Fourier series $f\left(e^{i\theta}\right) \sim \sum c_k e^{ik\theta}$, then*

$$(6.7) \qquad \sum_{k=-\infty}^{\infty} |c_k|^2 \leq \int_{-\pi}^{\pi} \left|f\left(e^{i\theta}\right)\right|^2 \frac{d\theta}{2\pi}.$$

From Bessel's inequality it follows in particular that $c_k \to 0$ as $k \to \pm\infty$. With this observation in hand, we are now ready to state and prove our main theorem on pointwise convergence of Fourier series.

Theorem. *Suppose $f\left(e^{i\theta}\right)$ is piecewise continuous (or square integrable), with Fourier series $f\left(e^{i\theta}\right) \sim \sum c_k e^{ik\theta}$. If $f\left(e^{i\theta}\right)$ is differentiable at θ_0, then the Fourier series of $f\left(e^{i\theta}\right)$ converges to $f(e^{i\theta_0})$ at $\theta = \theta_0$,*

$$f(e^{i\theta_0}) = \sum_{-\infty}^{\infty} c_k e^{ik\theta_0} = \lim_{m,n\to\infty} \sum_{k=-m}^{n} c_k e^{ik\theta_0}.$$

Though Fourier series had been studied intensively for well over a century, this ingenious proof was discovered only relatively recently, by P. Chernoff in 1980.

We consider first the special case in which $\theta_0 = 0$ and $e^{i\theta_0} = 1$. Define $g\left(e^{i\theta}\right) = \left[f\left(e^{i\theta}\right) - f(1)\right] / \left(e^{i\theta} - 1\right)$. The differentiability of $f\left(e^{i\theta}\right)$ at $\theta = 0$ implies that $g\left(e^{i\theta}\right)$ has a limit as $\theta \to 0$. Consequently, $g\left(e^{i\theta}\right)$ is also piecewise continuous. Denote the Fourier coefficients of $g\left(e^{i\theta}\right)$ by b_k, so that $g\left(e^{i\theta}\right) \sim \sum b_k e^{ik\theta}$. Bessel's inequality for $g\left(e^{i\theta}\right)$ shows that $b_k \to 0$ as $k \to \pm\infty$. Now we compute the c_k's in terms of the b_k's. Since $f\left(e^{i\theta}\right) = g\left(e^{i\theta}\right)\left(e^{i\theta} - 1\right) + f(1)$, we have

$$c_k = \int_{-\pi}^{\pi} g\left(e^{i\theta}\right)\left(e^{i\theta} - 1\right)e^{-ik\theta}\frac{d\theta}{2\pi} + f(1)\int_{-\pi}^{\pi} e^{-ik\theta}\frac{d\theta}{2\pi}.$$

When we express these integrals as Fourier coefficients of $g\left(e^{i\theta}\right)$, we obtain $c_k = b_{k-1} - b_k$ if $k \neq 0$, and $c_0 = b_{-1} - b_0 + f(1)$. Hence the series $\sum c_k$ telescopes, and we obtain

$$\sum_{k=-m}^{n} c_k = f(1) + \sum_{k=-m}^{n} (b_{k-1} - b_k) = f(1) + b_{-m-1} - b_n,$$

which tends to $f(1)$ as $m, n \to +\infty$. This proves the theorem when $\theta_0 = 0$.

The case when θ_0 is arbitrary is reduced to the above special case by a change of variable. Consider the function $h\left(e^{i\theta}\right) = f\left(e^{i(\theta+\theta_0)}\right) \sim \sum a_k e^{ik\theta}$, which is piecewise continuous and which is differentiable at $\theta = 0$.

The Fourier coefficient a_k of $h\left(e^{i\theta}\right)$ is

$$a_k = \int_{-\pi}^{\pi} f\left(e^{i(\theta+\theta_0)}\right) e^{-ik\theta} \frac{d\theta}{2\pi} = \int_{-\pi}^{\pi} f\left(e^{ik\varphi}\right) e^{-ik\varphi} e^{ik\theta_0} \frac{d\varphi}{2\pi} = c_k e^{ik\theta_0}.$$

Thus the Fourier series of $f\left(e^{i\theta}\right)$ evaluated at $\theta = \theta_0$ is $\sum c_k e^{ik\theta_0} = \sum a_k$, which is the same as the Fourier series of $h\left(e^{i\theta}\right)$ evaluated at $\theta = 0$. We have shown that the latter converges to $h(1) = f(e^{i\theta_0})$. This completes the proof.

Example. Consider the series in (6.5). By the convergence theorem, the series converges to $+1$ for $0 < \theta < \pi$ and to -1 for $-\pi < \theta < 0$. The convergence theorem does not give any information about the points $\theta = 0$ and $\theta = \pm\pi$, where the function is discontinuous. However, we see directly that the series converges to 0 at these points.

We aim now to establish a result on uniform convergence of Fourier series. We begin by showing that the Fourier series of a smooth function can be differentiated term by term.

Theorem. *Suppose $f\left(e^{i\theta}\right)$ is a continuously differentiable function of θ, with Fourier series $f\left(e^{i\theta}\right) \sim \sum c_k e^{ik\theta}$. Then the Fourier series of the derivative of $f\left(e^{i\theta}\right)$ is obtained by differentiating term by term,*

$$\frac{d}{d\theta} f\left(e^{i\theta}\right) \sim \sum ik c_k e^{ik\theta}.$$

To check this, we simply write down the expression for the kth Fourier coefficient of the derivative and we integrate by parts. The Fourier coefficient of the derivative is

$$\int_{-\pi}^{\pi} e^{-ik\theta} \frac{d}{d\theta} f\left(e^{i\theta}\right) \frac{d\theta}{2\pi} = e^{-ik\theta} f\left(e^{i\theta}\right)\Big|_{-\pi}^{\pi} + ik \int_{-\pi}^{\pi} f\left(e^{i\theta}\right) e^{-ik\theta} \frac{d\theta}{2\pi} = ik c_k.$$

If we combine this theorem with Bessel's inequality, we can show that the Fourier coefficients of an n-times continuously differentiable function tend to zero at least as rapidly as the nth power of $1/k$. The smoother the function, the more rapidly its Fourier coefficients decay.

Corollary. *If $f\left(e^{i\theta}\right)$ is an n-times continuously differentiable function of θ, with Fourier series $f\left(e^{i\theta}\right) \sim \sum c_k e^{ik\theta}$, then $\sum_{k=-\infty}^{\infty} k^{2n} |c_k|^2 < \infty$. Further, $k^n c_k \to 0$ as $k \to \pm\infty$.*

To see this, observe that the nth derivative of $f\left(e^{i\theta}\right)$ has Fourier series $\sum (ik)^n c_k e^{ik\theta}$ and apply Bessel's inequality. The second statement of the corollary follows immediately from the convergence of the series, since the terms of the series then tend to 0.

In particular, if $f\left(e^{i\theta}\right)$ is a twice continuously differentiable function of θ, then $k^2 c_k \to 0$ as $k \to \pm\infty$. Hence the series $\sum |c_k|$ converges, by comparison with $\sum 1/k^2$. By the Weierstrass M-test, the series $\sum c_k e^{ik\theta}$ then converges uniformly in θ. By the pointwise convergence theorem proved above, the sum of the series is $f\left(e^{i\theta}\right)$. We have proved the following.

Theorem. *Suppose $f\left(e^{i\theta}\right)$ is a twice continuously differentiable function of θ. Then the Fourier series of $f\left(e^{i\theta}\right)$ converges to $f\left(e^{i\theta}\right)$ uniformly in θ.*

Example. Define $g(\theta) = \theta^4 - 2\pi^2\theta^2$, $-\pi \le \theta \le \pi$. We check that $g(-\pi) = g(\pi)$, $g'(-\pi) = g'(\pi)$, and $g''(-\pi) = g''(\pi)$. Consequently, if we set $f\left(e^{i\theta}\right) = g(\theta)$, we obtain a twice continuously differentiable function of $e^{i\theta}$. By the theorem, the Fourier series of $f\left(e^{i\theta}\right)$ converges uniformly in θ. The explicit calculation of the Fourier series is left to the exercises (Exercise 4).

Example. The Fourier series in (6.5) does not converge uniformly, since the sum is not continuous.

Exercises for VI.6

1. Consider the continuous function $f(e^{i\theta}) = |\theta|$, $-\pi \le \theta \le \pi$. Find the complex Fourier series of $f\left(e^{i\theta}\right)$ and show that it can be expressed as a cosine series. Sketch the graphs of the first three partial sums of the cosine series. Discuss the convergence of the series. Does it converge uniformly? *Partial answer.* The cosine series is

$$|\theta| = \frac{\pi}{2} - \frac{4}{\pi}\left(\cos\theta + \frac{1}{3^2}\cos 3\theta + \frac{1}{5^2}\cos 5\theta + \cdots\right).$$

2. Let $f(e^{i\theta}) = \theta$, $-\pi < \theta \le \pi$ (the principal value of the argument). Find the complex Fourier series of $f\left(e^{i\theta}\right)$ and the sine series of $f\left(e^{i\theta}\right)$. Show that the complex Fourier series diverges at $\theta = \pm\pi$, while the sine series converges at $\pm\pi$. Differentiate the complex Fourier series term by term and determine where the differentiated series converges.

3. Consider the continuous function $f(e^{i\theta}) = \theta^2$, $-\pi \le \theta \le \pi$. Find the complex Fourier series of $f\left(e^{i\theta}\right)$ and show that it can be expressed as a cosine series. Discuss the convergence of the series. Does it converge uniformly? By substituting $\theta = 0$, show that

$$\frac{\pi^2}{12} = 1 - \frac{1}{2^2} + \frac{1}{3^2} - \frac{1}{4^2} + \cdots.$$

4. Consider the continuous function $f(e^{i\theta}) = \theta^4 - 2\pi^2\theta^2$, $-\pi \le \theta \le \pi$. Find the complex Fourier series of $f\left(e^{i\theta}\right)$ and show that it can be

expressed as a cosine series. Relate the Fourier series to the series of the function in Exercise 3.

5. Show that if $\sum c_k$ converges absolutely, then $\sum c_k e^{ik\theta}$ converges absolutely for each θ, and the series converges uniformly for $-\pi \leq \theta \leq \pi$.

6. Show that any function $f\left(e^{i\theta}\right)$ on the unit circle with absolutely convergent Fourier series has the form $f\left(e^{i\theta}\right) = g\left(e^{i\theta}\right) + \overline{h\left(e^{i\theta}\right)}$, where $g(z)$ and $h(z)$ are continuous functions on the unit circle that extend continuously to be analytic on the open unit disk.

7. If $f\left(e^{i\theta}\right) \sim \sum c_k e^{ik\theta}$, and the series converges uniformly to $f\left(e^{i\theta}\right)$, then

$$\int_{-\pi}^{\pi} \left|f\left(e^{i\theta}\right)\right|^2 \frac{d\theta}{2\pi} = \sum_{k=-\infty}^{\infty} |c_k|^2 .$$

Remark. This is called **Parseval's identity**. Formula (6.6) shows that Parseval's identity holds for a function $f\left(e^{i\theta}\right)$ if and only if the partial sums of the Fourier series of $f\left(e^{i\theta}\right)$ converge to $f\left(e^{i\theta}\right)$ in the sense of "mean-square" or "L^2-approximation."

8. By applying Parseval's identity to the piecewise constant function with series (6.5), show that

$$\frac{\pi^2}{8} = 1 + \frac{1}{3^2} + \frac{1}{5^2} + \frac{1}{7^2} + \cdots .$$

Use this identity and some algebraic manipulation to show that

$$\frac{\pi^2}{6} = 1 + \frac{1}{2^2} + \frac{1}{3^2} + \frac{1}{4^2} + \cdots .$$

9. By applying Parseval's identity to the function of Exercise 1, show that

$$\frac{\pi^4}{96} = 1 + \frac{1}{3^4} + \frac{1}{5^4} + \frac{1}{7^4} + \cdots .$$

Use this identity and some algebraic manipulation to show that

$$\frac{\pi^4}{90} = 1 + \frac{1}{2^4} + \frac{1}{3^4} + \frac{1}{4^4} + \cdots .$$

10. If $f(z)$ is analytic in some annulus containing the unit circle $|z| = 1$, with Laurent expansion $\sum a_k z^k$, then

$$\frac{1}{2\pi} \oint_{|z|=1} |f(z)|^2 |dz| = \sum_{k=-\infty}^{\infty} |a_k|^2 .$$

11. Let $f\left(e^{i\theta}\right)$ be a continuous function on the unit circle, with Fourier series $\sum c_k e^{ik\theta}$. Show that $f\left(e^{i\theta}\right)$ extends to be analytic on some annulus containing the unit circle if and only if there exist $r < 1$ and $C > 0$ such that $|c_k| \leq C r^{|k|}$ for $-\infty < k < \infty$.

12. Using the convergence theorem for Fourier series, prove that every continuous function on the unit circle in the complex plane can be approximated uniformly there by trigonometric polynomials, that is, by finite linear combinations of exponentials $e^{ik\theta}$, $-\infty < k < \infty$. *Strategy.* First approximate $f\left(e^{i\theta}\right)$ by a smooth function.

13. Let D be a domain bounded by a smooth boundary curve of length 2π. We parametrize the boundary of D by arc length s, so the boundary is given by a smooth periodic function $\gamma(s)$, $0 \leq s \leq 2\pi$. Let $\sum c_k e^{iks}$ be the Fourier series of $\gamma(s)$. (a) Show that $\sum k^2 |c_k|^2 = 1$. *Hint.* Apply Parseval's identity to $\gamma'(s)$ and use $|\gamma'(s)| = 1$ for a curve parameterized by arc length. (b) Show that the area of D is $\pi \sum k |c_k|^2$. *Hint.* Use Exercise IV.1.4. (c) Show that the area of D is $\leq \pi$, with equality if and only if D is a disk. *Remark.* This proves the **isoperimetric theorem**: *Among all smooth closed curves of a given length, the curve that surrounds the largest area is a circle.*

14. Show that

$$\int_{-\pi}^{\pi} \left| f\left(e^{i\theta}\right) - \sum_{k=-m}^{n} b_k e^{ik\theta} \right|^2 \frac{d\theta}{2\pi}$$

$$= \int_{-\pi}^{\pi} \left| f\left(e^{i\theta}\right) - \sum_{k=-m}^{n} c_k e^{ik\theta} \right|^2 \frac{d\theta}{2\pi} + \sum_{k=-m}^{n} |b_k - c_k|^2,$$

for any choice of complex numbers b_k, $-m \leq k \leq n$. *Remark.* This shows that the best mean-square approximant to $f\left(e^{i\theta}\right)$ by exponential sums $\sum_{-m}^{n} b_k e^{ik\theta}$, for fixed m and n, is the corresponding partial sum of the Fourier series.

15. Show that a continuously differentiable function on the unit circle has an absolutely convergent Fourier series. *Strategy.* Write the Fourier coefficients c_k of $f\left(e^{i\theta}\right)$ as $a_k b_k$, where $a_k = 1/ik$ and b_k is the Fourier coefficient of the derivative. Use Bessel's inequality and the Cauchy-Schwarz inequality $\left| \sum a_k \overline{\beta_k} \right| \leq \sqrt{\sum |a_k|^2} \sqrt{\sum |\beta_k|^2}$.

16. Let $f\left(e^{i\theta}\right)$ be a continuous function on the unit circle. Suppose that $f\left(e^{i\theta}\right)$ is piecewise continuously differentiable, in the sense that it has a continuous derivative except at a finite number of points, at each of which the derivative has limits from the left and from the right. Show that the Fourier series of $f\left(e^{i\theta}\right)$ is absolutely convergent. *Strategy.* Cancel the discontinuities of the derivative

using translates of the function in Exercise 3, whose Fourier series is absolutely convergent.

17. Let $f\left(e^{i\theta}\right)$ be piecewise continuously differentiable, in the sense that it is continuously differentiable except at a finite number of points, at each of which both the function and its derivative have limits from the left and from the right. Show that the Fourier series of $f\left(e^{i\theta}\right)$ converges at each point, to $f\left(e^{i\theta}\right)$ if the function is continuous at $e^{i\theta}$, and otherwise to the average of the limits of $f\left(e^{i\theta}\right)$ from the left and from the right. *Strategy.* Show that $f\left(e^{i\theta}\right) = f_1\left(e^{i\theta}\right) + \sum b_j h_j\left(e^{i\theta}\right)$, where $f_1\left(e^{i\theta}\right)$ satisfies the hypotheses of Exercise 15, and each $h_j\left(e^{i\theta}\right)$ is obtained from the function of Exercise 2 by a change of variable $\theta \mapsto \theta - \theta_j$.

VII

The Residue Calculus

Section 1 is devoted to the residue theorem and to techniques for evaluating residues. In the remaining sections we apply the residue theorem to evaluate various real integrals. This material provides a good training ground for the techniques of complex integration. The student who is anxious to move on can skip the final several sections of the chapter at first reading.

1. The Residue Theorem

Suppose z_0 is an isolated singularity of $f(z)$ and that $f(z)$ has Laurent series

$$f(z) = \sum_{n=-\infty}^{\infty} a_n(z - z_0)^n, \qquad 0 < |z - z_0| < \rho.$$

We define the **residue** of $f(z)$ at z_0 to be the coefficient a_{-1} of $1/(z - z_0)$ in this Laurent expansion,

$$(1.1) \qquad \operatorname{Res}[f(z), z_0] = a_{-1} = \frac{1}{2\pi i} \oint_{|z-z_0|=r} f(z) \, dz,$$

where r is any fixed radius satisfying $0 < r < \rho$.

Example. The definition yields immediately

$$\operatorname{Res}\left[\frac{1}{z}, 0\right] = 1, \qquad \operatorname{Res}\left[\frac{1}{(z - z_0)^2}, z_0\right] = 0.$$

Example. The partial fractions decomposition

$$\frac{1}{z^2 + 1} = \frac{1}{2i}\left[\frac{1}{z - i} - \frac{1}{z + i}\right] = \frac{1}{2i}\frac{1}{z - i} + [\text{analytic at } i]$$

yields

$$\operatorname{Res}\left[\frac{1}{z^2 + 1}, i\right] = \frac{1}{2i}.$$

The following residue theorem provides an important tool for evaluating complex line integrals. It extends Cauchy's theorem by allowing for a finite number of singularities inside the contour of integration. When there are no singularities present, the residue theorem reduces to Cauchy's theorem.

Theorem (Residue Theorem). *Let D be a bounded domain in the complex plane with piecewise smooth boundary. Suppose that $f(z)$ is analytic on $D \cup \partial D$, except for a finite number of isolated singularities $z_1, \dots, z_m$ in D. Then*

$$(1.2) \qquad \int_{\partial D} f(z)\,dz \;=\; 2\pi i \sum_{j=1}^{m} \mathrm{Res}\,[\,f(z), z_j\,].$$

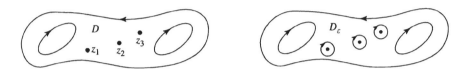

To see this, let D_ε be the domain obtained from D by punching out small disks U_j centered at z_j of radius ε. The formula (1.1) for the residue at z_j yields

$$\int_{\partial U_j} f(z)\,dz \;=\; 2\pi i \,\mathrm{Res}\,[\,f(z), z_j\,].$$

(This can be regarded as a special case of the residue theorem, for the domain U_j and a function with a singularity at z_j.) By Cauchy's theorem,

$$0 \;=\; \int_{\partial D_\varepsilon} f(z)\,dz \;=\; \int_{\partial D} f(z)\,dz \;-\; \sum_{j=1}^{m} \int_{\partial U_j} f(z)\,dz.$$

If we combine these two identities, we obtain (1.2).

We give four useful rules for calculating residues.

Rule 1. *If $f(z)$ has a simple pole at z_0, then*

$$\mathrm{Res}\,[\,f(z), z_0\,] \;=\; \lim_{z \to z_0} (z - z_0)f(z).$$

In this case the Laurent series of $f(z)$ is

$$f(z) \;=\; \frac{a_{-1}}{z - z_0} \;+\; [\text{analytic at } z_0]\,,$$

from which the rule follows immediately. Note that once we obtain an expression for $(z - z_0)f(z)$ as an analytic function, the limit is evaluated by simply plugging $z = z_0$ into the expression.

Example. From Rule 1 we have

$$\text{Res}\left[\frac{1}{z^2+1}, i\right] = \lim_{z\to i}\frac{z-i}{z^2+1} = \lim_{z\to i}\frac{1}{z+i} = \frac{1}{z+i}\bigg|_{z=i} = \frac{1}{2i}.$$

This method of obtaining the residue is faster than finding first the partial fractions decomposition. Rule 4 below is faster yet.

Rule 2. If $f(z)$ has a double pole at z_0, then

$$\text{Res}\left[f(z), z_0\right] = \lim_{z\to z_0}\frac{d}{dz}[(z-z_0)^2 f(z)].$$

In this case the Laurent expansion is

$$f(z) = \frac{a_{-2}}{(z-z_0)^2} + \frac{a_{-1}}{z-z_0} + a_0 + \cdots.$$

Thus

$$(z-z_0)^2 f(z) = a_{-2} + a_{-1}(z-z_0) + a_0(z-z_0)^2 + \cdots.$$

If we differentiate and then plug in $z = z_0$, we obtain Rule 2. Note that Rule 2 can be regarded as the formula for the coefficient of $z - z_0$ in the power series expansion of the analytic function $(z-z_0)^2 f(z)$.

Example. The function $1/(z^2+1)^2$ has double poles at $\pm i$. The residue at i is given by

$$\text{Res}\left[\frac{1}{(z^2+1)^2}, i\right] = \lim_{z\to i}\frac{d}{dz}\frac{1}{(z+i)^2} = \frac{-2}{(z+i)^3}\bigg|_{z=i} = \frac{1}{4i}.$$

Rule 3. If $f(z)$ and $g(z)$ are analytic at z_0, and if $g(z)$ has a simple zero at z_0, then

$$\text{Res}\left[\frac{f(z)}{g(z)}, z_0\right] = \frac{f(z_0)}{g'(z_0)}.$$

In this case $f(z)/g(z)$ has at most a simple pole at z_0. If we use Rule 1 and the definition of the derivative, we obtain for the residue

$$\lim_{z\to z_0}(z-z_0)\frac{f(z)}{g(z)} = \lim_{z\to z_0}\frac{f(z)}{(g(z)-g(z_0))/(z-z_0)} = \frac{f(z_0)}{g'(z_0)}.$$

Example. The partial fractions decomposition of the function $z^3/(z^2+1)$ was found in Section VI.4 to be

$$\frac{z^3}{z^2+1} = z - \frac{1}{2}\frac{1}{z-i} - \frac{1}{2}\frac{1}{z+i}.$$

From this we read off the residues at $\pm i$ to be both $-\frac{1}{2}$. The residues can also be obtained directly using Rule 3. The residue at i is given by

$$\text{Res}\left[\frac{z^3}{z^2+1}, i\right] = \frac{z^3}{2z}\bigg|_{z=i} = \frac{i^3}{2i} = -\frac{1}{2}.$$

The following special case of Rule 3 is particularly useful.

Rule 4. If $g(z)$ is analytic and has a simple zero at z_0, then

$$\text{Res}\left[\frac{1}{g(z)}, z_0\right] = \frac{1}{g'(z_0)}.$$

Example. If we apply Rule 4 to $1/(z^2+1)$, we obtain the residue even faster than before,

$$\text{Res}\left[\frac{1}{z^2+1}, i\right] = \frac{1}{2z}\bigg|_{z=i} = \frac{1}{2i}.$$

Exercises for VII.1

1. Evaluate the following residues.

(a) $\text{Res}\left[\dfrac{1}{z^2+4}, 2i\right]$ (d) $\text{Res}\left[\dfrac{\sin z}{z^2}, 0\right]$ (g) $\text{Res}\left[\dfrac{z}{\text{Log } z}, 1\right]$

(b) $\text{Res}\left[\dfrac{1}{z^2+4}, -2i\right]$ (e) $\text{Res}\left[\dfrac{\cos z}{z^2}, 0\right]$ (h) $\text{Res}\left[\dfrac{e^z}{z^5}, 0\right]$

(c) $\text{Res}\left[\dfrac{1}{z^5-1}, 1\right]$ (f) $\text{Res}\left[\cot z, 0\right]$ (i) $\text{Res}\left[\dfrac{z^n+1}{z^n-1}, e^{2\pi ki/n}\right]$

2. Calculate the residue at each isolated singularity in the complex plane of the following functions.

(a) $e^{1/z}$ (b) $\tan z$ (c) $\dfrac{z}{(z^2+1)^2}$ (d) $\dfrac{1}{z^2+z}$

3. Evaluate the following integrals, using the residue theorem.

(a) $\displaystyle\oint_{|z|=1} \frac{\sin z}{z^2}\, dz$ (c) $\displaystyle\oint_{|z|=2} \frac{z}{\cos z}\, dz$ (e) $\displaystyle\oint_{|z-1|=1} \frac{1}{z^8-1}\, dz$

(b) $\displaystyle\oint_{|z|=2} \frac{e^z}{z^2-1}\, dz$ (d) $\displaystyle\oint_{|z|=1} \frac{z^4}{\sin z}\, dz$ (f) $\displaystyle\oint_{|z-1/2|=3/2} \frac{\tan z}{z}\, dz$

4. Suppose $P(z)$ and $Q(z)$ are polynomials such that the zeros of $Q(z)$ are simple zeros at the points $z_1, \ldots, z_m$, and $\deg P(z) < \deg Q(z)$.

Show that the partial fractions decomposition of $P(z)/Q(z)$ is given by

$$\frac{P(z)}{Q(z)} = \sum_{j=1}^{m} \frac{P(z_j)}{Q'(z_j)} \frac{1}{z - z_j}.$$

5. Let $f(z)$ be a meromorphic function on the complex plane that is doubly periodic, and suppose that none of the poles of $f(z)$ lie on the boundary of the period parallelogram P constructed in Section VI.5. By integrating $f(z)$ around the boundary of P, show that the sum of the residues at the poles of $f(z)$ in P is zero. Conclude that there is no doubly periodic meromorphic function with only one pole, a simple pole, in the period parallelogram.

6. Consider the integral

$$\int_{\partial D_R} \frac{e^{\pi i(z-1/2)^2}}{1 - e^{-2\pi iz}} \, dz,$$

where D_R is the parallelogram with vertices $\pm(\frac{1}{2}) \pm (1+i)R$. (a) Use the residue theorem to show that the integral is $(1+i)/\sqrt{2}$. (b) By parameterizing the sides of the parallelogram, show that the integral tends to

$$(1+i) \int_{-\infty}^{\infty} e^{-2\pi t^2} \, dt$$

as $R \to \infty$. (c) Use (a) and (b) to show that

$$\int_{-\infty}^{\infty} e^{-s^2} \, ds = \sqrt{\pi}.$$

2. Integrals Featuring Rational Functions

The prototype for evaluation of an integral by means of contour integration is the derivation of the formula

(2.1) $$\int_{-\infty}^{\infty} \frac{dx}{1 + x^2} = \pi.$$

This integral can, of course, be evaluated using the usual integration formula featuring the inverse tangent function. To evaluate it using contour integration, we proceed as follows.

Let D_R be the half-disk in the upper half-plane bounded by the interval $[-R, R]$ on the real axis and the semicircular contour Γ_R of radius R in the upper half-plane. The function $1/(1 + z^2)$ has one pole in D_R, a simple

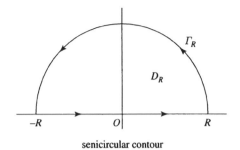

senicircular contour

pole at i with residue $1/2i$. The residue theorem yields

$$\int_{\partial D_R} \frac{dz}{1+z^2} \;=\; 2\pi i \, \mathrm{Res}\left[\frac{1}{1+z^2}, i\right] \;=\; 2\pi i \cdot \frac{1}{2i} \;=\; \pi.$$

Now,

$$\int_{\partial D_R} \frac{dz}{1+z^2} \;=\; \int_{-R}^{R} \frac{dx}{1+x^2} \;+\; \int_{\Gamma_R} \frac{dz}{1+z^2}.$$

On Γ_R we have

$$|1/(1+z^2)| \;\leq\; 1/(R^2-1) \;\sim\; 1/R^2,$$

while the length of Γ_R is πR. By the ML-estimate we have

$$\left|\int_{\Gamma_R} \frac{dz}{1+z^2}\right| \;\leq\; \frac{1}{R^2-1} \cdot \pi R \;\sim\; \frac{1}{R},$$

which tends to 0 as $R \to \infty$. Hence

$$\lim_{R\to\infty} \int_{-R}^{R} \frac{dx}{1+x^2} \;=\; \pi,$$

which is (2.1).

The same technique can be used to evaluate integrals of the form

$$(2.2) \qquad\qquad \int_{-\infty}^{\infty} \frac{P(x)}{Q(x)} \, dx,$$

where $P(z)$ and $Q(z)$ are polynomials, and $Q(z)$ has no zeros on the real axis. For convergence of the integral, we require that

$$(2.3) \qquad\qquad \deg Q(z) \;\geq\; \deg P(z) \;+\; 2.$$

The integral is evaluated by integrating $P(z)/Q(z)$ around the boundary of a half-disk in the upper half-plane, as above, and letting the radius tend to ∞. This yields the formula

$$(2.4) \qquad\qquad \int_{-\infty}^{\infty} \frac{P(x)}{Q(x)} \, dx \;=\; 2\pi i \sum \mathrm{Res}\left[\frac{P(z)}{Q(z)}, z_j\right],$$

summed over the poles z_j of $P(z)/Q(z)$ in the upper half-plane.

The same contour can be used to evaluate the integrals of rational functions times trigonometric functions. The typical integral has the form

$$\int_{-\infty}^{\infty} \frac{P(x)}{Q(x)} \cos(ax)\, dx\,,$$

where the polynomials $P(z)$ and $Q(z)$ have real coefficients and satisfy (2.3). To use the semicircular contour, we cannot use the function $\cos(az)$, because it behaves too badly in the upper half-plane. (As we have seen, the cosine function is essentially a hyperbolic cosine on the imaginary axis, which grows exponentially fast.) To obtain a tractable integral, we resort to a trick. The trick is to substitute e^{iz} for the cosine function in the contour integral, and to recover the cosine integral at the end by taking real parts. Since $|e^{i(x+iy)}| = e^{-y}$, the exponential function e^{iz} is bounded by 1 in the upper half-plane,

$$|e^{iz}| \leq 1, \qquad \operatorname{Im}(z) \geq 0.$$

Example. We show by contour integration that

$$(2.5) \qquad \int_{-\infty}^{\infty} \frac{\cos(ax)}{1+x^2}\, dx \;=\; \pi e^{-a}, \qquad a > 0.$$

Again we let D_R be the half-disk in the upper half-plane bounded by the interval $[-R, R]$ on the real axis and the semicircular contour Γ_R of radius R in the upper half-plane. This time we integrate the function $e^{iaz}/(1 + z^2)$ over the boundary of D_R. The function has only one pole in the upper half-plane, a simple pole at i, with residue calculated by Rule 3 to be

$$\operatorname{Res}\left[\frac{e^{iaz}}{1+z^2}, i\right] \;=\; \left.\frac{e^{iaz}}{2z}\right|_{z=i} \;=\; \frac{e^{-a}}{2i}.$$

Thus

$$\int_{\partial D_R} \frac{e^{iaz}}{1+z^2}\, dz \;=\; 2\pi i \cdot \frac{e^{-a}}{2i} \;=\; \pi e^{-a}.$$

Since $|e^{iaz}| \leq 1$ in the upper half-plane, the ML-estimate yields

$$\left| \int_{\Gamma_R} \frac{e^{iaz}}{1+z^2}\, dz \right| \;\leq\; \frac{1}{R^2-1} \cdot \pi R \;\sim\; \frac{1}{R}.$$

Again we have

$$\int_{\partial D_R} \frac{e^{iaz}}{1+z^2}\, dz \;=\; \int_{-R}^{R} \frac{e^{iax}}{1+x^2}\, dx + \int_{\Gamma_R} \frac{e^{iaz}}{1+z^2}\, dz\,.$$

Passing to the limit as $R \to \infty$, we obtain

$$\int_{-\infty}^{\infty} \frac{e^{iax}}{1+x^2}\, dx \;=\; \pi e^{-a}, \qquad a > 0.$$

Now we take the real parts of the integral, and we obtain (2.5). Note that if we take the imaginary part of the integral, we obtain

$$\int_{-\infty}^{\infty} \frac{\sin(ax)}{1+x^2} \, dx = 0,$$

which is no surprise, since the integrand is an odd function.

Exercises for VII.2

1. Show using residue theory that

$$\int_{-\infty}^{\infty} \frac{dx}{x^2 + a^2} = \frac{\pi}{a}, \qquad a > 0.$$

Remark. Check the result by evaluating the integral directly, using the arctangent function.

2. Show using residue theory that

$$\int_{-\infty}^{\infty} \frac{dx}{(x^2 + a^2)^2} = \frac{\pi}{2a^3}.$$

Remark. Check the result by differentiating the formula in the preceding exercise with respect to the parameter.

3. Show using residue theory that

$$\int_{-\infty}^{\infty} \frac{x^2 \, dx}{(x^2 + 1)^2} = \frac{\pi}{2}.$$

Remark. Check the result by combining the preceding two exercises.

4. Using residue theory, show that $\displaystyle\int_{-\infty}^{\infty} \frac{dx}{x^4 + 1} = \frac{\pi}{\sqrt{2}}.$

5. Using residue theory, show that $\displaystyle\int_{0}^{\infty} \frac{x^2}{x^4 + 1} \, dx = \frac{\pi}{2\sqrt{2}}.$

6. Show that $\displaystyle\int_{-\infty}^{\infty} \frac{x}{(x^2 + 2x + 2)(x^2 + 4)} \, dx = -\frac{\pi}{10}.$

7. Show that

$$\int_{-\infty}^{\infty} \frac{\cos(ax)}{x^4 + 1} \, dx = \frac{\pi}{\sqrt{2}} e^{-a/\sqrt{2}} \left(\cos \frac{a}{\sqrt{2}} + \sin \frac{a}{\sqrt{2}} \right), \qquad a > 0.$$

8. Show that

$$\int_{-\infty}^{\infty} \frac{\cos x}{(1 + x^2)^2} \, dx = \frac{\pi}{e}.$$

9. Show that

$$\int_{-\infty}^{\infty} \frac{\sin^2 x}{x^2 + 1}\, dx \;=\; \frac{\pi}{2}\left[1 - \frac{1}{e^2}\right].$$

10. Show that

$$\int_{-\infty}^{\infty} \frac{\cos(ax)}{x^2 + b^2}\, dx \;=\; \frac{\pi e^{-|a|b}}{b}, \qquad -\infty < a < \infty, \; b > 0.$$

For which complex values of the parameters a and b does the integral exist? Where does the integral depend analytically on the parameters?

11. Evaluate $\displaystyle\int_{-\infty}^{\infty} \frac{\cos x}{(x^2 + a^2)(x^2 + b^2)}\, dx$. Indicate the range of the parameters a and b.

12. Let $Q(z)$ be a polynomial of degree m with no zeros on the real line, and let $f(z)$ be a function that is analytic in the upper half-plane and across the real line. Suppose there is $b < m - 1$ such that $|f(z)| \le |z|^b$ for z in the upper half-plane, $|z| > 1$. Show that

$$\int_{-\infty}^{\infty} \frac{f(x)}{Q(x)}\, dx \;=\; 2\pi i \sum \operatorname{Res}\left[\frac{f(z)}{Q(z)}, z_j\right],$$

summed over the zeros z_j of $Q(z)$ in the upper half-plane.

3. Integrals of Trigonometric Functions

We have seen how complex contour integrals along a curve can be converted to garden-variety integrals by parameterizing the curve. Some definite integrals can be evaluated through the reverse process, by converting them to complex contour integrals and using the residue theorem. To illustrate this, we show how the integral

$$(3.1) \qquad\qquad \int_0^{2\pi} \frac{d\theta}{a + \cos\theta}, \qquad a > 1,$$

can be evaluated using residue calculus. The idea is to convert this to a complex contour integral around the unit circle. The usual parameterization $z = e^{i\theta}$ gives $dz = ie^{i\theta}\, d\theta = iz\, d\theta$, and we have

$$d\theta \;=\; \frac{dz}{iz}.$$

The trigonometric functions are easily expressible in terms of z on the unit circle. For $\cos\theta$ and $\sin\theta$ on the unit circle we have

$$\cos\theta = \frac{e^{i\theta} + e^{-i\theta}}{2} = \frac{z + 1/z}{2},$$

$$\sin\theta = \frac{e^{i\theta} - e^{-i\theta}}{2i} = \frac{z - 1/z}{2i}.$$

If we substitute the expression for $\cos\theta$ into the integral (3.1), we obtain

$$\int_0^{2\pi} \frac{d\theta}{a + \cos\theta} = \oint_{|z|=1} \frac{1}{a + \frac{1}{2}(z + 1/z)} \frac{dz}{iz} = \frac{2}{i} \oint_{|z|=1} \frac{dz}{z^2 + 2az + 1}.$$

The poles of the integrand are the two zeros of $z^2 + 2az + 1$, which are $-a \pm \sqrt{a^2 - 1}$. Only one of these roots is inside the unit circle, at $z_0 = -a + \sqrt{a^2 - 1}$. The residue at z_0 is calculated by Rule 3 above to be

$$\operatorname{Res}\left[\frac{1}{z^2 + 2az + 1}, z_0 \right] = \left. \frac{1}{2z + 2a} \right|_{z=z_0} = \frac{1}{2\sqrt{a^2 - 1}}.$$

Thus by the residue theorem

$$(3.2) \qquad \int_0^{2\pi} \frac{d\theta}{a + \cos\theta} = \frac{2}{i} \cdot 2\pi i \cdot \frac{1}{2\sqrt{a^2 - 1}} = \frac{2\pi}{\sqrt{a^2 - 1}}.$$

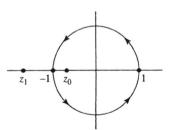

unit circle contour

Now consider the integral (3.1) with a replaced by a complex parameter w. We claim that

$$(3.3) \qquad \int_0^{2\pi} \frac{d\theta}{w + \cos\theta} = \frac{2\pi}{\sqrt{w^2 - 1}}, \qquad w \in \mathbb{C}\backslash[-1, 1],$$

where the right-hand side of the identity corresponds to the branch of $\sqrt{w^2 - 1}$ that is positive on the real interval $(1, +\infty)$. Indeed, the functions appearing on each side of the identity (3.3) are analytic for w in the slit plane $\mathbb{C}\backslash[-1, 1]$. The identity (3.2) shows that these two analytic functions coincide for $w \in (1, +\infty)$. By the uniqueness principle, the two analytic functions coincide for all $w \in \mathbb{C}\backslash[-1, 1]$. This establishes (3.3) also for complex values of the parameter w, except for w in the interval $[-1, 1]$, where the integral does not converge.

Exercises for VII.3

1. Show using residue theory that

$$\int_0^{2\pi} \frac{\cos\theta}{2 + \cos\theta}\, d\theta = 2\pi\left(1 - \frac{2}{\sqrt{3}}\right).$$

2. Show using residue theory that

$$\int_0^{2\pi} \frac{d\theta}{a + b\sin\theta} = \frac{2\pi}{\sqrt{a^2 - b^2}}, \qquad a > b > 0.$$

3. Show using residue theory that

$$\int_0^{\pi} \frac{\sin^2\theta}{a + \cos\theta}\, d\theta = \pi[a - \sqrt{a^2 - 1}], \qquad a > 1.$$

4. Show using residue theory that

$$\int_{-\pi}^{\pi} \frac{d\theta}{1 + \sin^2\theta} = \pi\sqrt{2}.$$

5. Show using residue theory that

$$\int_{-\pi}^{\pi} \frac{1 - r^2}{1 - 2r\cos\theta + r^2}\frac{d\theta}{2\pi} = 1, \qquad 0 \le r < 1.$$

Remark. The integrand is the Poisson kernel (Section X.1).

6. By expanding both sides of the identity

$$\frac{1}{2\pi}\int_0^{2\pi} \frac{d\theta}{w + \cos\theta} = \frac{1}{\sqrt{w^2 - 1}}$$

in a power series at ∞, show that

$$\frac{1}{2\pi}\int_0^{2\pi} \cos^{2k}\theta\, d\theta = \frac{(2k)!}{2^{2k}(k!)^2}, \qquad k \ge 0.$$

7. Show using residue theory that

$$\int_0^{2\pi} \frac{d\theta}{(w + \cos\theta)^2} = \frac{2\pi w}{(w^2 - 1)^{3/2}}, \qquad w \in \mathbb{C}\setminus[-1, 1].$$

Specify carefully the branch of the power function. Check your answer by differentiating the integral of $1/(w + \cos\theta)$ with respect to the parameter w.

4. Integrands with Branch Points

Integrals featuring x^a and $\log x$ can sometimes be evaluated using contour integration. It is important to specify carefully the branch of the function z^a or $\log z$ used in the complex integral. We illustrate by deriving the identity

$$(4.1) \qquad \int_0^\infty \frac{x^a}{(1+x)^2}\, dx \;=\; \frac{\pi a}{\sin(\pi a)}, \qquad -1 < a < 1.$$

This integral is easily seen to be 1 if $a = 0$, and we interpret the right-hand side also to be 1 at $a = 0$. We suppose then that $a \neq 0$.

We consider the branch of the function $z^a/(1+z)^2$ defined on the slit plane $\mathbb{C}\backslash[0, +\infty)$ by

$$(4.2) \qquad f(z) \;=\; \frac{r^a e^{ia\theta}}{(1+z)^2}, \qquad z = re^{i\theta},\; 0 < \theta < 2\pi.$$

We regard the slit $[0, +\infty)$ as having a top edge and a bottom edge, and we extend the function by continuity to each edge of the slit, so it is defined by (4.2) with $\theta = 0$ on the top edge, and by (4.2) with $\theta = 2\pi$ on the bottom edge. The values on the bottom edge are obtained from those on the top edge by multiplying by the phase factor $e^{2\pi ia}$.

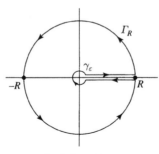

keyhole contour

For $\varepsilon > 0$ small and $R > 0$ large, we consider the keyhole domain D consisting of z in the slit plane $\mathbb{C}\backslash[0, \infty)$ satisfying $\varepsilon < |z| < R$. The function $f(z)$ has one pole in D, a double pole at $z = -1$. For the residue, Rule 2 gives

$$\mathrm{Res}\left[\frac{z^a}{(1+z)^2}, -1\right] \;=\; \frac{d}{dz} z^a \bigg|_{z=-1} \;=\; a\frac{z^a}{z}\bigg|_{z=-1} \;=\; -ae^{\pi ia}.$$

The residue theorem yields

$$(4.3) \qquad \int_{\partial D} f(z)\, dz \;=\; -2\pi iae^{\pi ia}.$$

The integral around ∂D breaks into the sum of four integrals, from ε to R along the top edge of the slit, around the circular contour Γ_R of radius R

in the counterclockwise direction, from R back to ε along the bottom edge of the slit, and around the circular contour γ_ε of radius ε in the clockwise direction,

$$\int_\varepsilon^R \frac{x^a}{(1+x)^2}\,dx + \int_{\Gamma_R} f(z)\,dz + \int_R^\varepsilon \frac{e^{2\pi i a}x^a}{(1+x)^2}\,dx + \int_{\gamma_\varepsilon} f(z)\,dz\,.$$

For the integrals over Γ_R and γ_ε, the ML-estimate gives

$$\left|\int_{\Gamma_R} \frac{z^a}{(1+z)^2}\,dz\right| \le \frac{R^a}{(R-1)^2}\cdot 2\pi R \sim R^{a-1},$$

$$\left|\int_{\gamma_\varepsilon} \frac{z^a}{(1+z)^2}\,dz\right| \le \frac{\varepsilon^a}{(1-\varepsilon)^2}\cdot 2\pi\varepsilon \sim \varepsilon^{a+1}.$$

Since $-1 < a < 1$, both these integrals tend to 0 as $R \to \infty$ and $\varepsilon \to 0$. If we reverse the direction of the integral from R to ε and use (4.3), we obtain in the limit

$$-2\pi i a e^{\pi i a} = (1 - e^{2\pi i a})\int_0^\infty \frac{x^a}{(1+x)^2}\,dx.$$

This yields the required identity,

$$\int_0^\infty \frac{x^a}{(1+x)^2}\,dx = \frac{-2\pi i a e^{\pi i a}}{1 - e^{2\pi i a}} = \frac{2\pi i a}{e^{\pi i a} - e^{-\pi i a}} = \frac{\pi a}{\sin(\pi a)}\,.$$

This identity can be extended to complex values of the parameter a. The function

$$g(w) = \int_0^\infty \frac{x^w}{(1+x)^2}\,dx$$

is analytic on the strip $\{-1 < \mathrm{Re}(w) < 1\}$, as is the function $\pi w/\sin(\pi w)$. We have shown that these two functions coincide on the interval $(-1, 1)$ where the strip meets the real axis. The uniqueness principle then shows that the functions coincide in the entire strip; that is, the identity (4.1) holds for all complex values of the parameter a satisfying $-1 < \mathrm{Re}(a) < 1$.

Exercises for VII.4

1. By integrating around the keyhole contour, show that

$$\int_0^\infty \frac{x^{-a}}{1+x}\,dx = \frac{\pi}{\sin(\pi a)}, \qquad 0 < a < 1.$$

2. By integrating around the boundary of a pie-slice domain of aperture $2\pi/b$, show that

$$\int_0^\infty \frac{dx}{1+x^b} = \frac{\pi}{b\sin(\pi/b)}, \qquad b > 1.$$

   Remark. Check the result by changing variable and comparing with Exercise 1.

3. By integrating around the keyhole contour, show that

$$\int_0^\infty \frac{\log x}{x^a(x+1)}\, dx = \frac{\pi^2 \cos(\pi a)}{\sin^2(\pi a)}, \qquad 0 < a < 1.$$

Remark. Check the result by differentiating the identity in Exercise 1.

4. For fixed $m \geq 2$, show by integrating around the keyhole contour that

$$\int_0^\infty \frac{x^{-a}}{(1+x)^m}\, dx = \frac{\pi a(a+1)\cdots(a+m-2)}{(m-1)!\,\sin(\pi a)}, \qquad 1 - m < a < 1.$$

Remark. The result can be obtained also by integrating the formula in Exercise 1 by parts.

5. By integrating a branch of $\dfrac{(\log z)^2}{(z+a)(z+b)}$ around the keyhole contour, show that

$$\int_0^\infty \frac{\log x}{(x+a)(x+b)}\, dx = \frac{(\log a)^2 - (\log b)^2}{2(a-b)}, \qquad a, b > 0, \ a \neq b.$$

6. Using residue theory, show that

$$\int_0^\infty \frac{x^a \log x}{(1+x)^2}\, dx = \frac{\pi \sin(\pi a) - a\pi^2 \cos(\pi a)}{\sin^2(\pi a)}, \qquad -1 < a < 1.$$

7. Show that

$$\int_0^\infty \frac{x^{a-1}}{1+x^b}\, dx = \frac{\pi}{b\sin(\pi a/b)}, \qquad 0 < a < b.$$

Determine for which complex values of the parameter a the integral exists (in the sense that the integral of the absolute value is finite), and evaluate it. Where does the integral depend analytically on the parameter a?

8. By integrating a branch of $(\log z)/(z^3 + 1)$ around the boundary of an indented sector of aperture $2\pi/3$, show that

$$\int_0^\infty \frac{\log x}{x^3+1}\, dx = \frac{-2\pi^2}{27}, \qquad \int_0^\infty \frac{1}{x^3+1}\, dx = \frac{2\pi}{3\sqrt{3}}.$$

Remark. Compare the results with those of Exercise 3 (after changing variable) and Exercise 2.

9. By integrating around an appropriate contour and using the results of Exercise 8, show that

$$\int_0^\infty \frac{(\log x)^2}{x^3 + 1}\, dx = \frac{10\pi^3}{81\sqrt{3}}.$$

10. By integrating a branch of $(\log z)/(z^3 - 1)$ around the boundary of an indented half-disk and using the result of Exercise 8, show that

$$\int_0^\infty \frac{\log x}{x^3 - 1}\, dx = \frac{4\pi^2}{27}.$$

5. Fractional Residues

Suppose z_0 is an isolated singularity of $f(z)$. For $\varepsilon > 0$ small, we consider the integral

$$\int_{C_\varepsilon} f(z)\, dz,$$

where C_ε is the arc of the circle $\{|z - z_0| = \varepsilon\}$ subtended by a sector of aperture α. If $\alpha = 2\pi$, then C_ε is the full circle, and the integral is equal to $2\pi i \operatorname{Res}[f(z), z_0]$. In general, there is no exact formula for the integral. However, if the isolated singularity of $f(z)$ is a simple pole, then we can calculate the limit of the integrals as $\varepsilon \to 0$.

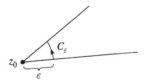

Theorem (Fractional Residue Theorem). *If z_0 is a simple pole of $f(z)$, and C_ε is an arc of the circle $\{|z - z_0| = \varepsilon\}$ of angle α, then*

$$(5.1) \qquad \lim_{\varepsilon \to 0} \int_{C_\varepsilon} f(z)\, dz = \alpha i \operatorname{Res}[f(z), z_0].$$

To see this, express $f(z) = A/(z - z_0) + g(z)$, where A is the residue of $f(z)$ at z_0 and $g(z)$ is analytic at z_0. Parameterizing the circle by $z = z_0 + \varepsilon e^{i\theta}$, and supposing $\theta_0 < \theta < \theta_0 + \alpha$ on the arc, we calculate

$$\int_{C_\varepsilon} \frac{A\, dz}{z - z_0} = iA \int_{\theta_0}^{\theta_0 + \alpha} d\theta = \alpha i A.$$

Since $g(z)$ is bounded near z_0 and the length of C_ε is at most $2\pi\varepsilon$, the ML-estimate shows that $\int_{C_\varepsilon} g(z) dz$ tends to 0 as $\varepsilon \to 0$. Consequently, $\int_{C_\varepsilon} f(z) dz$ tends to $\alpha i A$ as $\varepsilon \to 0$, which is (5.1).

Note that the angle α is taken to be negative if the arc of the circle is traversed in the negative direction, that is, in the clockwise direction.

To illustrate how the fractional residue theorem is used, we show that

(5.2)
$$\int_0^\infty \frac{\log x}{x^2 - 1}\, dx = \frac{\pi^2}{4}.$$

We choose the branch of $\log z$ that is real on the positive real axis (the principal branch of $\log z$), and we integrate $f(z) = (\log z)/(z^2 - 1)$ around the boundary ∂D of a domain D that is a half-disk of radius R in the upper half-plane, with indentations of radii ε and δ at the singularities of $f(z)$ at -1 and 0, respectively. Since the zeros of $\log z$ and $z^2 - 1$ at $z = 1$ are both simple, they cancel each other out, so that $f(z)$ is analytic at $z = 1$ and no indentation is required there. Since $f(z)$ is analytic on D, Cauchy's theorem yields

(5.3)
$$\int_{\partial D} f(z)\, dz = 0.$$

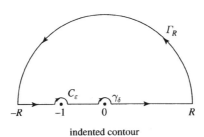

indented contour

Now the integral around ∂D breaks into the sum of six integrals, from δ to R along the positive real axis, around the semicircular contour Γ_R of radius R in the counterclockwise direction, from $-R$ to $-1-\varepsilon$ along the negative real axis, around a semicircular contour C_ε of radius ε in the clockwise direction around -1, from $-1 + \varepsilon$ to $-\delta$ along the negative real axis, and finally around a semicircular contour γ_δ of radius δ in the clockwise direction around 0. For the integrals over Γ_R and γ_δ, the ML-estimate gives

$$\left| \int_{\Gamma_R} \frac{\log z}{z^2 - 1}\, dz \right| \leq \frac{\sqrt{\log^2 R + \pi^2}}{R^2 - 1} \cdot \pi R \sim \frac{\log R}{R},$$

$$\left| \int_{\gamma_\delta} \frac{\log z}{z^2 - 1}\, dz \right| \leq \frac{\sqrt{\log^2 \delta + \pi^2}}{1 - \delta^2} \cdot \pi \delta \sim \delta |\log \delta|.$$

If we let $R \to \infty$ and $\delta \to 0$, these two contributions disappear, and (5.3) becomes

(5.4)
$$\int_0^\infty \frac{\log x}{x^2 - 1} \, dx + \int_{-\infty}^{-1-\varepsilon} \frac{\log |x| + \pi i}{x^2 - 1} \, dx$$

$$+ \int_{-1+\varepsilon}^0 \frac{\log |x| + \pi i}{x^2 - 1} \, dx + \int_{C_\varepsilon} \frac{\log z}{z^2 - 1} \, dz = 0.$$

The integral around C_ε is handled by the fractional residue formula. The angle α is $-\pi$, since C_ε is traversed halfway around the circle in the negative direction. The function $f(z)$ has a simple pole at $z = -1$, with residue computed by Rule 3 to be

$$\text{Res}\left[\frac{\log z}{z^2 - 1}, -1\right] = \left.\frac{\log z}{2z}\right|_{z=-1} = -\frac{1}{2}\log(-1) = -\frac{i\pi}{2}.$$

Thus the fractional residue formula (5.1) becomes

$$\lim_{\varepsilon \to 0} \int_{C_\varepsilon} \frac{\log z}{z^2 - 1} \, dz = i(-\pi)\left(-\frac{i\pi}{2}\right) = -\frac{\pi^2}{2}.$$

If we take the real part of (5.4) and let $\varepsilon \to 0$, we then obtain

$$\int_0^\infty \frac{\log x}{x^2 - 1} \, dx + \int_{-\infty}^0 \frac{\log |x|}{x^2 - 1} \, dx - \frac{\pi^2}{2} = 0.$$

After a change in variable $x \mapsto -x$ in the second integral, this becomes (5.2).

Exercises for VII.5

1. Use the keyhole contour indented on the lower edge of the axis at $x = 1$ to show that
$$\int_0^\infty \frac{\log x}{x^a(x - 1)} \, dx = \frac{2\pi^2}{1 - \cos(2\pi a)}, \qquad 0 < a < 1.$$

2. Show using residue theory that
$$\int_{-\infty}^\infty \frac{\sin(ax)}{x(x^2 + 1)} \, dx = \pi(1 - e^{-a}), \qquad a > 0.$$

   Hint. Replace $\sin(az)$ by e^{iaz}, and integrate around the boundary of a half-disk indented at $z = 0$.

3. Show using residue theory that
$$\int_{-\infty}^\infty \frac{\sin(ax)}{x(\pi^2 - a^2x^2)} \, dx = \frac{2}{\pi}, \qquad a > 0.$$

4. Show using residue theory that
$$\int_0^\infty \frac{1 - \cos x}{x^2} \, dx = \frac{\pi}{2}.$$

5. By integrating $(e^{\pm 2iz} - 1)/z^2$ over appropriate indented contours and using Cauchy's theorem, show that

$$\int_{-\infty}^{\infty} \frac{\sin^2 x}{x^2}\, dx \;=\; \pi.$$

6. By integrating a branch of $(\log z)/(z^3 - 1)$ around the boundary of an indented sector of aperture $2\pi/3$, show that

$$\int_0^{\infty} \frac{\log x}{x^3 - 1}\, dx \;=\; \frac{4\pi^2}{27}.$$

Remark. See also Exercise 4.10.

6. Principal Values

An integral $\int_a^b f(x)dx$ is **absolutely convergent** if the (proper or improper) integral $\int_a^b |f(x)|dx$ is finite. The integral is **absolutely divergent** if $\int_a^b |f(x)|dx = +\infty$. There is essentially only one way to assign a value to an absolutely convergent integral, while there may not be an obvious way to assign a value to an absolutely divergent integral. This is analogous to the dichotomy between absolutely and conditionally convergent series. Every rearrangement of an absolutely convergent series converges to the same value, while the rearrangements of a conditionally convergent series can converge to just about anything.

Example. The integral

$$\int_{-1}^{+1} \frac{1}{x}\, dx$$

is absolutely divergent, since $\int_{-1}^{+1}(1/|x|)dx = +\infty$. One natural way to assign a value to the integral is to take a limit as $\varepsilon \to 0$ of integrals over the two intervals $[-1, -\varepsilon]$ and $[\varepsilon, +1]$, which are obtained by excising a symmetric interval centered at the singularity 0. The negative contributions for $x < 0$ cancel out the positive contributions for $x > 0$, and we obtain what is called the "principal value" of the integral,

$$\mathrm{PV}\int_{-1}^{+1} \frac{1}{x}\, dx \;=\; \lim_{\varepsilon \to 0}\left(\int_{-1}^{-\varepsilon} + \int_{\varepsilon}^{1}\right)\frac{1}{x}\, dx \;=\; 0.$$

Observe, though, that if we excise an asymmetric interval, say from $-\varepsilon$ to $c\varepsilon$, and take a limit, instead of 0 we obtain

$$\lim_{\varepsilon \to 0}\left(\int_{-1}^{-\varepsilon} + \int_{c\varepsilon}^{1}\right)\frac{1}{x}\, dx \;=\; -\log c,$$

which can be any value at all.

Suppose that $f(x)$ is continuous for $a \leq x < x_0$ and for $x_0 < x \leq b$. We define the **principal value** of the integral $\int_a^b f(x)dx$ to be

$$(6.1) \qquad \mathrm{PV} \int_a^b f(x)\,dx \;=\; \lim_{\varepsilon \to 0} \left(\int_a^{x_0-\varepsilon} + \int_{x_0+\varepsilon}^b \right) f(x)\,dx,$$

provided that the limit exists. The principal value of the integral coincides with the usual value of the (proper or improper) integral if $f(x)$ is absolutely integrable. The same definition is used when the endpoints are infinite or when $f(x)$ is not continuous at the endpoints, provided that $f(x)$ is absolutely integrable over each of the intervals $(a, x_0 - \varepsilon)$ and $(x_0 + \varepsilon, b)$. If $f(x)$ has a finite number of discontinuities within (a,b), we define the principal value of the integral by dividing the interval (a,b) into subintervals, each containing one discontinuity of $f(x)$, and adding the principal values of the integrals corresponding to the subintervals.

Example. We illustrate by deriving the identity

$$\mathrm{PV} \int_{-\infty}^\infty \frac{1}{x^3-1}\,dx \;=\; -\frac{\pi}{\sqrt{3}},$$

Observe first that the integrand behaves like $1/(x-1)$ near $x = 1$, so that the integral is absolutely divergent. It is absolutely convergent on each interval $(-\infty, 1 - \varepsilon)$ and $(1 + \varepsilon, \infty)$, and the principal value is defined by

$$\mathrm{PV} \int_{-\infty}^\infty \frac{1}{x^3-1}\,dx \;=\; \lim_{\varepsilon \to 0} \left(\int_{-\infty}^{1-\varepsilon} + \int_{1+\varepsilon}^\infty \right) \frac{1}{x^3-1}\,dx.$$

We consider the function $f(z) = 1/(z^3 - 1)$, which has simple poles at the cube roots of unity. For $\varepsilon > 0$ small and $R > 0$ large, we consider the indented half-disk D in the upper half-plane, consisting of z in the upper half-plane satisfying $|z| < R$ and $|z - 1| > \varepsilon$. The function $f(z)$ has one pole in D, a simple pole at $z = e^{2\pi i/3}$. For the residue, Rule 4 gives

$$\mathrm{Res}\left[\frac{1}{z^3-1}, e^{2\pi i/3} \right] \;=\; \frac{1}{3z^2} \bigg|_{z=e^{2\pi i/3}} \;=\; \frac{e^{2\pi i/3}}{3}.$$

The residue theorem yields

$$(6.2) \qquad \int_{\partial D} f(z)\,dz \;=\; 2\pi i\,\frac{e^{2\pi i/3}}{3} \;=\; -\frac{\pi}{\sqrt{3}} - \frac{\pi}{3}i.$$

The integral around ∂D breaks into the sum of four integrals, from $-R$ to $1 - \varepsilon$, around a semicircular contour C_ε centered at 1 of radius ε in the clockwise direction, from $1+\varepsilon$ to R, and around the semicircular contour Γ_R

of radius R in the counterclockwise direction,

$$(6.3) \qquad \int_{\partial D} f(z)\, dz \;=\; \left(\int_{-R}^{1-\varepsilon} + \int_{C_\varepsilon} + \int_{1+\varepsilon}^{R} + \int_{\Gamma_R} \right) f(z)\, dz.$$

The ML-estimate gives

$$\left| \int_{\Gamma_R} \frac{1}{z^3 - 1}\, dz \right| \;\leq\; \frac{1}{R^3 - 1} \cdot \pi R \;\sim\; \frac{1}{R^2},$$

which tends to 0 as $R \to \infty$. We combine (6.2) and (6.3), and we pass to the limit as $R \to \infty$, to obtain

$$\left(\int_{-\infty}^{1-\varepsilon} + \int_{1+\varepsilon}^{\infty} \right) \frac{1}{x^3 - 1}\, dx \;+\; \int_{C_\varepsilon} \frac{1}{z^3 - 1}\, dz \;=\; -\frac{\pi}{\sqrt{3}} - \frac{\pi}{3} i.$$

From Rule 4, the residue of $f(z)$ at $z = 1$ is

$$\mathrm{Res}\left[\frac{1}{z^3 - 1}, 1 \right] \;=\; \left. \frac{1}{3z^2} \right|_{z=1} \;=\; \frac{1}{3}.$$

The fractional residue theorem, with angle $-\pi$, then yields

$$\lim_{\varepsilon \to 0} \int_{C_\varepsilon} \frac{1}{z^3 - 1}\, dz \;=\; -\frac{\pi}{3} i.$$

If we pass to the limit in (6.3) as $\varepsilon \to 0$, we then obtain

$$\mathrm{PV} \int_{-\infty}^{\infty} \frac{1}{x^3 - 1}\, dx \;-\; \frac{\pi}{3} i \;=\; -\frac{\pi}{\sqrt{3}} - \frac{\pi}{3} i.$$

If we add $\pi i / 3$ to both sides, we obtain our asserted identity.

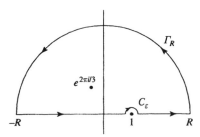

Principal value integrals are important in harmonic analysis. One of the most important integral operators in analysis is the singular integral operator called the **Hilbert transform**, defined by

$$(Hu)(t) \;=\; \mathrm{PV} \int_{-\infty}^{\infty} \frac{u(s)}{s - t}\, ds, \qquad -\infty < t < \infty,$$

where $u(s)$ is an integrable function on the real line. It turns out that the principal value exists for "almost all" real numbers t and defines a "good" operator $u \mapsto Hu$. If $u(s)$ is extended appropriately to be harmonic in the upper half-plane, then $(Hu)(s)$ is the boundary value function of its harmonic conjugate.

Exercises for VII.6

1. Integrate $1/(1 - x^2)$ directly, using partial fractions, and show that

$$\text{PV} \int_0^\infty \frac{dx}{1 - x^2} = 0.$$

Show that

$$\int_0^1 \frac{dx}{1 - x^2} = +\infty, \qquad \int_1^\infty \frac{dx}{1 - x^2} = -\infty.$$

2. Obtain the principal value in Exercise 1 by taking imaginary parts of the identity (5.4) in the preceding section and making a change of variable.

3. By integrating around the boundary of an indented half-disk in the upper half-plane, show that

$$\text{PV} \int_{-\infty}^\infty \frac{1}{(x^2 + 1)(x - a)} \, dx = -\frac{\pi a}{a^2 + 1}, \qquad -\infty < a < \infty.$$

4. Suppose $m \geq 2$ and $a_1 < a_2 < \cdots < a_m$. By integrating around the boundary of an indented half-disk in the upper half-plane, show that

$$\text{PV} \int_{-\infty}^\infty \frac{1}{(x - a_1)(x - a_2) \cdots (x - a_m)} \, dx = 0.$$

5. Show that

$$\text{PV} \int_0^\infty \frac{x^{a-1}}{x^b - 1} \, dx = -\frac{\pi}{b} \cot\left(\frac{\pi a}{b}\right), \qquad 0 < a < b.$$

Hint. For $b > 1$ one can integrate a branch of $z^{a-1}/(z^b - 1)$ around a sector of aperture $2\pi/b$, indented at $z = 1$ and at $z = e^{2\pi i/b}$.

6. By integrating a branch of $(\log z)/(z^b - 1)$ around an indented sector of aperture $2\pi/b$, show that for $b > 1$,

$$\int_0^\infty \frac{\log x}{x^b - 1} \, dx = \frac{\pi^2}{b^2 \sin^2(\pi/b)}, \qquad \text{PV} \int_0^\infty \frac{1}{x^b - 1} \, dx = -\frac{\pi}{b} \cot(\pi/b).$$

7. Suppose that $P(z)$ and $Q(z)$ are polynomials, $\deg Q(z) \geq \deg P(z) + 2$, and the zeros of $Q(z)$ on the real axis are all simple. Show that

$$\text{PV} \int_{-\infty}^\infty \frac{P(x)}{Q(x)} \, dx = 2\pi i \sum \text{Res}\left[\frac{P(z)}{Q(z)}, z_j\right] + \pi i \sum \text{Res}\left[\frac{P(z)}{Q(z)}, x_k\right],$$

summed over the poles z_j of $P(z)/Q(z)$ in the open upper half-plane and the poles x_k of $P(z)/Q(z)$ on the real axis. *Remark.* In other words, the principal value of the integral is $2\pi i$ times the sum of the residues in the upper half-plane, where we count the poles on the real axis as being half in and half out of the upper half-plane.

7. Jordan's Lemma

Another class of absolutely divergent integrals is made up of integrals of the form

$$\int_{-\infty}^{\infty} \frac{P(x)}{Q(x)} \sin x \, dx \,, \qquad \int_{-\infty}^{\infty} \frac{P(x)}{Q(x)} \cos x \, dx \,,$$

where $P(z)$ and $Q(z)$ are polynomials satisfying

$$\deg Q(z) \;=\; \deg P(z) \,+\, 1.$$

To evaluate such an integral we would replace the sine or cosine function by e^{iz} and integrate over the boundary of the usual half-disk in the upper half-plane. If $\deg Q(z) \geq \deg P(z) + 2$, then the ML-estimate shows that the integral over the semicircular piece Γ_R of the boundary tends to 0 as $R \to \infty$. If $\deg Q(z) = \deg P(z) + 1$, the integral over Γ_R still tends to 0, but this is not so obvious. It is a consequence of the following estimate.

Lemma (Jordan's Lemma). *If Γ_R is the semicircular contour $z(\theta) = Re^{i\theta}$, $0 \leq \theta \leq \pi$, in the upper half-plane, then*

$$(7.1) \qquad\qquad \int_{\Gamma_R} |e^{iz}||dz| \;<\; \pi.$$

For the parameterization $z(\theta) = Re^{i\theta}$ we have $|e^{iz}| = e^{-R\sin\theta}$ and $|dz| = R\,d\theta$, so the estimate (7.1) becomes

$$(7.2) \qquad\qquad \int_0^\pi e^{-R\sin\theta}d\theta \;<\; \frac{\pi}{R} \,.$$

In order to establish (7.2), note that $\sin\theta$ is concave down on the interval $0 \leq \theta \leq \pi/2$, so the graph of $\sin\theta$ lies above the straight line connecting its endpoints,

$$\sin\theta \;\geq\; 2\theta/\pi, \qquad 0 \leq \theta \leq \pi/2.$$

Thus

$$\int_0^\pi e^{-R\sin\theta}d\theta \;=\; 2\int_0^{\pi/2} e^{-R\sin\theta}d\theta \;\leq\; 2\int_0^{\pi/2} e^{-2R\theta/\pi}d\theta$$

$$=\; \frac{\pi}{R}\int_0^R e^{-t}dt \;<\; \frac{\pi}{R}\int_0^\infty e^{-t}dt \;=\; \frac{\pi}{R} \,.$$

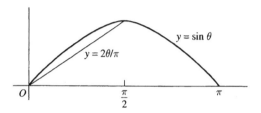

Example. As an application of Jordan's lemma, we show that

$$(7.3) \qquad \lim_{R \to \infty} \int_0^R \frac{\sin x}{x}\, dx = \frac{\pi}{2}.$$

Note that $(\sin z)/z$ is analytic at $z = 0$, so that $(\sin x)/x$ is absolutely integrable over any finite interval. It is not difficult to see (Exercise 1) that

$$\lim_{R \to \infty} \int_0^R \frac{|\sin x|}{x}\, dx = +\infty.$$

Thus the integral over $[0, \infty)$ is absolutely divergent. Since the integrand is an even function, (7.3) is equivalent to

$$(7.4) \qquad \lim_{R \to \infty} \int_{-R}^R \frac{\sin x}{x}\, dx = \pi.$$

We may think of the limit (7.4) as a principal value at infinity,

$$\lim_{R \to \infty} \int_{-R}^R \frac{\sin x}{x}\, dx = \mathrm{PV} \int_{-\infty}^{\infty} \frac{\sin x}{x}\, dx.$$

To evaluate the limit, we consider the function $f(z) = e^{iz}/z$, which has only one pole, a simple pole at $z = 1$ with residue 1. Let D be the indented half-disk consisting of points z in the upper half-plane satisfying $\varepsilon < |z| < R$. The integral around ∂D breaks into the sum of four integrals, from $-R$ to $-\varepsilon$, around a semicircular contour C_ε centered at 0 of radius ε in the clockwise direction, from ε to R, and around the semicircular contour Γ_R of radius R in the counterclockwise direction. Since $f(z)$ is analytic on D, we may apply Cauchy's theorem, to obtain

$$(7.5) \qquad 0 = \int_{\partial D} f(z)\, dz = \left(\int_{-R}^{-\varepsilon} + \int_{C_\varepsilon} + \int_\varepsilon^R + \int_{\Gamma_R} \right) f(z)\, dz.$$

The integral over C_ε is handled by the fractional residue formula, with angle $-\pi$,

$$\lim_{\varepsilon \to 0} \int_{C_\varepsilon} \frac{e^{iz}}{z}\, dz = -\pi i \operatorname{Res}\left[\frac{e^{iz}}{z}, 0 \right] = -\pi i.$$

Passing to the limit in (7.5) as $\varepsilon \to 0$, we obtain

$$0 = \mathrm{PV} \int_{-R}^R \frac{e^{ix}}{x}\, dx - \pi i + \int_{\Gamma_R} \frac{e^{iz}}{z}\, dz.$$

Taking the imaginary part of this identity, we obtain

$$\int_{-R}^{R} \frac{\sin x}{x}\, dx \; + \; \operatorname{Im} \int_{\Gamma_R} \frac{e^{iz}}{z}\, dz \; = \; \pi.$$

Now the integral over Γ_R is handled by Jordan's lemma,

$$\left| \int_{\Gamma_R} \frac{e^{iz}}{z}\, dz \right| \; \leq \; \frac{1}{R} \int_{\Gamma_R} |e^{iz}|\, |dz| \; < \; \frac{\pi}{R}.$$

Since this tends to 0 as $R \to \infty$, in the limit we obtain (7.4).

Exercises for VII.7

1. Show that

$$\int_0^\infty \frac{|\sin x|}{x}\, dx \; = \; +\infty.$$

   Hint: Show that the area under the mth arch of $|\sin x|/x$ is $\sim 1/m$.

2. Show that

$$\lim_{R \to \infty} \int_{-R}^{R} \frac{x^3 \sin x}{(x^2+1)^2}\, dx \; = \; \frac{\pi}{2e}.$$

3. Evaluate the limits

$$\lim_{R \to \infty} \int_{-R}^{R} \frac{x \sin(ax)}{x^2+1}\, dx, \qquad -\infty < a < +\infty.$$

   Show that they do not depend continuously on the parameter a.

4. By integrating $z^{a-1} e^{iz}$ around the boundary of a domain in the first quadrant bounded by the real and imaginary axes and a quarter-circle, show that

$$\lim_{R \to \infty} \int_0^{R} x^{a-1} \cos x\, dx \; = \; \Gamma(a) \cos(\pi a/2), \qquad 0 < a < 1,$$

$$\lim_{R \to \infty} \int_0^{R} x^{a-1} \sin x\, dx \; = \; \Gamma(a) \sin(\pi a/2), \qquad 0 < a < 1,$$

   where $\Gamma(a)$ is the gamma function defined by

$$\Gamma(a) \; = \; \int_0^\infty t^{a-1} e^{-t}\, dt.$$

   Remark: The formula for the sine integral holds also for $-1 < a < 0$. To see this, integrate by parts.

5. Show that

$$\lim_{R \to \infty} \int_0^{R} \sin(x^2)\, dx \; = \; \lim_{R \to \infty} \int_0^{R} \cos(x^2)\, dx \; = \; \frac{\sqrt{\pi}}{2\sqrt{2}},$$

by integrating e^{iz^2} around the boundary of the pie-slice domain determined by $0 < \arg z < \pi/4$ and $|z| < R$. *Remark.* These improper integrals are called the **Fresnel integrals**. The identities can also be deduced from the preceding exercise by changing variable.

8. Exterior Domains

An **exterior domain** is a domain D in the complex plane that includes all large z, that is, D includes all z such that $|z| \geq R$ for some R. The residue theorem is valid also for exterior domains, though the residue formula must take into account the point at ∞.

Theorem. *Let D be an exterior domain with piecewise smooth boundary. Suppose that $f(z)$ is analytic on $D \cup \partial D$, except for a finite number of isolated singularities $z_1, \ldots, z_m$ in D, and let a_{-1} be the coefficient of $1/z$ in the Laurent expansion $f(z) = \sum a_k z^k$ that converges for $|z| > R$. Then*

$$(8.1) \qquad \int_{\partial D} f(z)\, dz = -2\pi i\, a_{-1} + 2\pi i \sum_{j=1}^{m} \operatorname{Res}[f(z), z_j].$$

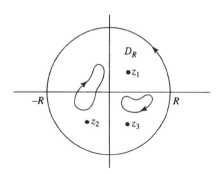

Here the boundary ∂D is oriented as usual, so that D lies on its left as ∂D is traversed in the positive direction. The formula is proved by applying the usual residue theorem to the bounded domain D_R consisting of $z \in D$ such that $|z| < R$. This yields

$$\int_{\partial D} f(z)\, dz + \int_{|z|=R} f(z)\, dz = \int_{\partial D_R} f(z)\, dz = 2\pi i \sum_{j=1}^{m} \operatorname{Res}[f(z), z_j].$$

If we substitute the Laurent series expansion for $f(z)$ into the integral over the circle $\{|z| = R\}$ and integrate term by term, we obtain $2\pi i a_{-1}$ for the integral. This is simply the integral formula for the Laurent coefficient a_{-1}. In any event, we move this summand to the right-hand side, picking up a minus sign, and we obtain (8.1).

Example. We show using contour integration that

$$\int_0^1 \frac{1}{\sqrt{x(1-x)}}\,dx \;=\; \pi.$$

Consider the function $1/\sqrt{z(1-z)}$, which has two analytic branches on the slit plane $\mathbb{C}\backslash[0,1]$. The branches are analytic at ∞, and the Laurent expansions are obtained in terms of a binomial series:

$$\frac{1}{\sqrt{z(1-z)}} \;=\; \pm\frac{i}{z}\left(1-\frac{1}{z}\right)^{-1/2}$$

$$= \;\pm\frac{i}{z}\left[1 + (-1/2)(-1/z) + \frac{(-1/2)(-3/2)}{2!}(-1/z)^2 + \cdots\right]$$

$$= \;\pm i\left[\frac{1}{z} + \frac{1}{2z^2} + \frac{3}{8z^3} + \cdots\right], \qquad |z| > 1.$$

Let $f(z)$ be the branch of the function that is positive on the top edge of the slit $[0,1]$. Since the phase factor of $f(z)$ is -1 at each of the branch points 0 and 1, $f(z)$ is negative on the bottom edge of the slit. By following the argument of $1/\sqrt{1-z}$ around a small semicircle from the top edge of $[0,1]$ to the interval $(1,+\infty)$, which increases by $\pi/2$, we see that the values $f(x)$ are positive imaginary for $x > 1$. Consequently, our branch $f(z)$ corresponds to the $+$ sign in the Laurent expansion, and the residue at ∞ is given by

$$\mathrm{Res}\,[\,f(z),\,\infty\,] \;=\; -a_{-1} \;=\; -i.$$

We integrate $f(z)$ around the **dogbone contour** Γ_ε (or **dumbbell contour**), traversing the top edge of the slit from ε to $1-\varepsilon$, a circle γ_ε centered at 1 of radius ε, the bottom edge of the slit from $1-\varepsilon$ to ε, and a circle C_ε centered at 0 of radius ε. If we apply the residue formula (8.1) to the exterior domain lying outside the dogbone contour, we obtain

$$(8.2) \qquad \int_{\Gamma_\varepsilon} f(z)\,dz \;=\; 2\pi i\,\mathrm{Res}\,[\,f(z),\,\infty\,] \;=\; 2\pi i(-i) \;=\; 2\pi.$$

The integrals over the circles γ_ε and C_ε are bounded by the ML-estimate

$$\left|\int_{C_\varepsilon} f(z)\,dz\right| \;\le\; \frac{2}{\sqrt{\varepsilon}}\cdot 2\pi\varepsilon \;\sim\; \sqrt{\varepsilon},$$

$$\left|\int_{\gamma_\varepsilon} f(z)\,dz\right| \;\le\; \frac{2}{\sqrt{\varepsilon}}\cdot 2\pi\varepsilon \;\sim\; \sqrt{\varepsilon}.$$

Since the values of $f(z)$ on the bottom edge of the slit are minus those on the top edge, while the direction is reversed, the integral along the bottom edge is equal to the integral along the top edge. Thus we obtain

$$\int_{\Gamma_\varepsilon} f(z)\,dz \;\rightarrow\; 2\int_0^1 \frac{1}{\sqrt{x(1-x)}}\,dx$$

as $\varepsilon \to 0$. If we combine this with (8.2), we obtain the desired identity.

dogbone contour

Note that it was not necessary to determine the choice of sign in the Laurent expansion of $f(z)$. If we had carried along the $\pm$ sign, we would have arrived at a value for the integral of $\pm \pi$. Since the integrand is positive, the value $+\pi$ must be the correct choice. Working backwards, we see that our choice of the $+$ sign in the Laurent expansion of $f(z)$ was correct.

Suppose now that $f(z)$ is analytic for $|z| \geq R$, with Laurent expansion

$$f(z) = \sum_{n=-\infty}^{\infty} a_n z^n, \qquad |z| \geq R.$$

We define the **residue of $f(z)$ at ∞** to be

$$(8.3) \qquad \qquad \operatorname{Res}[f(z), \infty] = -a_{-1}.$$

If D_R is the exterior domain $\{|z| > R\}$, this definition is equivalent to

$$\int_{\partial D_R} f(z)\, dz = 2\pi i \operatorname{Res}[f(z), \infty].$$

The orientation of the circle $\{|z| = R\}$ with respect to D_R is clockwise, and this accounts for the minus sign. With this definition of residue at ∞, formula (8.1) becomes

$$(8.4) \qquad \int_{\partial D} f(z)\, dz = 2\pi i \operatorname{Res}[f(z), \infty] + 2\pi i \sum_{j=1}^{m} \operatorname{Res}[f(z), z_j].$$

Thus the residue theorem for exterior domains is the same as for bounded domains, except that we must take the residue at ∞ into account.

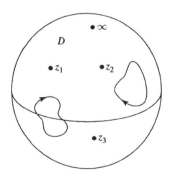

Note that the function $f(z)$ can be analytic at ∞ and still have a nonzero residue at ∞. For instance, the function $1/z$ is analytic at ∞, while

$$\operatorname{Res}\left[\frac{1}{z}, \infty\right] = -1.$$

There is a mathematical explanation for this state of affairs. The explanation is that in order to have a coordinate-free definition of residues, the residue should not be associated with the function $f(z)$ but with the differential $f(z)dz$. Even though $f(z)$ may be analytic at ∞, the differential $f(z)dz$ need not be analytic at ∞, as can be seen by making the usual change of variable $w = 1/z$. (See Exercise 13.)

Exercises for VII.8

1. Evaluate the residue at ∞ of the following functions.

(a) $\dfrac{z}{z^2 - 1}$ (c) $\dfrac{z^3 + 1}{z^2 - 1}$ (e) $z^n e^{1/z}, \quad n = 0, \pm 1, \ldots$

(b) $\dfrac{1}{(z^2 + 1)^2}$ (d) $\dfrac{z^9 + 1}{z^6 - 1}$ (f) $\sqrt{\dfrac{z - a}{z - b}}$

   Note. There are two possibilities for (f), one for each branch of the square root.

2. Show by integrating around the dogbone contour that

$$\int_0^1 \frac{x^4}{\sqrt{x(1 - x)}}\, dx = \frac{35\pi}{128}.$$

3. Fix an integer n, positive or negative. Determine for which complex values of the parameter a the integral

$$\int_0^1 \frac{x^n}{x^a(1 - x)^{1-a}}\, dx$$

   converges, and evaluate it.

4. Show that

$$\int_{-1}^1 \frac{\sqrt{1 - x^2}}{1 + x^2}\, dx = \pi\left(\sqrt{2} - 1\right).$$

5. Show that the sum of the residues of a rational function on the extended complex plane is equal to zero.

6. Find the residue of $z/(z^2 + 1)$ at each pole in the extended complex plane, and check that the sum of the residues is zero.

7. Find the sum of the residues of $(3z^4 + 2z + 1)/(8z^5 + 5z^2 + 2)$ at its poles in the (finite) complex plane.

8. Fix $n \geq 1$ and $k \geq 0$. Find the residue of $z^k/(z^n - 1)$ at ∞ by expanding $1/(z^n - 1)$ in a Laurent series. Find the residue of $z^k/(z^n - 1)$ at each finite pole, and verify that the sum of all the residues is zero.

9. Show that if $f(z)$ is analytic at ∞, then
$$\text{Res}[f(z), \infty] = -\lim_{z \to \infty} z(f(z) - f(\infty)).$$

10. Let D be an exterior domain. Suppose that $f(z)$ is analytic on $D \cup \partial D$ and at ∞. Show that
$$\frac{1}{2\pi i} \int_{\partial D} \frac{f(\zeta)}{\zeta - z} \, d\zeta = f(z) - f(\infty), \qquad z \in D.$$

11. If $f(x)$ is not integrable at ∞, we define the principal value
$$\text{PV} \int_{-\infty}^{\infty} f(x) \, dx = \lim_{R \to \infty} \text{PV} \int_{-R}^{R} f(x) \, dx.$$
Show that
$$\text{PV} \int_{-\infty}^{\infty} \frac{1}{x - a} \, dx = \begin{cases} i\pi, & \text{Im } a > 0, \\ 0, & \text{Im } a = 0, \\ -i\pi, & \text{Im } a < 0. \end{cases}$$

12. Suppose that $P(z)$ and $Q(z)$ are polynomials such that the degree of $Q(z)$ is strictly greater than the degree of $P(z)$. Suppose that the zeros $x_1, \ldots, x_m$ of $Q(z)$ on the real axis are all simple, and set $x_0 = \infty$. Show that
$$\text{PV} \int_{-\infty}^{\infty} \frac{P(x)}{Q(x)} \, dx = 2\pi i \sum \text{Res}\left[\frac{P(z)}{Q(z)}, z_j\right] + \pi i \sum \text{Res}\left[\frac{P(z)}{Q(z)}, x_k\right],$$
summed over the poles z_j of $P(z)/Q(z)$ in the open upper half-plane and summed over the x_k's including ∞. *Hint.* Use the preceding exercise. See also Exercise 6.7.

13. Show that the analytic differential $f(z)dz$ transforms under the change of variable $w = 1/z$ to $-f(1/w)dw/w^2$. Show that the residue of $f(z)$ at $z = \infty$ coincides with that of $-f(1/w)/w^2$ at $w = 0$.

VIII

The Logarithmic Integral

In this chapter we discuss the argument principle and develop several of its consequences. In Section 1 we derive the argument principle from the residue theorem, and we use the argument principle to locate the zeros of analytic functions. Sections 2 through 5 can be viewed as a study of how the zeros of an analytic function depend on various types of parameters. Sections 6 and 7 are devoted to winding numbers of closed paths and the jump theorem for the Cauchy integral. The jump theorem yields an easy proof of the Jordan curve theorem in the smooth case, and a proof of the full Jordan curve theorem is laid out in the exercises. In Section 8 we introduce simply connected domains and we characterize these in several ways. While the material in this chapter is of fundamental importance for the Riemann mapping theorem in Chapter XI and for various further developments, the student can skip to Chapter IX immediately after Sections 1 and 2.

1. The Argument Principle

Suppose $f(z)$ is analytic on a domain D. For a curve γ in D such that $f(z) \neq 0$ on γ, we refer to

$$(1.1) \qquad \frac{1}{2\pi i} \int_\gamma \frac{f'(z)}{f(z)} \, dz \;=\; \frac{1}{2\pi i} \int_\gamma d\log f(z)$$

as the **logarithmic integral** of $f(z)$ along γ. Thus the logarithmic integral measures the change of $\log f(z)$ along the curve γ. It can be used to count zeros and poles of meromorphic functions. The following theorem is one version of the argument principle.

Theorem. *Let D be a bounded domain with piecewise smooth boundary ∂D, and let $f(z)$ be a meromorphic function on D that extends to be analytic on ∂D, such that $f(z) \neq 0$ on ∂D. Then*

$$(1.2) \qquad \frac{1}{2\pi i} \int_{\partial D} \frac{f'(z)}{f(z)} \, dz \;=\; N_0 - N_\infty,$$

where N_0 is the number of zeros of $f(z)$ in D and N_∞ is the number of poles of $f(z)$ in D, counting multiplicities.

The formula (1.2) is simply the residue theorem for $f'(z)/f(z)$. This function is analytic on $D \cup \partial D$ except possibly at the zeros and poles of $f(z)$ in D. Let z_0 be a zero or pole of $f(z)$, and let N be the order of $f(z)$ at z_0, that is, N is the order of the zero if z_0 is a zero, and N is minus the order of the pole if z_0 is a pole. Then

$$f(z) = (z - z_0)^N g(z),$$

where $g(z)$ is analytic at z_0 and $g(z_0) \neq 0$. So

$$\frac{f'(z)}{f(z)} = \frac{N(z - z_0)^{N-1}g(z)}{(z - z_0)^N g(z)} + \frac{(z - z_0)^N g'(z)}{(z - z_0)^N g(z)} = \frac{N}{z - z_0} + [\text{analytic}].$$

Thus $f'(z)/f(z)$ has a simple pole at z_0, with residue N. If now we sum the N's over the zeros and poles, we find that the sum of the residues of $f'(z)/f(z)$ in D is $N_0 - N_\infty$. Thus (1.2) follows from the residue theorem.

Now we look more carefully at the logarithmic integral (1.1), which we express in the form

$$\frac{1}{2\pi i} \int_\gamma d\log f(z) = \frac{1}{2\pi i} \int_\gamma d\log |f(z)| + \frac{1}{2\pi} \int_\gamma d\arg(f(z)).$$

The differential $d\log|f(z)|$ is exact. If we parametrize the curve γ by $\gamma(t) = x(t) + iy(t)$, $a \le t \le b$, then

$$\int_\gamma d\log|f(z)| = \log|f(\gamma(b))| - \log|f(\gamma(a))|,$$

which depends only on the initial point $\gamma(a)$ and the endpoint $\gamma(b)$ of the curve. In particular, $\int_\gamma d\log|f(z)| = 0$ if γ is a closed curve.

The differential $d\arg f(z)$ is closed but not exact. Its integral is computed by choosing a continuous single-valued determination of $\arg f(\gamma(t))$ for $a \le t \le b$. Then for this determination,

(1.3) $$\int_\gamma d\arg(f(z)) = \arg f(\gamma(b)) - \arg f(\gamma(a)).$$

This quantity (1.3) is referred to as the **increase in the argument of** $f(z)$ **along** γ. It is defined for any (continuous) path γ in D providing there are no zeros or poles on the path. Since any two continuous determinations of $\arg f(\gamma(t))$ differ by a constant, the increase in the argument given by (1.3) is independent of the continuous determination.

If γ is a concatenation of curve segments, the increase in the argument of $f(z)$ along γ is obtained by adding the increases along the various curve segments. If the direction of the curve γ is reversed, the increase in the argument is replaced by its negative. If γ is a closed curve, so that $\gamma(a) = \gamma(b)$, the determinations of $\arg f(\gamma(b))$ and $\arg f(\gamma(a))$ differ by an integral multiple of 2π, so that the increase in the argument of $f(z)$ around γ is an integral multiple of 2π.

Example. The increase in the argument of $f(z) = (z - z_0)^n$ counterclockwise around the circle $|z - z_0| = \rho$ is determined as follows. We parametrize the circle by $\gamma(t) = z_0 + \rho e^{it}$, $0 \le t \le 2\pi$. Then $f(\gamma(t)) = \rho^n e^{int}$, and a continuous determination for the argument of $f(\gamma(t))$ is given by $\arg f(\gamma(t)) = nt$, $0 \le t \le 2\pi$. Thus the increase in the argument of $f(z)$ around γ is

$$\int_\gamma d\arg(f(z)) \;=\; nt \Big|_{t=0}^{t=2\pi} \;=\; 2\pi n.$$

Now we return to a bounded domain D, whose boundary ∂D consists of a finite number of piecewise-smooth closed curves with the usual orientation, so that D lies on the left as we traverse the curves in ∂D in the positive direction. We define the **increase in the argument of $f(z)$ around the boundary of D** to be the sum of its increases around the closed curves in ∂D. The argument principle can be restated as follows.

Theorem. *Let D be a bounded domain with piecewise smooth boundary ∂D, and let $f(z)$ be a meromorphic function on D that extends to be analytic on ∂D, such that $f(z) \ne 0$ on ∂D. Then the increase in the argument of $f(z)$ around the boundary of D is 2π times the number of zeros minus the number of poles of $f(z)$ in D,*

$$(1.4) \qquad\qquad \int_{\partial D} d\arg(f(z)) \;=\; 2\pi(N_0 - N_\infty).$$

Before illustrating the theorem with an example, we note that it is easy to track the argument of $f(z)$ along a segment of a curve in a domain where a single-valued determination of the argument function is available. In this situation, the increase in the argument is obtained simply by evaluating the branch of the argument function at the initial and terminal values of $f(z)$. Suppose, for instance, that the values $f(\gamma(t))$ start from the positive real axis at $t = a$, they lie in the upper half-plane for $a < t < b$, and they hit the negative real axis at $t = b$. Then the increase in the argument of $f(\gamma(t))$ for $a \le t \le b$ is π, no matter how wildly the curve might wiggle in the upper half-plane, since it can be evaluated by evaluating a branch of the argument function, in this case the principal branch, at the endpoints. By the same token, if the curve $f(\gamma(t))$ goes from the positive real axis at $t = a$ through the lower half-plane for $a < t < b$ and terminates on the negative real axis, then the increase of the argument of $f(\gamma(t))$ along the curve is $-\pi$, no matter how erratically the curve behaves in the lower half-plane.

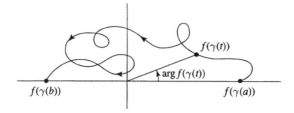

Example. We show how to use the argument principle to find the number of zeros of the polynomial $p(z) = z^6 + 9z^4 + z^3 + 2z + 4$ in the first quadrant $\{0 < \arg z < \pi/2\}$ and in the second quadrant $\{\pi/2 < \arg z < \pi\}$. First note that $p(x) > 0$ for $-\infty < x < \infty$. This is clear if $x \geq 0$, and it follows from the estimates $4 + x^3 + 2x > 0$ for $-1 \leq x \leq 0$ and $9x^4 + x^3 + 2x > 0$ for $x \leq -1$. Since there are no real zeros of $p(z)$, and since the coefficients of $p(z)$ are real, the zeros come in complex conjugate pairs, and there are three zeros in the upper half-plane. To determine the number of zeros in the first quadrant, we estimate the increase of $\arg p(z)$ around the boundary of the quarter disk D_R of a large radius R, consisting of the points z in the first quadrant satisfying $|z| < R$. Since the increase in the argument of $p(z)$ around ∂D_R is 2π times the number of zeros in D_R, any reasonable approximation will yield the exact value. We break ∂D_R into three paths. Along the real axis from 0 to R we have $p(x) > 0$, so that the argument is constant, and the increase is zero. Along the quarter-circle Γ_R defined by $|z| = R$ and $0 \leq \arg z \leq \pi/2$, the term z^6 dominates, and $\arg p(z) \approx 6 \arg z$. The increase in $\arg p(z)$ along Γ_R is thus approximately $6(\pi/2) = 3\pi$. To determine the increase in $\arg p(z)$ down the imaginary axis from iR to 0, we substitute $z = iy$ and we follow the values of

$$p(iy) = -y^6 + 9y^4 + 4 + i(-y^3 + 2y)$$

as y decreases from R to 0. At the initial point of this segment we have $\operatorname{Re} p(iR) \approx -R^6$ and $\operatorname{Im} p(iR) \approx -R^3$, so that $p(iR)$ lies in the third quadrant and has argument approximately $-\pi$. At the terminal point we have $p(0) = 4$ on the positive real axis. To see how $\arg p(iy)$ behaves, we determine where the curve $p(iy)$ crosses the x-axis. The curve crosses the real axis for parameter values y that satisfy $-y^3 + 2y = 0$. The solutions of this cubic are $y = 0$ and $y = \pm\sqrt{2}$. There is only one crossing point corresponding to $y > 0$, at $y = \sqrt{2}$. The crossing point is at $p\left(i\sqrt{2}\right) = 32$, which lies on the positive real axis. Thus $p(iy)$ remains in the lower half-plane as the parameter y decreases from $y = R$ to $y = \sqrt{2}$, and it hits the positive real axis at $y = \sqrt{2}$. Thus we see that the increase in the argument of $p(iy)$ as y decreases from R to $\sqrt{2}$ is approximately π. Since $p(iy)$ is in the upper half-plane for y between $\sqrt{2}$ and 0, and since the starting and ending points for this segment of the curve are on the positive real axis, the increase in the argument of $p(iy)$ as y decreases from $\sqrt{2}$ to 0 is zero. Thus the total increase of $\arg p(z)$ around ∂D_R is approximately $3\pi + \pi = 4\pi$,

hence it is exactly 4π. By the argument principle, $p(z)$ has two zeros in
the first quadrant. We have seen that there are no zeros on the imaginary
axis, so the remaining zero of $p(z)$ in the upper half-plane lies in the second
quadrant.

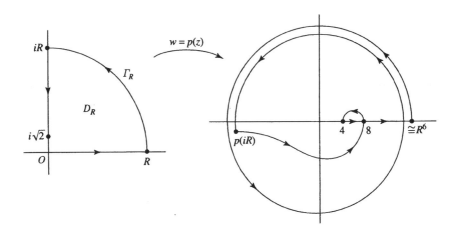

Exercises for VIII.1

1. Show that $z^4 + 2z^2 - z + 1$ has exactly one root in each quadrant.

2. Find the number of zeros of the polynomial $p(z) = z^4 + z^3 + 4z^2 + 3z + 2$ in each quadrant.

3. Find the number of zeros of the polynomial $p(z) = z^6 + 4z^4 + z^3 + 2z^2 + z + 5$ in the first quadrant $\{\operatorname{Re} z > 0,\ \operatorname{Im} z > 0\}$.

4. Find the number of zeros of the polynomial $p(z) = z^9 + 2z^5 - 2z^4 + z + 3$ in the right half-plane.

5. For a fixed real number α, find the number of zeros of $z^4 + z^3 + 4z^2 + \alpha z + 3$ satisfying $\operatorname{Re} z < 0$. (Your answer depends on α.)

6. For a fixed real number α, find the number of solutions of $z^5 + 2z^3 - z^2 + z = \alpha$ satisfying $\operatorname{Re} z > 0$.

7. For a fixed complex number λ, show that if m and n are large integers, then the equation $e^z = z + \lambda$ has exactly $m + n$ solutions in the horizontal strip $\{-2\pi i m < \operatorname{Im} z < 2\pi i n\}$.

8. Show that if $\operatorname{Re} \lambda > 1$, then the equation $e^z = z + \lambda$ has exactly one solution in the left half-plane.

9. Show that if $f(z)$ is analytic in a domain D, and if γ is a closed curve in D such that the values of $f(z)$ on γ lie in the slit plane

$\mathbb{C}\backslash(-\infty, 0]$, then the increase in the argument of $f(z)$ around γ is zero.

2. Rouché's Theorem

There is a general principle to the effect that the number of zeros of an analytic function on a domain does not change if we make a small change in the function. In other words, the number of zeros of an analytic function remains constant under small perturbations. One version of this principle is as follows.

Theorem (Rouché's Theorem). *Let D be a bounded domain with piecewise smooth boundary ∂D. Let $f(z)$ and $h(z)$ be analytic on $D \cup \partial D$. If $|h(z)| < |f(z)|$ for $z \in \partial D$, then $f(z)$ and $f(z) + h(z)$ have the same number of zeros in D, counting multiplicities.*

The hypothesis implies that $f(z) \neq 0$ on ∂D, and that $f(z) + h(z) \neq 0$ on ∂D. From $f(z) + h(z) = f(z)[1 + h(z)/f(z)]$, we obtain

$$(2.1) \qquad \arg(f(z) + h(z)) = \arg(f(z)) + \arg\left(1 + \frac{h(z)}{f(z)}\right).$$

Since $|h(z)/f(z)| < 1$, the values of $1 + h(z)/f(z)$ lie in the right half-plane, and the increase of $\arg(1 + h(z)/f(z))$ around a closed boundary curve is 0. From (2.1) we see that $\arg(f(z) + h(z))$ and $\arg f(z)$ have the same increase around ∂D. By the argument principle, the functions have the same number of zeros in D.

Example. We use Rouché's theorem to find the number of zeros of the polynomial $p(z) = z^6 + 9z^4 + z^3 + 2z + 4$ that lie inside the unit circle. To apply Rouché's theorem, we seek to express $p(z)$ in the form

$$p(z) = \text{BIG} + \text{little},$$

where the function "BIG" dominates "little" on the unit circle, and where it is apparent how many zeros "BIG" has inside the unit circle. Our choice for "BIG" in this case is $f(z) = 9z^4$, which has four zeros inside the unit circle, all at the origin. The function "little" is $h(z) = z^6 + z^3 + 2z + 4$, which satisfies $|h(z)| < |f(z)|$ for $|z| = 1$. By Rouché's theorem, $p(z) = f(z) + h(z)$ also has four zeros inside the unit circle.

Example. We wish to find all solutions of the equation $e^z = 1 + 2z$ that satisfy $|z| < 1$. One obvious solution is $z = 0$. To determine whether there are other solutions, we apply Rouché's theorem to count the number of zeros of $e^z - 1 - 2z$ in the unit disk. In this case, the term $f(z) = -2z$ is the dominant term on the unit circle. We take $h(z) = e^z - 1$, and we estimate

$$|h(z)| = |e^z - 1| = \left| z + \frac{z^2}{2!} + \frac{z^3}{3!} + \cdots \right| \leq |z| + \frac{|z|^2}{2!} + \frac{|z|^3}{3!} + \cdots.$$

If $|z| = 1$, the right-hand side is $e - 1 \approx 1.7 < 2$. Thus $|h(z)| < |f(z)|$ for $|z| = 1$. Since $f(z)$ has only one zero in the unit disk, $f(z) + h(z)$ also has only one zero, and the equation has no solutions in the unit disk other than $z = 0$.

There is a simple proof of the fundamental theorem of algebra based on Rouché's theorem. Let

$$p(z) = z^m + a_{m-1}z^{m-1} + \cdots + a_0$$

be a monic polynomial of degree $m \geq 1$. For $|z|$ large, the term z^m dominates. We take $f(z) = z^m$ and $h(z) = a_{m-1}z^{m-1} + \cdots + a_0$. If R is large, we have $|h(z)| < |f(z)|$ for $|z| = R$. Consequently, $p(z) = f(z) + h(z)$ has the same number of zeros in the disk $|z| < R$ as $f(z) = z^m$, which is m.

Exercises for VIII.2

1. Show that $2z^5 + 6z - 1$ has one root in the interval $0 < x < 1$ and four roots in the annulus $\{1 < |z| < 2\}$.

2. How many roots does $z^9 + z^5 - 8z^3 + 2z + 1$ have between the circles $\{|z| = 1\}$ and $\{|z| = 2\}$?

3. Show that if m and n are positive integers, then the polynomial

$$p(z) = 1 + z + \frac{z^2}{2!} + \cdots + \frac{z^m}{m!} + 3z^n$$

has exactly n zeros in the unit disk.

4. Fix a complex number λ such that $|\lambda| < 1$. For $n \geq 1$, show that $(z-1)^n e^z - \lambda$ has n zeros satisfying $|z - 1| < 1$ and no other zeros in the right half-plane. Determine the multiplicity of the zeros.

5. For a fixed λ satisfying $|\lambda| < 1$, show that $(z-1)^n e^z + \lambda(z+1)^n$ has n zeros in the right half-plane, which are all simple if $\lambda \neq 0$.

6. Let $p(z) = z^6 + 9z^4 + z^3 + 2z + 4$ be the polynomial treated in the example in this section. (a) Determine which quadrants contain the four zeros of $p(z)$ that lie inside the unit circle. (b) Determine which quadrants contain the two zeros of $p(z)$ that lie outside the unit circle. (c) Show that the two zeros of $p(z)$ that lie outside the unit circle satisfy $\{|z \pm 3i| < 1/10\}$.

7. Let $f(z)$ and $g(z)$ be analytic functions on the bounded domain D that extend continuously to ∂D and satisfy $|f(z) + g(z)| < |f(z)| + |g(z)|$ on ∂D. Show that $f(z)$ and $g(z)$ have the same number of

zeros in D, counting multiplicity. *Remark.* This is a variant of Rouché's theorem, in which the hypotheses are symmetric in $f(z)$ and $g(z)$. Rouché's theorem is obtained by setting $h(z) = -f(z) - g(z)$. For the solution of the exercise, see Exercise 9 in the preceding section.

8. Let D be a bounded domain, and let $f(z)$ and $h(z)$ be meromorphic functions on D that extend to be analytic on ∂D. Suppose that $|h(z)| < |f(z)|$ on ∂D. Show by example that $f(z)$ and $f(z) + h(z)$ can have different numbers of zeros on D. What *can* be said about $f(z)$ and $f(z) + h(z)$? Prove your assertion.

9. Let $f(z)$ be a continuously differentiable function on a domain D Suppose that for all complex constants a and b, the increase in the argument of $f(z) + az + b$ around any small circle in D on which $f(z) + az + b \neq 0$ is nonnegative. Show that $f(z)$ is analytic.

3. Hurwitz's Theorem

The argument principle (logarithmic integral) provides a tool for studying the behavior of the zeros of a convergent sequence of analytic functions. The zeros of the functions in the sequence converge to the zeros of the limit function, in a sense made precise by the following theorem.

Theorem (Hurwitz's Theorem). *Suppose $\{f_k(z)\}$ is a sequence of analytic functions on a domain D that converges normally on D to $f(z)$, and suppose that $f(z)$ has a zero of order N at z_0. Then there exists $\rho > 0$ such that for k large, $f_k(z)$ has exactly N zeros in the disk $\{|z - z_0| < \rho\}$, counting multiplicity, and these zeros converge to z_0 as $k \to \infty$.*

For the proof, let $\rho > 0$ be sufficiently small so that the closed disk $\{|z - z_0| \leq \rho\}$ is contained in D and so that $f(z) \neq 0$ for $0 < |z - z_0| \leq \rho$. Choose $\delta > 0$ such that $|f(z)| \geq \delta$ on the circle $|z - z_0| = \rho$. Since $f_k(z)$ converges uniformly to $f(z)$ for $|z - z_0| \leq \rho$, for k large we have $|f_k(z)| > \delta/2$ for $|z - z_0| = \rho$, and further, $f_k'(z)/f_k(z)$ converges uniformly to $f'(z)/f(z)$ for $|z - z_0| = \rho$. Hence the integrals converge,

$$\frac{1}{2\pi i} \int_{|z-z_0|=\rho} \frac{f_k'(z)}{f_k(z)} \, dz \longrightarrow \frac{1}{2\pi i} \int_{|z-z_0|=\rho} \frac{f'(z)}{f(z)} \, dz \,.$$

Now, the expression on the left is the number N_k of zeros of $f_k(z)$ in the disk $\{|z - z_0| < \rho\}$, while the expression on the right is the number N of zeros of $f(z)$ in the disk. Since $N_k \to N$, in fact $N_k = N$ for k large, that is, $f_k(z)$ has N zeros satisfying $|z - z_0| < \rho$. The same argument works for smaller $\rho > 0$, so the zeros of $f_k(z)$ must accumulate at z_0.

An alternative proof of Hurwitz's theorem can be based on Rouché's theorem, applied to the "little" function $f_k(z) - f(z)$ and the "big" function $f(z)$. See Exercise 2.

We say that a function is **univalent** on a domain D if it is analytic and one-to-one on D. In other words, the univalent functions on D are the conformal maps of D to other domains. If we apply the preceding theorem to univalent functions, where $N = 1$, we obtain the following, which is sometimes also referred to as Hurwitz's theorem.

Theorem. *Suppose $\{f_k(z)\}$ is a sequence of univalent functions on a domain D that converges normally on D to a function $f(z)$. Then either $f(z)$ is univalent or $f(z)$ is constant.*

Indeed, suppose the limit function $f(z)$ is not constant. Suppose z_0 and ζ_0 satisfy $f(z_0) = f(\zeta_0) = w_0$. Then z_0 and ζ_0 are zeros of finite order for $f(z) - w_0$. By the preceding theorem, there are sequences $z_k \to z_0$ and $\zeta_k \to \zeta_0$ such that $f_k(z_k) - w_0 = 0$, and $f_k(\zeta_k) - w_0 = 0$. Since $f_k(z)$ is univalent, we have $z_k = \zeta_k$, and in the limit $z_0 = \zeta_0$. Thus $f(z)$ is univalent.

Example. For each $k \geq 1$, the function $f_k(z) = z/k$ is univalent on the complex plane. The sequence $\{f_k(z)\}$ converges normally to the constant function $f(z) \equiv 0$. Thus a normal limit of univalent functions need not be univalent, and both alternatives of the theorem occur.

Exercises for VIII.3

1. Let $\{f_k(z)\}$ be a sequence of analytic functions on D that converges normally to $f(z)$, and suppose that $f(z)$ has a zero of order N at $z_0 \in D$. Use Rouché's theorem to show that there exists $\rho > 0$ such that for k large, $f_k(z)$ has exactly N zeros counting multiplicity on the disk $\{|z - z_0| < \rho\}$.

2. Let $\mathcal{S}$ be the family of univalent functions $f(z)$ defined on the open unit disk $\{|z| < 1\}$ that satisfy $f(0) = 0$ and $f'(0) = 1$. Show that $\mathcal{S}$ is closed under normal convergence, that is, if a sequence in $\mathcal{S}$ converges normally to $f(z)$, then $f \in \mathcal{S}$. *Remark.* It is also true, but more difficult to prove, that $\mathcal{S}$ is a compact family of analytic functions, that is, every sequence in $\mathcal{S}$ has a normally convergent subsequence.

4. Open Mapping and Inverse Function Theorems

Let $f(z)$ be a meromorphic function on a domain D. We say that $f(z)$ **attains the value** w_0 m **times** at z_0 if $f(z) - w_0$ has a zero of order m at z_0. We make the usual modifications to cover the cases $z_0 = \infty$ and

$w_0 = \infty$, so that $f(z)$ attains a finite value w_0 m times at $z_0 = \infty$ if $f(1/z) - w_0$ has a zero of order m at $z = 0$, and $f(z)$ attains the value ∞ m times at z_0 if z_0 is a pole of $f(z)$ of order m.

The number of times that $f(z)$ attains a value w_0 on D is obtained by adding the number of times it attains the value w_0 at the various points of D.

Example. The polynomial $z^m + 1$ attains the value $w = 1$ m times at $z = 0$, and it attains the value $w = \infty$ m times at $z = \infty$.

Example. A polynomial $f(z)$ of degree $m \geq 1$ attains each value $w \in \mathbb{C}^*$ m times on $\mathbb{C}^*$, always counting multiplicity. In fact, for fixed finite w, $f(z) - w$ is a polynomial of degree m, which has m zeros counting multiplicity. Since $f(z)$ has a pole of order m at ∞, it attains the value $w = \infty$ m times at $z = \infty$.

We can use the logarithmic integral to study the dependence on w of the number of times that an analytic or meromorphic function attains the value w. The technique is the same as that used in the preceding section, except that instead of a sequence of functions we treat functions that depend continuously on a parameter. The basic idea is as follows.

Let $f(z)$ be a nonconstant analytic function on a domain D. Let $z_0 \in D$, $w_0 = f(z_0)$, and assume that $f(z) - w_0$ has a zero of order m at z_0. Since the zeros of $f(z) - w_0$ are isolated, we can choose $\rho > 0$ so that $f(z) - w_0 \neq 0$ for $0 < |z - z_0| \leq \rho$. Let $\delta > 0$ satisfy $|f(z) - w_0| \geq \delta$ for $|z - z_0| = \rho$. The integral

$$(4.1) \qquad N(w) = \frac{1}{2\pi i} \int_{|z - z_0| = \rho} \frac{f'(z)}{f(z) - w} \, dz, \qquad |w - w_0| < \delta,$$

is then defined, and it depends analytically on w. Since $N(w)$ is the number of zeros of $f(z) - w$ in the disk $\{|z - z_0| < \rho\}$, the analytic function $N(w)$ is integer-valued, hence constant. Since $N(w_0) = m$, we obtain $N(w) = m$ for $|w - w_0| < \delta$. Thus each value w satisfying $|w - w_0| < \delta$ is assumed exactly m times, counting multiplicity, by $f(z)$ in the disk $\{|z - z_0| < \rho\}$.

As a first consequence of this analysis, we show that nonconstant analytic functions are "open mappings."

Theorem (Open Mapping Theorem for Analytic Functions). *If $f(z)$ is analytic on a domain D, and $f(z)$ is not constant, then $f(z)$ maps open sets to open sets, that is, $f(U)$ is open for each open subset U of D.*

Indeed, let $w_0 \in f(U)$, say $w_0 = f(z_0)$. If we apply the above discussion to $f(z)$, we obtain a disk centered at z_0 of radius ρ and contained in U, such that the values of $f(z)$ on this disk include the disk centered at w_0 of radius δ. This implies that $f(U)$ is open.

Now consider the case where $f(z) - w_0$ has a simple zero at z_0. In this case, $N(w_0) = 1$, and we conclude that $N(w) = 1$ for $|w - w_0| < \delta$. Thus every value w satisfying $|w - w_0| < \delta$ is taken on exactly once by $f(z)$ in the disk $\{|z - z_0| < \rho\}$. This allows us to define an inverse function $f^{-1}(w)$ for $|w - w_0| < \delta$. Further, the residue theorem allows us to derive an explicit formula for the inverse function.

Theorem (Inverse Function Theorem). *Suppose $f(z)$ is analytic for $|z - z_0| \leq \rho$ and satisfies $f(z_0) = w_0$, $f'(z_0) \neq 0$, and $f(z) \neq w_0$ for $0 < |z - z_0| \leq \rho$. Let $\delta > 0$ be chosen such that $|f(z) - w_0| \geq \delta$ for $|z - z_0| = \rho$. Then for each w such that $|w - w_0| < \delta$, there is a unique z satisfying $|z - z_0| < \rho$ and $f(z) = w$. Writing $z = f^{-1}(w)$, we have*

$$(4.2) \qquad f^{-1}(w) = \frac{1}{2\pi i} \int_{|\zeta - z_0| = \rho} \frac{\zeta f'(\zeta)}{f(\zeta) - w} \, d\zeta, \qquad |w - w_0| < \delta.$$

It remains only to establish (4.2). Fix w such that $|w - w_0| < \delta$, and set $z = f^{-1}(w)$, so that $f(z) = w$. The function $\zeta f'(\zeta)/[f(\zeta) - w]$ is an analytic function of ζ for $|\zeta - z_0| \leq \rho$, except for a simple pole at $\zeta = z$ with residue

$$\operatorname{Res}\left[\frac{\zeta f'(\zeta)}{f(\zeta) - w}, z\right] = \lim_{\zeta \to z} \frac{(\zeta - z)\zeta f'(\zeta)}{f(\zeta) - w} = z.$$

The residue theorem yields (4.2) immediately.

We remark that this gives an alternative proof of the existence of a locally defined inverse for $f(z)$ when $f'(z) \neq 0$, which depends on the residue theorem rather than on the inverse function theorem. This procedure also gives an explicit disk on which the inverse function is defined. The explicit formula (4.2) shows that the inverse function is analytic.

Exercises for VIII.4

1. Suppose D is a bounded domain with piecewise smooth boundary. Let $f(z)$ be meromorphic and $g(z)$ analytic on D. Suppose that both $f(z)$ and $g(z)$ extend analytically across the boundary of D, and that $f(z) \neq 0$ on ∂D. Show that

$$\frac{1}{2\pi i} \oint_{\partial D} g(z) \frac{f'(z)}{f(z)} \, dz = \sum_{j=1}^{n} m_j g(z_j),$$

where $z_1, \ldots, z_n$ are the zeros and poles of $f(z)$, and m_j is the order of $f(z)$ at z_j.

2. Let $f(z)$ be a meromorphic function on the complex plane that is doubly periodic. Suppose that the zeros and poles of $f(z)$ are at the points $z_1, \ldots, z_n$ and at their translates by periods of $f(z)$, and

suppose no z_j is a translate by a period of another z_k. Let m_j be the order of $f(z)$ at z_j. Show that $\sum m_j z_j$ is a period of $f(z)$. *Hint.* Integrate $z f'(z)/f(z)$ around the boundary of the fundamental parallelogram P constructed in Section VI.5.

3. Let $\{f_k(z)\}$ be a sequence of analytic functions on a domain D that converges normally to $f(z)$. Suppose that $f_k(z)$ attains each value w at most m times (counting multiplicity) in D. Show that either $f(z)$ is constant, or $f(z)$ attains each value w at most m times in D.

4. Let $f(z)$ be an analytic function on the open unit disk $\mathbb{D} = \{|z| < 1\}$. Suppose there is an annulus $U = \{r < |z| < 1\}$ such that the restriction of $f(z)$ to U is one-to-one. Show that $f(z)$ is one-to-one on $\mathbb{D}$.

5. Let $f(z) = p(z)/q(z)$ be a rational function, where $p(z)$ and $q(z)$ are polynomials that are relatively prime (no common zeros). We define the **degree** of $f(z)$ to be the larger of the degrees of $p(z)$ and $q(z)$. Denote the degree of $f(z)$ by d. (a) Show that each value $w \in \mathbb{C}$, $w \neq f(\infty)$, is assumed d times by $f(z)$ on $\mathbb{C}$. (b) Show that $f(z)$ attains each value $w \in \mathbb{C}^*$ d times on $\mathbb{C}^*$ (as always, counting multiplicity).

6. Let $f(z)$ be a meromorphic function on the complex plane, and suppose there is an integer m such that $f^{-1}(w)$ has at most m points for all $w \in \mathbb{C}$. Show that $f(z)$ is a rational function.

7. Let $F(z, w)$ be a continuous function of z and w that depends analytically on z for each fixed w, and let $F_1(z, w)$ denote the derivative of $F(z, w)$ with respect to z. Suppose $F(z_0, w_0) = 0$, and $F_1(z_0, w_0) \neq 0$. Choose ρ such that $F(z, w_0) \neq 0$ for $0 < |z - z_0| \leq \rho$.
   (a) Show that there exists $\delta > 0$ such that if $|w - w_0| < \delta$, there is a unique $z = g(w)$ satisfying $|z - z_0| < \rho$ and $F(z, w) = 0$.
   (b) Show that

$$g(w) = \frac{1}{2\pi i} \int_{|\zeta - z_0| = \rho} \frac{\zeta F_1(\zeta, w)}{F(\zeta, w)} \, d\zeta, \qquad |w - w_0| < \delta.$$

   (c) Suppose further that $F(z, w)$ is analytic in w for each fixed z, and let $F_2(z, w)$ denote the derivative of $F(z, w)$ with respect to w. Show that $g(w)$ is analytic, and

$$g'(w) = -F_2(g(w), w)/F_1(g(w), w).$$

   (d) Derive the inverse function theorem given in this section, together with the usual formula for the derivative of the inverse function, as a corollary of (a), (b), and (c).

Remark. This is the **implicit function theorem for analytic functions**. Note that a specific formula is given for the function $g(w)$ defined implicitly by $F(g(w), w) = 0$.

8. Let D be a bounded domain, and let $f(z)$ be a continuous function on $D \cup \partial D$ that is analytic on D. Show that $\partial(f(D)) \subseteq f(\partial D)$, that is, the boundary of the open set $f(D)$ is contained in the image under $f(z)$ of the boundary of D.

5. Critical Points

Let $f(z)$ be a nonconstant analytic function on a domain D. A point z_0 is called a **critical point** of $f(z)$ if $f'(z_0) = 0$. The value $f(z_0) = w_0$ is called a **critical value** of $f(z)$. We define the **order** of the critical point z_0 to be the order of zero of $f'(z)$ at z_0. Since the critical points are the zeros of the nonconstant analytic function $f'(z)$, critical points are isolated. Our aim is to understand the behavior of an analytic function near a critical point, and to understand the behavior of the inverse function near the corresponding critical value.

Example. The function $f(z) = z^m + c$ has a critical point of order $m - 1$ at $z = 0$, with critical value c. The inverse function for $f(z)$ is given by the m branches of $(w - c)^{1/m}$.

Suppose z_0 is a critical point of $f(z)$ of order $m - 1$ with critical value w_0. Then $f(z) - w_0$ has a zero of order m at z_0, and we can proceed as before by considering the logarithmic integral (4.1). This time the constant value of $N(w)$ is m. Thus for $\rho > 0$ small and for w near w_0, there are m points $z_1(w), \ldots, z_m(w)$, repeated according to multiplicity, such that $f(z)$ attains the value w in the disk $\{|z - z_0| < \rho\}$ precisely at the points $z_1(w), \ldots, z_m(w)$. We wish to see how the points $z_j(w)$ depend on w. To do this, we will use a simple procedure, based only on the existence of an analytic inverse for an analytic function at a noncritical point. The procedure is to make a change of variable in order to show that the behavior of $f(z)$ near z_0 is effectively the same as the behavior of the analytic function ζ^m at the critical point $\zeta = 0$. We proceed as follows.

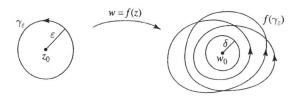

Since $f(z) - w_0$ has order m at z_0, we can factor $f(z) - w_0 = (z - z_0)^m h(z)$, where $h(z)$ is analytic at z_0, and $h(z_0) \neq 0$. Near z_0 we define an analytic branch of $h(z)^{1/m}$, and then for $g(z) = (z - z_0)h(z)^{1/m}$ we have

$$f(z) - w_0 = g(z)^m, \qquad f(z) = g(z)^m + w_0.$$

Moreover, $g(z)$ is analytic at z_0 and has a simple zero there, so $g'(z_0) \neq 0$ and $g(z)$ is one-to-one near z_0. Thus $f(z)$ is represented as the composition of three functions, the univalent function $g(z)$, followed by the map $\zeta \mapsto \zeta^m$, followed by a translation $\xi \mapsto \xi + w_0$. Since we understand the power function $\xi = \zeta^m$ and the translation $w = \xi + w_0$ very well, and since $g(z)$ is univalent near z_0, we can draw from this a clear picture of the behavior of $f(z)$. For instance, the set where ζ^m is real consists of m straight line segments passing through 0, dividing the ζ-plane near $\zeta = 0$ into $2m$ sectors of aperture π/m, on which $\text{Im}(\zeta^m)$ is alternately positive and negative. The set where $f(z) - w_0$ is real is the image under $g^{-1}(\zeta)$ of these line segments and hence consists of m "analytic" curves through z_0. They divide the z-plane near z_0 into $2m$ pie-slice domains with vertex at z_0 and aperture π/m, on which $\text{Im}(f(z) - w_0)$ is alternately positive and negative.

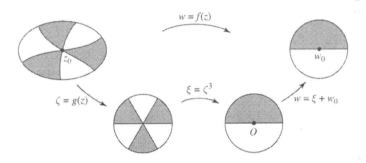

A point w near w_0, $w \neq w_0$, has m distinct preimages $z_1(w), \ldots, z_m(w)$. They are the m branches of $(w - w_0)^{1/m}$ composed with $g^{-1}(\zeta)$. If we make a branch cut and consider the principal branch $(w - w_0)^{1/m}$ on the slit disk $\{|w - w_0| < \delta\} \backslash (w_0 - \delta, w_0]$, the other branches are of the form $e^{2\pi ij/m}(w - w_0)^{1/m}$, and the preimages of w are given by the composition of $g^{-1}(\zeta)$ and these branches,

$$z_j(w) = g^{-1}\left(e^{2\pi ij/m}(w - w_0)^{1/m}\right), \qquad 1 \le j \le m.$$

In particular, for $w \neq w_0$, the preimages $z_j(w)$ are distinct, and each $z_j(w)$ depends analytically on w. The values of $z_j(w)$ at the "top edge" of the slit coincide with the values of $z_{j+1}(w)$ at the "bottom edge" of the slit, so that $z_j(w)$ is continued analytically to $z_{j+1}(w)$ when w follows a circle around w_0 in the positive direction. In the same way that we constructed a Riemann surface for the inverse function of ζ^m, we can construct a Riemann surface for the inverse function $z(w)$ of $w = f(z)$. We create m copies of the slit disk, and we define $z(w)$ to be $z_j(w)$ on the jth copy. Then we glue

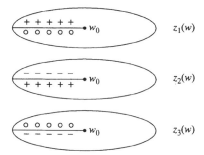

together the edges of the branch cuts so that the function $z(w)$ becomes continuous on the surface.

Now consider the function

$$(z - z_1(w)) \cdots (z - z_m(w)) = z^m + a_{m-1}(w)z^{m-1} + \cdots + a_1(w)z + a_0(w),$$

where

$$a_0(w) = (-1)^m z_1(w) \cdots z_m(w),$$

$$\vdots$$

$$a_{m-1}(w) = -\sum_{j=1}^{m} z_j(w).$$

If we continue $a_k(w)$ around a circle centered at w_0, the values of the $z_j(w)$'s are permuted and we return to the same value for $a_k(w)$. Thus $a_k(w)$ is analytic in the punctured disk $0 < |w - w_0| < \varepsilon$. Evidently, $a_k(w)$ is bounded as $w \to w_0$. By Riemann's theorem on removable singularities, $a_k(w)$ is analytic at w_0. In fact, the value $a_k(w_0)$ is just the coefficient of z^k in the expansion of $(z - z_0)^m$ in powers of z. Thus we have found a polynomial equation

(5.1) $z^m + a_{m-1}(w)z^{m-1} + \cdots + a_1(w)z + a_0(w) = 0,$

with analytic coefficients $a_0(w), \ldots, a_{m-1}(w)$, that reduces to the equation $(z - z_0)^m = 0$ at $w = w_0$ and that has solutions given precisely by $z_1(w), \ldots, z_m(w)$.

We summarize our results as follows.

Theorem. Let $f(z)$ be analytic at z_0. Suppose z_0 is a critical point of order $m - 1$ for $f(z)$, with critical value $f(z_0) = w_0$. Let $\rho > 0$ satisfy $f(z) \neq w_0$ for $0 < |z - z_0| \leq \rho$, and let $\delta > 0$ satisfy $|f(z) - w_0| \geq \delta$ for $|z - z_0| = \rho$. Then for each w such that $0 < |w - w_0| < \delta$, the equation $f(z) = w$ has exactly m distinct solutions $z_1(w), \ldots, z_m(w)$ in the disk $\{|z - z_0| < \rho\}$. The functions $z_1(w), \ldots, z_m(w)$ can be chosen to depend analytically on w in the slit disk $\{|w - w_0| < \delta\}\backslash(w_0 - \delta, w_0]$. They form the m branches of an analytic function $z = z(w)$ on an m-sheeted Riemann *surface over the punctured disk* $\{0 < |w - w_0| < \delta\}$. *The* $z_j(w)$*'s are the solutions of a polynomial equation* (5.1) *with coefficients* $a_j(w)$ *that are analytic for* $|w - w_0| < \delta$.

We have dealt with the solutions of the equation $f(z) - w = 0$ near a solution point (z_0, w_0). A similar state of affairs holds for the solutions of an equation $F(z, w) = 0$, where $F(z, w)$ depends analytically on both z and w in a neighborhood of a solution point (z_0, w_0), though in this case the solution set is a finite union of surfaces passing through (z_0, w_0). This is the simplest example of an **analytic variety**, which is defined to be the set of common zeros of a family of analytic functions on some domain in complex n-space.

Exercises for VIII.5

1. Find the critical points and critical values of $f(z) = z + 1/z$. Sketch the curves where $f(z)$ is real. Sketch the regions where $\text{Im} f(z) > 0$ and where $\text{Im} f(z) < 0$.

2. Suppose $g(z)$ is analytic at $z = 0$, with power series $g(z) = 2 + iz^4 + \mathcal{O}(z^5)$. Sketch and label the curves passing through $z = 0$ where $\text{Re} g(z) = 2$ and where $\text{Im} g(z) = 0$.

3. Find the critical points and critical values of $f(z) = z^2 + 1$. Sketch the set of points z such that $|f(z)| \le 1$, and locate the critical points of $f(z)$ on the sketch.

4. Suppose that $f(z)$ is analytic at z_0. Show that if the set of z such that $\text{Re} f(z) = \text{Re} f(z_0)$ consists of just one curve passing through z_0, then $f'(z_0) \ne 0$. Show also that if the set of z such that $|f(z)| = |f(z_0)|$ consists of just one curve passing through z_0, then $f'(z_0) \ne 0$.

5. How many critical points, counting multiplicity, does a polynomial of degree m have in the complex plane? Justify your answer.

6. Find and plot the critical points and critical values of $f(z) = z^2 + 1$ and of its iterates $f(f(z)) = (z^2 + 1)^2 + 1$ and $f(f(f(z)))$. *Suggestion.* Use the chain rule.

7. Let $f(z)$ be a polynomial of degree m. How many (finite) critical points does the N-fold iterate $f \circ \cdots \circ f$ (N times) have? Describe them in terms of the critical points of $f(z)$.

8. We define a pole z_0 of $f(z)$ to be a **critical point of $f(z)$ of order k** if z_0 is a critical point of $1/f(z)$ of order k. We define $z = \infty$ to be a **critical point of $f(z)$ of order k** if $w = 0$ is a critical point of $g(w) = f(1/w)$ of order k. Show that with this definition, a point

$z_0 \in \mathbb{C}^*$ is a critical point of order k for a meromorphic function $f(z)$ if and only if there are open sets U containing z_0 and V containing $w_0 = f(z_0)$ such that each $w \in V$, $w \neq w_0$, has exactly $k + 1$ preimages in U. *Remark.* We say that $f(z)$ is a $(k + 1)$-**sheeted covering of** $f^{-1}(V \setminus \{w_0\}) \cap U$ **over** $V \setminus \{w_0\}$.

9. Show that a polynomial of degree m, regarded as a meromorphic function on C^*, has a critical point of order $m - 1$ at $z_0 = \infty$.

10. Locate the critical points and critical values in the extended complex plane of the polynomial $f(z) = z^4 - 2z^2$. Determine the order of each critical point. Sketch the set of points z such that $\operatorname{Im} f(z) \geq 0$.

11. Show that if f is a rational function, and if g is a fractional linear transformation, then f and $g \circ f$ have the same critical points in the extended complex plane $\mathbb{C}^*$. What can be said about the critical values of $g \circ f$? What can be said about the critical points and critical values of $f \circ g$?

12. Let $f(z) = p(z)/q(z)$ be a rational function of degree d, so that $p(z)$ and $q(z)$ are relatively prime, and d is the larger of the degrees of $p(z)$ and $q(z)$. (See Exercise 4.5.) Show that $f(z)$ has $2d - 2$ critical points, counting multiplicity, in the extended complex plane $\mathbb{C}^*$. *Hint.* If $\deg p \neq \deg q$, then the number of critical points of $f(z)$ in the finite plane $\mathbb{C}$ is $\deg(qp' - q'p) = \deg p + \deg q - 1$, while the order of the critical point at ∞ is $|\deg p - \deg q| - 1$.

13. Show that the set of solution points (w, z) of the equation $z^2 - 2(\cos w)z + 1 = 0$ consists of the graphs of two entire functions $z_1(w)$ and $z_2(w)$ of w. Specify the entire functions, and determine where their graphs meet. *Remark.* The solution set forms a reducible one-dimensional analytic variety in $\mathbb{C}^2$.

14. Let $a_0(w), \dots, a_{m-1}(w)$ be analytic in a neighborhood of $w = 0$ and vanish at $w = 0$. Consider the monic polynomial in z whose coefficients are analytic functions in w,

$$P(z, w) = z^m + a_{m-1}(w)z^{m-1} + \cdots + a_0(w), \qquad |w| < \delta.$$

Suppose that for each fixed w, $0 < |w| < \delta$, there are m distinct solutions of $P(z, w) = 0$.

(a) Show that the m roots of the equation $P(z, w) = 0$ determine analytic functions $z_1(w), \dots, z_m(w)$ in the slit disk $\{|w| < \delta\} \setminus (-\delta, 0]$. *Hint.* Use the implicit function theorem (Exercise 4.7).

(b) Glue together branch cuts to form an m-sheeted (possibly disconnected) surface over the punctured disk $\{0 < |w| < \delta\}$ on which the branches $z_j(w)$ determine a continuous function.

(c) Suppose that the indices are arranged so that for some fixed k, $1 \leq k \leq n$, the continuation of $z_j(w)$ once around $w = 0$ is $z_{j+1}(w)$ for $1 \leq j \leq k-1$, while the continuation of $z_k(w)$ once around $w = 0$ is $z_1(w)$. Show that $Q(z, w) = (z - z_1(w)) \cdots (z - z_k(w))$ determines a polynomial in z whose coefficients are analytic functions of w for $|w| < \delta$. Show further that the polynomial $Q(z, w)$ is an irreducible factor of $P(z, w)$, and that all irreducible factors of $P(z, w)$ arise from subsets of the $z_j(w)$'s in this way.

(d) Show that if $f(z)$ is an analytic function that has a zero of order m at $z = 0$, and $z_1(w), \ldots, z_m(w)$ are the solutions of the equation $w = f(z)$, then the polynomial $P(z, w) = (z - z_1(w)) \cdots (z - z_m(w))$ is irreducible.

15. Consider monic polynomials in z of the form

$$P(z, w) = z^m + a_{m-1}(w)z^{m-1} + \cdots + a_0(w),$$

where the functions $a_0(w), \ldots, a_{m-1}(w)$ are defined and meromorphic in some disk centered at $w = 0$. Let $P_0(z, w)$ and $P_1(z, w)$ be two such polynomials, and consider the following algorithm. Using the division algorithm, find polynomials $A_2(z, w)$ and $P_2(z, w)$ such that $P_0(z, w) = A_2(z, w)P_1(z, w) + P_2(z, w)$ and the degree of $P_2(z, w)$ is less than the degree of $P_1(z, w)$. Continue in this fashion, finding polynomials $A_{j+1}(z, w)$ and $P_{j+1}(z, w)$ such that

$$P_{j-1}(z, w) = A_{j+1}(z, w)P_j(z, w) + P_{j+1}(z, w)$$

and $\deg P_{j+1}(z, w) < \deg P_j(z, w)$, until eventually we reach

$$P_{\ell+1}(z, w) = 0, \qquad P_{\ell-1}(z, w) = A_\ell(z, w)P_\ell(z, w).$$

Let $D(z, w)$ be the monic polynomial in z obtained by dividing $P_\ell(z, w)$ by the coefficient of the highest power of z.

(a) Show that $D(z, w)$ is the greatest common divisor of $P_0(z, w)$ and $P_1(z, w)$, in the sense that $D(z, w)$ divides both $P_0(z, w)$ and $P_1(z, w)$, and each polynomial that divides both $P_0(z, w)$ and $P_1(z, w)$ also divides $D(z, w)$.

(b) Show that there are polynomials $A(z, w)$ and $B(z, w)$ such that $D = AP_0 + BP_1$.

(c) Show that if $P_0(z, w)$ and $P_1(z, w)$ are relatively prime (that is, $D(z, w) = 1$), then there is $\varepsilon > 0$ such that for each fixed w, $0 < |w| < \varepsilon$, the polynomials $P_0(z, w)$ and $P_1(z, w)$ have no common zeros.

(d) Show that any polynomial $P(z, w)$ as above can be factored as a product of irreducible polynomials, and the factorization is unique up to the order of the factors and multiplication of a factor by a meromorphic function in w.

(e) Show that if the coefficients of $P(z, w)$ are analytic at $w = 0$, then the irreducible factors of $P(z, w)$ can be chosen so that their coefficients are analytic at $w = 0$.

(f) Show that if $P(z, w)$ is irreducible, then there is $\varepsilon > 0$ such that for each fixed w, $0 < |w| < \varepsilon$, the roots of $P(z, w)$ are distinct.

(g) Show that the results of Exercise 14(a)–(c) hold without the supposition that the solutions of $P(z, w) = 0$ are distinct.

6. Winding Numbers

Let $\gamma(t)$, $a \leq t \leq b$, be a closed path in D. We define the **trace of** γ to be the image $\Gamma = \gamma([a, b])$ of γ. In this section and the next it will be important to distinguish between the parameterized path γ and its trace Γ.

For $z_0 \notin \Gamma$, we define the **winding number** $W(\gamma, z_0)$ of γ around z_0 to be the increase in the argument of $z - z_0$ around γ, normalized by dividing by 2π. If γ is piecewise smooth, the winding number is the integer

$$(6.1) \quad W(\gamma, z_0) = \frac{1}{2\pi i} \int_\gamma \frac{dz}{z - z_0} = \frac{1}{2\pi} \int_\gamma d \arg(z - z_0), \qquad z_0 \notin \Gamma.$$

We may think of a cameraman standing at z_0 and videotaping a child emerging through a door, romping around, and returning to the door. The winding number of γ around z_0 is the number of revolutions the cameraman makes as the child runs around the path γ, counting counterclockwise as positive and clockwise as negative.

The winding number around z_0 is defined not only for piecewise smooth curves but also for any closed path γ. It is determined by choosing values $h(t)$ of $\arg(\gamma(t) - z_0)$ that vary continuously with t for $a \leq t \leq b$, and setting

$$W(\gamma, z_0) = \frac{1}{2\pi}[h(b) - h(a)].$$

If there is a single-valued determination $\psi(z)$ of $\arg(z - z_0)$ defined on Γ, we can take $h(t) = \psi(\gamma(t))$, and then $h(b) = h(a)$, so that the winding number of γ around z_0 is 0.

Example. The closed curve γ represented by the figure loops twice around z_2 but only once around z_1, and so $W(\gamma, z_2) = 2$, $W(\gamma, z_1) = 1$. For z_0 lying to the left of the curve, there is a single-valued determination of the argument function defined for $z \in \Gamma$, namely the principal branch $\operatorname{Arg}(z - z_0)$. Consequently $W(\gamma, z_0) = 0$.

The representation of the winding number as a Cauchy integral in (6.1) shows that $W(\gamma, \zeta)$ depends analytically on ζ for $\zeta \notin \Gamma$. Since the function is integer-valued, it is constant on each component of the complement of Γ. Since the Cauchy integral in (6.1) tends to 0 as $z_0 \to \infty$, the integer

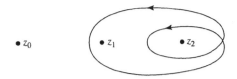

$W(\gamma, \zeta)$ must be 0 on the unbounded component of the complement of Γ. This can be seen also by noting that there is a single-valued determination of $\arg(z - \zeta)$ defined for $z \in \Gamma$ when ζ is near ∞. We state our observations as a theorem.

Theorem. *Let* $\gamma(t)$, $a \leq t \leq b$, *be a closed path in the complex plane, and let* $\Gamma = \gamma([a, b])$ *be its trace. The winding number* $W(\gamma, \zeta)$ *is constant on each connected component of* $\mathbb{C} \backslash \Gamma$. *Further,* $W(\gamma, \zeta) = 0$ *for all* ζ *in the unbounded component of* $\mathbb{C} \backslash \Gamma$.

Various of the theorems involving Cauchy integrals around closed curves can be rephrased in terms of winding numbers. We have seen that if D is a star-shaped domain, and if $f(z)$ is analytic on D, then $\int_\gamma f(z)\, dz = 0$ around any closed curve γ in D. This statement extends to arbitrary domains, provided that γ does not wind around a component of the complement of D.

Theorem. *If* $f(z)$ *is analytic on a domain* D, *then* $\int_\gamma f(z)\, dz = 0$ *for each closed path* γ *in* D *such that* $W(\gamma, \zeta) = 0$ *for all* $\zeta \in \mathbb{C} \backslash D$.

The first step of the proof is to find a bounded domain U whose boundary consists of a finite number of piecewise smooth closed curves, such that U and its boundary are contained in D, and such that U contains the trace Γ of γ. To do this, we choose $\delta > 0$ so small that every point of Γ has distance at least 4δ from any point of $\mathbb{C} \backslash D$, and we divide the complex plane into a grid of closed squares of side length δ. (See the figure.) Let K be the union of all the (closed) squares in the grid that contain a point with distance less than δ from Γ, and let $U = K \backslash \partial K$. Then U is a domain with boundary $\partial U = \partial K$ consisting of the sides of squares in K that are adjacent to a square in the grid not in K. By following along these sides, we see that ∂U is a finite union of closed curves consisting of consecutive sides of squares. Two boundary curves may intersect at a point where two squares in U are kitty-corner from each other, but these intersections can be eliminated by adjoining to U a smaller square with vertex in the corner. Finally, we adjoin to U the squares from any component of $\mathbb{C}^* \backslash U$ that does not meet $\mathbb{C}^* \backslash D$. Then each component of $\mathbb{C}^* \backslash U$ meets $\mathbb{C}^* \backslash D$.

Green's theorem is proved (completely rigorously) for domains built in this way from contiguous squares, so that the Cauchy integral theorem is

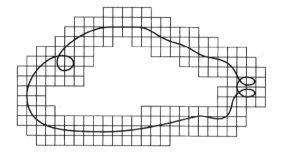

at our disposal. If $z \in \Gamma$, we have

$$f(z) = \frac{1}{2\pi i} \int_{\partial U} \frac{f(\zeta)}{\zeta - z} \, d\zeta.$$

If we integrate around γ and then interchange the order of integration, we obtain

$$\int_\gamma f(z) \, dz = \frac{1}{2\pi i} \int_{\partial U} \left[\int_\gamma \frac{1}{\zeta - z} \, dz \right] f(\zeta) \, d\zeta = -\int_{\partial U} W(\gamma, \zeta) f(\zeta) \, d\zeta.$$

The winding number $W(\gamma, \zeta)$ is constant on each connected component of $\mathbb{C} \backslash \Gamma$, and it is zero for $\zeta \in \mathbb{C} \backslash D$. Since each component of $\mathbb{C} \backslash U$ meets $\mathbb{C} \backslash D$, it is zero for $\zeta \in \mathbb{C} \backslash U$, hence also for $\zeta \in \partial U$. Hence the integral is also zero.

As an analogue of the Cauchy integral formula we have the following.

Theorem. Let $f(z)$ be analytic on a domain D, and let γ be a closed path in D with trace $\Gamma = \gamma([a, b])$. If $W(\gamma, \zeta) = 0$ for all $\zeta \in \mathbb{C} \backslash D$, then

$$\frac{1}{2\pi i} \int_\gamma \frac{f(z)}{z - z_0} \, dz = W(\gamma, z_0) f(z_0), \qquad z_0 \in D \backslash \Gamma.$$

The proof is simple. We apply the preceding version of the Cauchy integral theorem to the analytic function $g(z) = (f(z) - f(z_0))/(z - z_0)$. From $\int_\gamma g(z) \, dz = 0$, we obtain

$$\frac{1}{2\pi i} \int_\gamma \frac{f(z)}{z - z_0} \, dz = \frac{1}{2\pi i} \int_\gamma \frac{f(z_0)}{z - z_0} \, dz = W(\gamma, z_0) f(z_0).$$

Winding numbers can be defined for a finite collection of closed curves, by simply adding the winding numbers for the various curves in the collection. (See the exercises for Section 8.) Thus for instance, if D is a bounded domain with piecewise smooth boundary, we define the winding number $W(\partial D, \zeta)$ of ∂D around $\zeta \notin \partial D$ to be the sum of the winding numbers of the various component curves of ∂D, appropriately oriented. From Cauchy's theorem and the Cauchy integral representation formula, we then have

$$W(\partial D, \zeta) \;=\; \frac{1}{2\pi i} \int_{\partial D} \frac{dz}{z-\zeta} \;=\; \begin{cases} 1, & \zeta \in D, \\ 0, & \zeta \notin D \cup \partial D. \end{cases}$$

Thus the winding number of ∂D around points ζ outside of $D \cup \partial D$ is 0, and the winding number jumps to $+1$ as ζ crosses over ∂D into D. This is a special case of the jump theorem for Cauchy integrals treated in the next section.

Exercises for VIII.6

1. Sketch the closed path $\gamma(t) = e^{it} \sin(2t)$, $0 \le t \le 2\pi$, and determine the winding number $W(\gamma, \zeta)$ for each point ζ not on the path.

2. Sketch the closed path $\gamma(t) = e^{-2it} \cos t$, $0 \le t \le 2\pi$, and determine the winding number $W(\gamma, \zeta)$ for each point ζ not on the path.

3. Let $f(z)$ be analytic on an open set containing a closed path γ, and suppose $f(z) \neq 0$ on γ. Show that the increase in $\arg f(z)$ around γ is $2\pi W(f \circ \gamma, 0)$.

4. Let D be a domain, and suppose z_0 and z_1 lie in the same connected component of $\mathbb{C}\backslash D$. (a) Show that the increase in the argument of $f(z) = (z - z_0)(z - z_1)$ around any closed curve in D is an even multiple of 2π. (b) Show that $(z - z_0)(z - z_1)$ has an analytic square root in D. (c) Show by example that $(z - z_0)(z - z_1)$ does not necessarily have an analytic cube root in D.

5. Show that if γ is a piecewise smooth closed curve in the complex plane, with trace Γ, and if $z_0 \notin \Gamma$, then

$$\int_\gamma \frac{1}{(z - z_0)^n}\, dz \;=\; 0, \qquad n \ge 2.$$

6. Let γ be a closed path in a domain D such that $W(\gamma, \zeta) = 0$ for all $\zeta \notin D$. Suppose that $f(z)$ is analytic on D except possibly at a finite number of isolated singularities $z_1, \dots, z_m \in D\backslash\Gamma$. Show that

$$\int_\gamma f(z)\, dz \;=\; 2\pi i \sum W(\gamma, z_k)\, \mathrm{Res}[f, z_k].$$

Hint. Consider the Laurent decomposition at each z_k, and use Exercise 5.

7. Evaluate

$$\frac{1}{2\pi i} \int_\gamma \frac{dz}{z(z^2 - 1)},$$

where γ is the closed path indicated in the figure. *Hint.* Either use Exercise 6, or proceed directly with partial fractions.

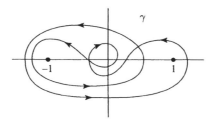

8. Let $\gamma(t)$ and $\sigma(t)$, $a \le t \le b$, be closed paths. (a) Show that if $\zeta \in \mathbb{C}$ does not lie on the straight line segment between $\gamma(t)$ and $\sigma(t)$, for $a \le t \le b$, then $W(\sigma, \zeta) = W(\gamma, \zeta)$. (b) Show that if $|\sigma(t) - \gamma(t)| < |\zeta - \gamma(t)|$ for $a \le t \le b$, then $W(\sigma, \zeta) = W(\gamma, \zeta)$.

9. Let $f(z)$ be a continuous complex-valued function on the complex plane such that $f(z)$ is analytic for $|z| < 1$, $f(z) \ne 0$ for $|z| \ge 1$, and $f(z) \to 1$ as $z \to \infty$. Show that $f(z) \ne 0$ for $|z| < 1$.

10. Let K be a nonempty closed bounded subset of the complex plane, and let $f(z)$ be a continuous complex-valued function on the complex plane that is analytic on $\mathbb{C}\backslash K$ and at ∞. Show that every value attained by $f(z)$ on $\mathbb{C}^* = \mathbb{C} \cup \{\infty\}$ is attained by $f(z)$ somewhere on K, that is, $f(\mathbb{C}^*) = f(K)$.

7. The Jump Theorem for Cauchy Integrals

Suppose γ is a piecewise smooth curve with trace Γ, and let $f(z)$ be a continuous function on Γ. We define the **Cauchy integral of** $f(z)$ along γ to be the function $F(\zeta)$ defined off Γ by

(7.1)
$$F(\zeta) = \frac{1}{2\pi i} \int_\gamma \frac{f(z)}{z - \zeta}\, dz, \qquad \zeta \in \mathbb{C}\backslash\Gamma.$$

We have seen that the Cauchy integral $F(\zeta)$ is analytic on $\mathbb{C}\backslash\Gamma$, and $F(\zeta)$ vanishes at ∞.

Example. If γ is a closed curve parameterizing the (positively oriented) boundary Γ of a domain D, and if $f(z)$ is analytic on D and across Γ, then the Cauchy integral $F(\zeta)$ of $f(z)$ around γ is given by

$$F(\zeta) = \begin{cases} f(\zeta), & \zeta \in D, \\ 0, & \zeta \in \mathbb{C}\backslash(D \cup \Gamma). \end{cases}$$

This follows from Cauchy's theorem if $\zeta \in \mathbb{C}\backslash(D \cup \Gamma)$, and from the Cauchy integral formula if $\zeta \in D$.

Example. If γ is a closed path, the Cauchy integral of the constant function $f(z) \equiv 1$ around γ is the winding number $F(\zeta) = W(\gamma, \zeta)$ of γ around ζ.

Now suppose that the continuous function $f(z)$ on Γ is analytic at some point $z_0 \in \Gamma$. We suppose that there is a small disk U centered at z_0 such that $f(z)$ is analytic on $U \cup \partial U$ and such that γ passes through U in a piecewise smooth subarc γ_0 traveling from $z_1 \in \partial U$ to another point $z_2 \in \partial U$, so that $\gamma \backslash \gamma_0$ does not enter U, while the trace Γ_0 of γ_0 divides U into two connected open domains. As we traverse γ_0 in the positive direction, one of these domains U_+ lies on the right side of γ_0, the other U_- on the left side. We break the integral for $F(\zeta)$ into two pieces

$$F(\zeta) = \frac{1}{2\pi i} \int_{\gamma_0} \frac{f(z)}{z - \zeta} \, dz + \frac{1}{2\pi i} \int_{\gamma \backslash \gamma_0} \frac{f(z)}{z - \zeta} \, dz = F_0(\zeta) + G(\zeta).$$

The second integral $G(\zeta)$ is analytic on U, while the first integral $F_0(\zeta)$ is analytic on $U \backslash \Gamma_0$. Let γ_+ denote the arc of ∂U from z_1 to z_2 that lies in the boundary of U_+, and let γ_- denote the arc of ∂U from z_1 to z_2 that lies in the boundary of U_-, as in the next figure. We define

$$F_+(\zeta) = \frac{1}{2\pi i} \int_{\gamma_-} \frac{f(z)}{z - \zeta} \, dz + G(\zeta), \qquad \zeta \in U.$$

By Cauchy's theorem,

$$\frac{1}{2\pi i} \int_{\gamma_0} \frac{f(z)}{z - \zeta} \, dz = \frac{1}{2\pi i} \int_{\gamma_-} \frac{f(z)}{z - \zeta} \, dz, \qquad \zeta \in U_+,$$

and consequently, $F(\zeta) = F_+(\zeta)$ for $\zeta \in U_+$. Similarly, we define

$$F_-(\zeta) = \frac{1}{2\pi i} \int_{\gamma_+} \frac{f(z)}{z - \zeta} \, dz + G(\zeta), \qquad \zeta \in U,$$

and then from Cauchy's theorem we obtain $F(\zeta) = F_-(\zeta)$ for $\zeta \in U_-$. If we integrate along the arc γ_+ of ∂U from z_1 to z_2, then backwards along the arc γ_- of ∂U from z_2 to z_1, we obtain the integral around the boundary ∂U. This and Cauchy's theorem yield

$$F_-(\zeta) - F_+(\zeta) = \frac{1}{2\pi i} \int_{\partial U} \frac{f(z)}{z - \zeta} \, dz = f(\zeta), \qquad \zeta \in U.$$

We have proved the following.

Theorem (Jump Theorem for Cauchy Integrals). *Suppose that Γ is a smooth curve that passes through z_0, and $f(z)$ is a continuous function on Γ that is analytic at z_0. Let U be a small disk containing z_0 such that $f(z)$ is analytic on U and such that Γ divides U into the two components U_+ and U_- as above. Then there are analytic functions $F_+(\zeta)$ and $F_-(\zeta)$ on U satisfying*

$$F_-(\zeta) - F_+(\zeta) = f(\zeta), \qquad \zeta \in U,$$

and such that the Cauchy integral $F(\zeta)$ of $f(z)$, defined by (7.1), satisfies

$$F(\zeta) \;=\; \begin{cases} F_+(\zeta), & \zeta \in U_+, \\ F_-(\zeta), & \zeta \in U_-. \end{cases}$$

Thus as ζ crosses Γ from right to left (from U_+ to U_-), the values of the Cauchy integral $F(\zeta)$ jump by $f(\zeta)$,

$$F_-(\zeta) \;=\; F_+(\zeta) + f(\zeta), \qquad \zeta \in U.$$

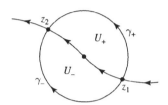

Example. In the first example above, where γ parameterizes the boundary Γ of a domain D and $f(z)$ is analytic on D and across Γ, the Cauchy integral of $f(z)$ jumps from 0 outside Γ to $f(\zeta)$ inside Γ.

Now consider the special case where $f(z) \equiv 1$ and γ is a closed curve. In this case, the Cauchy integral (7.1) is the winding number $W(\gamma, \zeta)$, and we are led to the following jump theorem.

Theorem (Jump Theorem for Winding Numbers). *Let γ be a closed path with trace Γ, and let $z_0 = \gamma(t_0) \in \Gamma$. Suppose that $\gamma(t)$ is continuously differentiable for t near t_0, with $\gamma'(t_0) \neq 0$. Let U be a small disk centered at z_0 that is divided into two components $U_\pm$ by a segment γ_0 on γ as above, and suppose that $\gamma \backslash \gamma_0$ does not enter U. Then as $\zeta \in U$ crosses Γ from right to left (from U_+ to U_-), the winding number $W(\gamma, \zeta)$ jumps by $+1$,*

$$W(\gamma, \zeta_-) \;=\; W(\gamma, \zeta_+) + 1, \qquad \zeta_- \in U_-, \; \zeta_+ \in U_+.$$

This does not follow immediately from the preceding jump theorem, since the path $\gamma \backslash \gamma_0$ is not necessarily smooth. However, the winding number can be expressed as

$$W(\gamma, \zeta) \;=\; \frac{1}{2\pi i} \int_{\gamma_0} \frac{dz}{z - \zeta} + G(\zeta),$$

where $G(\zeta)$ is the increase of the argument of $z - \zeta$ along $\gamma \backslash \gamma_0$. If we apply the jump theorem to the Cauchy integral along γ_0, and we observe that $G(\zeta)$ is analytic on U and hence does not jump, we obtain the jump theorem for the winding number.

Recall (Section III.1) that a closed path Γ in the complex plane is **simple** if it is parameterized by a continuous function $\gamma(t)$, $a \le t \le b$, such that $\gamma(s) \ne \gamma(t)$ for $a \le s < t < b$. We refer to a simple closed path also as a **simple closed curve**. By regarding the interval $[a, b]$ with its endpoints identified as a circle, we can think of a simple closed curve as a continuous one-to-one image of a circle. The Jordan curve theorem asserts that a simple closed curve in the complex plane divides the plane into exactly two connected components, one bounded (the "inside") and the other unbounded (the "outside"). While the theorem may seem intuitively plausible, a glance at the figure for Exercise 6 should convince one that the theorem is not completely obvious. In fact, Jordan did not give a rigorous proof of the Jordan curve theorem, and the known proofs of the theorem require a fair amount of effort.

Möbius band

To appreciate the sort of difficulty that might arise, consider the Möbius band, obtained from a long thin strip of paper by flipping over one end and then gluing the two ends together. If one cuts the strip along the median line with a scissors, the resulting surface is connected. Thus the median line, which is a simple closed curve on the surface, does not divide the surface into two connected components. The median line has only one "side."

The proof of the Jordan curve theorem for piecewise smooth curves is substantially easier than the proof for arbitrary simple closed curves. We give a proof based on the jump theorem for the winding number. This idea forms the basis for a proof in the general case, which is laid out in the exercises.

Suppose γ is a simple closed curve such that $\gamma(t)$ is continuously differentiable on some parameter interval. Then we can find z_0 on γ and a disk U centered at z_0 such that the smooth segment divides U into two pieces as above. (This requires a nonzero tangent vector and some elementary analysis at the level of the implicit function theorem.) Since the winding number $W(\gamma, \zeta)$ jumps as ζ crosses the segment of γ in U, it attains at least two values on $\mathbb{C} \backslash \Gamma$. Since it is constant on each component of $\mathbb{C} \backslash \Gamma$, the complement of Γ has at least two connected components.

If the entire curve γ is smooth, or even just piecewise smooth, we can cover Γ by a finite number of open disks U as above, each of which is divided into two pieces by γ. By tracking successive components of $U \backslash \Gamma$ corresponding to consecutive U's, we see that each point in any $U \backslash \Gamma$ can be

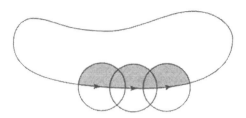

joined by a path in the $U\backslash\Gamma$'s to one or the other component of any other $U\backslash\Gamma$. It follows that each point of $\mathbb{C}\backslash\Gamma$ can be joined by a path in $\mathbb{C}\backslash\Gamma$ to one or the other side of some fixed U, and consequently $\mathbb{C}\backslash\Gamma$ has at most two components. We have proved the following.

Theorem (Jordan Curve Theorem for Smooth Curves). *Let γ be a piecewise smooth simple closed curve in the complex plane $\mathbb{C}$, and let $\Gamma = \gamma([a, b])$ be its trace. Then Γ divides the complex plane into two connected components, one bounded and the other unbounded, each of which has Γ as its boundary. Further, $W(\gamma, \zeta) = 0$ if ζ is in the unbounded component of $\mathbb{C}\backslash\Gamma$, and $W(\gamma, \zeta) = \pm 1$ if ζ is in the bounded component of $\mathbb{C}\backslash\Gamma$, where the choice of sign depends on the orientation of γ.*

Exercises for VIII.7

1. Let $f(z)$ be an entire function, and suppose $g(\zeta)$ is analytic for ζ in the open upper and lower half-planes and across the interval $(-1, 1)$ on the real line. Suppose that

$$\int_{-1}^{1} \frac{f(x)}{x - \zeta}\, dx = g(\zeta)$$

   for ζ in the upper half-plane. What is the value of the integral when ζ is in the lower half-plane? Justify your answer carefully.

2. Show that

$$\int_{-1}^{1} \frac{dx}{x - \zeta} = \mathrm{Log}\left(\frac{\zeta - 1}{\zeta + 1}\right), \qquad \zeta \in \mathbb{C}\backslash[-1, +1].$$

   (Note that we use the principal branch of the logarithm here.) Reconcile this result with your solution to Exercise 1.

3. Find the Cauchy integrals of the following functions around the unit circle $\Gamma = \{|z| = 1\}$, positively oriented. (a) z, (b) $\frac{1}{z}$, (c) $x = \mathrm{Re}(z)$, (d) $y = \mathrm{Im}(z)$.

4. Suppose $f(z)$ is analytic on an annulus $\{\rho < |z| < \sigma\}$, and let $f(z) = f_0(z) + f_1(z)$ be the Laurent decomposition of $f(z)$. (See Section VI.1.) Fix r between ρ and σ, and let $F(\zeta)$ be the Cauchy integral of $f(z)$ around the circle $|z| = r$. Show that $f_0(\zeta) = F(\zeta)$

for $|\zeta| < r$, and $f_1(\zeta) = -F(\zeta)$ for $|\zeta| > r$. Show further that $f_0(\zeta) = F_-(\zeta)$ and $f_1(\zeta) = -F_+(\zeta)$. *Remark.* The formula $f(z) = f_0(z) + f_1(z)$ reflects the jump theorem for the Cauchy integral of $f(z)$ around circles $|z| = r$.

5. Let γ be a piecewise smooth curve, and let $F(\zeta)$ be the Cauchy integral (7.1) of a continuous function $f(z)$ on γ. Show that if $g(z)$ is a smooth function on the complex plane that is zero off some bounded set, then

$$\int_\gamma g(z)f(z)\,dz \;=\; 2i \iint_{\mathbb{C}} \frac{\partial g}{\partial \bar{z}}\, F(z)\,dx\,dy.$$

Hint. Recall Pompeiu's formula (Section IV.8).

6. Determine whether the point z lies inside or outside. Explain.

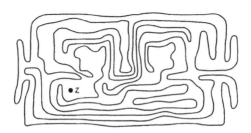

7. A **simple arc** Γ in $\mathbb{C}$ is the image of a continuous one-to-one function $\gamma(t)$ from a closed interval $[a, b]$ to the complex plane. Show that a simple arc Γ in $\mathbb{C}$ has a connected complement, that is, $\mathbb{C}\backslash\Gamma$ is connected. You may use the Tietze extension theorem, that a continuous real-valued function on a closed subset of the complex plane can be extended to a continuous real-valued function on the entire complex plane. *Hint.* Suppose z_0 belongs to a bounded component of $\mathbb{C}\backslash\Gamma$. Find a continuous determination $h(z)$ of $\log(z - z_0)$ on Γ, extend $h(z)$ to a continuous function on $\mathbb{C}^*$, and define $f(z) = z - z_0$ on the component of $\mathbb{C}\backslash\Gamma$ containing z_0, and $f(z) = e^{h(z)}$ on the remainder of $\mathbb{C}^*\backslash\Gamma$. Consider the increase in the argument of $f(z)$ around circles centered at z_0.

8. Prove the Jordan curve theorem for a simple closed curve γ by filling in the following proof outline.
   (a) Show that each component of $\mathbb{C}\backslash\Gamma$ has boundary Γ. *Hint.* For $z_0 = \gamma(t_0) \in \Gamma$, apply the preceding exercise to the simple arc $\Gamma\backslash\gamma(I)$, where I is a small open parameter interval containing t_0.
   (b) Prove the Jordan curve theorem in the case where γ contains a straight line segment.

(c) Show that for any $z_0 = \gamma(t_0) \in \Gamma$, any small disk D_0 containing z_0, and any component U of $\mathbb{C}\backslash\Gamma$, there are points $z_1 = \gamma(t_1)$ and $z_2 = \gamma(t_2)$ such that the image of the parameter segment between t_1 and t_2 is contained in D_0 and such that z_1 and z_2 can be joined by a broken line segment in $U \cap D_0$.

(d) With notation as in (b), let σ be the simple closed curve obtained by replacing the segment of γ in D_0 between z_0 and z_1 by the broken line segment in $U \cap D_0$ between them, and let τ be the simple closed curve in D_0 obtained by following the segment of γ in D_0 from z_0 to z_1 and returning to z_0 along the broken line segment. Show that $W(\tau, \zeta) = 0$ and $W(\gamma, \zeta) = W(\sigma, \zeta)$ for $\zeta \in \mathbb{C}\backslash\Gamma$, $\zeta \notin D_0$.

(e) Using (b) and (d), show that $\mathbb{C}\backslash\Gamma$ has at least two components and that $W(\gamma, \zeta) = \pm 1$ for ζ in each bounded component of $\mathbb{C}\backslash\Gamma$.

(f) By taking U in (c) to be a bounded component of $\mathbb{C}\backslash\Gamma$, show that $W(\gamma, \zeta) = 0$ for ζ in any other component of $\mathbb{C}\backslash\Gamma$.

8. Simply Connected Domains

A domain in the plane is "simply connected" if it has no "holes." Disks and rectangles are simply connected. More generally, star-shaped domains are simply connected. Annuli, punctured disks, and the punctured plane have "holes" and are *not* simply connected. Our aim in this section is to make more precise the notion of "simple connectedness."

Let $\gamma(t)$, $a \leq t \leq b$, be a closed path in a domain D. We say that γ is **deformable to a point** if there are closed paths $\gamma_s(t)$, $a \leq t \leq b$, $0 \leq s \leq 1$, in D such that $\gamma_s(t)$ depends continuously on both s and t, $\gamma_0 = \gamma$, and $\gamma_1(t) \equiv z_1$ is the constant path at some point $z_1 \in D$. The domain D is **simply connected** if every closed path in D can be deformed to a point.

Example. If D is a star-shaped domain with respect to z_0, then any closed path γ in D can be continuously deformed to the point z_0 by pulling the path to z_0 along straight line segments. We set

$$\gamma_s(t) = sz_0 + (1-s)\gamma(t), \qquad a \leq t \leq b, \ 0 \leq s \leq 1,$$

and then the closed paths γ_s deform $\gamma_0 = \gamma$ to the point path $\gamma_1 \equiv z_0$. Thus D is simply connected.

If a closed path can be deformed to a point, then it can be deformed to a point in such a way that each path in the deformation starts and ends at the same fixed point. That is the content of the following lemma.

Lemma. *Let* $\gamma(t)$, $0 \le t \le 1$, *be a closed path in* D, *with* $z_0 = \gamma(0) = \gamma(1)$. *Suppose that* γ *can be deformed continuously to a point in* D. *Then there is a continuous family of closed paths* γ_s, $0 \le s \le 1$, *such that* $\gamma_0 = \gamma$, γ_1 *is the constant path at* z_0, *and each path* γ_s *starts and ends at* z_0.

Suppose that γ is deformed to a point by closed paths σ_s, $0 \le s \le \frac{1}{2}$. Thus $\sigma_0(t) = \gamma(t)$ for $0 \le t \le 1$, $\sigma_{1/2}(t) = z_1$ for $0 \le t \le 1$, and $\sigma_s(0) = \sigma_s(1)$ for $0 \le s \le \frac{1}{2}$. We deform γ to the point z_0 by means of paths γ_s, $0 \le s \le 1$, that all start and terminate at z_0, in two stages. In the first stage we deform γ to a path $\gamma_{1/2}$ that follows the starting points $\sigma_s(0)$ of σ_s from z_0 to z_1, and then reverses direction and follows them backward from z_1 to z_0. Such a deformation γ_s is given explicitly for $0 \le s < \frac{1}{2}$ by

$$\gamma_s(t) = \begin{cases} \sigma_t(0), & 0 \le t \le s, \\ \sigma_s((t-s)/(1-2s)), & s \le t \le 1-s, \\ \sigma_{1-t}(1), & 1-s \le t \le 1. \end{cases}$$

For the second stage we deform $\gamma_{1/2}$ to the point path at z_0 by running partway along $\gamma_{1/2}$, pausing, then returning, with increasing pauses. Such a deformation γ_s, $\frac{1}{2} \le s \le 1$, is given explicitly by

$$\gamma_s(t) = \begin{cases} \sigma_t(0), & 0 \le t \le 1-s, \\ \sigma_{1-s}(0), & 1-s \le t \le s, \\ \sigma_{1-t}(0), & s \le t \le 1. \end{cases}$$

The definition of the function $\gamma_s(t)$, regarded as a function on the square, is suggested in the figure. On the horizontal intervals in the bottom half, $\gamma_s(t)$ is a reparametrization of $\sigma_s(t)$, while on each horizontal interval in the top half and on each vertical interval $\gamma_s(t)$ is constant.

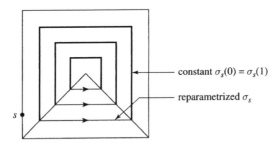

constant $\sigma_s(0) = \sigma_s(1)$

reparametrized σ_s

Another topological notion we require is connectedness. We have already defined what it means for an open subset of the complex plane to be connected, namely, that any two points can be joined by a path. In the context of general topological spaces this concept is defined to be "path connectedness." For arbitrary topological spaces there is another definition of connectedness, which we will use in this section. It is equivalent to path connectedness for open sets in the plane.

Let X be a subset of the extended complex plane $\mathbb{C}^* = \mathbb{C} \cup \{\infty\}$. We say that X is **connected** if every continuous integer-valued function on X is constant. Thus X is *not* connected if and only if there is a subset E of X such that both E and $X \backslash E$ are nonempty, and no sequence in E has a limit in $X \backslash E$, nor does any sequence in $X \backslash E$ have a limit in E. If there is such a set E, then the function that is 1 on E and 0 on $X \backslash E$ is a continuous integer-valued function on X that is not constant. Conversely, if there is a nonconstant continuous integer-valued function on X, we take E to be one of the level sets of the function, and we see that X is not connected. Note that if X is a closed subset of $\mathbb{C}^*$ that is not connected, we can take E above such that $\infty \notin E$, and then E is a compact subset of the complex plane with positive distance from $X \backslash E$, that is, there is $\delta > 0$ such that $|z - w| \geq \delta$ for all $z \in E$ and $w \in X \backslash E$.

connected   not connected

For domains in the plane many conditions can be formulated that are equivalent to simple connectivity. We give several that are useful for our purposes.

Theorem. *The following properties are equivalent, for a domain D in the complex plane:*

  (i) *D is simply connected,*
 (ii) *every closed differential on D is exact,*
(iii) *for each $z_0 \in \mathbb{C} \backslash D$, there is an analytic branch of $\log(z - z_0)$ defined on D,*
(iv) *each closed curve γ in D has winding number $W(\gamma, z_0) = 0$ about all points $z_0 \in \mathbb{C} \backslash D$,*
 (v) *the complement of D in the extended complex plane $\mathbb{C}^*$ is connected.*

In practice, the condition (v) is usually the easiest to check for a given domain D. It is usually easy to spot any set E in $\mathbb{C} \backslash D$ that has positive distance from the rest of $\mathbb{C} \backslash D$. We think of such sets E as representing holes in D.

Example. If D is an annulus $\{r < |z| < s\}$, we can take E to be the closed disk $\{|z| \leq r\}$. Then the rest of $\mathbb{C} \backslash D$ is the exterior set $\{|z| \geq s\}$, every point of which has distance at least $s - r$ from E. Consequently, $\mathbb{C} \backslash D$ is not connected, and D is not simply connected. The set $\{|z| \leq r\}$ is a hole in D. Similarly, if D is the punctured disk $\{0 < |z| < 1\}$ or the punctured plane $\mathbb{C} \backslash \{0\}$, we can take E to be the puncture $\{0\}$, and we see that D is not simply connected.

Now we turn to the proof of the theorem. The implications that are trivial or easy consequences of our previous discussions are (i) $\Rightarrow$ (ii) $\Rightarrow$ (iii) $\Leftrightarrow$ (iv) and (v) $\Rightarrow$ (iv). New ideas are required to establish the implications (iv) $\Rightarrow$ (v) and (iv) $\Rightarrow$ (i).

Proof that (i) $\Longrightarrow$ (ii): This follows from our discussion of deformation of paths in Section III.2. Let $P\,dx + Q\,dy$ be a closed differential on D. According to the discussion in Section III.2, $\int_\gamma P\,dx + Q\,dy = \int_\sigma P\,dx + Q\,dy$ for any closed paths γ and σ in D such that γ can be deformed continuously to σ. Since the integral over a one-point path is zero, the simple connectivity of D implies that $\int_\gamma P\,dx + Q\,dy = 0$ for any closed path in D. Consequently, $P\,dx + Q\,dy$ is independent of path, hence exact.

Proof that (ii) $\Longrightarrow$ (iii): If every closed differential on D is exact, then in particular for fixed $z_0 \in \mathbb{C}\backslash D$ the differential $d\arg(z - z_0)$ is exact. We write $d\arg(z - z_0) = d\,h(z)$ for some smooth function $h(z)$, normalized so that $h(z_1)$ is a value of $\arg(z_1 - z_0)$ for some fixed point $z_1 \in D$. Then $h(z)$ is a continuous determination of $\arg(z - z_0)$ in D, and $\log|z - z_0| + ih(z)$ is an analytic branch of $\log(z - z_0)$ in D.

Proof that (iii) $\Longleftrightarrow$ (iv): If there is an analytic branch $f(z)$ of $\log(z - z_0)$, then

$$W(\gamma, z_0) = \frac{1}{2\pi i}\int_\gamma df = 0$$

for any closed path γ in D. Conversely, if $W(\gamma, z_0) = 0$ for any closed path γ in D, then the analytic differential $dz/(z - z_0)$ is independent of path, and there is an analytic function $f(z)$ such that $df = dz/(z - z_0)$. An analytic branch of $\log(z - z_0)$ is obtained by adding an appropriate constant to $f(z)$.

Proof that (v) $\Longrightarrow$ (iv): Let γ be a closed path in D. The winding number $W(\gamma, \zeta)$ is a continuous integer-valued function on $\mathbb{C}\backslash D$ that is zero for large ζ. If we set it equal to 0 for $\zeta = \infty$, we obtain a continuous integer-valued function on $\mathbb{C}^*\backslash D$ that vanishes at ∞. Since $\mathbb{C}^*\backslash D$ is connected, the function is identically zero. Hence $W(\gamma, \zeta) = 0$ for $\zeta \in \mathbb{C}\backslash D$.

Proof that (iv) $\Longrightarrow$ (v): If $\mathbb{C}^*\backslash D$ is not connected, then there are a closed bounded subset E of $\mathbb{C}\backslash D$ and $\delta > 0$ such that $|z - w| \geq 4\delta$ for every point $z \in E$ and every point $w \in F = \mathbb{C}\backslash(D \cup E)$. Fix a point $z_0 \in E$, and cover the complex plane by a grid of squares of side length δ so that z_0 lies in the interior of one of the squares. Let K be the union of all the (closed) squares in the grid that contain a point of distance at most δ from E. Then K is a finite union of closed squares, K is disjoint from F, and the boundary ∂K of K does not contain points of either E or F. Thus ∂K is a finite union of straight line segments contained in D. Further, if we denote by K_j the squares from the grid in K, then

$$\int_{\partial K} d\arg(z - z_0) = \sum \int_{\partial K_j} d\arg(z - z_0) = 2\pi,$$

since the only summand that is not zero is the integral over the boundary of the square containing z_0. Now, ∂K is a union of a finite number of closed paths in D. Thus there is a closed path γ in D obtained by following consecutively sides of squares in the grid, for which $\int_\gamma d \arg(z - z_0) \neq 0$. Consequently, $W(\gamma, z_0) \neq 0$.

 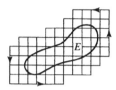

Proof that (iv) $\implies$ (i): Our proof is based on the "northeast corner" argument. We can deform any closed γ in D to a path that follows the sides of a grid of squares of small side length $\delta > 0$. Thus we fix δ, and we consider closed paths in D that are obtained by following consecutively sides of squares in the grid, so that the path can be represented as a sequence of sides, each with a direction. We proceed by induction on the number of consecutive sides in the path. Thus we assume that γ is represented by a sequence of sides of the grid, and that any closed path formed by a shorter sequence of sides in the grid can be deformed in D to a point. We must show that γ can be deformed in D to a point.

Suppose first that the closed path γ intersects itself, so that γ is obtained by traversing successively two shorter closed paths, each of which starts and terminates at a vertex z_0 of the grid. By our induction hypothesis, each of the shorter paths can be deformed in D to a point. By the lemma, each of the paths can be deformed in D to the point z_0 by means of closed paths that begin and end at z_0. By traversing successively the paths in the two deformations, we can then deform the path γ in D to the point z_0.

We may suppose, then, that the closed path γ does not intersect itself. In the path γ consider the vertices furthest to the right, for which $\text{Re}(z)$ is the largest, and among these let z_{NE} be the highest vertex, for which $\text{Im}(z)$ is the largest. Reversing the direction of the path if necessary, we can assume that the path follows the horizontal edge from $z_{\text{NE}} - \delta$ to z_{NE}, where it makes a right turn and proceeds down to $w_{\text{NE}} = z_{\text{NE}} - im\delta$, then makes another right turn and proceeds to $w_{\text{NE}} - \delta$, as in the figure. Let R be the open rectangle with vertices $z_{\text{NE}} - \delta$, z_{NE}, w_{NE}, and $w_{\text{NE}} - \delta$. Since $W(\gamma, \zeta) = 0$ for ζ in the unbounded component of the complement of γ, we have $W(\gamma, \zeta) = 0$ just to the right of the segment from z_{NE} to w_{NE}. By the jump theorem for the winding number (Section 3), we have $W(\gamma, \zeta) = -1$ for ζ to the left of the vertical segment, hence for all $\zeta \in R$. It follows that R is contained in D. Further, by continuity we have $W(\gamma, \zeta) = -1$ at all points of the left vertical side of R that are not on γ, so that none of the points of the boundary of R belong to $\mathbb{C} \backslash D$, and ∂R is also contained in D. Let σ be the path obtained from γ by replacing the segments of γ consisting of the three edges of the rectangle R by the vertical edge of R from $z_{NE} - \delta$ to $w_{NE} - \delta$. Then γ can be deformed to σ in D, simply by deforming the three sides of the rectangle to the fourth side. Since σ is shorter than γ, σ can be deformed in D to a point. Thus γ can also be deformed in D to a point. This completes the proof of the implication and of the theorem.

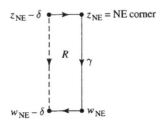

In Chapter XI we will prove the Riemann mapping theorem, which asserts that every simply connected domain in the plane except the plane itself can be mapped conformally onto the open unit disk. What we will actually establish in Section XI.3 is that if $D \neq \mathbb{C}$ and if (iii) holds, then there is a conformal map of D onto the open unit disk. From this it follows easily that a domain D satisfying (iii) is simply connected (see Exercise 7), and also that D has a connected complement. Thus the proof of the Riemann mapping theorem will provide us with an alternative method to show that (iii) implies (i) and (v). The proof given above has the advantage of being "elementary" in nature, whereas the proof of the Riemann mapping theorem depends on the arguably nonelementary notion of compactness of families of analytic functions.

Exercises for VIII.8

1. Which of the following domains in $\mathbb{C}$ are simply connected? Justify your answers. (a) $D = \{\operatorname{Im} z > 0\}\backslash[0, i]$, the upper half-plane with a vertical slit from 0 to i. (b) $D = \{\operatorname{Im} z > 0\}\backslash[i, 2i]$, the upper half-plane with a vertical slit from i to $2i$. (c) $D = \mathbb{C}\backslash[0, +\infty)$, the complex plane slit along the positive real axis. (d) $D = \mathbb{C}\backslash[-1, 1]$, the complex plane with an interval deleted.

2. Show that a domain D in the extended complex plane $\mathbb{C}^* = \mathbb{C}\cup\{\infty\}$ is simply connected if and only if its complement $\mathbb{C}^*\backslash D$ is connected. *Hint.* If $D \neq \mathbb{C}^*$, move a point in the complement of D to ∞. If $D = \mathbb{C}^*$, first deform a given closed path to one that does not cover the sphere, then deform it to a point by pulling along arcs of great circles.

3. Which of the following domains in $\mathbb{C}^*$ are simply connected? Justify your answers. (a) $D = \mathbb{C}^*\backslash[-1, 1]$, the extended complex plane with

an interval deleted, (b) $D = \mathbb{C}^*\backslash\{-1,0,1\}$, the thrice-punctured sphere.

4. Show that a domain D in the complex plane is simply connected if and only if any analytic function $f(z)$ on D that does not vanish at any point of D has an analytic logarithm on D. *Hint.* If $f(z) \neq 0$ on D, consider the function

$$G(z) = \int_{z_0}^{z} \frac{f'(w)}{f(w)} dw.$$

5. Show that a domain D is simply connected if and only if any analytic function $f(z)$ on D that does not vanish at any point of D has an analytic square root on D. Show that this occurs if and only if for any point $z_0 \notin D$ the function $z - z_0$ has an analytic square root on D.

6. Show that a domain D is simply connected if and only if each continuous function $f(z)$ on D that does not vanish at any point of D has a continuous logarithm on D.

7. Let E be a closed connected subset of the extended complex plane $\mathbb{C}^*$. Show that each connected component of $\mathbb{C}^*\backslash E$ is simply connected.

8. Show that simple connectivity is a "topological property," that is, if U and V are domains, and φ is a continuous map of U onto V such that φ^{-1} is also continuous, then U is simply connected if and only if V is simply connected.

9. Suppose that $f(z)$ is analytic on a domain D, and $f'(z)$ has no zeros on D. Suppose also that $f(D)$ is simply connected, and that there is a branch $g(w)$ of f^{-1} that is analytic at $w_0 = f(z_0)$ and that can be continued analytically along any path in $f(D)$ starting at w_0. Show that $f(z)$ is one-to-one on D.

10. We define an **integral 1-cycle** in D to be an expression of the form $\sigma = \sum k_j \gamma_j$, where $\gamma_1, \ldots, \gamma_m$ are closed paths in D and $k_1, \ldots, k_m$ are integers. We define the **winding number of σ about ζ** to be $W(\sigma, \zeta) = \sum k_j W(\gamma_j, \zeta)$, $\zeta \in \mathbb{C}\backslash D$. Show that if $h(\zeta)$ is a continuous integer-valued function on $\mathbb{C}^*\backslash D$ such that $h(\infty) = 0$, then there is an integral 1-cycle σ on D such that $W(\sigma, \zeta) = h(\zeta)$ for all $\zeta \in \mathbb{C}\backslash D$.

11. An integral 1-cycle σ is **homologous to zero** in D if $W(\sigma, \zeta) = 0$ whenever $\zeta \notin D$. Let U be a bounded domain whose boundary consists of a finite number of piecewise smooth closed curves $\gamma_1, \ldots, \gamma_m$, oriented positively with respect to U, such that U together with its

boundary is contained in D. Show that the 1-cycle $\partial U = \sum \gamma_j$ is homologous to zero in D.

12. Let D be a domain in $\mathbb{C}$ such that $\mathbb{C}^* \backslash D$ consists of $m + 1$ disjoint closed connected sets. Show that there are m piecewise smooth closed curves $\gamma_1, \ldots, \gamma_m$ such that every integral 1-cycle σ can be expressed uniquely in the form $\sigma = \sigma_0 + \sum k_j \gamma_j$, where the k_j's are integers and σ_0 is homologous to zero in D. *Remark.* The γ_j's form a **homology** basis for D.

IX

The Schwarz Lemma and Hyperbolic Geometry

This short chapter is devoted to the Schwarz lemma, which is a simple consequence of the power series expansion and the maximum principle. The Schwarz lemma is proved in Section 1, and it is used in Section 2 to determine the conformal self-maps of the unit disk. In Section 2 we formulate the Schwarz lemma to be invariant under the conformal self-maps of the unit disk, thereby obtaining Pick's lemma. This leads in Section 3 to the hyperbolic metric and hyperbolic geometry of the unit disk.

1. The Schwarz Lemma

The Schwarz lemma is easy to prove, yet it has far-reaching consequences.

Theorem (Schwarz Lemma). *Let $f(z)$ be analytic for $|z| < 1$. Suppose $|f(z)| \leq 1$ for all $|z| < 1$, and $f(0) = 0$. Then*

$$(1.1) \qquad |f(z)| \leq |z|, \qquad |z| < 1.$$

Further, if equality holds in (1.1) at some point $z_0 \neq 0$, then $f(z) = \lambda z$ for some constant λ of unit modulus.

For the proof, we factor $f(z) = zg(z)$, where $g(z)$ is analytic, and we apply the maximum principle to $g(z)$. Let $r < 1$. If $|z| = r$, then $|g(z)| = |f(z)|/r \leq 1/r$. By the maximum principle, $|g(z)| \leq 1/r$ for all z satisfying $|z| \leq r$. If we let $r \to 1$, we obtain $|g(z)| \leq 1$ for all $|z| < 1$. This yields (1.1). If $|f(z_0)| = |z_0|$ for some $z_0 \neq 0$, then $|g(z_0)| = 1$, and by the strict maximum principle, $g(z)$ is constant, say $g(z) = \lambda$. Then $f(z) = \lambda z$.

An analogous estimate holds in any disk. If $f(z)$ is analytic for $|z - z_0| < R$, $|f(z)| \leq M$, and $f(z_0) = 0$, then

$$(1.2) \qquad |f(z)| \leq \frac{M}{R} |z - z_0|, \qquad |z - z_0| < R,$$

with equality only when $f(z)$ is a multiple of $z - z_0$. This can be proved directly, based on the factorization $f(z) = (z - z_0)g(z)$. It can also be obtained from (1.1) by scaling in both the z-variable and the w-variable, $w = f(z)$, and by translating the center of the disk to z_0, as follows. The change of variable $\zeta \mapsto R\zeta + z_0$ maps the unit disk $\{|\zeta| < 1\}$ onto the disk $\{|z - z_0| < R\}$. If we define $h(\zeta) = f(R\zeta + z_0)/M$, then $h(\zeta)$ is analytic on the open unit disk and satisfies $|h(\zeta)| \leq 1$ and $h(0) = 0$. The estimate $|h(\zeta)| \leq |\zeta|$ becomes (1.2).

The Schwarz lemma gives an explicit estimate for the "modulus of continuity" of an analytic function. It shows that a uniformly bounded family of analytic functions is "equicontinuous" at each point. We will return in Chapter XI to treat the ideas of equicontinuity and compactness for families of analytic functions.

There is an infinitesimal version of the Schwarz lemma.

Theorem. Let $f(z)$ be analytic for $|z| < 1$. If $|f(z)| \leq 1$ for $|z| < 1$, and $f(0) = 0$, then

$$(1.3) \qquad\qquad |f'(0)| \leq 1,$$

with equality if and only if $f(z) = \lambda z$ for some constant λ with $|\lambda| = 1$.

The estimate (1.3) follows by taking $z \to 0$ in the Schwarz lemma. For the case of equality, we consider the factorization $f(z) = zg(z)$ used in the proof of the Schwarz lemma, and we observe that $g(0) = f'(0)$. If $|f'(0)| = 1$, we then have $|g(0)| = 1$, and we conclude as before from the strict maximum principle that $g(z)$ is constant. Hence $f(z) = \lambda z$.

Note that the estimate (1.3) is the same as the Cauchy estimate for $f'(0)$ derived in Section IV.4, without the hypothesis that $f(0) = 0$. See also Exercise 7.

Exercises for IX.1

1. Let $f(z)$ be analytic and satisfy $|f(z)| \leq M$ for $|z - z_0| < R$. Show that if $f(z)$ has a zero of order m at z_0, then

$$|f(z)| \leq \frac{M}{R^m} |z - z_0|^m, \qquad |z - z_0| < R.$$

Show that equality holds at some point $z \neq z_0$ only when $f(z)$ is a constant multiple of $(z - z_0)^m$.

2. Suppose that $f(z)$ is analytic and satisfies $|f(z)| \leq 1$ for $|z| < 1$. Show that if $f(z)$ has a zero of order m at z_0, then $|z_0|^m \geq |f(0)|$. *Hint.* Let $\psi(z) = (z - z_0)/(1 - \overline{z_0}z)$, which is a fractional linear transformation mapping the unit disk onto itself, and show that $|f(z)| \leq |\psi(z)|^m$.

3. Suppose that $f(z)$ is analytic for $|z| \leq 1$, and suppose that $1 < |f(z)| < M$ for $|z| = 1$, while $f(0) = 1$. Show that $f(z)$ has a zero in the unit disk, and that any such zero z_0 satisfies $|z_0| > 1/M$. *Hint.* For the second assertion, consider $\psi(f(z))$, where $\psi(w)$ is a fractional linear transformation mapping 1 to 0 and the circle $\{|w| = M\}$ to the unit circle. Or use Exercise 2.

4. Suppose that $f(z)$ is analytic for $|z| < 1$ and satisfies $f(0) = 0$ and $\mathrm{Re}\, f(z) < 1$. (a) Show that $|f(z)| \leq 2|z|/(1 - |z|)$. *Hint.* Consider the composition of $f(z)$ and the fractional linear transformation mapping the half-plane $\{\mathrm{Re}\, w < 1\}$ onto the unit disk. (b) Show that $|f'(0)| \leq 2$. (c) For fixed z_0 with $0 < |z_0| < 1$, determine for which functions $f(z)$ there is equality in (a). (d) Determine for which functions $f(z)$ there is equality in (b). (e) By scaling the estimates in (a) and (b), obtain sharp estimates for $|g(z)|$ and $|g'(0)|$ for functions $g(z)$ analytic for $|z| < R$ and satisfying $g(0) = 0$ and $\mathrm{Re}\, g(z) < C$.

5. Suppose that $f(z)$ is analytic and satisfies $|f(z)| \leq 1$ for $|z| < 1$. Show that if $|f(0)| \geq r$, then $|f(z)| \geq (r - |z|)/(1 - r|z|)$ for $|z| < r$. Determine for which functions $f(z)$ equality holds at some point z_0 with $|z_0| < r$.

6. Let $f(z)$ be a conformal map of the open unit disk onto a domain D. Show that the distance from $f(0)$ to the boundary of D is estimated by $\mathrm{dist}(f(0), \partial D) \leq |f'(0)|$.

7. Suppose that $f(z) = \sum_{k=0}^{\infty} a_k z^k$ is analytic for $|z| < 1$ and satisfies $|f(z)| \leq M$.
   (a) Show that $\sum_{k=0}^{\infty} |a_k|^2 \leq M^2$. *Hint.* Integrate $|f(z)|^2$ around a circle of radius r.
   (b) Show using (a) that $|f'(0)| \leq M$, with equality only if $f(z)$ is a constant multiple of z. *Remark.* It is not assumed that $f(0) = 0$.
   (c) Show that $|f^{(k)}(0)| \leq k!M$, with equality only if $f(z)$ is a constant multiple of z^k.

8. Suppose that $f(z)$ is analytic for $|z| < 1$ and satisfies $|f(z)| < 1$, $f(0) = 0$, and $|f'(0)| < 1$. Let $r < 1$. Show that there is a constant $c < 1$ such that $|f(z)| \leq c|z|$ for $|z| \leq r$. Show that the nth iterate $f_n(z) = f(f(\cdots f(z) \cdots)) = f(f_{n-1}(z))$ of $f(z)$ satisfies $|f_n(z)| \leq c^n |z|$ for $|z| \leq r$. Deduce that $f_n(z)$ converges to zero normally on the open unit disk $\mathbb{D}$.

2. Conformal Self-Maps of the Unit Disk

We denote by $\mathbb{D}$ the open unit disk in the complex plane, $\mathbb{D} = \{|z| < 1\}$. A **conformal self-map of the unit disk** is an analytic function from $\mathbb{D}$ to itself that is one-to-one and onto. The composition of two conformal self-maps is again a conformal self-map, and the inverse of a conformal self-map is a conformal self-map. The conformal self-maps form what is called a "group," with composition as the group operation. The group identity is the identity map $g(z) = z$.

For fixed angle φ, the rotation $z \mapsto e^{i\varphi}z$ is a conformal self-map of $\mathbb{D}$ that fixes the origin, and these are the only conformal self-maps that leave 0 fixed.

Lemma. *If $g(z)$ is a conformal self-map of the unit disk $\mathbb{D}$ such that $g(0) = 0$, then $g(z)$ is a rotation, that is, $g(z) = e^{i\varphi}z$ for some fixed φ, $0 \le \varphi \le 2\pi$.*

To see this, we apply the Schwarz lemma to $g(z)$ and to its inverse. Since $g(0) = 0$ and $|g(z)| < 1$, the Schwarz lemma applies, and $|g(z)| \le |z|$. If we apply the Schwarz lemma also to $g^{-1}(w)$, we obtain $|g^{-1}(w)| \le |w|$, which for $w = g(z)$ becomes $|z| \le |g(z)|$. Thus $|g(z)| = |z|$. Since $g(z)/z$ has constant modulus, it is constant. Hence $g(z) = \lambda z$ for a unimodular constant λ.

Theorem. *The conformal self-maps of the open unit disk $\mathbb{D}$ are precisely the fractional linear transformations of the form*

$$(2.1) \qquad f(z) = e^{i\varphi} \frac{z - a}{1 - \bar{a}z}, \qquad |z| < 1,$$

where a is complex, $|a| < 1$, and $0 \le \varphi \le 2\pi$.

Define $g(z) = (z - a)/(1 - \bar{a}z)$. Since $g(z)$ is a fractional linear transformation, it is a conformal self-map of the extended complex plane, and it maps circles to circles. From

$$|e^{i\theta} - a| = |e^{-i\theta} - \bar{a}| = |1 - \bar{a}e^{i\theta}|, \qquad 0 \le \theta \le 2\pi,$$

we see that $|g(z)| = 1$ for $z = e^{i\theta}$, so that $g(z)$ maps the unit circle to itself. Since $g(a) = 0$, $g(z)$ must map the open unit disk to itself. Consequently, $g(z)$ is a conformal self-map of the unit disk, and so is $f(z)$ defined by (2.1). Let $h(z)$ be an arbitrary conformal self-map of $\mathbb{D}$, and set $a = h^{-1}(0)$. Then $h \circ g^{-1}$ is a conformal self-map of $\mathbb{D}$, and $(h \circ g^{-1})(0) = h(a) = 0$. By the lemma, $(h \circ g^{-1})(w) = e^{i\varphi}w$ for some fixed φ, $0 \le \varphi \le 2\pi$. Writing $w = g(z)$, we obtain $h(z) = e^{i\varphi}g(z)$, and $h(z)$ has the form (2.1).

The parameters a and $e^{i\varphi}$ are uniquely determined by the conformal self-map $f(z)$ of $\mathbb{D}$. The parameter a is $f^{-1}(0)$, and since

$$f'(z) = e^{i\varphi} \frac{1 - |a|^2}{(1 - \bar{a}z)^2}, \qquad |z| < 1,$$

the parameter φ is uniquely specified (modulo 2π) as the argument of $f'(0)$. Thus there is a one-to-one correspondence between points of the parameter space $\mathbb{D} \times \partial\mathbb{D}$ and conformal self-maps of the open unit disk.

Next we take a giant step by proving a form of the Schwarz lemma that is invariant under conformal self-maps of the open unit disk.

Theorem (Pick's Lemma). *If $f(z)$ is analytic and satisfies $|f(z)| < 1$ for $|z| < 1$, then*

$$(2.2) \qquad |f'(z)| \leq \frac{1 - |f(z)|^2}{1 - |z|^2}, \qquad |z| < 1.$$

If $f(z)$ is a conformal self-map of $\mathbb{D}$, then equality holds in (2.2); otherwise, there is strict inequality for all $|z| < 1$.

To prove (2.2), our strategy is to transport z and $f(z)$ to 0 using conformal self-maps, and to apply the Schwarz lemma to the resulting composition. Fix $z_0 \in \mathbb{D}$ and set $w_0 = f(z_0)$. Let $g(z)$ and $h(z)$ be conformal self-maps of $\mathbb{D}$ mapping 0 to z_0 and w_0 to 0, respectively, say

$$g(z) = \frac{z + z_0}{1 + \overline{z_0}z}, \qquad h(w) = \frac{w - w_0}{1 - \overline{w_0}w}.$$

Then $h \circ f \circ g$ maps 0 to 0. The estimate (1.3) and the chain rule yield

$$(2.3) \qquad |(h \circ f \circ g)'(0)| = |h'(w_0)f'(z_0)g'(0)| \leq 1,$$

hence $|f'(z_0)| \leq 1/|g'(0)||h'(w_0)|$. Substituting $g'(0) = 1 - |z_0|^2$ and $h'(w_0) = 1/(1 - |w_0|^2)$, we obtain (2.2).

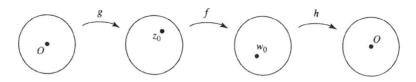

If $f(z)$ is a conformal self-map of $\mathbb{D}$, then so is $h \circ f \circ g$, so we have equality in (2.3), which yields equality in (2.2). Conversely, suppose that $f(z)$ is an analytic function from $\mathbb{D}$ to $\mathbb{D}$ such that equality holds in (2.2) at one point z_0. Then the calculations above give $|(h \circ f \circ g)'(0)| = 1$. According to Section 1, $h \circ f \circ g$ is multiplication by a unimodular constant, hence a conformal self-map of $\mathbb{D}$. Composing by h^{-1} on the left and by g^{-1} on the right, we conclude that f is a conformal self-map of $\mathbb{D}$.

Exercises for IX.2

1. A **finite Blaschke product** is a rational function of the form

$$B(z) = e^{i\varphi} \left(\frac{z - a_1}{1 - \bar{a}_1 z} \right) \cdots \left(\frac{z - a_n}{1 - \bar{a}_n z} \right),$$

where $a_1, \ldots, a_n \in \mathbb{D}$ and $0 \le \varphi \le 2\pi$. Show that if $f(z)$ is continuous for $|z| \le 1$ and analytic for $|z| < 1$, and if $|f(z)| = 1$ for $|z| = 1$, then $f(z)$ is a finite Blaschke product.

2. Show that $f(z) = (1 + 3z^2)/(3 + z^2)$ is a finite Blaschke product.

3. Suppose $f(z)$ is analytic for $|z| < 3$. If $|f(z)| \le 1$, and $f(\pm i) = f(\pm 1) = 0$, what is the maximum value of $|f(0)|$? For which functions is the maximum attained?

4. For fixed $z_0, z_1 \in \mathbb{D}$, find the maximum value of $|f(z_1) - f(z_0)|$ among all analytic functions $f(z)$ on the open unit disk $\mathbb{D}$ satisfying $|f(z)| < 1$. Determine for which such functions the maximum value is attained. *Hint.* Consider first the case where $z_0 = r > 0$ and $z_1 = -r$, and show that the maximum is $2r$, attained only for $f(z) = \lambda z$, $|\lambda| = 1$.

5. Show that any conformal self-map of the upper half-plane has the form

$$f(z) = \frac{az + b}{cz + d}, \qquad \operatorname{Im} z > 0,$$

where a, b, c, d are real numbers satisfying $ad - bc = 1$. When do two such coefficient choices for a, b, c, d determine the same conformal self-map of the upper half-plane?

6. Show that the conformal maps of the upper half-plane onto the open unit disk are of the form

$$f(z) = e^{i\varphi} \frac{z - a}{z - \bar{a}}, \qquad \operatorname{Im} a > 0, \ 0 \le \varphi \le 2\pi.$$

Show that a and $e^{i\varphi}$ are uniquely determined by the conformal map.

7. Show that every conformal self-map of the complex plane $\mathbb{C}$ has the form $f(z) = az + b$, where $a \ne 0$. *Hint.* The isolated singularity of $f(z)$ at ∞ must be a simple pole.

8. Show that every conformal self-map of the Riemann sphere $\mathbb{C}^*$ is given by a fractional linear transformation.

9. Show that any conformal self-map of the punctured unit disk $\{0 < |z| < 1\}$ is a rotation $z \mapsto e^{i\varphi} z$.

10. Show that any conformal self-map of the punctured complex plane $\{0 < |z| < \infty\}$ is either a multiplication $z \mapsto az$, or such a multiplication followed by the inversion $z \mapsto 1/z$.

11. Let $D = \mathbb{C}\backslash\{a_1, \dots, a_m\}$ be the complex plane with m punctures. Show that any conformal self-map of D is a fractional linear transformation that permutes $\{a_1, \dots, a_m, \infty\}$.

12. Determine the conformal self-maps of the following domains D: (a) $D = \mathbb{C}\backslash\{0, 1\}$, (b) $D = \mathbb{C}\backslash\{-1, 0, 1\}$, (c) $D = \mathbb{C}\backslash\{-1, 0, 2\}$.

13. Suppose $f(z)$ is an analytic function from the open unit disk $\mathbb{D}$ to itself that is not the identity map z. Show that $f(z)$ has at most one fixed point in $\mathbb{D}$. *Hint.* Make a change of variable with a conformal self-map of $\mathbb{D}$ to place the fixed point at 0.

14. Suppose $f(z)$ is an analytic function from the open unit disk $\mathbb{D}$ to itself that is not a conformal self-map, and denote by $f_n(z)$ the nth iterate of $f(z)$. Show that if $f(z)$ has a fixed point $z_0 \in \mathbb{D}$, then $f_n(z)$ converges to z_0 for each $z \in \mathbb{D}$. Show that for each $r < 1$, the convergence is uniform for $|z| \leq r$. *Hint.* See Exercise 1.8.

15. We say that two conformal self-maps f and g of $\mathbb{D}$ are **conjugate** if there is a conformal self-map h of $\mathbb{D}$ such that $g = h \circ f \circ h^{-1}$. (See the exercises for Section II.7.) Let f be a conformal self-map of $\mathbb{D}$ that is not the identity map z. (a) Show that either f has two fixed points on $\partial\mathbb{D}$, counting multiplicity, or f has one fixed point in $\mathbb{D}$. (b) Show that f has a fixed point in $\mathbb{D}$ if and only if f is conjugate to a rotation $g(z) = e^{i\varphi}z$. (c) Show that rotations by different angles are not conjugate. (d) Show that f has two distinct fixed points on $\partial\mathbb{D}$ if and only if f is conjugate to $g(z) = (z - s)/(1 - sz)$ for some s satisfying $0 < s < 1$. (e) Show that g's for different s's are not conjugate. (f) Show that any two conformal self-maps of $\mathbb{D}$ with one fixed point on $\partial\mathbb{D}$ (of multiplicity two) are conjugate.

3. Hyperbolic Geometry

Suppose $w = f(z)$ is a conformal self-map of the open unit disk $\mathbb{D}$. From Pick's lemma we then have equality in (2.2),

$$\left| \frac{dw}{dz} \right| = \frac{1 - |w|^2}{1 - |z|^2}.$$

In differential form this becomes

$$\frac{|dw|}{1 - |w|^2} = \frac{|dz|}{1 - |z|^2},$$

which means that if γ is any smooth curve in $\mathbb{D}$, and $w = f(z)$ is a conformal self-map of $\mathbb{D}$, then

(3.1)
$$\int_{f \circ \gamma} \frac{|dw|}{1 - |w|^2} = \int_\gamma \frac{|dz|}{1 - |z|^2}.$$

Thus to obtain a length function that is invariant under conformal self-maps of $\mathbb{D}$, we are led to make the following definition. We define the **length of γ in the hyperbolic metric** by

(3.2)
$$\text{hyperbolic length of } \gamma = 2 \int_\gamma \frac{|dz|}{1 - |z|^2}.$$

The factor 2 is a harmless factor, which is often omitted. (It adjusts the metric so that its curvature is -1.) The identity (3.1) shows that $f \circ \gamma$ has the same hyperbolic length as γ for any conformal self-map $f(z)$ of $\mathbb{D}$. Thus hyperbolic lengths are invariant under conformal self-maps of $\mathbb{D}$.

We define the **hyperbolic distance** $\rho(z_0, z_1)$ from z_0 to z_1 to be the infimum (greatest lower bound) of the hyperbolic lengths of all piecewise smooth curves in $\mathbb{D}$ from z_0 to z_1. Since conformal self-maps of $\mathbb{D}$ preserve the hyperbolic lengths of curves, they also preserve hyperbolic distances; that is, for any conformal self-map $w = f(z)$ of $\mathbb{D}$,

$$\rho(f(z_0), f(z_1)) = \rho(z_0, z_1), \qquad z_0, z_1 \in \mathbb{D}.$$

Theorem. *For any two distinct points z_0, z_1 in the open unit disk $\mathbb{D}$, there is a unique shortest curve in $\mathbb{D}$ from z_0 to z_1 in the hyperbolic metric, namely, the arc of the circle passing through z_0 and z_1 that is orthogonal to the unit circle.*

The paths of shortest hyperbolic length between points are called **hyperbolic geodesics**. The hyperbolic geodesics play the role that straight lines play in the Euclidean geometry of the plane. They satisfy all the axioms of Euclidean geometry except the parallel axiom (that through each point not on a given line there passes a unique straight line through the point and parallel to the given line).

hyperbolic geodesics

For a proof of the theorem, let $w = f(z)$ be a conformal self-map of $\mathbb{D}$ such that $f(z_0) = 0$. By multiplying by a unimodular constant, we can arrange that $f(z_1) = r > 0$. Since $f(z)$ preserves hyperbolic lengths, and since $f(z)$ maps circles orthogonal to the unit circle onto circles orthogonal to the unit circle, it suffices to show that the straight line segment from 0 to r is a unique path of shortest hyperbolic length from 0 to r. For this, let $\gamma(t) = x(t) + iy(t)$, $0 \leq t \leq 1$, be a piecewise smooth path in $\mathbb{D}$ from 0 to r. Then $\alpha(t) = \text{Re}(\gamma(t)) = x(t)$ defines a path in $\mathbb{D}$ from 0 to r along the real axis, and

$$\int_\alpha \frac{|dz|}{1 - |z|^2} = \int_0^1 \frac{|dx(t)|}{1 - x(t)^2} \leq \int_0^1 \frac{|dx(t)|}{1 - |\gamma(t)|^2} \leq \int_\gamma \frac{|dz|}{1 - |z|^2}.$$

If $y(t) \neq 0$ for some t, then $|\gamma(t)| > |x(t)|$, and the first inequality above is strict. In this case, the path $\alpha(t)$ on the real axis is strictly shorter than the path $\gamma(t)$. Further, if $\alpha(t)$ is decreasing on some interval, we could reduce the integral by deleting a parameter interval over which $\alpha(t)$ starts and ends at the same value. We conclude that the integral is a minimum exactly when $\gamma(t)$ is real and nondecreasing, in which case the path is the straight line segment from 0 to r.

We turn now to an important reinterpretation of Pick's lemma.

Theorem. *Every analytic function $w = f(z)$ from the open unit disk $\mathbb{D}$ to itself is a contraction mapping with respect to the hyperbolic metric ρ,*

$$(3.3) \qquad \rho(f(z_0), f(z_1)) \leq \rho(z_0, z_1), \qquad z_0, z_1 \in \mathbb{D}.$$

Further, there is strict inequality for all points $z_0, z_1 \in \mathbb{D}$, $z_0 \neq z_1$, unless $f(z)$ is a conformal self-map of $\mathbb{D}$, in which case there is equality for all $z_0, z_1 \in \mathbb{D}$.

To see this, let γ be the geodesic from z_0 to z_1. Then $f \circ \gamma$ is a curve from $f(z_0)$ to $f(z_1)$. Pick's lemma and the definition of the hyperbolic metric yield

$$\rho(f(z_0), f(z_1)) \leq 2 \int_{f \circ \gamma} \frac{|dw|}{1 - |w|^2} = 2 \int_\gamma \frac{|f'(z)| \, |dz|}{1 - |f(z)|^2}$$

$$\leq 2 \int_\gamma \frac{|dz|}{1 - |z|^2} = \rho(z_0, z_1).$$

If $f(z)$ is not a conformal self-map of $\mathbb{D}$, there is strict inequality in Pick's lemma, and we obtain strict inequality in this estimate, hence in (3.3).

The hyperbolic distance from 0 to z can be computed explicitly. It is

$$\rho(0, z) = 2 \int_0^{|z|} \frac{dt}{1 - t^2} = \int_0^{|z|} \left[\frac{1}{1 - t} + \frac{1}{1 + t} \right] dt = \log\left(\frac{1 + |z|}{1 - |z|} \right).$$

This shows that the hyperbolic distance from 0 to z tends to $+\infty$ when z tends to the boundary of the unit disk.

A **geodesic triangle** is an area bounded by three hyperbolic geodesics. Since the hyperbolic geodesics and the angles between them are preserved

hyperbolic triangles

by conformal self-maps of $\mathbb{D}$, we can map any geodesic triangle to a triangle with vertex at 0 and with the same angles between sides. For a geodesic triangle with vertex at 0, two of the sides are radial segments, and the third is an arc of a circle lying inside the Euclidean triangle with the two radii as sides. From this representation we see that the sum of the angles of any geodesic triangle is strictly less than π, which is the sum of the angles of the corresponding Euclidean triangle.

In connection with complex analysis, we have now been in contact with three spaces with strikingly different geometries. The first space is the complex plane $\mathbb{C}$ with the usual Euclidean metric $|dz|$. In the Euclidean plane, the geodesics are straight lines, and the sum of the angles of a geodesic triangle is exactly equal to π. The second space is the open unit disk $\mathbb{D}$ with the hyperbolic metric $2|dz|/(1-|z|^2)$. For the hyperbolic disk, the geodesics are arcs of circles orthogonal to the unit circle, and the sum of angles of a geodesic triangle is strictly less than π.

The third space is the extended complex plane $\mathbb{C}^* = \mathbb{C} \cup \{\infty\}$ with the spherical metric, which can be introduced in a manner completely analogous to the hyperbolic metric. Recall (Section I.3) that the chordal metric induced on $\mathbb{C}$ by the Euclidean metric of the sphere via the stereographic projection is given explicitly by

$$\text{chordal distance from } z \text{ to } w = \frac{2|z-w|}{\sqrt{1+|z|^2}\sqrt{1+|w|^2}}.$$

The infinitesimal form of this metric is $2|dz|/(1+|z|^2)$. If γ is a path in $\mathbb{C}^*$, its **length in the spherical metric** is

$$\text{spherical length of } \gamma = 2\int_\gamma \frac{|dz|}{1+|z|^2} = 2\int \frac{|\gamma'(t)|}{1+|\gamma(t)|^2}\, dt.$$

This is the length of the corresponding path on the unit sphere in $\mathbb{R}^3$. The **distance from z_1 to z_2 in the spherical metric** is defined to be the infimum of the spherical lengths of the paths joining z_1 to z_2. Since the chordal metric is invariant under rotations of the sphere, so is the spherical metric, and consequently, the lengths of paths and the distances between points in the spherical metric are invariant under rotations. It is not difficult to show that the geodesics in the spherical metric correspond to great circles on the sphere, and the sum of the angles of a geodesic triangle is strictly greater than π.

Each of these three spaces is homogeneous, in the sense that any prescribed point can be transported to any other by an "isometry." Thus for each of these spaces, any scalar quantity that is invariant under isometries is constant. It turns out that a notion of "scalar curvature" can be associated to each of the spaces (see the exercises), and the curvature is invariant under isometries, so that in each case the curvature is constant. The Euclidean plane has constant zero curvature, the sphere has constant positive curvature, and the hyperbolic disk has constant negative curvature. The curvature can be related to the area and the sum of angles of geodesic triangles (Gauss-Bonnet formula). We summarize these properties in tabular form.

Geometry	Euclidean	Spherical	Hyperbolic
Infinitesimal length	$|dz|$	$\dfrac{2|dz|}{1+|z|^2}$	$\dfrac{2|dz|}{1-|z|^2}$
Oriented isometries	$e^{i\varphi}z + b$	rotations	conformal self-maps
Curvature	0	$+1$	-1
Geodesics	lines	great circles	circles $\perp$ unit circle
Angles of triangle	$=\pi$	$>\pi$	$<\pi$
Disk circumference	$2\pi\rho$	$2\pi\rho - \dfrac{\pi\rho^3}{3} + \mathcal{O}(\rho^5)$	$2\pi\rho + \dfrac{\pi\rho^3}{3} + \mathcal{O}(\rho^5)$

Euclidean disk   spherical disk   hyperbolic disk

Exercises for IX.3

1. Show by direct computation that $|w'(z)| = (1 - |w|^2)/(1 - |z|^2)$ for any conformal self-map $w = f(z)$ of $\mathbb{D}$.

2. A **hyperbolic disk** centered at $z_0 \in \mathbb{D}$ of radius $\rho > 0$ consists of all $z \in \mathbb{D}$ such that $\rho(z, z_0) < \rho$. (a) Show that the hyperbolic disk centered at 0 of radius ρ is a Euclidean disk of radius $r = (e^\rho - 1)/(e^\rho + 1)$. (b) Show that any hyperbolic disk is a Euclidean disk.

3. Denote by $c(z, \rho)$ and $r(z, \rho)$ the Euclidean center and Euclidean radius of the hyperbolic disk centered at z of hyperbolic radius ρ. (a) For fixed ρ, show that $r(z, \rho)/(1 - |z|)$ tends to a constant $A > 0$ as $|z| \to 1$. (b) For fixed ρ, show that $|z - c(z, \rho)|/r(z, \rho)$ tends to a constant B, $0 < B < 1$, as $|z| \to 1$.

4. Show that the circumference of a hyperbolic disk of radius ρ is $2\pi \sinh \rho$. *Hint.* Show first that the hyperbolic circumference of a Euclidean disk of radius r centered at 0 is $4\pi r/(1 - r^2)$.

5. We define the **hyperbolic area** of a subset E of $\mathbb{D}$ to be

$$4 \iint_E \frac{dx \, dy}{(1 - |z|^2)^2} .$$

Show that the hyperbolic area is invariant under conformal self-maps of $\mathbb{D}$. Show that the hyperbolic area of a hyperbolic disk of radius ρ is given by

$$2\pi(\cosh \rho - 1) = \pi\rho^2 + \frac{\pi}{12}\rho^4 + \mathcal{O}(\rho^6).$$

6. Establish the following, for the spherical metric. (a) The circumference of a spherical disk of radius ρ is $2\pi \sin \rho$, $0 < \rho < \pi$. (b) The area of a spherical disk of radius ρ is given by

$$2\pi(1 - \cos \rho) = \pi\rho^2 - \frac{\pi}{12}\rho^4 + \mathcal{O}(\rho^6).$$

(c) The geodesics in the spherical metric correspond to great circles on the sphere. *Hint.* It suffices to show that the shortest curve from 0 to ε in the spherical metric is the straight line segment joining them.

7. Show that an isometry of the hyperbolic disk $\mathbb{D}$ is either a conformal self-map of $\mathbb{D}$ or the composition of a conformal self-map and the reflection $z \mapsto \bar{z}$.

8. Let $f(z) = (az+b)/(cz+d)$, where $ad - bc = 1$. Show that $f(z)$ is an isometry in the spherical metric if and only if the matrix $\begin{pmatrix} a & b \\ c & d \end{pmatrix}$ is unitary.

9. Show that the function $f(z) = z^2$ is strictly contracting with respect to the hyperbolic metric on any subdisk $\{|z| \leq r\}$, $0 < r < 1$, and that any branch of the square root function is strictly expanding, by establishing the following. (a) For fixed r, $0 < r < 1$, show that

$$\rho(z^2, \zeta^2) \leq \frac{2r}{1 + r^2}\rho(z, \zeta), \qquad |z|, |\zeta| \leq r.$$

When does equality hold? (b) Show that the constant $2r/(1 + r^2)$ in (a) is sharp. (c) For fixed s, $0 < s < 1$, show that

$$\rho\left(\pm\sqrt{z}, \pm\sqrt{\zeta}\right) \geq \frac{1 + s}{2\sqrt{s}}\rho(z, \zeta), \qquad |z|, |\zeta| \leq s.$$

10. Show that

$$d(z,w) \;=\; \left| \frac{z-w}{1-\bar{w}z} \right|, \qquad |z|, |w| < 1,$$

satisfies the triangle inequality, that is, $d(z,w) \le d(z,\zeta) + d(\zeta,w)$ for all $z, \zeta, w \in \mathbb{D}$. *Remark.* This can be regarded as the analogue of the chordal metric for the sphere (defined in Section I.3). Except for the constant factor 2, the hyperbolic metric is the infinitesimal version of the metric function $d(z,w)$.

11. Show that the metric function $d(z,w)$ defined in the preceding exercise satisfies

$$d(f(z), f(w)) \le d(z,w), \qquad |z|, |w| < 1,$$

for any analytic function $f(z)$ from $\mathbb{D}$ to $\mathbb{D}$. Show that equality obtains whenever $f(z)$ is a conformal self-map of $\mathbb{D}$, and otherwise there is strict inequality for all $z \ne w$.

12. A conformal map $g(z)$ of a domain D onto the open unit disk $\mathbb{D}$ induces the metric ρ_D on D defined by

$$d\rho_D(z) = \frac{2|g'(z)|}{1 - |g(z)|^2} |dz|, \qquad z \in D.$$

Show that ρ_D is independent of the conformal map $g(z)$ of D onto $\mathbb{D}$. *Remark.* The metric ρ_D is called the **hyperbolic metric** of the simply connected domain D.

13. Show that the hyperbolic metric of the upper half-plane $\mathbb{H}$ is given by

$$d\rho_{\mathbb{H}}(z) = \frac{|dz|}{y}, \qquad z = x+iy, \; y > 0.$$

What are the geodesics in the hyperbolic metric? Illustrate with a sketch.

14. Show that the horizontal strip $S = \{-\pi/2 < \operatorname{Im} z < \pi/2\}$ has hyperbolic metric

$$d\rho_S(z) = \frac{|dz|}{\cos y}, \qquad z = x + iy, \; -\pi/2 < y < \pi/2.$$

Sketch the hyperbolic geodesics that are orthogonal to the vertical interval $\{iy : -\pi/2 < y < \pi/2\}$.

15. The **curvature** of the metric $\sigma(z)|dz|$ is defined to be

$$\kappa(z) \;=\; -\frac{1}{\sigma(z)^2} \left(\frac{\partial^2}{\partial x^2} + \frac{\partial^2}{\partial y^2} \right) \log \sigma(z).$$

Find the curvature of each of the spherical, the hyperbolic, and the
Euclidean metrics.

16. (Wolff-Denjoy Theorem.) Let $f(z)$ be an analytic function from $\mathbb{D}$
    to $\mathbb{D}$. Let $f_n(z)$ denote the nth iterate of $f(z)$, and let K_r denote
    the closed disk $\{|z| \le r\}$.
    (a) Show that if $f(z)$ is not a conformal self-map of $\mathbb{D}$, then for
        any $r < 1$ there is a constant $c < 1$ such that $\rho(f(z), f(w)) \le
        c\rho(z, w)$ for $z, w \in K_r$.
    (b) Show that if the image $f(\mathbb{D})$ is contained in K_r for some $r < 1$,
        then the iterates $f_n(z)$ converge uniformly on $\mathbb{D}$ to a fixed point
        for $f(z)$.
    (c) Show that if $f(z)$ is not a conformal self-map of $\mathbb{D}$, and if there
        is $r < 1$ such that the iterates of some point $z_0 \in \mathbb{D}$ visit K_r
        infinitely often, then the iterates $f_n(z)$ converge normally on $\mathbb{D}$
        to a fixed point of $f(z)$. *Hint.* First find the fixed point.
    (d) Show that if the iterates of some point $z_0 \in \mathbb{D}$ tend to the unit
        circle $\partial\mathbb{D}$, then there is a point $\zeta \in \partial\mathbb{D}$ (the **Wolff-Denjoy
        point**) such that the iterates $f_n(z)$ converge normally on $\mathbb{D}$
        to ζ. *Hint.* Suppose $z_0 = 0$. Define $g_\varepsilon(z) = (1 - \varepsilon)f(z)$, let z_ε
        be the fixed point of $g_\varepsilon(z)$, and let D_ε be the hyperbolic disk
        centered at z_ε with 0 on its boundary. Show that the limit D
        of the D_ε's is a Euclidean disk that is invariant under $f(z)$ and
        whose boundary meets $\partial\mathbb{D}$ in exactly one point.

X

Harmonic Functions and the Reflection Principle

In Section 1 we introduce the Poisson kernel function and we develop the Poisson integral representation for harmonic functions on the open unit disk. The Poisson kernel is the analogue for harmonic functions of the Cauchy kernel for analytic functions, and the Poisson integral formula solves the Dirichlet problem for the unit disk. In Section 2 we use this solution to characterize harmonic functions by the mean value property. This characterization is the analogue of Morera's theorem characterizing analytic functions. In Section 3 we apply the characterization of harmonic functions to establish the Schwarz reflection principle for harmonic functions. The reflection principle plays a key role in the study of boundary behavior of conformal maps.

1. The Poisson Integral Formula

We wish to extend a given continuous complex-valued function $h\left(e^{i\theta}\right)$ on the unit circle continuously to the closed unit disk $\{|z| \leq 1\}$ so as to be harmonic on the interior of the disk. Any such extension is unique, since the difference of two such extensions is zero on the boundary, hence zero on the entire disk, by the maximum principle. Our strategy is to derive a formula for the extension in the case that $h\left(e^{i\theta}\right)$ is a trigonometric polynomial, and then to show that the formula provides an extension even when $h\left(e^{i\theta}\right)$ is only continuous.

We start with the trigonometric monomial $e^{ik\theta}$. A harmonic extension is given explicitly by $r^{|k|}e^{ik\theta}$. If $k \geq 0$ this extension is $r^k e^{ik\theta} = z^k$, which is analytic. If $k < 0$ this extension is $r^{-k}e^{ik\theta} = \bar{z}^{(-k)}$, which is conjugate-analytic hence harmonic. Proceeding by linearity, we see that the trigonometric polynomial $h\left(e^{i\theta}\right) = \sum_{k=-N}^{N} a_k e^{ik\theta}$ has the (unique) harmonic extension

$$\tilde{h}\left(re^{i\theta}\right) = \sum_{k=-N}^{N} a_k r^{|k|} e^{ik\theta}.$$

We capture the coefficient a_m by multiplying $h\left(e^{i\theta}\right)$ by $e^{-im\theta}$ and integrating. The orthogonality relations for complex exponentials (Section VI.6) yield

$$a_m = \int_{-\pi}^{\pi} h\left(e^{i\theta}\right) e^{-im\theta} \frac{d\theta}{2\pi}.$$

Substituting this expression into the formula for $\tilde{h}\left(re^{i\theta}\right)$, we obtain

$$\tilde{h}(re^{i\theta}) = \sum_{k=-\infty}^{\infty} \left(\int_{-\pi}^{\pi} h\left(e^{i\varphi}\right) e^{-ik\varphi} \frac{d\varphi}{2\pi} \right) r^{|k|} e^{ik\theta}$$

$$= \int_{-\pi}^{\pi} h\left(e^{i\varphi}\right) \left[\sum_{k=-\infty}^{\infty} r^{|k|} e^{-ik\varphi} e^{ik\theta} \right] \frac{d\varphi}{2\pi}.$$

In order to simplify this expression, we introduce the **Poisson kernel function** defined by

$$(1.1) \qquad P_r(\theta) = \sum_{k=-\infty}^{\infty} r^{|k|} e^{ik\theta}.$$

For each fixed $\rho < 1$, this series converges uniformly for $r \le \rho$ and $-\pi \le \theta \le \pi$, by the Weierstrass M-test, since then $\left| r^{|k|} e^{ik\theta} \right| \le \rho^{|k|}$. In terms of the Poisson kernel, the formula for the harmonic extension $\tilde{h}\left(re^{i\theta}\right)$ becomes

$$(1.2a) \qquad \tilde{h}(re^{i\theta}) = \int_{-\pi}^{\pi} h\left(e^{i\varphi}\right) P_r(\theta - \varphi) \frac{d\varphi}{2\pi}, \qquad re^{i\theta} \in \mathbb{D}.$$

If we make a change of variable $\varphi \mapsto \theta - \varphi$ and use the 2π-periodicity of $P_r(\theta)$, we obtain an alternative form of (1.2a),

$$(1.2b) \qquad \tilde{h}(re^{i\theta}) = \int_{-\pi}^{\pi} h\left(e^{i(\theta-\varphi)}\right) P_r(\varphi) \frac{d\varphi}{2\pi}, \qquad re^{i\theta} \in \mathbb{D}.$$

Now we look at the Poisson kernel function $P_r(\theta)$ more closely. Setting $z = re^{i\theta}$ and setting $j = -k$ when $k < 0$ in (1.1), we obtain

$$P_r(\theta) = 1 + \sum_{k=1}^{\infty} z^k + \sum_{j=1}^{\infty} \bar{z}^j.$$

Summing these two geometric series, we obtain

$$(1.3) \qquad P_r(\theta) = 1 + \frac{z}{1-z} + \frac{\bar{z}}{1-\bar{z}}, \qquad z = re^{i\theta} \in \mathbb{D}.$$

Putting this over the common denominator

$$|1 - z|^2 = (1 - z)(1 - \bar{z}) = 1 + r^2 - 2r \cos\theta,$$

we obtain

(1.4) $$P_r(\theta) = \frac{1 - |z|^2}{|1 - z|^2} = \frac{1 - r^2}{1 + r^2 - 2r \cos\theta}, \qquad z = re^{i\theta} \in \mathbb{D}.$$

Since the Poisson kernel is 2π-periodic, we focus on its behavior on the period interval $-\pi \leq \theta \leq \pi$. From the Poisson integral representation formula (1.2b) for the constant function $h \equiv 1$, we have

(1.5) $$\int_{-\pi}^{\pi} P_r(\theta) \frac{d\theta}{2\pi} = 1.$$

Three other properties of the Poisson kernel, which follow immediately from the formula (1.4), are

(1.6) $$P_r(\theta) > 0, \qquad -\pi \leq \theta \leq \pi,$$

(1.7) $$P_r(-\theta) = P_r(\theta), \qquad -\pi \leq \theta \leq \pi,$$

(1.8) $P_r(\theta)$ is increasing for $-\pi \leq \theta \leq 0$ and decreasing for $0 \leq \theta \leq \pi$.

Since $P_r(\delta) \to 0$ as $r \to 1$ for each fixed $\delta > 0$, we obtain from (1.8) the following key property of the Poisson kernel:

(1.9) for fixed $\delta > 0$, $\max\{ P_r(\theta) : \delta \leq |\theta| \leq \pi \} \to 0$ as $r \to 1$.

These five properties, (1.5) through (1.9), are fundamental for understanding the Poisson kernel. Properties (1.5) and (1.6) together show that each $P_r(\theta)d\theta/2\pi$ is a "probability measure," that is, a positive "mass distribution" with unit total mass. Properties (1.5), (1.6), and (1.9) show that the family of functions $P_r(\theta)$ form an "approximate identity," in the sense that the mass of the probability measure $P_r(\theta)d\theta/2\pi$ concentrates at the point $\theta = 0$ as $r \to 1$. Identity (1.7) reflects the symmetry of the Poisson kernel.

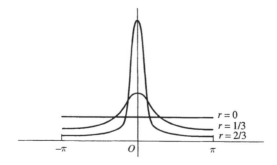

Another important property of the Poisson kernel function is that $P_r(\theta)$ is a harmonic function of $z = re^{i\theta}$ for $z \in \mathbb{D}$. In fact, (1.3) yields an explicit

representation of the Poisson kernel function as the real part of an analytic function,

$$P_r(\theta) = 1 + 2 \operatorname{Re}\left(\frac{z}{1-z}\right) = \operatorname{Re}\left(\frac{1+z}{1-z}\right), \qquad z = re^{i\theta} \in \mathbb{D}.$$

The Poisson integral formula (1.2a) becomes

$$\tilde{h}(re^{i\theta}) = \int_{-\pi}^{\pi} h\left(e^{i\varphi}\right) \operatorname{Re}\left(\frac{1+re^{i(\theta-\varphi)}}{1-re^{i(\theta-\varphi)}}\right) \frac{d\varphi}{2\pi}.$$

Thus

$$(1.10) \qquad \tilde{h}(z) = \int_{-\pi}^{\pi} h\left(e^{i\varphi}\right) \operatorname{Re}\left(\frac{e^{i\varphi}+z}{e^{i\varphi}-z}\right) \frac{d\varphi}{2\pi}, \qquad z \in \mathbb{D}.$$

If we substitute for $h\left(e^{i\theta}\right)$ a real-valued trigonometric polynomial $u\left(e^{i\theta}\right)$ in this formula, we may take real parts after integrating, and we obtain

$$(1.11) \qquad \tilde{u}(z) = \operatorname{Re} \int_{-\pi}^{\pi} u(e^{i\varphi}) \frac{e^{i\varphi}+z}{e^{i\varphi}-z} \frac{d\varphi}{2\pi}, \qquad z \in \mathbb{D}.$$

The integral depends analytically on the parameter z. Thus (1.11) expresses the Poisson integral $\tilde{u}(z)$ as the real part of an explicit analytic function.

Now we change our point of view. Instead of focusing on trigonometric polynomials, we consider an arbitrary continuous complex-valued function $h\left(e^{i\theta}\right)$ on the unit circle. We define the **Poisson integral** $\tilde{h}(z)$ of $h\left(e^{i\theta}\right)$ to be the function on the open unit disk $\mathbb{D}$ given by

$$(1.12) \qquad \tilde{h}(z) = \int_{-\pi}^{\pi} h\left(e^{i\varphi}\right) P_r(\theta - \varphi) \frac{d\varphi}{2\pi}, \qquad z = re^{i\theta} \in \mathbb{D}.$$

This is just (1.2a), and it can be rewritten as (1.2b). Formula (1.10) still holds, and for real-valued continuous functions $u\left(e^{i\theta}\right)$, formula (1.11) holds.

The correspondence $h \mapsto \tilde{h}$ is linear, that is, the Poisson integral of $c_1 h_1 + c_2 h_2$ is $c_1 \widetilde{h_1} + c_2 \widetilde{h_2}$. Also, observe that the maximum principle holds, in the sense that if $|h\left(e^{i\theta}\right)| \leq M$ for all θ, then $|\tilde{h}(z)| \leq M$ for all $z \in \mathbb{D}$. This follows from (1.5) and (1.6), that is, from the fact that the Poisson kernel determines a probability measure.

Theorem. Let $h\left(e^{i\theta}\right)$ be a continuous function on the unit circle. Then the Poisson integral $\tilde{h}(z)$ defined by (1.12) is a harmonic function on the open unit disk that has boundary values $h\left(e^{i\theta}\right)$, that is, $\tilde{h}(z)$ tends to $h(\zeta)$ as $z \in \mathbb{D}$ tends to $\zeta \in \partial\mathbb{D}$.

The harmonicity can be checked by differentiating under the integral sign in (1.12). It can also be seen by decomposing $h\left(e^{i\theta}\right)$ into its real and imaginary parts, $h\left(e^{i\theta}\right) = u\left(e^{i\theta}\right) + iv\left(e^{i\theta}\right)$. Formula (1.11) shows that

$\tilde{u}(z)$ and $\tilde{v}(z)$ are harmonic, and hence $\tilde{h}(z) = \tilde{u}(z) + i\tilde{v}(z)$ is harmonic for $|z| < 1$.

Since the statement about the boundary values is important, we sketch two proofs. The first proof depends upon the fact that any continuous function on the unit circle can be approximated uniformly by trigonometric polynomials. (One way to see this is to approximate the continuous function by a smooth function and then to appeal to the result proved in Section VI.6 that the Fourier series of any smooth function on the unit circle converges uniformly to the function.) The second proof illustrates how the properties of an approximate identity are typically used.

The first proof boils down to the fact that a uniform limit of continuous functions is continuous. It runs as follows. Let $h\left(e^{i\theta}\right)$ be continuous, and let $\varepsilon > 0$. Let $g\left(e^{i\theta}\right) = \sum a_k e^{ik\theta}$ be a trigonometric polynomial such that $\left|h\left(e^{i\theta}\right) - g\left(e^{i\theta}\right)\right| \le \varepsilon$ for all $e^{i\theta} \in \partial\mathbb{D}$. Then $\left|\tilde{h}(z) - \tilde{g}(z)\right| \le \varepsilon$ for all $z \in \mathbb{D}$. Now, $\tilde{g}\left(re^{i\theta}\right) = \sum a_k r^{|k|} e^{ik\theta}$ attains the boundary values $g\left(e^{i\theta}\right)$ continuously on $\partial\mathbb{D}$. Hence the composite function equal to $h\left(e^{i\theta}\right)$ on $\partial\mathbb{D}$ and $\tilde{h}\left(re^{i\theta}\right)$ on $\mathbb{D}$ can be approximated to within ε by a continuous function. Consequently, the values $\tilde{h}(z)$ cluster to within ε of $h\left(e^{i\theta}\right)$ as $z \in \mathbb{D}$ tends to $e^{i\theta} \in \partial\mathbb{D}$. Since $\varepsilon > 0$ is arbitrary, $\tilde{h}(z)$ tends to $h\left(e^{i\theta}\right)$ as $z \in \mathbb{D}$ tends to $e^{i\theta} \in \partial\mathbb{D}$.

The second proof depends only on properties (1.5), (1.6), and (1.9) of the Poisson kernel. Again let $\varepsilon > 0$. Choose $M > 0$ such that $\left|h\left(e^{i\theta}\right)\right| \le M$. Since $h\left(e^{i\theta}\right)$ is uniformly continuous, we can choose $\delta > 0$ so small that $\left|h\left(e^{i\theta}\right) - h\left(e^{i\varphi}\right)\right| < \varepsilon$ whenever $|\theta - \varphi| < \delta$. Using (1.5) we have

$$\tilde{h}(z) - h\left(e^{i\theta}\right) = \int_{-\pi}^{\pi} \left[h\left(e^{i(\theta-\varphi)}\right) - h\left(e^{i\theta}\right)\right] P_r(\varphi) \frac{d\varphi}{2\pi} .$$

Now we take absolute values, break the integral cleverly into pieces, and make the obvious estimate on each piece, to obtain

$$\left|\tilde{h}\left(re^{i\theta}\right) - h\left(e^{i\theta}\right)\right| \le \left(\int_{-\delta}^{\delta} + \int_{\delta \le |\varphi| \le \pi}\right) \left|h\left(e^{i(\theta-\varphi)}\right) - h\left(e^{i\theta}\right)\right| P_r(\varphi) \frac{d\varphi}{2\pi}$$

$$\le \int_{-\delta}^{\delta} \varepsilon P_r(\varphi) \frac{d\varphi}{2\pi} + 2M \max_{\delta \le |\varphi| \le \pi} P_r(\varphi) \int_{\delta \le |\varphi| \le \pi} \frac{d\varphi}{2\pi}$$

$$\le \varepsilon + 2M \max_{\delta \le |\varphi| \le \pi} P_r(\varphi).$$

By (1.9), the summand on the right tends to 0 as $r \to 1$. Thus again the values $\tilde{h}(z)$ cluster to within ε of $h\left(e^{i\theta}\right)$ as $z \to e^{i\theta}$, this for any $\varepsilon > 0$, so $\tilde{h}(z) \to h\left(e^{i\theta}\right)$ as $z \to e^{i\theta}$.

Exercises for X.1

1. Show that the Fourier coefficients of $P_r(\theta)$ are $c_k = r^{|k|}$, $-\infty < k < \infty$. (Here we fix r, $0 \le r < 1$, and we regard $P_r(\theta)$ as a function on the circle.)

2. Let $R > 0$, and let $h\left(Re^{i\theta}\right)$ be a continuous function on the circle $\{|z| = R\}$. Show that the function

$$\tilde{h}(z) = \int_{-\pi}^{\pi} \frac{R^2 - r^2}{R^2 + r^2 - 2rR\cos(\theta - \varphi)} h\left(Re^{i\varphi}\right) \frac{d\varphi}{2\pi}, \qquad |z| < R,$$

   is harmonic on the disk $\{|z| < R\}$ and has boundary values $h\left(Re^{i\theta}\right)$ on the boundary circle.

3. Suppose that $f(z) = u(z) + iv(z)$ is analytic for $|z| < 1$ and that $u(z)$ extends to be continuous on the closed disk $\{|z| \le 1\}$. Show that

$$f(z) = \int_0^{2\pi} u(e^{i\varphi}) \frac{e^{i\varphi} + z}{e^{i\varphi} - z} \frac{d\varphi}{2\pi} + iv(0), \qquad |z| < 1.$$

   Remark. This is the **Schwarz formula**, expressing an analytic function in terms of the boundary values of its real part.

4. Let $\{f_n(z) = u_n(z) + iv_n(z)\}$ be a sequence of analytic functions on the open unit disk $\mathbb{D}$ such that $u_n(z)$ extends continuously to $\partial\mathbb{D}$, $u_n\left(e^{i\theta}\right)$ converges uniformly on $\partial\mathbb{D}$ to $u\left(e^{i\theta}\right)$, and $v_n(0)$ converges. Show that $f_n(z)$ converges normally on $\mathbb{D}$ to an analytic function $f(z)$ whose real part is $\tilde{u}(z)$. *Hint.* Use the Schwarz formula.

5. Let $h\left(e^{i\theta}\right)$ be a piecewise continuous function (or an integrable function) on the unit circle. Show that the Poisson integral $\tilde{h}(z)$ tends to $h(\zeta)$ as $z \in \mathbb{D}$ tends to any point ζ of the unit circle at which $h\left(e^{i\theta}\right)$ is continuous.

6. A function $f(z)$, $z \in \mathbb{D}$, is said to have **radial limit** L at $\zeta \in \partial\mathbb{D}$ if $f(r\zeta) \to L$ as r increases to 1. Let $h\left(e^{i\theta}\right)$ be a piecewise continuous function on the unit circle. Show that $\tilde{h}(z)$ has a radial limit at each $\zeta \in \partial\mathbb{D}$, equal to the average of the limits of $h\left(e^{i\theta}\right)$ at ζ from each side.

7. For each $t > 0$, define the kernel function $C_t(s)$ on the real line by

$$C_t(s) = \frac{t}{\pi} \frac{1}{s^2 + t^2}, \qquad -\infty < s < \infty.$$

   For $h(\xi)$ a bounded piecewise continuous function on the real line, define a function $\tilde{h}(s + it)$ on the open upper half-plane $\mathbb{H}$ by

$$\tilde{h}(s + it) = \int_{-\infty}^{\infty} C_t(s - \xi)h(\xi)\,d\xi, \qquad s + it \in \mathbb{H}.$$

(a) Sketch the graph of $C_t(s)$ for $t = 1$, $t = 0.1$, and $t = 0.01$. (b) Show that $C_t(s) > 0$ and $\int_{-\infty}^{\infty} C_t(s)\,ds = 1$. (Thus $C_t(s)ds$ is a probability measure.) (c) Show that for each $\delta > 0$, $\int_{\{|s|>\delta\}} C_t(s)\,ds \to 0$ as $t \to 0$. (Thus $C_t(s)ds$ is an approximate identity.) (d) Show that $\tilde{h}(s + it)$ is a bounded harmonic function in the upper half-plane. (e) Show that $\tilde{h}(s + it) \to h(\xi_0)$ as $s + it \in \mathbb{H}$ tends to ξ_0, whenever $h(\xi)$ is continuous at ξ_0. *Remark.* The kernel function $C_t(s)$ is the **Poisson kernel** for the upper half-plane.

8. Let $h\left(e^{i\varphi}\right)$ be a continuous complex-valued function on $\partial\mathbb{D}$, and let $\tilde{h}(z)$ be its Poisson integral.
   (a) Show that

$$\frac{\partial^m}{\partial z^m}\tilde{h}(z) = m! \int_{-\pi}^{\pi} h(e^{i\varphi}) \frac{e^{i\varphi}}{(e^{i\varphi} - z)^{m+1}} \frac{d\varphi}{2\pi}, \qquad z \in \mathbb{D},$$

$$\frac{\partial^m}{\partial \bar{z}^m}\tilde{h}(z) = m! \int_{-\pi}^{\pi} h(e^{i\varphi}) \frac{e^{-i\varphi}}{(e^{-i\varphi} - \bar{z})^{m+1}} \frac{d\varphi}{2\pi}, \qquad z \in \mathbb{D}.$$

Hint. Use the identity

$$P_r(\theta - \varphi) = 1 + \frac{ze^{-i\varphi}}{1 - ze^{-i\varphi}} + \frac{\bar{z}e^{i\varphi}}{1 - \bar{z}e^{i\varphi}}, \qquad z = re^{i\theta},$$

and justify differentiating under the integral sign.
   (b) Show that if $\left|h\left(e^{i\theta}\right)\right| \le M$ on $\partial\mathbb{D}$, then

$$\left|\frac{\partial^m}{\partial z^m}\tilde{h}(z)\right| \le m!M \int_{-\pi}^{\pi} \frac{1}{|e^{i\varphi} - \rho|^{m+1}} \frac{d\varphi}{2\pi}, \qquad |z| \le \rho,$$

with similar estimates for the $\bar{z}$-derivatives.
   (c) Show that if $h\left(e^{i\theta}\right) = e^{-i\theta}\left(e^{i\theta} - z\right)^{m+1}/|e^{i\theta} - z|^{m+1}$, then equality holds in (b) when $|z| = \rho$.

9. Let $\{h_n(e^{i\varphi})\}$ be a sequence of continuous functions on the unit circle $\partial\mathbb{D}$ that converges uniformly to $h(e^{i\varphi})$. Show that for each fixed $\rho < 1$, the partial derivatives of $\tilde{h}_n(z)$ converge to the corresponding partial derivatives of $\tilde{h}(z)$ uniformly for $|z| \le \rho$. *Remark.* It suffices to show this for the partial derivatives $\partial^m/\partial z^m$ and $\partial^m/\partial \bar{z}^m$. See Exercise IV.8.4.

2. Characterization of Harmonic Functions

Recall (Section III.4) that a continuous function $h(z)$ on a domain D has the mean value property if for each $z_0 \in D$, the value $h(z_0)$ is the average of $h(z)$ over any small circle centered at z_0. We have seen that harmonic functions have the mean value property. Here we prove the converse.

Theorem. *Let $h(z)$ be a continuous function on a domain D. Then $h(z)$ is harmonic on D if and only if $h(z)$ has the mean value property on D.*

The proof depends on solving the Dirichlet problem for disks in D. Roughly speaking, the **Dirichlet problem** for a domain U with boundary ∂U is to extend a given function $f(\zeta)$ on ∂U to a harmonic function $\tilde{f}(z)$ on U, so that the harmonic extension $\tilde{f}(z)$ has boundary values $f(\zeta)$. Typically we assume that $f(\zeta)$ is a continuous function on ∂U, and we ask that $\tilde{f}(z)$ attain the boundary values $f(\zeta)$ continuously, in the sense that $\tilde{f}(z)$ tends to $f(\zeta)$ as $z \in U$ tends to $\zeta \in \partial U$. In the preceding section we showed that the Poisson integral solves the Dirichlet problem for the open unit disk $\mathbb{D}$. By making a change of variable $z \mapsto az + b$, we can solve the Dirichlet problem on any disk. An explicit solution is obtained by changing variable in the Poisson integral formula. (See Exercise 1.2.)

So let U be any open disk in D whose boundary is also contained in D. Since the Dirichlet problem is solvable for U, there is a harmonic function $g(z)$ on U that has boundary values $h(z)$ on ∂U. Then $h(z) - g(z)$ is a continuous function on U that has the mean value property on U, and $h(z) - g(z)$ tends to 0 on ∂U. By the maximum principle for functions with the mean value property (Section III.4), $h(z) - g(z) = 0$ on U. Consequently, $h(z) = g(z)$ on U, and $h(z)$ is harmonic on U. Since this holds for all U, $h(z)$ is harmonic on D.

The characterization of harmonic functions by the mean value property is the analogue for harmonic functions of Morera's theorem for analytic functions. The hypothesis involves an integration, which is easier to perform than differentiation. The characterization can be exploited in the same way as the characterization of analytic functions by Morera's theorem. For instance, it can be used to give a simple proof that the limit of a uniformly convergent sequence of harmonic functions is harmonic, since it is easy to check that the uniform limit of functions with the mean value property has the mean value property. It can also be used (Exercise 1) to show that an integral of a function that depends harmonically on a parameter also depends harmonically on the parameter.

It is rather striking that a continuous function with the mean value property automatically has partial derivatives of all orders.

Exercises for X.2

1. Let $g(t, z)$ be a continuous function defined for $a \leq t \leq b$ and z in a domain D, and suppose that $g(t, z)$ is a harmonic function of z for each fixed t. Show that

$$G(z) \;=\; \int_a^b g(t, z)\, dt, \qquad z \in D,$$

is harmonic on D.

2. Assume that $u(x, y)$ is a twice continuously differentiable function on a domain D.

(a) For $(x_0, y_0) \in D$, let $A_\varepsilon(x_0, y_0)$ be the average of $u(x, y)$ on the circle centered at (x_0, y_0) of radius ε. Show that

$$\lim_{\varepsilon \to 0} \frac{A_\varepsilon(x_0, y_0) - u(x_0, y_0)}{\varepsilon^2} = \frac{1}{4} \Delta u(x_0, y_0),$$

where Δ is the Laplacian operator (Section II.5).

(b) Let $B_\varepsilon(x_0, y_0)$ be the area average of $u(x, y)$ on the disk centered at (x_0, y_0) of radius ε. Show that

$$\lim_{\varepsilon \to 0} \frac{B_\varepsilon(x_0, y_0) - u(x_0, y_0)}{\varepsilon^2} = \frac{1}{8} \Delta u(x_0, y_0).$$

3. For fixed $\rho > 0$, define $h(z) = e^{i\rho \operatorname{Im} z}$. Show that if ρ is a zero of the Bessel function $J_1(z)$, then $\int_\gamma h(z) dz = 0$ for all circles γ of radius 1. *Suggestion.* See the Schlömilch formula (Exercise VI.1.3).

4. Suppose that $r_1, r_2 > 0$ are such that r_2/r_1 is a quotient of two positive zeros of the Bessel function $J_1(z)$. Show that there is a continuous function $g(z)$ on the complex plane such that $\int_\gamma g(z) dz = 0$ for all circles of radius r_1 and for all circles of radius r_2, yet $g(z)$ is not analytic. Use the preceding exercise together with the fact (easily derived from the differential equation in Section V.4) that $J_1(z)$ has zeros on the positive real axis. *Remark.* The condition is sharp. If r_2/r_1 is not the quotient of two positive zeros of the Bessel function $J_1(z)$, then any continuous function $f(z)$ on the complex plane such that $\int_\gamma f(z) dz = 0$ for all circles of radius r_1 and r_2 is analytic, by a theorem of L. Zalcman. There is an analogous result for harmonic functions and the mean value property. According to a theorem of J. Delsarte, if r_1, $r_2 > 0$ are such that r_2/r_1 is not the quotient of two complex numbers for which $J_0(z) = 1$, then any continuous function on the complex plane that has the mean value property for circles of radius r_1 and r_2 is harmonic.

3. The Schwarz Reflection Principle

Suppose a given function is analytic on a domain that abuts on one side of an analytic curve. The Schwarz reflection principle asserts that under certain conditions, the analytic function extends analytically across the curve, and further, there is a formula for the extension. The formula reflects the function analytically to the mirror image of the domain, and the nub of the problem is to establish analyticity on the axis of reflection.

At its core, the Schwarz reflection principle is a theorem about harmonic functions. We begin by reflecting harmonic functions across the real line.

First note that if $u(z)$ is harmonic in a domain D, then $u^*(z) = u(\bar{z})$ is harmonic on the reflected domain $D^* = \{\bar{z} : z \in D\}$. Indeed, if $u(x, y)$ has the mean value property on circles, then so does $u(x, -y)$. Another way to see this is to note that the change of variables $(x, y) \mapsto (x, -y)$ does not change the Laplacian operator Δ.

Theorem. *Let D be a domain that is symmetric with respect to the real axis, and let $D^+ = D \cap \{\operatorname{Im} z > 0\}$ be the part of D in the open upper half-plane. Let $u(z)$ be a real-valued harmonic function on D^+ such that $u(z) \to 0$ as $z \in D^+$ tends to any point of $D \cap \mathbb{R}$. Then $u(z)$ extends to be harmonic on D, and the extension satisfies*

$$(3.1) \qquad u(\bar{z}) = -u(z), \qquad z \in D.$$

For the proof, note that D is the disjoint union of $D^+ = D \cap \{\operatorname{Im} z > 0\}$, its reflection $D^- = D \cap \{\operatorname{Im} z < 0\}$, and $D \cap \mathbb{R}$. We extend $u(z)$ to D by defining $u(z) = -u(\bar{z})$ for $z \in D^-$, and setting $u(z) = 0$ for $z \in D \cap \mathbb{R}$. Then $u(z)$ is continuous on D, $u(z)$ is harmonic on D^+ and on D^-, and (3.1) holds. We claim that the extended function has the mean value property. Fix $z_0 \in D$. If z_0 is in D^+ or in D^-, then $u(z)$ has the mean value property for small disks centered at z_0, since $u(z)$ is harmonic near z_0. If $z_0 \in D \cap \mathbb{R}$, then by (3.1) the average value of $u(z)$ over the top half of a circle centered at z_0 cancels the average over the bottom half of the circle, and the average of $u(z)$ over the circle is 0, which coincides with $u(z_0)$. Thus $u(z)$ has the mean value property on D, and $u(z)$ is harmonic on D.

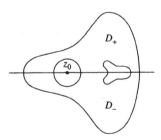

There is a corresponding result for analytic functions. First note that if $f(z)$ is analytic on a domain in the upper half-plane, then $g(z) = \overline{f(\bar{z})}$ is analytic on the reflected domain in the lower half-plane. This can be checked using the Cauchy-Riemann equations or a power series expansion. We can think of $g(z)$ as the composition of an anticonformal map $z \mapsto \bar{z}$, a conformal map $f(z)$ (except where $f'(z) = 0$), and an anticonformal map $w \mapsto \bar{w}$; hence it is conformal.

Theorem. *Let D be a domain that is symmetric with respect to the real axis, and let $D^+ = D \cap \{\operatorname{Im} z > 0\}$. Let $f(z)$ be an analytic function on D^+ such that $\operatorname{Im} f(z) \to 0$ as $z \in D^+$ tends to $D \cap \mathbb{R}$. Then $f(z)$ extends to be analytic on D, and the extension satisfies*

$$(3.2) \qquad\qquad f(\bar{z}) = \overline{f(z)}, \qquad z \in D.$$

For the proof, write $f(z) = u(z) + iv(z)$ in D^+. By the preceding theorem, $v(z)$ extends to be harmonic on D and satisfies $v(\bar{z}) = -v(z)$, $z \in D$. Fix a point $x_0 \in D \cap \mathbb{R}$, and let D_0 be a disk in D centered at x_0. Since $v(z)$ is harmonic on D_0, it has a harmonic conjugate on D_0, and consequently $f(z)$ extends to be analytic on D_0. The function $\overline{f(\bar{z})}$ is also analytic on D_0, and it coincides with $f(z)$ on the real axis, so by the uniqueness principle it coincides with $f(z)$ in D_0. Hence (3.2) holds for $z \in D_0$. If we now extend $f(z)$ to $D^- = D \cap \{\operatorname{Im} z < 0\}$ by setting $f(z) = \overline{f(\bar{z})}$, $z \in D^-$, the extended $f(z)$ is analytic on D^- and coincides with the analytic continuation of $f(z)$ across $D \cap \mathbb{R}$ from D^+, and so is analytic on all of D. Further, it satisfies (3.2) for all $z \in D$.

The above theorem is somewhat easier to prove if it is assumed that $f(z)$ extends continuously from D^+ to $D \cap \mathbb{R}$ and is real-valued there. In this case, formula (3.2) defines a continuous extension of $f(z)$ to all of D, which is analytic on D^-, and we can use Morera's theorem to show that it is also analytic across $D \cap \mathbb{R}$.

We define a curve γ to be an **analytic curve** if every point of γ has an open neighborhood U for which there is a conformal map $\zeta \mapsto z(\zeta)$ of a disk D centered on the real line $\mathbb{R}$ onto U, such that the image of $D \cap \mathbb{R}$ coincides with $U \cap \gamma$. We also refer to such a γ as an **analytic arc**. The conjugation $\zeta \mapsto \bar{\zeta}$ induces a map $z \mapsto z^*$ of U onto itself, by $z(\zeta)^* = z(\bar{\zeta})$. The map is an "involution" in the sense that the map followed by itself is the identity, $z^{**} = z$. Further, $z^* = z$ if and only if $z \in \gamma$. The map $\zeta \mapsto \bar{\zeta}$ interchanges the top half and the bottom half of $D \backslash \mathbb{R}$, which are the two components of $D \backslash \mathbb{R}$, so the map $z \mapsto z^*$ interchanges the two components of $U \backslash \gamma$. We refer to these two components as the **neighborhoods of the sides of** γ, and we refer to the map $z \mapsto z^*$ as the **reflection across** γ. Since the map $\zeta \mapsto \bar{\zeta}$ is anticonformal, and the composition of an anticonformal map and conformal maps is anticonformal, the reflection $z \mapsto z^*$ is anticonformal.

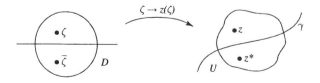

The reflection $z \mapsto z^*$ is unique, in the following sense. If there is an open neighborhood of a point $z_0 \in \gamma \cap U$ and an anticonformal map $z \mapsto z^+$

that fixes each point of γ in the neighborhood, then $z^+ = z^*$. Indeed, the map $z \mapsto (z^+)^*$ is the composition of two anticonformal maps, so it is conformal, hence analytic (Section V.8). Since it fixes points of γ, the uniqueness principle shows that it is the identity map. Thus $(z^+)^* = z$, and $z^+ = (z^+)^{**} = (z^{+*})^* = z^*$.

We can extend the involution $z \mapsto z^*$ to a neighborhood of the entire analytic arc γ, by covering γ with small coordinate sets U and using the uniqueness principle to see that the involutions agree on overlapping U's.

Example. The map $z(\zeta) = (\zeta-i)/(\zeta+i)$ sends the extended real line onto the unit circle. It induces a reflection across the unit circle,

$$z^* = (\bar{\zeta} - i)/(\bar{\zeta} + i) = \overline{(\zeta+i)/(\zeta-i)} = 1/\bar{z} = z/|z|^2.$$

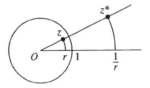

From this formula we see that the reflected point z^* lies on the same ray through the origin as z. The reflection $z \mapsto z^*$ maps the circle centered at 0 of radius r to that of radius $1/r$, sending $re^{i\theta}$ to $e^{i\theta}/r$.

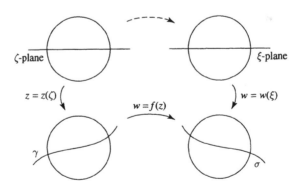

A more general version of the Schwarz reflection principle permits us to reflect harmonic and analytic functions across analytic curves. The idea is that if $w = f(z)$ is defined and analytic on one side of an analytic arc γ, and if the values w tend to another analytic arc σ as z tends to γ, then $f(z)$ extends analytically across γ, and the extension satisfies $f(z^*) = f(z)^*$ in a neighborhood of γ. (Here we use the same notation for both reflections.) It is a straightforward matter to justify the general version by coordinatizing both analytic arcs and reducing to the case of reflection across the real line.

Rather than state a precise theorem, we derive a version of the principle that will be used later.

Let D be a domain. An analytic arc $\gamma \subset \partial D$ is a **free analytic boundary arc** of D if every point of γ is contained in a disk U such that $U \backslash \gamma$ has two components, one contained in D and the other disjoint from D.

Example. Suppose D is obtained from the open disk $\{|z| < 2\}$ by excising a sequence of disjoint closed disks from the top half accumulating on the interval $[-1, 1]$, as in the figure. Then the boundary circle of each excised disk is a free analytic boundary arc of D, but the analytic arc $(-1, 1)$ is not a free analytic boundary arc.

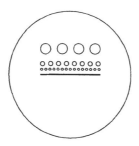

Theorem. *Let D be a domain, and let γ be a free analytic boundary arc of D. Let $f(z)$ be analytic on D. If $|f(z)| \to 1$ as $z \in D$ tends to γ, then $f(z)$ extends to be analytic in a neighborhood of γ, and the extension satisfies*

$$(3.3) \qquad\qquad f(z^*) \; = \; 1/\overline{f(z)}$$

in a neighborhood of γ, where $z \mapsto z^*$ is the reflection across γ.

For the proof, fix a point $z_0 \in \gamma$, and consider a conformal map $\zeta \mapsto z(\zeta)$ of a disk D_0 centered on $\mathbb{R}$ onto an open set containing z_0 so that $D_0 \cap \mathbb{R}$ corresponds to $U \cap \gamma$. Then $g(\zeta) = \log f(z(\zeta))$ is defined and analytic on one side of $D_0 \cap \mathbb{R}$. Further, $\operatorname{Re} g(\zeta) = \log |f(z(\zeta))| \to 0$ as $\zeta \to D_0 \cap \mathbb{R}$. By the Schwarz reflection principle, $g(\zeta)$ extends to be analytic on D_0, as does $f(z(\zeta)) = e^{g(\zeta)}$. Consequently, $f(z)$ extends analytically to U, and the extension is unique. To see that the extension satisfies (3.3), we argue as follows. The function $h(z) = 1/\overline{f(z^*)}$ is the composition of $f(z)$ and the two anticonformal maps $z \mapsto z^*$ and $w \mapsto 1/\bar{w}$. Thus it is conformal wherever $f'(z) \neq 0$. By the equivalence of analyticity and conformality (Section IV.8), $h(z)$ is analytic wherever $f'(z) \neq 0$, and it is continuous, hence analytic also at the isolated points where $f'(z) = 0$. Further, if $z \in \gamma$, then $z^* = z$, and $|f(z)| = 1$, so $h(z) = f(z)$ on γ. By the uniqueness principle, $h(z) = f(z)$ everywhere, and $f(z)$ satisfies (3.3).

Exercises for X.3

1. Show that the reflection in the circle $\{|z - z_0| = R\}$ is given by $z^* = z_0 + R^2(z - \bar{z_0})/|z - z_0|^2$.

2. Show that a reflection in a circle maps circles in the plane to circles.

3. What happens to angles between curves when they are reflected in an analytic arc?

4. Suppose the curve γ passing through 0 is the graph of a function $y = h(x)$ that can be expressed as a convergent power series $h(x) = \sum_{k=1}^{\infty} a_k x^k$, $-r < x < r$, where the a_k's are real. (a) Show that $z = \zeta + ih(\zeta)$ can be solved for $\zeta = \zeta(z)$ as an analytic function of z for $|z| < \varepsilon$. (b) Show that γ is an analytic arc. (c) Show that the reflection through γ is given by $z^* = 2\overline{\zeta(z)} - \bar{z}$.

5. Determine the reflection z^* of z across the parabola $y = x^2$. Expand z^* in a power series, in powers of $\bar{z}$. Determine the radius of convergence of the series. Try to explain graphically why the radius of convergence is finite.

6. Let $f(z)$ be an entire function whose modulus is constant on some circle. Show that $f(z) = c(z - z_0)^n$ for some $n \geq 0$ and some constant c, where z_0 is the center of the circle.

7. Show that if $f(z)$ is meromorphic for $|z| < 1$, and $|f(z)| \to 1$ as $|z| \to 1$, then $f(z)$ is a rational function. Show further that $f(z)$ is the quotient of two finite Blaschke products. (For the definition of finite Blaschke product, see Exercise IX.2.1.)

8. The **modulus of an annulus** $\{a < |z - z_0| < b\}$ is defined to be $(1/2\pi) \log(b/a)$. (a) Show that any conformal map from one annulus centered on the origin to another such annulus extends to a conformal self-map of the punctured plane. (b) Show that there is a conformal map of one annulus onto another if and only if the annuli have the same moduli. (c) Show that any conformal self-map of the annulus $\{a < |z| < b\}$ is either a rotation $z \mapsto e^{i\varphi}z$ or a rotation followed by the inversion $z \mapsto ab/z$.

9. Let γ be an analytic curve passing through the origin and tangent to the imaginary axis at 0. Suppose that $\gamma(t) = it + b_2 t^2 + b_3 t^3 + \cdots$, $-\delta < t < \delta$, where the b_n's are real. Show that there is a conformal map $\varphi(z)$ defined in a neighborhood of 0 such that $\varphi(0) = 0$, $\varphi'(0) = 1$, and $\varphi(z)$ maps the angle between the positive real axis and the segment of γ in the upper half-plane to the angle between the positive real and imaginary axes if and only if γ coincides with its reflection $\bar{\gamma}$ in the real axis. Show that this occurs if and only if $b_n = 0$ for n odd.

10. Let γ be an analytic curve passing through the origin and making an angle θ_0, $0 < \theta_0 < \pi/2$, with the positive real axis. Suppose γ is the image of an interval $(-\delta, \delta)$ under $h(z) = z + i(b_1 z + b_2 z^2 + \cdots)$, where the b_j's are real. Define

$$g(z) = \overline{h(\,\overline{h^{-1}(z)}\,)}, \qquad \lambda = e^{-2i\theta_0}.$$

(a) Show that $g(z)$ is analytic at $z = 0$, $g'(0) = \lambda$, and $g(z) = \bar{z}$ for $z \in \gamma$.

(b) Show that if $\varphi(z)$ is analytic at $z = 0$, $\varphi(0) = 0$, $\varphi'(0) = 1$, and $\varphi(z)$ maps the angle between the positive real axis and the segment of γ in the first quadrant to the angle between the positive real axis and the straight line segment at angle θ_0, then $\varphi(g(z)) = \lambda \overline{\varphi(z)}$. *Remark.* This equation for $\varphi(z)$ is **Schröder's equation**.

(c) Show by plugging in power series that any (normalized) solution $\varphi(z)$ of Schröder's equation is unique, if it exists, and is given by $\varphi(z) = z + c_2 z^2 + c_3 z^3 + \cdots$, where

$$c_n = \frac{1}{\lambda^n - \lambda} A_n(\lambda, a_2, \ldots, a_n, c_2, \ldots, c_{n-1}), \qquad n \geq 2,$$

A_n is a polynomial, and the a_n's are the power series coefficients of $g(z)$. *Remark.* The problem of estimating the c_n's to determine whether the series converges is called a "small denominator problem ." As the powers λ^{n-1} return sporadically near to 1, the denominators become sporadically small.

(d) Show that if $\varphi(z)$ is a solution of Schröder's equation (analytic at 0 and satisfying $\varphi(0) = 0$, $\varphi'(0) = 1$), and if $\varphi(x)$ is real when x is real, then $\varphi(z)$ maps the angle between the positive real axis and γ to the angle between the straight line segments at angles 0 and θ_0.

XI

Conformal Mapping

In this chapter we will be concerned with conformal maps from domains onto the open unit disk. One of our goals is the celebrated Riemann mapping theorem: Any simply connected domain in the complex plane, except the entire complex plane itself, can be mapped conformally onto the open unit disk. We begin in Section 1 by reviewing and enlarging our repertoire of conformal maps onto the open unit disk, or equivalently, onto the upper half-plane. In Section 2 we state and discuss the Riemann mapping theorem. Before embarking on the proof, we give some applications to the conformal mapping of polygons in Section 3 and to fluid dynamics in Section 4. In Section 5 we develop some prerequisite material concerning compactness of families of analytic functions, which is at a deeper level than the analysis used up to this point. The proof of the Riemann mapping theorem follows in Section 6.

1. Mappings to the Unit Disk and Upper Half-Plane

Recall that a **conformal map** of a domain D onto a domain V is an analytic function $\varphi(z)$ from D to V that is one-to-one and onto. The composition of two conformal maps is a conformal map, and the inverse of a conformal map is a conformal map. In this section we seek to find conformal maps from various domains D onto the open unit disk $\mathbb{D}$. Such a map $\varphi(z)$ is never unique, as we can compose it with any conformal self-map of $\mathbb{D}$. However, if $\psi(z)$ is any other conformal map of D onto $\mathbb{D}$, then $g = \psi \circ \varphi^{-1}$ is a conformal self-map of $\mathbb{D}$ for which $g \circ \varphi = \psi$. Thus once one conformal map φ of D onto $\mathbb{D}$ is known, the others are precisely the maps of the form $\psi = g \circ \varphi$, where g is a conformal self-map of $\mathbb{D}$.

We have seen in Section IX.2 that the conformal self-maps of the open unit disk have the form

$$g(z) = \lambda \frac{z - a}{1 - \bar{a}z}, \qquad z \in \mathbb{D},$$

where $|a| < 1$ and $|\lambda| = 1$. The conformal self-maps of $\mathbb{D}$ are determined by three real parameters, namely, the real and imaginary parts of a, and the argument of the unimodular constant λ. Consequently, a conformal map of D onto $\mathbb{D}$ is uniquely determined by specifying three real parameters. Specifying the image of a fixed point of D counts as two real parameters, corresponding to real and imaginary parts. Specifying the image of a boundary point of D counts as one real parameter.

The fractional linear transformation $w = (z - i)/(z + i)$ maps the open upper half-plane $\mathbb{H}$ onto the open unit disk $\mathbb{D}$. The problem of finding a conformal map of a given domain onto $\mathbb{D}$ is equivalent to finding one onto $\mathbb{H}$. We can go back and forth between $\mathbb{D}$ and $\mathbb{H}$ by composing with the above map and its inverse $z = i(1 + w)/(1 - w)$. The complex analyst does not distinguish between the half-plane and the unit disk, and works in whichever space the calculations are simpler.

We turn to three specific classes of domains: sectors, strips, and lunes.

Sectors. A sector can be mapped onto a half-plane with the aid of the power function z^β, for an appropriate choice of β. Any sector with vertex at 0 can be rotated by the map $z \mapsto \lambda z$, $|\lambda| = 1$, to a sector of the form $D = \{0 < \arg z < \alpha\}$, where $\alpha \le 2\pi$. Since z^β multiplies angles by β, we set $\beta = \pi/\alpha$, to obtain a map $\zeta = z^{\pi/\alpha}$ of the sector D onto the upper half-plane. If we follow this by the map $w = (\zeta - i)/(\zeta + i)$ of $\mathbb{H}$ onto $\mathbb{D}$, we obtain a conformal map

$$w = \varphi(z) = \frac{z^{\pi/\alpha} - i}{z^{\pi/\alpha} + i}, \qquad z \in D,$$

mapping the sector D onto the open unit disk. Under this map, the vertex of the sector at $z = 0$ corresponds to $w = -1$, and the other "vertex" at $z = \infty$ corresponds to $w = +1$. Rays emanating from the origin are mapped to arcs of circles from -1 to $+1$ in $\mathbb{D}$, and the circular arcs $\{|z| = \text{constant}\}$ are mapped to their orthogonal trajectories, which are arcs of circles from the bottom half of the unit circle to the top half.

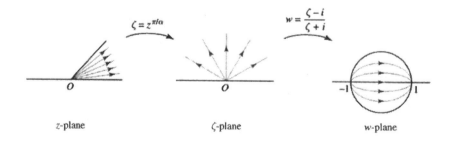

z-plane ζ-plane w-plane

Exercise. Find a conformal map of the sector $\{-\pi < \arg z < \pi/2\}$ onto the open unit disk.

Solution. The rotation $\zeta = e^{\pi i}z = -z$ maps the given sector to the sector $\{0 < \arg \zeta < 3\pi/2\}$ in the ζ-plane. The power function $\xi = \zeta^{2/3}$ maps this to the upper half of the ξ-plane. We compose this with the map $w = (\xi - i)/(\xi + i)$ onto the open unit disk, and we obtain

$$w = \frac{(-z)^{2/3} - i}{(-z)^{2/3} + i} = \frac{e^{2\pi i/3} z^{2/3} - i}{e^{2\pi i/3} z^{2/3} + i},$$

where we take the branch of $z^{2/3}$ satisfying $-2\pi/3 < \arg\left(z^{2/3}\right) < \pi/3$ on the sector.

Strips. An arbitrary strip can be mapped to a horizontal strip by a rotation $z \mapsto \lambda z$. The exponential function e^z maps horizontal strips to sectors. Thus an appropriate power $e^{\alpha z}$ maps a horizontal strip onto a half-plane. A further rotation maps the half-plane onto the upper half-plane, which can be then mapped to the unit disk.

Exercise. Find a conformal map of the vertical strip $\{-1 < \operatorname{Re} z < 1\}$ onto the open unit disk $\mathbb{D}$. What are the images of vertical and horizontal lines in the strip under the map?

Solution. The preliminary rotation $\zeta = iz$ maps the vertical strip to the horizontal strip $\{-1 < \operatorname{Im} z < 1\}$. The exponential map $\xi = e^{\alpha \zeta}$ maps the horizontal strip onto the sector $\{-\alpha < \arg \xi < \alpha\}$. Thus if we take $\alpha = \pi/2$, we map onto the right half of the ξ-plane. Multiplication by i maps this onto the upper half-plane, and we obtain finally

$$w = \frac{i\xi - i}{i\xi + i} = \frac{\xi - 1}{\xi + 1} = \frac{e^{\pi \zeta/2} - 1}{e^{\pi \zeta/2} + 1} = \frac{e^{\pi i z/2} - 1}{e^{\pi i z/2} + 1}.$$

Under this map, $z = -i\infty$ is mapped to $w = +1$ and $z = +i\infty$ is mapped to $w = -1$. Vertical lines in the strip in the z-plane are rotated to horizontal lines in the ζ-plane, which are carried by the exponential map to rays in the ξ-plane. These are carried to circles in the unit disk in the w-plane passing through ± 1. Horizontal lines in the strip are mapped to the orthogonal trajectories of the circles passing through ± 1. These orthogonal trajectories are also arcs of circles, passing from the bottom half of the unit circle to the top half.

Note that the solution map for this problem is not unique, and different maps will give different images of horizontal and vertical lines. However, once the images of $-i\infty$ and $+i\infty$ are known, then the vertical lines in the strip will be mapped to arcs of circles through these two image points, and horizontal lines in the strip will be mapped to the arcs of circles that are their orthogonal trajectories.

Exercise. Find a conformal map of the vertical strip $\{-1 < \operatorname{Re} z < 1\}$ onto the open unit disk $\mathbb{D}$ that maps $-i\infty$ to -1 and $+i\infty$ to i.

Solution. The desired map is obtained by composing the map just constructed with a conformal self-map $g(w) = \lambda(w - a)/(1 - \bar{a}w)$ of the unit disk. Since $z = -i\infty$ goes to $w = +1$, we require that $g(+1) = -1$, and since $z = i\infty$ goes to $w = -1$, we require that $g(-1) = i$. Thus

$$\lambda \frac{1-a}{1-\bar{a}} = -1, \qquad \lambda \frac{-1-a}{1+\bar{a}} = i.$$

If we eliminate the unimodular constant λ and simplify, we obtain $|a|^2 + (\bar{a} - a) = 1$, which is the equation of a circle centered at $+1$ of radius $\sqrt{2}$. Thus the solution is not unique, as we can choose a to be any point on this circle satisfying $|a| < 1$. For instance, we could take $a = 1 - \sqrt{2}$, which yields $\lambda = -1$. Choosing the point a on the circle corresponds to specifying the third parameter. Note that there is no solution for which additionally the point $z = 0$ is mapped to 0. We would ordinarily not expect there to be such a map, since this would correspond to specifying a total of four parameters, and we have only three at our disposal.

Lunar domains. Suppose the domain D has boundary consisting of two curves, each of which is an arc of a circle or a straight line segment. Let z_0 and z_1 be the endpoints of the curves. We assume $z_0 \neq z_1$. Then there is a fractional linear transformation $\zeta = g(z)$ mapping z_0 to 0 and z_1 to ∞. Since fractional linear transformations map circles to circles, the images of the two arcs lie on circles passing through 0 and ∞, and so are rays from 0 to ∞. Consequently, $\zeta = g(z)$ maps D onto a sector in the ζ-plane. This reduces the problem to mapping the sector onto $\mathbb{H}$ or $\mathbb{D}$.

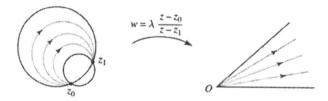

Exercise. Find a conformal map of the part of the upper half-plane outside the unit circle onto the entire upper half-plane mapping -1 to -1, i to 0, and $+1$ to $+1$.

Solution. We think of D as a lunar domain in the extended complex plane. One of the bounding arcs is the straight line segment from $+1$ through ∞ to -1, and the other bounding arc is the top half of the unit circle. The map $\zeta(z) = (z - 1)/(z + 1)$ sends the two vertices -1 and $+1$ of the lune to ∞ and 0, respectively, and it maps D to the first quadrant. Under this map, i goes to i. The first quadrant is mapped to the upper half-plane by $\xi = \zeta^2$. Thus the composition $\xi = (z - 1)^2/(z + 1)^2$ maps D onto the upper half-plane, and it sends -1 to ∞, i to -1, and $+1$ to 0. Finally, we compose with the fractional linear transformation that maps the upper half-plane to itself and sends ∞ to -1, -1 to 0, and 0 to 1. This map is found to be $w = (1 + \xi)/(1 - \xi)$. The final solution is then

$$(1.1) \quad w = \frac{1 + (z-1)^2/(z+1)^2}{1 - (z-1)^2/(z+1)^2} = \frac{(z+1)^2 + (z-1)^2}{(z+1)^2 - (z-1)^2} = \frac{1}{2}\left(z + \frac{1}{z}\right),$$

which has already appeared in Exercise II.6.6.

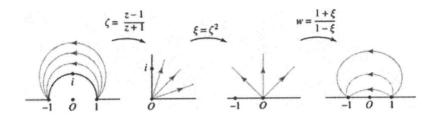

Exercises for XI.1

1. Find a conformal map of the sector $\{|\arg z| < \pi/3\}$ onto the open unit disk mapping 0 to -1 and ∞ to $+1$. Sketch the images of radial lines and of arcs of circles centered at 0. Is the map unique?

2. Find a conformal map of the slit plane $\mathbb{C}\backslash(-\infty, 0]$ onto the open unit disk satisfying $w(0) = i$, $w(-1 + 0i) = +1$, $w(-1 - 0i) = -1$. What are the images of circles centered at 0 under the map? Sketch them.

3. For fixed $A > 0$, find the conformal map $w(z)$ of the open unit disk $\{|z| < 1\}$ onto the vertical strip $\{-A < \operatorname{Re} w < A\}$ that satisfies $w(0) = 0$ and $w'(0) > 0$. Sketch the curves in the disk that correspond to vertical and horizontal lines in the strip.

4. Find a conformal map $w(z)$ of the strip $\{\operatorname{Im} z < \operatorname{Re} z < \operatorname{Im} z + 2\}$ onto the upper half-plane such that $w(0) = 0$, $w(z) \to +1$ as $\operatorname{Re} z \to -\infty$, and $w(z) \to -1$ as $\operatorname{Re} z \to +\infty$. Sketch the images of the straight lines $\{\operatorname{Re} z = \operatorname{Im} z + c\}$ in the strip. What is the image of the median line $\{\operatorname{Re} z = \operatorname{Im} z + 1\}$ of the strip?

5. Find a conformal map $w(z)$ of the right half-disk $\{\operatorname{Re} z > 0, |z| < 1\}$ onto the upper half-plane that maps $-i$ to 0, $+i$ to ∞, and 0 to -1. What is $w(1)$?

6. Let $w = g(z)$ be the conformal map of the right half-disk $\{\operatorname{Re} z > 0, |z| < 1\}$ onto the entire unit disk that fixes the points $\pm i$ and $+1$. (a) Without computing $g(z)$ explicitly, show that $g(\bar{z}) = \overline{g(z)}$. *Hint.* Argue that $h(z) = \overline{g(\bar{z})}$ is another conformal map satisfying the same conditions, and appeal to uniqueness. (b) Use symmetry to show that $g(0) = -1$. (In other words, use part (a).) (c) Find

$g(z)$ as a composition of explicit conformal maps, and use this to check that $g(0) = -1$.

7. Find the conformal map of the pie-slice domain $\{|\arg z| < \alpha, |z| < 1\}$ onto the open unit disk such that $w(0) = -1$, $w(+1) = +1$, and $w(e^{i\alpha}) = i$. It is enough to express $w(z)$ as a composition of specific conformal maps.

8. For fixed b in the interval $(-1, 1)$, find all conformal maps of the unit disk slit along the interval $[-1, b]$ onto the entire unit disk that map b to -1 and leave $+1$ fixed. It is enough to express them as a composition of specific conformal maps.

9. For fixed $a > 0$, let D be the domain obtained by slitting the upper half-plane along the vertical interval from $z = 0$ to $z = ia$.
   (a) Find a conformal map $w(z)$ of D onto the entire upper half-plane such that $w(z) \sim z$ as $|z| \to \infty$. *Hint.* Consider the preliminary map $\zeta = z^2$.
   (b) Describe how the map can be used to model the flow of water over a vertical metal sheet lying in a flat river bed, perpendicular to the flow of the water. Give a rough sketch of the streamlines of the flow.

10. Find the conformal map $w = f(z)$ of the exterior domain $\{|z| > 1\} \cup \{\infty\}$ onto $\mathbb{C}^* \backslash [-1, +1]$ such that $f(\infty) = \infty$ and the argument of $f(z)/z$ tends to α as $z \to \infty$, where α is a fixed real number. Sketch the images of circles $\{|z| = r\}$, for $r > 1$, and the images of the intervals $(-\infty, -1]$ and $[1, \infty)$. What is the inverse map? (Specify the branch.) *Hint.* For $\alpha = 0$, use the map $(z + 1/z)/2$ treated in (1.1) above. For the general case, do a preliminary rotation. See also the exercises in Section II.6.

11. Show that the half-strip $\{-\pi/2 < \operatorname{Re} z < \pi/2, \operatorname{Im} z > 0\}$ is mapped conformally by $w = \sin z$ onto the upper half-plane. Sketch the images of horizontal and vertical lines.

12. Find a conformal map of the half-strip in Exercise 11 onto the open unit disk that maps $-\pi/2$ to $-i$, $\pi/2$ to i, and 0 to $+1$. Where does ∞ go under this map?

2. The Riemann Mapping Theorem

This section is devoted to a preliminary discussion of the Riemann mapping theorem. The proof will be postponed to the end of the chapter. The Riemann mapping theorem was first established in the following generality by W.F. Osgood in 1900.

Theorem (Riemann Mapping Theorem). *If D is a simply connected domain in the complex plane, and D is not the entire complex plane, then there is a conformal map of D onto the open unit disk $\mathbb{D}$.*

Concerning the hypothesis, recall from Section VIII.8 that a domain is simply connected if every closed curve in the domain can be deformed to a point in the domain. Several characterizations of simply connected domains were given in Section VIII.8. Roughly speaking, a domain in the plane is simply connected if it has no "holes." Disks are simply connected, and annuli are not.

We say that two domains are **conformally equivalent** if there is a conformal map of one onto the other. Thus the Riemann mapping theorem asserts that any simply connected domain in the complex plane $\mathbb{C}$ either coincides with $\mathbb{C}$ or is conformally equivalent to $\mathbb{D}$.

We refer to a conformal map $w = \varphi(z)$ of D onto $\mathbb{D}$ as the **Riemann map** of D onto $\mathbb{D}$. It is unique, up to postcomposing with a conformal self-map of $\mathbb{D}$. To specify the Riemann map uniquely we must specify three real parameters. One way to do this is to specify $\varphi(z_0) = 0$ and $\varphi'(z_0) > 0$ for some $z_0 \in D$.

Each of the hypotheses in the Riemann mapping theorem is necessary. The complex plane $\mathbb{C}$ cannot be mapped conformally onto any bounded domain, because according to Liouville's theorem the only bounded analytic functions on $\mathbb{C}$ are the constants. Since any closed curve in $\mathbb{D}$ can be deformed to a point, any closed curve in any domain conformally equivalent to $\mathbb{D}$ can also be deformed to a point, by composing a deformation in $\mathbb{D}$ with the inverse of the Riemann map. Thus any domain conformally equivalent to $\mathbb{D}$ is simply connected.

For a simply connected domain D in the Riemann sphere (the extended complex plane), there are three possibilities. Indeed, if the domain is not the entire Riemann sphere, we can move a point in the complement of the domain to ∞ by a fractional linear transformation and reduce to the case of a simply connected domain in the plane, where there are two possibilities. The Riemann mapping theorem then yields the following.

Corollary. *A simply connected domain in the Riemann sphere is either the entire Riemann sphere, or it is conformally equivalent to the complex plane, or it is conformally equivalent to the open unit disk.*

Suppose now that D is a simply connected domain with Riemann map $\varphi(z)$ mapping D onto $\mathbb{D}$. We consider the behavior of $\varphi(z)$ as z approaches the boundary of D. For each fixed $\varepsilon > 0$, the set $\{|\varphi(z)| \leq 1 - \varepsilon\}$ is a compact subset of D, which is at a positive distance from ∂D. It follows that $|\varphi(z)| \to 1$ as $z \to \partial D$. Hence the harmonic function $\log|\varphi(z)|$ tends to 0 as z tends to the boundary of D. The Schwarz reflection principle for harmonic functions (Section X.3) then implies that $\log|\varphi(z)|$ extends harmonically across any free analytic boundary arc of D. Consequently, $\varphi(z)$ extends analytically across any free analytic boundary arc of D, and in fact we have the following.

Theorem. *Let D be a simply connected domain in $\mathbb{C}$, $D \neq \mathbb{C}$. Then the Riemann map $\varphi(z)$ of D onto $\mathbb{D}$ extends analytically across any free analytic boundary arc γ of D, and $\varphi(z)$ maps γ one-to-one onto an arc of $\partial\mathbb{D}$. The extended function satisfies $\varphi'(z) \neq 0$ for $z \in \gamma$, and $\varphi(z^*) = 1/\overline{\varphi(z)}$ for z in a neighborhood of γ, where $z \mapsto z^*$ is the reflection across γ. Disjoint free analytic boundary arcs of D are mapped by $\varphi(z)$ to disjoint arcs of $\partial\mathbb{D}$.*

For the formula for the analytic extension, see Section X.3. From the formula and the fact that $|\varphi(z)| < 1$ on one side of γ in D, we see that $|\varphi(z)| > 1$ on the other side of γ. Thus near any point of γ the locus where $|\varphi(z)| = 1$ consists only of points of the one analytic curve γ, so no point of γ can be a critical point of $\varphi(z)$. Let $z_0 \in \gamma$, and let $\zeta_0 = \varphi(z_0)$. If U_ε is a small disk centered at z_0, then $\varphi(U_\varepsilon)$ includes all points of $\mathbb{D}$ in a sufficiently small disk centered at ζ_0. Since $\varphi(z)$ is one-to-one on D, no other point of D is mapped by $\varphi(z)$ to this disk centered at ζ_0. Thus no other point of γ, and no point lying on any other free analytic boundary arc of D, is mapped to ζ_0. It follows that $\varphi(z)$ is one-to-one on γ, and further, the image of γ is disjoint from the image of any other free analytic boundary arc of D.

Exercises for XI.2

1. Show that no two of the domains $\mathbb{C}^*$, $\mathbb{C}$, and $\mathbb{D}$ are conformally equivalent.

2. Let $\varphi(z)$ be a conformal map from a domain D onto the open unit disk $\mathbb{D}$. For $0 < r < 1$, let D_r be the set of $z \in D$ such that $|\varphi(z)| < r$. Find a conformal map of D_r onto $\mathbb{D}$.

3. Let D be a domain in the complex plane whose complement $\mathbb{C}^* \backslash D$ in the extended complex plane consists of a finite number of disjoint closed connected sets, not all of which are points. Show that D can be mapped conformally onto a bounded domain whose boundary consists of a finite number of points and simple closed analytic curves.

3. The Schwarz-Christoffel Formula

Suppose that D is a polygonal domain, bounded by a finite number of straight line segments, and $w = g(z)$ is a conformal map of the upper half-plane onto D. The Schwarz-Christoffel formula is a differential equation that $g(z)$ must satisfy. Even though the differential equation cannot be solved in closed form except in very special cases, it provides a basis for computation of conformal maps onto arbitrary domains by approximating them by polygons and solving the associated differential equations numerically.

We begin by studying the behavior of a conformal map at a corner. It will be convenient to work in the upper half-plane instead of the unit disk.

Let D be a simply connected domain that has a corner at w_0, so that the part of D near w_0 has the form $D_\delta = \{w_0 + re^{i\theta} : 0 < r < \delta, \theta_1 < \theta < \theta_0\}$. Thus the part of ∂D near w_0 consists of two straight line segments terminating at w_0 and subtending an interior angle of $\theta_0 - \theta_1 = \pi\alpha$, where $0 < \alpha < 2$. Let $z = \varphi(w)$ be the Riemann map of D onto the upper half-plane $\mathbb{H}$. It maps the two sides of the corner analytically onto two disjoint intervals on the extended real line. By composing with a conformal self-map of $\mathbb{H}$, we can assume that the side at angle θ_0 is mapped to a finite interval $I = (a_0, b_0)$, and the side at angle θ_1 to a finite interval $J = (a_1, b_1)$, ordered so that $b_0 \leq a_1$. The image under $\varphi(w)$ of the circular arc $\{|w - w_0| = \varepsilon\} \cap D$ is a curve γ_ε in $\mathbb{H}$ from I to J. Since the circular arcs tend to $w_0 \in \partial D$, the curves γ_ε tend to the boundary of $\mathbb{H}$ as $\varepsilon \to 0$, in such a way that the initial point of γ_ε increases to b_0 and the terminal point decreases to a_1. It follows that the curves tend to the closed interval $[b_0, a_1]$ as $\varepsilon \to 0$. Thus the inverse function $w = g(z)$ of $\varphi(w)$ tends to the vertex w_0 as $z \in \mathbb{H}$ tends to the interval $[b_0, a_1]$. Now, the interval cannot have any interior points, or else $g(z) - w_0$ would tend to zero on the interval, hence reflect analytically across, hence be identically zero, which is absurd. Thus $b_0 = a_1$. We conclude that $\varphi(w)$ extends continuously to the corner w_0, and that $\varphi(w)$ maps the two segments in ∂D abutting at a corner w_0 to two contiguous intervals on the real line meeting at the image $\varphi(w_0)$ of the corner.

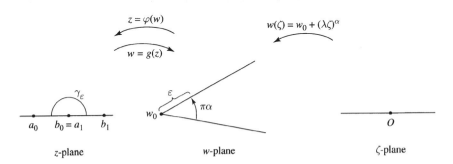

z-plane   w-plane ζ-plane

Now we consider more carefully the behavior of the inverse map $w = g(z)$ of the upper half-plane onto the domain D, mapping the point $a_0 \in \mathbb{R}$ to the corner w_0, and mapping intervals on each side of a_0 onto two straight line segments in ∂D terminating at w_0. Let $\lambda = e^{i\theta_1}$, a unimodular constant.

The function $w(\zeta) = w_0 + \lambda \zeta^\alpha$ maps a semidisk $\{|\zeta| < \varepsilon, \operatorname{Im} \zeta > 0\}$ onto the part of D at the corner. We may express ζ as a function of z, by $\lambda^{1/\alpha} \zeta(z) = (g(z) - w_0)^{1/\alpha}$. Since $\operatorname{Im} \zeta(z) \to 0$ as $\operatorname{Im} z \to 0$, the Schwarz reflection principle shows that $\zeta(z)$ is analytic at $z = a_0$, and further, a_0 cannot be a critical point of $\zeta(z)$. Thus $\lambda^{1/\alpha} \zeta(z) = (z - a_0) f(z)$ where $f(z)$ is analytic at a_0 and $f(a_0) \neq 0$. Let $h(z)$ be an analytic branch of $f(z)^\alpha$, so that $\lambda \zeta^\alpha = (z - a_0)^\alpha h(z)$ and $h(a_0) \neq 0$. Then from $g(z) = w_0 + \lambda \zeta^\alpha$ we obtain

$$g(z) = w_0 + (z - a_0)^\alpha h(z),$$

where $h(z)$ is analytic at a_0 and $h(a_0) \neq 0$. If we differentiate once, we eliminate w_0. If we differentiate again, we obtain

$$g'(z) = (z - a_0)^{\alpha-1} \left(\alpha h(z) + \mathcal{O}(z - a_0) \right),$$
$$g''(z) = (z - a_0)^{\alpha-2} \left(\alpha(\alpha - 1) h(z) + \mathcal{O}(z - a_0) \right),$$

and this allows us to eliminate the branch point by dividing,

$$(3.1) \qquad \frac{g''(z)}{g'(z)} = \frac{\alpha - 1}{z - a_0} + \text{analytic}, \qquad |z - a_0| < \varepsilon, \ \operatorname{Im} z > 0.$$

Thus while $g(z)$ has a branch point at a_0, the combination of derivatives $g''(z)/g'(z)$ extends meromorphically to a_0 and has a simple pole there with residue $\alpha - 1$.

Something similar happens if the point ∞ is mapped to the corner w_0 by $g(z)$. We parametrize the corner by ζ as before, and apply the Schwarz reflection principle, to see that $\lambda^{1/\alpha} \zeta(z) = (g(z) - w_0)^{1/\alpha}$ depends analytically on z at $z = \infty$. In this case, $\lambda^{1/\alpha} \zeta(z) = f(z)/z$, where $f(z)$ is analytic at ∞ and $f(\infty) \neq 0$. Then $g(z) = w_0 + h(z)/z^\alpha$, where $h(z)$ is analytic at ∞ and $h(\infty) \neq 0$. A calculation similar to the one above shows that

$$(3.2) \qquad \frac{g''(z)}{g'(z)} = -\frac{\alpha + 1}{z} + \cdots, \qquad |z| > R, \ \operatorname{Im} z > 0,$$

is analytic at ∞ and vanishes there. We have proved the following theorem on corners.

Theorem. *Let D be a simply connected domain with a corner at $w_0 \in \partial D$ with interior angle $\alpha\pi$, $0 < \alpha < 2$. Suppose $w = g(z)$ is a conformal map of $\mathbb{H}$ onto D. Then there are two contiguous intervals of the extended real line that are mapped analytically by $g(z)$ onto the sides of the corner, and $g(z)$ extends continuously to map the common endpoint a_0 of the intervals to the vertex w_0 of the corner. If a_0 is finite, then $g''(z)/g'(z)$ extends to be meromorphic at a_0 with a simple pole and with residue $\alpha - 1$. If $a_0 = \infty$, then $g''(z)/g'(z)$ extends to be analytic at ∞ and vanishes there.*

The preceding theorem also applies to a virtual corner, that is, to a point lying on straight line segments in ∂D. Such a point can be considered a corner with angle π. If a_0 is finite, the residue at a_0 is zero, and $g''(z)/g'(z)$ is analytic at a_0, as it is also in the case that $a_0 = \infty$.

If D is a polygon, we can carry the analysis further. In this case, the preceding analysis shows that the conformal map $g(z)$ of $\mathbb{H}$ onto D extends continuously to map the extended real line one-to-one onto ∂D.

Theorem. *Suppose $g(z)$ is a conformal map of the upper half-plane onto a (bounded) polygon D. Let $a_1 < \cdots < a_m$ be the points of $\mathbb{R}$ mapped to vertices of D. (It may be that ∞ is also mapped to a vertex.) Suppose D has interior angle $\alpha_j \pi$ at the vertex $w_j = g(a_j)$, $1 \le j \le m$. Then*

$$(3.3) \qquad \frac{g''(z)}{g'(z)} = \frac{\alpha_1 - 1}{z - a_1} + \cdots + \frac{\alpha_m - 1}{z - a_m},$$

and there are constants A and B such that

$$(3.4) \qquad g'(z) = A(z - a_1)^{\alpha_1 - 1} \cdots (z - a_m)^{\alpha_m - 1},$$

$$(3.5) \qquad g(z) = A \int_{z_0}^{z} (t - a_1)^{\alpha_1 - 1} \cdots (t - a_m)^{\alpha_m - 1} dt + B.$$

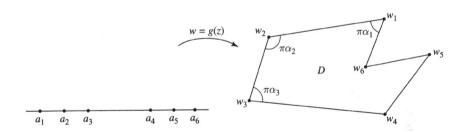

The Schwarz reflection principle shows that $g(z)$ reflects analytically across each interval on $\mathbb{R}$ between consecutive a_j's, and also across the intervals $(-\infty, a_1)$ and $(a_m, +\infty)$. The reflected function across (a_j, a_{j+1}) maps the lower half-plane conformally onto the polygon obtained by reflecting D across the straight line passing through w_j and w_{j+1}. (The extensions obtained by reflecting across different intervals may not coincide.) Thus $g'(z)$ and $g''(z)$ extend analytically across each interval to the lower half-plane, and since $g(z)$ has no critical points on the intervals, the extended $g'(z)$ has no zeros. Thus $g''(z)/g'(z)$ extends analytically across each of the intervals to the lower half-plane. The form (3.1) of $g''(z)/g'(z)$ at the a_j's shows that the reflections of $g''(z)/g'(z)$ across contiguous intervals coincide. Thus $g''(z)/g'(z)$ extends to be meromorphic on the entire complex plane, with a simple pole at each a_j with residue $\alpha_j - 1$ and with no other poles. The theorem on corners (which applies to virtual corners) shows that $g''(z)/g'(z)$ is analytic at ∞ and vanishes there. Thus $g''(z)/g'(z)$ is a meromorphic function on the extended complex plane, hence rational, and (3.3) is its partial fractions decomposition (Section VI.4). Integrating (3.3), we obtain

$$\log g'(z) = \sum (\alpha_j - 1) \log(z - a_j) + \text{ constant},$$

and this yields (3.4) for some constant A. One more integration yields (3.5).

Both (3.4) and (3.5) are referred to as the Schwarz-Christoffel formula. In order to clarify the formula, we make a series of remarks.

1. Each of the singularities in (3.5) is integrable, since $\alpha_j > 0$.

2. If the angle at w_j is π, so that we have only an apparent vertex, then $\alpha_j - 1 = 0$, and the corresponding factor disappears from the integral (as it should).

3. We can allow an angle of 2π, corresponding to the tip of a slit. In this case we think of the slit as two intervals, one for each side of the slit.

4. We can allow intervals in ∂D to overlap, and we can allow one point to serve as a multiple vertex, as in the figures.

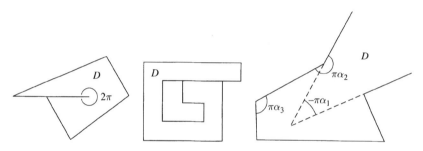

5. We can allow ∞ to be a vertex, provided that we count the corresponding angle as negative. For instance, suppose $g(z)$ maps a_1 to the vertex $w_1 = \infty$, with angle $\pi\alpha_1$, where $-2 < \alpha_1 < 0$. The map to the ζ-plane is then $\zeta(z) = g(z)^{1/\alpha_1} = (z - a_1)\psi(z)$, and we obtain $g(z) = (z - a_1)^{\alpha_1} h(z)$, where $h(z)$ is analytic at a_1 and $h(a_1) \neq 0$. We have (3.1) as before, and we obtain the same Schwarz-Christoffel formula (3.5). This time the factor $(z - a_1)^{\alpha_1}$ of $g'(z)$ is not integrable at a_1, corresponding to the fact that $g(a_1) = \infty$.

6. We can allow ∞ to be a vertex with angle zero. This corresponds to the "end" of an infinite strip. In this case the map to the ζ-plane is given by $\zeta(z) = e^{\beta g(z)}$ for some β, and

$$g(z) = \frac{1}{\beta} \log \zeta = \frac{1}{\beta} \log(z - a_1) + h(z),$$

where $h(z)$ is analytic at a_1. A calculation reveals that (3.1) again holds. We are led to the same Schwarz-Christoffel formula (3.5), with $\alpha_1 = 0$ for the vertex at ∞.

7. If $a_0 = \infty$ is mapped to a vertex $w_0 = \infty$, the Schwarz-Christoffel formula (3.5) holds, in which there is no factor corresponding to ∞. The

calculations are similar to those above. The result can also be obtained by letting the parameters in the Schwarz-Christoffel formula tend to ∞. (See Exercises 3 and 4.)

Example. Let D be the domain obtained by cutting a vertical slit in the upper half-plane from 0 to ia on the imaginary axis. (See Exercise 1.9.) Let $w = g(z)$ be the Schwarz-Christoffel map of the upper half-plane onto D that sends 0 to ia and ± 1 to the two corners at 0. This determines $g(z)$ uniquely, since we have specified three real parameters. In the Schwarz-Christoffel formula, the consecutive points are $a_1 = -1$ with angle $\pi/2$, $a_2 = 0$ with angle 2π, and $a_3 = +1$ with angle $\pi/2$. Thus (3.4) becomes

$$g'(z) = A(z-1)^{-1/2}z(z+1)^{-1/2} = Az/\sqrt{z^2-1}.$$

This can be integrated in closed form, and after matching boundary points to determine A and the integration constant, we obtain $g(z) = a\sqrt{z^2-1}$, where the branch of the square root is the one that is positive on the interval $(1, \infty)$.

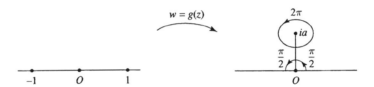

Example. Suppose D is a triangle, with vertices at 0, 1, and a point w_0 in the upper half-plane, and interior angles $\alpha\pi$, $\beta\pi$, and $\gamma\pi$, where $\alpha + \beta + \gamma = 1$. Fix a point $a > 1$, and let $w = g(z)$ map the upper half-plane conformally onto D, with $g(0) = 0$, $g(1) = 1$, and $g(a) = w_0$. Since we have specified three real parameters, $g(z)$ is uniquely determined. Since $g(0) = 0$, the Schwarz-Christoffel formula becomes

(3.6)
$$g(z) = A \int_0^z t^{\alpha-1}(1-t)^{\beta-1}(a-t)^{\gamma-1}dt.$$

The constant A is determined by the condition $g(1) = 1$.

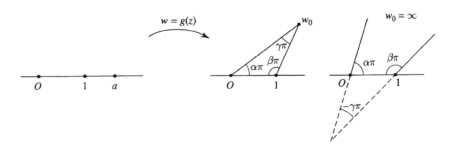

Exercises for XI.3

1. Derive the formula (3.2) for the case that $g(z)$ maps ∞ to the corner w_0 of D.

2. Let $\varphi(w)$ be the Riemann map of a simply connected domain D onto $\mathbb{D}$. Suppose w_0 is a corner of D with interior angle $\alpha\pi$. Set $w(\zeta) = w_0 + \lambda\zeta^\alpha$, where $|\lambda| = 1$. (a) Use the Schwarz reflection principle to show that $\varphi(w(\zeta))$ is analytic at $\zeta = 0$. (b) Show that $\varphi(w)$ has a series expansion $\sum b_j(w - w_0)^{j/\alpha}$ that converges uniformly for $w \in D$, $|w - w_0| < \varepsilon$. (c) Use (b) to show that $\varphi(w)$ has a limit as $w \in D$ tends to w_0.

3. Denote by $g_a(z)$ the conformal map of $\mathbb{H}$ onto the triangle with vertices 0, 1, and w_0 given by (3.6), so that $g_a(0) = 0$, $g_a(1) = 1$, and $g_a(a) = w_0$. Show that $g_a(z)$ converges as $a \to +\infty$, uniformly for $|z| \le R$. Express the limit as an integral, and relate it to the Schwarz-Christoffel map.

4. Fix α such that $0 < \alpha < 1$, and let $R > 0$. Denote by $f_R(z)$ the conformal map of $\mathbb{H}$ onto the triangle with vertices 0, 1, and $Re^{i\alpha\pi}$ given by (3.6), so that $f_R(0) = 0$, $f_R(1) = 1$, and $f_R(a) = Re^{i\alpha\pi}$ Show that $f_R(z)$ converges as $R \to \infty$, uniformly for $|z - a| \ge \varepsilon$. Express the limit as an integral, and relate it to the Schwarz-Christoffel map.

5. For fixed $0 < k < 1$, show that

$$w(z) = \int_0^z \frac{dt}{\sqrt{1 - t^2}\sqrt{1 - k^2 t^2}}$$

maps the upper half-plane conformally onto a rectangle. Sketch the rectangle, using the notation

$$K = \int_0^1 \frac{dt}{\sqrt{1 - t^2}\sqrt{1 - k^2 t^2}}, \qquad K' = \int_1^{1/k} \frac{dt}{\sqrt{t^2 - 1}\sqrt{1 - k^2 t^2}}.$$

Show that the inverse $z = z(w)$ of the function defined above extends to be a meromorphic function $h(w)$ on the entire complex plane. Show that $h(w)$ is an odd function, and that $h(w)$ is doubly periodic with periods $4K$ and $2K'i$. Locate the zeros and the poles of $h(w)$, and indicate their orders. *Remark.* The integrals appearing above are **elliptic integrals**. The function $h(w)$ is the **Jacobian elliptic function** denoted classically by $\mathrm{sn}(w)$. It maps the torus formed by identifying the opposite edges of the period rectangle two-to-one onto the extended complex plane.

6. What is the Schwarz-Christoffel formula (3.4) for the conformal map $g(z)$ of the upper half-plane onto the horizontal strip $\{0 < \mathrm{Im}(w) <$

a} that maps 0 to $-\infty$ and ∞ to $+\infty$? By integrating, show that
$g(z) = g(1) + \dfrac{a}{\pi} \operatorname{Log} z.$

7. Use the Schwarz-Christoffel formula to find a conformal map $w = g(z)$ of the upper half-plane onto the vertical half-strip $\{\operatorname{Im} w > 0, \, -a < \operatorname{Re} w < a\}$ such that $g(1) = a$, $g(-1) = -a$, and $g(0) = 0$. Compare your result with that of Exercise 1.11.

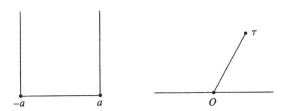

8. Let τ be a complex number such that $\operatorname{Im}\tau > 0$, and let D_τ be the domain obtained from the upper half-plane by deleting the straight line segment from 0 to τ. (a) Find an integral expression for conformal maps $w = g(z)$ of the upper half-plane onto D such that $g(0) = \tau$ and $g(\infty) = \infty$. (b) Show directly that if $C > 0$ and $0 < \sigma < 1$, the function $g(z) = C(z - \sigma)^\sigma (z + 1 - \sigma)^{1-\sigma}$ maps the upper half-plane conformally onto the domain D_τ for appropriate τ. (c) How are the maps in (a) and (b) related?

9. Let D be the step domain that is the union of the half-plane $\{\operatorname{Im} w > 1\}$ and the first quadrant $\{\operatorname{Re} w > 0, \operatorname{Im} w > 0\}$. Find a conformal map $w = g(z)$ of the upper half-plane onto D satisfying $g(-1) = i$, $g(+1) = 0$, and $g(\infty) = \infty$.

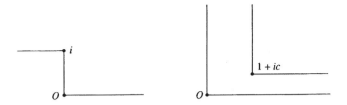

10. For $c > 0$, find an integral expression for a conformal map $w = g(z)$ of the upper half-plane onto the corridor with a turn described by

$$D = \{0 < \operatorname{Re} w < 1, \operatorname{Im} w > 0\} \cup \{0 < \operatorname{Im} w < c, \operatorname{Re} w > 0\}.$$

In the case $c = 1$, find explicitly the conformal map $g(z)$ that satisfies $g(0) = 0$, $g(1) = +\infty$, and $g(\infty) = 1 + i$.

4. Return to Fluid Dynamics

We return to fluid dynamics, to see how conformal mapping techniques can be applied to solve fluid flow problems. Following the notation of Section III.6, we denote the complex velocity potential by $f(z) = \phi(z) + i\psi(z)$, where $\phi(z)$ is the velocity potential, and $\psi(z)$ is the stream function whose level sets are the flow lines of the fluid. The velocity vector field is $\mathbf{V}(z) = \overline{f'(z)}$. The flow is tangent to the boundary along boundary curves on which there are no sources or sinks. Thus $\psi(z)$ is a harmonic function on D that is constant on each boundary curve with no sources or sinks.

Example. The stream function $\psi(z) = \operatorname{Im} z$ on the upper half-plane $\mathbb{H}$ corresponds to a flow parallel to the x-axis with no sources or sinks on the axis. The stream function $\psi(z) = \operatorname{Arg} z$ on $\mathbb{H}$ corresponds to a flow from a source at the origin. We may superimpose the flows, to obtain the stream function

$$\psi(z) \;=\; c_1 \operatorname{Im} z + c_2 \operatorname{Arg} z, \qquad \operatorname{Im} z > 0,$$

which corresponds to a flow near the bank of a large river with a pipe discharging an effluent at one point on the riverbank.

As in Section III.6, our strategy for solving a flow problem on a domain D is to find a conformal map $h(w)$ from D to the upper half of the z-plane, in order to transfer the boundary problem to $\mathbb{H}$. The transferred problem involves finding a harmonic function $\psi(z)$ on $\mathbb{H}$ that is constant on the intervals of the real axis corresponding to the boundary arcs of D where there are no sinks or sources. The composed function $\psi(h(w))$ is then the stream function for the flow in D. In the case that D is a polygon domain, the Schwarz-Christoffel formula gives the conformal map $w = g(z)$ from $\mathbb{H}$ onto D, and this must be solved for the inverse function $z = h(w)$ if a solution in closed form is required.

Example. Consider the flow near the bank of a large river with a small parallel tributary. We represent the configuration by the upper half of the w-plane with a horizontal slit terminating at the point ib on the positive imaginary axis, $D = \mathbb{H} \backslash (ib - \infty, ib]$. Consider the conformal map $w = g(z)$ of $\mathbb{H}$ onto D such that -1 corresponds to the tip of the slit, 0 to the source of the tributary, and $i\infty$ to $i\infty$. The three conditions $g(-1) = ib$, $g(0) = -\infty$, and $g(i\infty) = i\infty$ determine $g(z)$ uniquely. The function $g(z)$ maps the interval $(-\infty, -1]$ to the top edge of the slit, the interval $(-1, 0)$ to the bottom edge reversing direction, and the interval $(0, \infty)$ to the real axis $(-\infty, \infty)$. The Schwarz-Christoffel formula (3.4) becomes $g'(z) = A(z+1)z^{-1}$, corresponding to angle 2π at -1 and 0 at 0. The term involving $-\infty$ does not enter. Thus $g(z) = A(z + \operatorname{Log} z) + B$. The constants

A and B are determined by the boundary conditions, with final result

$$g(z) = \frac{b}{\pi}(1 + z + \text{Log}\, z), \qquad \text{Im}\, z > 0.$$

A flow function corresponding to a source in the tributary in the w-plane then corresponds to a flow function corresponding to a source at $z = 0$ in the z-plane. This is given by $\text{Arg}\, z = \text{Arg}\, h(w)$, where $h(w)$ is the inverse of the conformal map $g(z)$. We superimpose this upon a flow in the river, to obtain a flow function of the form

$$\psi(w) = c_1 \text{Arg}\, h(w) + c_2 \text{Im}\, h(w),$$

where the constants c_1 and c_2 are adjusted to the magnitudes of the flows.

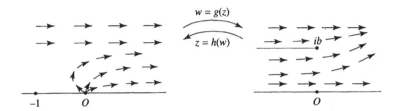

Exercises for XI.4

1. Let D be the domain obtained from the upper half-plane by deleting the part of the imaginary axis above ib. Find a conformal map $w = g(z)$ of the upper half-plane onto D such that $g(0) = -\infty$, $g(\infty) = +\infty$, and $g(-1) = ib$. Solve for $z = h(w)$. With $\psi(w) = \arg(h(w))$ as a stream function, use the map to sketch the streamlines of the flow of water in a river as it passes under a vertical gate. *Hint.* In the Schwarz-Christoffel formula, remember to take the angles at ∞ to be negative. An alternative method is to use an inversion to map the infinite vertical slit onto a finite slit from 0 to ia. The answer is

$$g(z) = \frac{b}{2}\left(\sqrt{z} - \frac{1}{\sqrt{z}}\right), \qquad h(w) = 1 + 2\frac{w^2}{b^2} + 2\frac{w}{b}\sqrt{\frac{w^2}{b^2} + 1}.$$

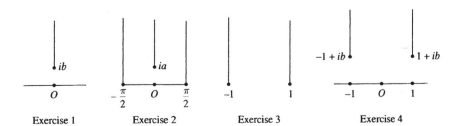

2. Let D be the domain obtained from the vertical half-strip $\{\operatorname{Im} z > 0, \, |\operatorname{Re} z| < \pi/2\}$ by deleting the part of the imaginary axis above ia. Find a conformal map $w = g(z)$ of the upper half-plane onto D. Solve for $z = h(w)$. Use the map to sketch the streamlines of the flow of water in a channel with a hairpin turn. *Hint.* Use the inverse sine function to map the half-strip onto an upper half-plane (see Exercise I.11), and refer to Exercise 1.

3. Find a conformal map $w = g(z)$ of the upper half-plane onto the plane with two vertical slits from ± 1 to $\pm 1 + i\infty$ deleted. Use the map to sketch roughly the flow of water near the mouth of a channel that empties into the middle of a large reservoir. Express the stream function for the flow in terms of the inverse $z = h(w)$ of $g(z)$.

4. Let D be the upper half-plane with two vertical slits from $\pm 1 + ib$ to $\pm 1 + i\infty$, and let $w = g(z)$ be the conformal map of the upper half-plane onto D satisfying $w(0) = 0$, $w(-1) = -\infty$, and $w(+1) = +\infty$. Show that the map is symmetric about the imaginary axis, so that $w(i\infty) = i\infty$. Express the map as an integral. Suppose in your integral formula that the point $a > 1$ is mapped to the end $1 + ib$ of the slit. Sketch the streamlines of the flow in the w-plane that corresponds to the stream function $\psi(z) = \arg(z - a) - \arg(z + a)$ in the z-plane. What are its sources and sinks?

5. Compactness of Families of Functions

We have reached a turning point. Our proof of the Riemann mapping theorem requires some technical ingredients at a higher level of mathematical sophistication than what we have used heretofore. In this section we summarize some of the ideas involved in the Arzelà-Ascoli theorem, and we use it to prove Montel's theorem (thesis grade). Montel's theorem is one of the key ingredients of the proof of the Riemann mapping theorem given in the next section. For an expanded treatment of the background material, see any good introduction to mathematical analysis.

Let E be a subset of the complex plane $\mathbb{C}$, and let $\mathcal{F}$ be a family of complex-valued functions on E. We say that $\mathcal{F}$ is **equicontinuous** at a point $z_0 \in E$ if for any $\varepsilon > 0$, there is $\delta > 0$ such that if $z \in E$ satisfies $|z - z_0| < \delta$, then $|f(z) - f(z_0)| < \varepsilon$ for all $f \in \mathcal{F}$. This coincides with the "ε-δ" definition of continuity at a point z_0, except that the same δ must serve simultaneously for *all* functions f in the family $\mathcal{F}$.

We say that the family $\mathcal{F}$ is **uniformly bounded** on E if there is a constant $M > 0$ such that $|f(z)| \le M$ for all $z \in E$ and all $f \in \mathcal{F}$.

One condition guaranteeing equicontinuity of a family of functions is that the derivatives of the functions in the family be uniformly bounded. Suppose, for instance, that $\mathcal{F}$ is a family of analytic functions on a domain D, and suppose there is $M > 0$ such that $|f'(z)| \leq M$ for all $z \in D$ and all $f \in \mathcal{F}$. Integrating along the straight line segment from z_0 to a nearby point z, we obtain

$$|f(z) - f(z_0)| = \left| \int_{z_0}^{z} f'(\zeta)\, d\zeta \right| \leq M|z - z_0|, \qquad f \in \mathcal{F},$$

for z near z_0. This estimate implies that the family $\mathcal{F}$ is equicontinuous at z_0.

Recall that a subset E of the complex plane $\mathbb{C}$ is **compact** if it is closed and bounded. We are now prepared to state one version of the **Arzelà-Ascoli theorem**.

Theorem. *Let E be a compact subset of $\mathbb{C}$, and let $\mathcal{F}$ be a family of continuous complex-valued functions on E that is uniformly bounded. Then the following are equivalent.*
(1) The family $\mathcal{F}$ is equicontinuous at each point of E.
(2) Each sequence of functions in $\mathcal{F}$ has a subsequence that converges uniformly on E.

In the sequel we will use only the forward implication, that (1) implies (2). The proof that (1) implies (2) proceeds in outline as follows. Let $\{f_n\}$ be a sequence in $\mathcal{F}$. Let $\{z_j\}$ be a sequence of points in E that is dense in E. First one uses a standard diagonalization argument to find a subsequence $\{f_{n_k}\}$ such that $f_{n_k}(z_j)$ converges for each j. Then one uses the equicontinuity of the f_{n_k}'s and the density of the z_j's to show that $\{f_{n_k}\}$ converges uniformly on E. The proof of the converse, that (2) implies (1), is an easy exercise in "proof by contradiction." For more details, see any good introduction to mathematical analysis.

The proof of the Arzelà-Ascoli theorem actually extends to a quite general situation. The theorem is valid for a family of functions on any compact metric space, and there is a version of the theorem for arbitrary compact topological spaces. The theorem also holds for functions whose ranges lie in any compact metric space. In particular, the theorem holds for functions from a compact set E to the extended complex plane $\mathbb{C}^* = \mathbb{C} \cup \{\infty\}$, provided that we use the spherical metric to measure distances in the range space. To be more precise, we denote by $\sigma(z, w)$ the spherical distance from z to w, as in Section IX.3. We say that a sequence of functions $\{f_n\}$ on E **converges uniformly to f in the spherical metric** if $\sigma(f_n(z), f_m(z))$ tends to 0 uniformly for $z \in E$ as $n, m \to \infty$. A family $\mathcal{F}$ is **equicontinuous with respect to the spherical metric** at $z_0 \in E$ if for any $\varepsilon > 0$ there is $\delta > 0$ such that if $z \in E$ satisfies $|z - z_0| < \delta$, then $\sigma(f(z), f(z_0)) < \varepsilon$ for all $f \in \mathcal{F}$. When we combine the Arzelà-Ascoli theorem with the diagonalization argument used above, we obtain the following version of the theorem, which we record for use when we return to this circle of ideas in Chapter XIV.

Theorem. *Let D be a domain in the complex plane, and let $\mathcal{F}$ be a family of continuous functions from D to the extended complex plane $\mathbb{C}^*$. Then the following are equivalent.*

(1) *Any sequence in $\mathcal{F}$ has a subsequence that converges uniformly on compact subsets of D in the spherical metric.*

(2) *The family $\mathcal{F}$ is equicontinuous at each point of D, with respect to the spherical metric.*

For now, our applications to complex analysis will be based on the following preliminary version of Montel's theorem, which was obtained by P. Montel in his thesis. We will prove a stronger version in Chapter XII.

Theorem. *Suppose $\mathcal{F}$ is a family of analytic functions on a domain D such that $\mathcal{F}$ is uniformly bounded on each compact subset of D. Then every sequence in $\mathcal{F}$ has a subsequence that converges normally on D, that is, uniformly on each compact subset of D.*

If $z_0 \in D$, there is $r > 0$ such that the closed disk $\{|z - z_0| \leq r\}$ is contained in D. The family $\mathcal{F}$ is uniformly bounded on the disk. By the Cauchy estimates, the derivatives of the functions in $\mathcal{F}$ are uniformly bounded on the smaller disk $\{|z - z_0| \leq r - \varepsilon\}$. Hence $\mathcal{F}$ is equicontinuous at each $z_0 \in D$. To complete the proof we combine the Arzelà-Ascoli theorem with a diagonalization argument. Let E_n be the set of $z \in D$ such that $|z| \leq n$ and the distance from z to ∂D is at least $1/n$. Then E_n is compact, the E_n's increase to D, and any compact subset of D is contained in some E_n. Let $\{f_n\}$ be a sequence in $\mathcal{F}$. By the Arzelà-Ascoli theorem, there is a subsequence $f_{11}, f_{12}, f_{13}, \ldots$ that converges uniformly on E_1. This has a further subsequence $f_{21}, f_{22}, f_{23}, \ldots$ that converges uniformly on E_2, and so on. The diagonal sequence $f_{11}, f_{22}, f_{33}, \ldots$ is then a subsequence of the original sequence that converges uniformly on each E_n, hence uniformly on each compact subset of D.

One typical application of Montel's theorem is to guarantee the existence of extremal functions for "compact" extremal problems. To illustrate the principle, we consider the extremal problem of maximizing the derivative at a prescribed point over a compact family of analytic functions.

Let D be a domain, and fix a point $z_0 \in D$. Let $\mathcal{F}$ be the family of analytic functions $f(z)$ on D such that $|f(z)| \leq 1$ on D. The extremal problem is to maximize $|f'(z_0)|$ among all functions $f \in \mathcal{F}$. The **extremal value** for the problem is

$$A = \sup \{|f'(z_0)| : f \in \mathcal{F}\}.$$

Since the functions in $\mathcal{F}$ are uniformly bounded on D, their derivatives are uniformly bounded at z_0, and A is finite. A function $G \in \mathcal{F}$ such that $|G'(z_0)| = A$ is an **extremal function** for the problem. The existence of an extremal function follows almost immediately from Montel's theorem.

If $\{f_n(z)\}$ is any sequence of functions in $\mathcal{F}$ such that $|f_n'(z_0)| \to A$, then by Montel's theorem, the f_n's have a subsequence that converges normally on D to an analytic function $G(z)$. On account of the normal convergence, $|G(z)| \le 1$ and $|G'(z_0)| = A$.

Theorem. *Let D be a domain in the complex plane on which there is a nonconstant bounded analytic function, and let $z_0 \in D$. Then there is an analytic function $G(z)$ on D such that $|G(z)| \le 1$ for $z \in D$, and $|f'(z_0)| \le |G'(z_0)|$ for any analytic function $f(z)$ on D satisfying $|f(z)| \le 1$ on D. Further, $G(z_0) = 0$ and $G'(z_0) \ne 0$.*

It remains only to establish the final statement. Let $h(z)$ be a non-constant bounded analytic function on D. Then $h(z) - h(z_0)$ has a zero of finite order, say of order N, at z_0. If $\varepsilon > 0$ is tiny, the function $f(z) = \varepsilon(h(z) - h(z_0))/(z - z_0)^{N-1}$ satisfies $|f(z)| \le 1$ and $f'(z_0) \ne 0$. Consequently, $A = |G'(z_0)| > 0$. Finally, $g(z) = (G(z) - G(z_0))/(1 - \overline{G(z_0)}z)$ is the composition of $G(z)$ and a conformal self-map of Δ, so that $g \in \mathcal{F}$. From

$$|G'(z_0)| \ge |g'(z_0)| = \frac{|G'(z_0)|}{1 - |G(z_0)|^2},$$

we deduce that $G(z_0) = 0$.

The extremal function $G(z)$ is called the **Ahlfors function** of D. The Ahlfors function depends on z_0. However, it can be shown (Exercise 9) that the Ahlfors function corresponding to a fixed point $z_0 \in D$ is unique, up to multiplication by a unimodular constant. The extremal value $A = |G'(z_0)|$ can be regarded as the best constant for which the Schwarz lemma (infinitesimal version) holds with respect to $z_0 \in D$. We will show in the next section that when D is simply connected, the Ahlfors function maps D conformally onto the open unit disk.

Exercises for XI.5

1. Let $\{f_n(z)\}$ be a uniformly bounded sequence of analytic functions on a domain D, and let $z_0 \in D$. Suppose that for each $m \ge 0$, $f_n^{(m)}(z_0) \to 0$ as $n \to \infty$. Show that $f_n(z) \to 0$ normally on D.

2. Let $\{f_n(z)\}$ be a sequence of analytic functions on a domain D, and let $z_0 \in D$. Suppose $\operatorname{Re} f_n(z) \ge -C$ for all $z \in D$, and $f_n^{(m)}(z_0) \to 0$ as $n \to \infty$ for each $m \ge 0$. Show that $f_n(z) \to 0$ normally on D. *Hint.* Use Problem 1, which is a special case.

3. Let $f(z)$ be a bounded analytic function on the horizontal strip $\{-1 < \operatorname{Im} z < 1\}$, such that $f(x)$ tends to 0 as x tends to $+\infty$. Show that for any $\varepsilon > 0$, $f(x + iy)$ tends to 0 as x tends to $+\infty$, uniformly for $-1 + \varepsilon < y < 1 - \varepsilon$.

4. Let $f(z)$ be a bounded analytic function on the upper half-plane.
   (a) Show that if $f(z)$ tends to 0 as z tends to ∞ along the imaginary axis, then for any $\varepsilon > 0$, $f(z)$ tends to 0 uniformly as z tends to ∞ through the sector $\{\varepsilon < \arg z < \pi - \varepsilon\}$. (b) Show that if $f(z)$ tends to a limit L as $z = x_0 + iy$ tends to $x_0 \in \mathbb{R}$ along a vertical line, then $f(z)$ tends uniformly to L as z tends to x_0 through any cone of the form $\{\varepsilon < \arg(z - x_0) < \pi - \varepsilon\}$.

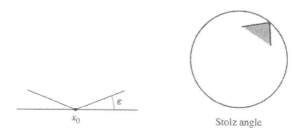

Stolz angle

5. Let $f(z)$ be a bounded analytic function on the open unit disk, and fix $e^{i\theta}$ on the unit circle. Show that if $f(re^{i\theta})$ has a limit L as r tends to 1, then $f(z)$ tends to L uniformly as z tends to $e^{i\theta}$ through any cone with vertex at $e^{i\theta}$, centered on the radius and with aperture strictly less than π. *Remark.* Such an approach sector is called a **Stolz angle** at $e^{i\theta}$.

6. Let D be a bounded domain, and let $f(z)$ be an analytic function from D into D. Denote by $f_n(z)$ the nth iterate of $f(z)$. Suppose that z_0 is an attracting fixed point for $f(z)$, that is, $f(z_0) = z_0$ and $|f'(z_0)| < 1$. Show that $f_n(z)$ converges uniformly on compact subsets of D to z_0.

7. Let D be a bounded domain, and let $f(z)$ be an analytic function from D into D. Show that if $z_0 \in D$ is a fixed point for $f(z)$, then $|f'(z_0)| \le 1$.

8. Let $\{f_n(z)\}$ be a sequence of analytic functions on a domain D. Suppose that $\iint_D |f_n(z)|\, dx\, dy \le 1$ for $n \ge 1$.
   (a) Show that $\{f_n(z)\}$ has a subsequence that converges normally to an analytic function $f(z)$ on D. *Hint.* To estimate $f(z)$, use the mean value property with respect to area (see Exercise III.4.1).
   (b) Show that $\iint_D |f(z)|\, dx\, dy \le 1$.
   (c) Show that if $\iint_D |f_n(z) - f_m(z)|\, dx\, dy \to 0$ as $m, n \to \infty$, then $\iint_D |f_n(z) - f(z)|\, dx\, dy \to 0$ as $n \to \infty$.

9. Let D be a domain in the complex plane, and let $\mathcal{F}$ be the family of analytic functions $f(z)$ on D such that $|f(z)| \le 1$ for $z \in D$. A

function $g(z)$ is an **extreme point** of $\mathcal{F}$ if it is not the midpoint of a line segment in $\mathcal{F}$, that is, if whenever $g = (g_0 + g_1)/2$ where $g_0, g_1 \in \mathcal{F}$, then $g_0 = g_1 = g$.

(a) Show that the function $g \in \mathcal{F}$ is an extreme point of $\mathcal{F}$ if and only if the only analytic function $h(z)$ satisfying $|g(z) \pm h(z)| \leq 1$ is $h = 0$.

(b) Show that the function $g \in \mathcal{F}$ is an extreme point of $\mathcal{F}$ if and only if the only analytic function $h(z)$ satisfying $|g(z)| + |h(z)| \leq 1$ is $h = 0$. *Hint.* First show that if $|a \pm b| \leq 1$, then $|a| + |b^2|/2 \leq 1$.

(c) Suppose that there is a nonconstant bounded analytic function on D, and fix $z_0 \in D$. Show that the Ahlfors function $G(z)$ associated with $z_0 \in D$ is an extreme point of $\mathcal{F}$.

(d) Show that the Ahlfors function $G(z)$ associated with $z_0 \in D$ is unique, up to multiplication by a unimodular constant.

(e) Let $L = \sum_{j,k} a_{jk} L_{jk}$ be a finite linear combination of linear functionals of the form $L_{jk}(f) = f^{(k)}(z_j)$, where the z_j's are fixed points of D. Suppose there is a nonconstant bounded analytic function f on D such that $L(f) \neq 0$. Show that the extremal problem of maximizing $\operatorname{Re} L(f)$ over all $f \in \mathcal{F}$ has a unique extremal function $\varphi(z)$, which is an extreme point of $\mathcal{F}$.

6. Proof of the Riemann Mapping Theorem

Now we are ready for the proof of the Riemann mapping theorem. The key ingredients are Montel's theorem, Hurwitz's theorem, some elementary mapping properties of the square root function, and the fact that $\sqrt{z}$ is expanding on the open unit disk. We begin with the elementary mapping properties of the square root function.

Suppose that D is simply connected and that $D \neq \mathbb{C}$. Choose $a \in \mathbb{C} \backslash D$. By the characterization of simple connectivity (Section VIII.8), there is an analytic branch $g(z)$ of $\log(z - a)$ in D. Then $h(z) = e^{g(z)/2}$ is an analytic branch of $\sqrt{z-a}$ in D, and $h(z)^2 = z - a \neq 0$ in D. If $h(z_1) = h(z_2)$, then $z_1 = h(z_1)^2 + a = h(z_2)^2 + a = z_2$. Thus $h(z)$ is univalent, and $h(z)$ maps D conformally onto $h(D)$. Finally, note that if $w_0 \in h(D)$, then $-w_0 \notin h(D)$. Indeed, if $w_0 = h(z_0)$ and $-w_0 = h(z_1)$ for $z_0, z_1 \in D$, then $z_0 = h(z_0)^2 + a = w_0^2 + a = h(z_1)^2 + a = z_1$, which is impossible. We summarize.

Lemma. *Let D be a simply connected domain. Suppose $a \notin D$, and let $h(z)$ be an analytic branch of $\sqrt{z-a}$ in D. Then $h(z)$ is univalent on D, and further, $h(D)$ is disjoint from $-h(D)$.*

We remark that on the basis of the lemma we can show that any simply connected domain $D \neq \mathbb{C}$ is conformally equivalent to a subdomain of the unit disk, where we have hyperbolic geometry at our disposal. Indeed, if the disk $|w - w_0| \leq \varepsilon$ is contained in $h(D)$, then its negative is disjoint from $h(D)$, so that $|h(z) + w_0| > \varepsilon$ for $z \in D$, and $\varepsilon/(h(z) + w_0)$ maps D conformally onto a domain contained in the unit disk.

Our proof depends at least conceptually on hyperbolic geometry. Let $\rho(\zeta, \xi)$ be the hyperbolic metric in $\mathbb{D}$ (Section IX.3). The function ζ^2, regarded as an analytic function from $\mathbb{D}$ to $\mathbb{D}$, is a contraction with respect to the hyperbolic metric, and it is a strict contraction on bounded subsets $\{|\zeta| \leq 1 - \varepsilon\}$ of the hyperbolic disk, in the sense that there is $c < 1$, depending on ε, such that $\rho(\zeta^2, \xi^2) \leq c\rho(\zeta, \xi)$ for $|\zeta|, |\xi| \leq 1 - \varepsilon$. Thus any branch of its inverse $\sqrt{\zeta}$ is expanding with respect to the hyperbolic metric, and it is strictly expanding on bounded subsets of the hyperbolic disk, in the sense that for any $r < 1$, there is $C > 1$ such that

$$\rho\left(\pm\sqrt{\zeta}, \pm\sqrt{\xi}\right) \geq C\rho(\zeta, \xi), \qquad |\zeta|, |\xi| \leq r.$$

(For a precise value for C, see Exercise IX.3.9.) The expanding property of $\sqrt{\zeta}$ is the key to the following lemma.

Lemma. *Let D be a simply connected subdomain of $\mathbb{D}$ such that $0 \in D$. If $D \neq \mathbb{D}$, then there is a conformal map $\psi(\zeta)$ of D onto a subdomain of $\mathbb{D}$ such that $\psi(0) = 0$ and $|\psi'(0)| > 1$.*

Let $b \in \mathbb{D}\backslash D$, and let g be the conformal self-map of $\mathbb{D}$ that maps b to 0. Then $g(D)$ is a simply connected domain in the unit disk that does not contain 0. Hence there is an analytic branch $h(\xi)$ of $\sqrt{\xi}$ on $g(D)$, and further, as shown above, h is univalent on $g(D)$. Let f be a conformal self-map of $\mathbb{D}$ that maps $h(g(0))$ to 0, and set $\psi = f \circ h \circ g$. Then $\psi(\zeta)$ is a conformal map of D onto a subdomain of $\mathbb{D}$, and $\psi(0) = 0$. Since f and g are isometries in the hyperbolic metric, and h is a strict expansion near $g(0)$, ψ is a strict expansion near 0, that is, there is $C > 1$ such that $\rho(\psi(\zeta), 0) \geq C\rho(\zeta, 0)$ for $|\zeta| < \varepsilon$. Since $\rho(\psi(\zeta), 0) \sim 2|\psi(\zeta)|$ and $\rho(\zeta, 0) \sim 2|\zeta|$ as $\zeta \to 0$, we obtain

$$C \leq \frac{\rho(\psi(\zeta), 0)}{\rho(\zeta, 0)} \sim \frac{|\psi(\zeta)|}{|\zeta|} \to |\psi'(0)|,$$

and $|\psi'(0)| > 1$. This inequality can also be checked by expressing f and g explicitly and differentiating (Exercise 1).

Now we complete the proof of the Riemann mapping theorem by producing the Riemann map as an extremal function for an extremal problem similar to the problem considered in Section 2. We assume that D is a simply connected domain, $D \neq \mathbb{C}$. Fix $z_0 \in D$, and let $\mathcal{F}$ be the family of univalent(!) functions $f(z)$ on D such that $|f(z)| < 1$ for $z \in D$ and

$f(z_0) = 0$. The family $\mathcal{F}$ is nonempty, since if $h(z)$ maps D conformally onto a bounded domain, then the function $f(z) = \varepsilon(h(z) - h(z_0))$ is in $\mathcal{F}$ for $\varepsilon > 0$ small. We consider the extremal problem of maximizing $|f'(z_0)|$ over $f \in \mathcal{F}$. As before, set

$$A = \sup\{|f'(z_0)| : f \in \mathcal{F}\} > 0,$$

and let $\{f_n(z)\}$ be a sequence of functions in $\mathcal{F}$ such that $|f_n'(z_0)| \to A$. By Montel's theorem, the f_n's have a subsequence that converges normally on D to an analytic function $\varphi(z)$. Clearly, $|\varphi(z)| \leq 1$, $\varphi(z_0) = 0$, and $|\varphi'(z_0)| = A$. The functions f_n are univalent, so by Hurwitz's theorem (Section VIII.2), either φ is constant or φ is univalent. Since $\varphi'(z_0) \neq 0$, φ is nonconstant, and consequently, φ is univalent, mapping D onto a subdomain of $\mathbb{D}$. We claim that $\varphi(D) = \mathbb{D}$. Otherwise, we could apply the preceding Lemma to the domain $\varphi(D)$ and find a conformal map $\psi(\zeta)$ of $\varphi(D)$ onto a subdomain of $\mathbb{D}$ such that $\psi(0) = 0$ and $|\psi'(0)| > 1$. Then $\psi \circ \varphi \in \mathcal{F}$ would satisfy $|(\psi \circ \varphi)'(z_0)| = |\psi'(0)\varphi'(z_0)| > A$, contradicting the definition of A. This completes the proof.

We will sketch a different proof of the Riemann mapping theorem in Section XV.5, based on the solvability of the Dirichlet problem.

Exercises for XI.6

1. Find explicitly the functions $f(\zeta)$ and $g(\zeta)$ used in the proof of the lemma. Show by computing the derivative that $|(f \circ h \circ g)'(0)| > 1$.

2. Let $\varphi(z)$ be the Riemann map of a simply connected domain D onto the open unit disk, normalized by $\varphi(z_0) = 0$ and $\varphi'(z_0) > 0$. Show that if $f(z)$ is any analytic function on D such that $|f(z)| \leq 1$ for $z \in D$, then $|f'(z_0)| \leq \varphi'(z_0)$, with equality only when $f(z)$ is a constant multiple of $\varphi(z)$. *Remark.* This shows that $\varphi(z)$ is the Ahlfors function of D corresponding to z_0.

3. Let $\varphi(z)$ be the Riemann map of a simply connected domain D onto the open unit disk, normalized by $\varphi(z_0) = 0$ and $\varphi'(z_0) > 0$. Show that if $f(z)$ is any analytic function on D such that $|f(z)| \leq 1$ for $z \in D$, then $\operatorname{Re} f'(z_0) \leq \varphi'(z_0)$, with equality only when $f(z) = \varphi(z)$.

4. Let $\{D_m\}$ be an increasing sequence of simply connected domains, and let φ_m be the Riemann map of D_m onto the open unit disk $\mathbb{D}$, normalized so that $\varphi_m(z_0) = 0$ and $\varphi_m'(z_0) > 0$ for some fixed $z_0 \in D_1$. Let D be the union of the D_m's. Show that if D is the entire complex plane, then the φ_m's are eventually defined on any disk $\{|z| \leq R\}$ and converge there uniformly to 0. Otherwise, D is simply connected and the φ_m's are eventually defined on each compact subset of D and converge there uniformly to the Riemann map φ of D onto $\mathbb{D}$ satisfying $\varphi(z_0) = 0$ and $\varphi'(z_0) > 0$.

5. Let $\{D_m\}$ be a decreasing sequence of simply connected domains, and suppose $w_0 \in D_m$ for all m. Let $g_m(z)$ be the conformal map of the open unit disk $\mathbb{D}$ onto D_m, normalized so that $g_m(0) = w_0$ and $g'_m(0) > 0$. Show that $g_m(z)$ converges normally on $\mathbb{D}$ to a function $g(z)$. If the distance from w_0 to the boundary of D_m tends to 0, then $g(z)$ is the constant function w_0, and otherwise, $g(z)$ maps $\mathbb{D}$ conformally onto some simply connected domain D. Describe D in terms of the D_m's.

6. Show that the function $\zeta \mapsto \zeta^2$ is not a strict contraction of the hyperbolic disk, that is, show that there is no constant $c < 1$ such that $\rho\left(\zeta^2, \xi^2\right) \le c\rho(\zeta, \xi)$ for all $\zeta, \xi \in \mathbb{D}$. *Remark.* See Exercise IX.3.9.

7. Suppose $f : \mathbb{D} \to \mathbb{D}$ is an analytic function from the unit disk into itself with a fixed point at $z_0 \in \mathbb{D}$. Show that the stretching at z_0 of $f(z)$ in the hyperbolic metric is the same as the stretching at z_0 of $f(z)$ in the Euclidean metric,

$$\lim_{z \to z_0} \frac{\rho(f(z), z_0)}{\rho(z, z_0)} = \lim_{z \to z_0} \frac{|f(z) - z_0|}{|z - z_0|} = |f'(z_0)|.$$

XII

Compact Families of Meromorphic Functions

In Sections 1 and 2 we treat normal families of meromorphic functions. These are families that are sequentially compact when regarded as functions with values in the extended complex plane. We give two characterizations of normal families, Marty's theorem in Section 1 and the Zalcman lemma in Section 2. From the latter characterization we deduce Montel's theorem on compactness of families of meromorphic functions that omit three points, and we also prove the Picard theorems. Sections 3 and 4 constitute an introduction to iteration theory and Julia sets. In Section 3 we proceed far enough into the theory to see how Montel's theorem enters the picture and to indicate the fractal nature (self-similarity) of Julia sets. In Section 4 we relate the connectedness of Julia sets to the orbits of critical points. In Section 5 we introduce the Mandelbrot set, which has been called the "most fascinating and complicated subset of the complex plane."

1. Marty's Theorem

In this section we consider the convergence of sequences of meromorphic functions. Our aim is to give conditions guaranteeing compactness of families of meromorphic functions. Since functions may now assume the value ∞, we must modify our notion of convergence. We do this by regarding meromorphic functions on a domain D as functions from D to the extended complex plane $\mathbb{C}^* = \mathbb{C} \cup \{\infty\}$, which we identify with a sphere via stereographic projection (Section I.3). When we identify the extended complex plane with the sphere, we refer to it also as the **Riemann sphere**. We will use distances on the sphere to measure distances between function values.

Let $\sigma(z, w)$ denote the distance from z to w in the spherical metric. Recall from Section IX.3 that $\sigma(z, w)$ is the length of the arc of the great circle on the Riemann sphere joining the points corresponding to z and w. The spherical metric is invariant under the transformations corresponding to rotations of the sphere. Since the inversion $z \mapsto 1/z$ corresponds to a rotation of the sphere (by 180° around the x-axis), we have

(1.1) $$\sigma(z, w) = \sigma(1/z, 1/w), \qquad z, w \in \mathbb{C}.$$

On any fixed bounded subset of the complex plane, the spherical metric is equivalent to the Euclidean metric,

$$\frac{1}{B_R}|z - w| \leq \sigma(z, w) \leq B_R|z - w|, \qquad |z|, |w| \leq R,$$

while on any fixed subset of the complex plane at a positive distance from the origin, the spherical distance from z to w is comparable to the Euclidean distance between the inverse points $1/z$ and $1/w$,

$$\frac{1}{C_\varepsilon}\left|\frac{1}{z} - \frac{1}{w}\right| \leq \sigma(z, w) \leq C_\varepsilon\left|\frac{1}{z} - \frac{1}{w}\right|, \qquad |z|, |w| \geq \varepsilon.$$

We say that a sequence $\{f_n(z)\}$ of meromorphic functions on a domain D **converges normally** to $f(z)$ on D if the sequence converges uniformly on compact subsets of D to $f(z)$ in the spherical metric, that is, $\sigma(f_n(z), f(z))$ converges to zero uniformly on each compact subset of D as $n \to \infty$. On account of (1.1), we see that $\{f_n(z)\}$ converges normally to $f(z)$ if and only if $\{1/f_n(z)\}$ converges normally to $1/f(z)$. Here we declare $1/0 = \infty$ and $1/\infty = 0$, and we further allow the possibility of the constant function that is identically equal to ∞.

Since the spherical and Euclidean metrics are equivalent on bounded subsets of the complex plane, the definition of normal convergence given above is consistent with our earlier definition of a normally convergent sequence of analytic functions. However, now we allow the possibility of the limit being ∞.

Theorem. *If the sequence $\{f_n(z)\}$ of meromorphic functions on a domain D converges normally to $f(z)$, then either $f(z)$ is meromorphic on D or $f(z) \equiv \infty$. If a sequence $\{f_n(z)\}$ of analytic functions on D converges normally to $f(z)$, then either $f(z)$ is analytic on D or $f(z) \equiv \infty$.*

For the first statement, note that every point z for which $f(z) \neq \infty$ has a neighborhood on which the f_n's are eventually uniformly bounded and converge uniformly in the Euclidean metric. Thus $f(z)$ is analytic on the set where $|f(z)| < \infty$. By the same token, since $1/f_n(z)$ converges normally to $1/f(z)$, also $1/f(z)$ is analytic wherever $f(z) \neq 0$. Either $1/f(z)$ is identically zero, in which case $f(z) \equiv \infty$; or else the zeros of $1/f(z)$ are isolated in D and they are poles of $f(z)$, in which case $f(z)$ is meromorphic on D.

For the second statement, suppose that the f_n's are analytic with limit $f(z)$ satisfying $f(z_0) = \infty$ at some point. Then $1/f_n(z)$ is a meromorphic function with no zeros near z_0, and $1/f_n(z_0) \to 0$. Hurwitz's theorem

(Section VIII.3) shows that $1/f_n(z)$ converges uniformly to zero on some disk centered at z_0, and consequently, $1/f(z) = 0$ on the disk. By the uniqueness principle, $1/f(z) = 0$ on D, and $f(z) = \infty$ on D.

We define a family $\mathcal{F}$ of meromorphic functions on D to be a **normal family** if every sequence in $\mathcal{F}$ has a subsequence that converges normally on D. In view of the preceding theorem, a family $\mathcal{F}$ of analytic functions on D is a normal family if and only if every sequence in $\mathcal{F}$ has either a subsequence that converges uniformly on compact subsets of D to an analytic function or a subsequence that converges uniformly on compact subsets of D to ∞. The thesis version of Montel's theorem (Section XI.5) can be formulated in terms of normal families. It asserts that any family of analytic functions that is uniformly bounded on each compact set is a normal family.

The Arzelà-Ascoli theorem (Section XI.5) provides a characterization of normal families of meromorphic functions. A family $\mathcal{F}$ of meromorphic functions is a normal family if and only if the family $\mathcal{F}$ is equicontinuous, regarded as functions from D, with the Euclidean metric, to the extended complex plane $\mathbb{C}^*$, with the spherical metric. In order to exploit this condition, we study the spherical metric in more detail.

Recall (Section IX.3) that the spherical metric is $2|dz|/(1 + |z|^2)$. The spherical length of a curve $\gamma(t)$, $a \leq t \leq b$, is given by

$$\text{spherical length of } \gamma \;=\; \int_\gamma \frac{2|dz|}{1 + |z|^2} \;=\; \int_a^b \frac{2|\gamma'(t)|\,dt}{1 + |\gamma(t)|^2}.$$

If now $w = f(z)$ is a meromorphic function, then the spherical length of the image curve $f \circ \gamma$ is given by

$$\text{spherical length of } f \circ \gamma \;=\; \int_\gamma \frac{2|f'(z)|}{1 + |f(z)|^2}\,|dz|.$$

This leads us to define the **spherical derivative** of $f(z)$ to be

$$f^\sharp(z) \;=\; \frac{2|f'(z)|}{1 + |f(z)|^2}.$$

The spherical length of $f \circ \gamma$ is then simply

$$\text{spherical length of } f \circ \gamma \;=\; \int_\gamma f^\sharp(z)\,|dz|.$$

By taking γ to be a short straight line segment from z to $z + \Delta z$, we see that $f^\sharp(z)$ measures the expansion of $f(z)$, regarded as a map from the complex plane with the usual Euclidean metric to the extended complex plane with the spherical metric.

Since the spherical metric is invariant under the inversion $z \mapsto 1/z$, by (1.1), the spherical derivative is also invariant under the inversion,

(1.2) $$(1/f)^\sharp = f^\sharp.$$

This identity can also be checked by direct calculation (Exercise 1). In dealing with the spherical metric on the range of $f(z)$, (1.2) allows us to replace $f(z)$ near its poles by $g(z) = 1/f(z)$, thereby reducing to the situation where $f(z)$ is analytic. This observation is the key to the following lemma and to Marty's theorem.

Lemma. *If* $f_k(z) \to f(z)$ *normally on* D, *then* $f_k^\sharp(z) \to f^\sharp(z)$ *uniformly on compact subsets of* D.

Indeed, if f is analytic at z_0, then $f_k' \to f'$ uniformly in some neighborhood of z_0, so $f_k^\sharp \to f^\sharp$ uniformly in some neighborhood of z_0. If f is not analytic at z_0, then $1/f$ is analytic at z_0 and $1/f_k \to 1/f$ normally. Again $f_k^\sharp = (1/f_k)^\sharp$ converges uniformly to $f^\sharp = (1/f)^\sharp$ in some neighborhood of z_0.

Theorem (Marty's Theorem). *A family* $\mathcal{F}$ *of meromorphic functions is normal on a domain* D *if and only if the spherical derivatives* $\{f^\sharp(z) : f \in \mathcal{F}\}$ *are uniformly bounded on each compact subset of* D.

Suppose the spherical derivatives are uniformly bounded near $z_0 \in D$, say $f^\sharp(z) \le C$ for $|z - z_0| < r$ and $f \in \mathcal{F}$. If $|z_1 - z_0| < r$, and γ is the straight line segment from z_0 to z_1, then the spherical distance from $f(z_0)$ to $f(z_1)$ is estimated by

$$\sigma(f(z_0), f(z_1)) \le \int_\gamma f^\sharp(z)\,|dz| \le C|z_0 - z_1|.$$

Since this estimate is independent of the function $f \in \mathcal{F}$, regarded as functions from the Euclidean to the spherical metric the family $\mathcal{F}$ is equicontinuous at z_0. The Arzelà-Ascoli theorem then implies that $\mathcal{F}$ is normal. On the other hand, if the spherical derivatives of the functions in $\mathcal{F}$ are not uniformly bounded on compact sets, there are $f_k \in \mathcal{F}$ such that the maximum of $f_k^\sharp$ over some compact set tends to $+\infty$. By the lemma, $\{f_k\}$ cannot have a normally convergent subsequence, so $\mathcal{F}$ is not normal.

The definition of a normal family of meromorphic functions on a domain D can be extended to include domains D that contain ∞. We say that a family $\mathcal{F}$ of meromorphic functions on a domain D in the extended complex plane is a **normal family of meromorphic functions** on D if $\mathcal{F}$ is a normal family on $D\backslash\{\infty\}$, and if further the functions $g(w) = f(1/w)$, $1/w \in D$, form a normal family of meromorphic functions in some disk containing $w = 0$. Thus the notion of a normal family is "local." Marty's theorem remains valid for domains containing ∞.

Exercises for XII.1

1. Verify the identity $(1/f)^\sharp(z) = f^\sharp(z)$ by direct calculation.

2. Show that $(g \circ f)^\sharp(z) = g^\sharp(f(z))|f'(z)|$. Interpret the identity in terms of stretching with respect to Euclidean and spherical metrics.

3. Show that the functions $f_\varepsilon(z) = z/(z+\varepsilon)$, $0 < \varepsilon \le 1$, form a normal family of meromorphic functions on $\mathbb{C}^*\backslash\{0\}$. Show that the functions $g_\varepsilon(w) = f_\varepsilon(1/w)$ form a normal family of meromorphic functions on $\mathbb{C}$. Find $\lim f_\varepsilon^\sharp(z)$ and $\lim g_\varepsilon^\sharp(w)$ as $\varepsilon \to 0$.

4. Show that the functions $z^3/(z+\varepsilon)$, $0 < \varepsilon \le 1$, form a normal family of meromorphic functions on $\mathbb{C}^*\backslash\{0\}$, but they do *not* form a normal family of meromorphic functions on $\mathbb{C}$.

5. Fix $M > 1$, and let $\{f_n(z)\}$ be a sequence of meromorphic functions on a domain D. Show that $f_n(z)$ converges normally to $f(z)$ if and only if $f_n(z)$ converges uniformly to $f(z)$ on compact subsets of the open set $\{|f(z)| < M\}$, and $1/f_n(z)$ converges uniformly to $1/f(z)$ on compact subsets of the open set $\{|f(z)| > 1/M\}$.

6. Let E be a compact connected subset of the complex plane that contains more than one point. Show that the family of meromorphic functions on a domain D that omits E (that is, with range in $\mathbb{C}^*\backslash E$) is a normal family of meromorphic functions.

7. Let $g(z)$ be a nonconstant analytic function on a domain U, let $V = g(U)$, and let $\mathcal{F}$ be a family of meromorphic functions on V. Show that the family $\mathcal{F}$ is normal on V if and only if the family of compositions $\{f \circ g : f \in \mathcal{F}\}$ is normal on U.

8. Let $\{f_n(z)\}$ be a sequence of rational functions that converges normally to $f(z)$ on the extended complex plane $\mathbb{C}^*$. Show that $f_n(z)$ has the same degree as $f(z)$ for n large. (See Exercise VIII.4.5 for the definition of the degree of a rational function.)

9. Let $\mathcal{F}$ be a family of meromorphic functions on a domain D. Suppose there is an increasing function $\psi(t)$, $t \ge 0$, such that $|f'(z)| \le \psi(|f(z)|)$ for all $z \in D$ and $f \in \mathcal{F}$. Show that $\mathcal{F}$ is normal on D. *Remark*. This is **Royden's theorem**. For the proof, assume that $\psi(t)$ is smooth and increasing, and $\psi(t) \ge 1+t^2$. Consider the metric in which the distance from z to w is the infimum of $\int 1/\psi(|\zeta|)\,|d\zeta|$ over all smooth paths from z to w. Show that this metric is equivalent to the spherical metric.

10. Show that the family of analytic functions on a domain satisfying $|f'(z)| \le e^{|f(z)|}$ is normal.

2. Theorems of Montel and Picard

As an application of Marty's theorem, we give a proof of Montel's theorem characterizing normal families of meromorphic functions. Another proof of Montel's theorem will be outlined at the end of Chapter XVI, based on a modular function that arises as a covering map. We begin with the following lemma. Since the converse is trivial, it actually provides a characterization of normal families.

Theorem (Zalcman's Lemma). *Suppose $\mathcal{F}$ is a family of meromorphic functions on a domain D that is not normal. Then there are points $z_n \in D$ converging to a point of D, numbers $\rho_n > 0$ converging to 0, and functions $f_n \in \mathcal{F}$ such that the scaled functions $g_n(\zeta) = f_n(z_n + \rho_n \zeta)$ converge normally to a nonconstant meromorphic function $g(\zeta)$ on $\mathbb{C}$ satisfying $g^\sharp(0) = 1$ and $g^\sharp(\zeta) \le 1$ for $\zeta \in \mathbb{C}$.*

The proof is tricky. By Marty's theorem, there are sequences $\{w_n\}$ in a compact subset of D and $f_n \in \mathcal{F}$ such that $f_n^\sharp(w_n) \to +\infty$. We can assume that $w_n \to 0 \in D$, and we assume for convenience that the closed unit disk $\{|z| \le 1\}$ is contained in D. Define

$$R_n = \max_{|z| \le 1} f_n^\sharp(z)(1 - |z|).$$

Since $w_n \to 0$ and $f_n^\sharp(w_n) \to \infty$, we have $R_n \to \infty$. Suppose $f_n^\sharp(z)(1 - |z|)$ attains its maximum at the point z_n,

$$R_n = f_n^\sharp(z_n)(1 - |z_n|).$$

Since $f_n^\sharp(z_n) \ge R_n$, also $f_n^\sharp(z_n) \to \infty$. Define

$$\rho_n = 1/f_n^\sharp(z_n);$$

then $\rho_n \to 0$. The disk centered at z_n of radius $1 - |z_n| = \rho_n R_n$ is contained in the unit disk and hence in D. We parametrize it with a parameter ζ, by

$$\zeta \mapsto z_n + \rho_n \zeta, \qquad |\zeta| < R_n.$$

Define the scaled function $g_n(\zeta)$ by

$$g_n(\zeta) = f_n(z_n + \rho_n \zeta), \qquad |\zeta| < R_n.$$

Since $R_n \to \infty$, these functions are eventually defined on any compact set. From the chain rule $(f \circ h)^\sharp(\zeta) = f^\sharp(h(\zeta))|h'(\zeta)|$ (Exercise 1.2) with $h(\zeta) = z_n + \rho_n \zeta$, we obtain

$$g_n^\sharp(\zeta) = \rho_n f_n^\sharp(z_n + \rho_n \zeta), \qquad |\zeta| < R_n.$$

Now fix $R > 0$. If n is so large that $R_n > R$, then $g_n(\zeta)$ is defined on the disk $\{|\zeta| < R\}$, and since $f_n^\sharp(z_n + \rho_n\zeta)(1 - |z_n + \rho_n\zeta|) \le R_n$ we have

$$g_n^\sharp(\zeta) \le \rho_n \frac{R_n}{1 - |z_n + \rho_n\zeta|} \le \frac{\rho_n R_n}{1 - |z_n| - \rho_n R}$$

$$= \frac{\rho_n R_n}{\rho_n R_n - \rho_n R} = \frac{1}{1 - R/R_n}, \qquad |\zeta| < R.$$

By Marty's theorem, the g_n's for n large form a normal family on the disk $\{|z| < R\}$. Passing to a subsequence, we can then assume that $\{g_n\}$ converges normally on $\mathbb{C}$ to a meromorphic function $g(\zeta)$. Since for each fixed R, $1/(1 - R/R_n) \to 1$, the estimate on $g_n^\sharp(\zeta)$ shows that $g^\sharp(\zeta) \le 1$ for all $\zeta \in \mathbb{C}$. Since $g_n^\sharp(0) = \rho_n f_n^\sharp(z_n) = 1$, $g^\sharp(0) = 1$.

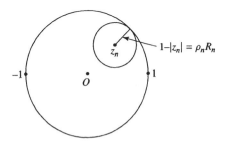

Suppose that $f(z)$ is meromorphic for $0 < |z - z_0| < r$. A value $w_0 \in \mathbb{C}^*$ is an **omitted value** of $f(z)$ at z_0 if there is $\delta > 0$ such that $f(z) \ne w_0$ for $0 < |z - z_0| < \delta$. Thus w_0 is *not* an omitted value of $f(z)$ at z_0 if and only if there is a sequence, $z_n \to z_0$, $z_n \ne z_0$, such that $f(z_n) = w_0$. An omitted value of $f(z)$ at ∞ is defined similarly.

Example. The function $e^{1/z}$ omits the two values 0 and ∞ at $z_0 = 0$. The function e^z omits the two values 0 and ∞ at ∞.

Theorem (Montel's Theorem). *A family $\mathcal{F}$ of meromorphic functions on a domain D that omits three values is normal.*

Since normality is a local property, we can assume that the domain D is the open unit disk $\{|z| < 1\}$. By composing the functions in $\mathcal{F}$ with a fractional linear transformation, we can assume that the omitted values are 0, 1, and ∞. Since the functions in $\mathcal{F}$ are then analytic and nonzero on the unit disk, they have analytic roots of all orders. Let $\mathcal{F}_k$ be the family of all 2^kth roots of functions in $\mathcal{F}$. Evidently $\mathcal{F}_k$ is normal if and only if $\mathcal{F}$ is normal. The functions in $\mathcal{F}_k$ omit the values 0, ∞, and all 2^kth roots of unity.

We will argue by contradiction. Suppose $\mathcal{F}$ is *not* normal. Then $\mathcal{F}_k$ is not normal. Let $G_k(\zeta)$ be the entire function from the Zalcman lemma, so that $G_k^\sharp(\zeta) \le 1$, $G_k^\sharp(0) = 1$, and G_k is a limit of restrictions of functions in $\mathcal{F}_k$, appropriately scaled. Since the functions in $\mathcal{F}_k$ omit the 2^kth roots of unity, so do their scaled restrictions, and from Hurwitz's theorem (Section VIII.3), so does any nonconstant normal limit. Thus G_k omits the 2^kth roots of unity. Now, Marty's theorem shows that $\{G_k\}$ is a normal family. Let G be any normal limit of a subsequence of the G_k's. Then $G^\sharp(0) = 1$, so that G is nonconstant, and by Hurwitz's theorem again, G omits all 2^kth roots of unity, this for all $k \geq 1$. Since G is an open mapping, G omits the unit circle. Thus either $|G| < 1$ or $|G| > 1$. Applying Liouville's theorem to G in the first case and to $1/G$ in the second, we conclude that G is constant. This contradiction establishes the theorem.

The following theorem is a substantial generalization of the Casorati-Weierstrass theorem.

Theorem (Picard's Big Theorem). *Suppose $f(z)$ is meromorphic on a punctured neighborhood $\{0 < |z - z_0| < \delta\}$ of z_0. If $f(z)$ omits three values at z_0, then $f(z)$ extends to be meromorphic at z_0.*

For the proof, we may assume that $z_0 = 0$ and that $f(z)$ omits the values 0 and ∞ on the punctured disk. Then $f(z)$ is analytic on the punctured disk. Let $\{\varepsilon_n\}$ be a sequence that decreases to 0, and define

$$g_n(z) \;=\; f(\varepsilon_n z), \qquad 0 < |z| < \delta\,.$$

Then $\{g_n\}$ omits three values, including 0 and ∞. By Montel's theorem, $\{g_n\}$ is a normal family. Passing to a subsequence, we can assume that $g_n(z)$ converges normally to $g(z)$ for $0 < |z| < \delta$. Assume first that $g(z)$ is not identically ∞. Then $g(z)$ is analytic for $0 < |z| < \delta$. Fix ρ, $0 < \rho < \delta$, and choose M such that $|g(z)| < M$ for $|z| = \rho$. Then for large n we have $|g_n(z)| < M$ for $|z| = \rho$, and consequently, $|f(z)| < M$ for $|z| = \varepsilon_n \rho$. By the maximum principle, $|f(z)| < M$ for $\varepsilon_n \rho \leq |z| \leq \rho$, this for all large n. These annuli increase to a punctured neighborhood of 0 on which $|f(z)| < M$. By Riemann's theorem on removable singularities, $f(z)$ extends to be analytic at 0, and the theorem is proved in this case. If, on the other hand, $g(z) \equiv \infty$, then we apply the above argument to $1/f(z)$. The argument shows that $1/f(z)$ extends analytically to 0, so in this case $f(z)$ extends meromorphically to have a pole at 0.

Theorem (Picard's Little Theorem). *A nonconstant entire function assumes every value in the complex plane with at most one exception.*

In other words, an entire function that omits two (finite) values is constant. To see this, we regard ∞ as an isolated singularity of the function. The function omits two finite values and the value ∞ at the isolated singularity at ∞. By Picard's big theorem, ∞ is a removable singularity or pole. Thus the function is a polynomial, and since any nonconstant polynomial assumes all finite values, the function is constant.

Example. The entire function e^z omits the value 0. The entire function $\sin z$ does not omit any (finite) value.

Example. The entire function ze^z does not omit any (finite) value. However, it does omit the value 0 at ∞.

Exercises for XII.2

1. Show that $e^{1/z} + e^{-1/z}$ omits only the value ∞ at $z = 0$.

2. Show that the meromorphic function $e^{1/z} + 1/(1 - e^{1/z})$ on $\mathbb{C}\backslash\{0\}$ does not omit any value at $z = 0$.

3. Show that for $\alpha > 0$ and $0 < \beta < 1$, there are constants $C(\alpha, \beta) > 0$ with the following property. If $\mathcal{F}$ is a family of analytic functions on the open unit disk that omits the values 0 and 1, and if $|f(0)| < \alpha$ for all $f \in \mathcal{F}$, then $|f(z)| \leq C(\alpha, \beta)$ for all $f \in \mathcal{F}$ and $|z| \leq \beta$. *Remark.* This is **Schottky's theorem**.

4. Let $f(z)$ be analytic on the punctured disk $\{0 < |z| < 1\}$, and define $f_n(z) = f(z/n)$, $n \geq 1$. Show that $\{f_n(z)\}$ is a normal family on the punctured disk if and only if the singularity of $f(z)$ at 0 is removable or a pole.

5. Let E_0, E_1, E_2 be three disjoint compact subsets of the Riemann sphere, and let $\mathcal{F}$ be a family of meromorphic functions on a domain D such that each $f \in \mathcal{F}$ omits at least one point of each of the three sets. Show that $\mathcal{F}$ is a normal family.

6. Let $\mathcal{G}$ be the family of univalent analytic functions on a fixed domain D. (a) Show that $\mathcal{G}$ is not normal. (b) Show that the family of functions in $\mathcal{G}$ that omit 0 is normal. (c) Show that the family of derivatives of functions in $\mathcal{G}$ is normal. (d) Show that for fixed $z_0 \in D$ and $M > 0$ the family of functions $f \in \mathcal{G}$ satisfying $|f'(z_0)| \leq M$ is normal.

7. Let $\mathcal{S}$ denote the family of univalent functions $f(z)$ on the open unit disk $\mathbb{D}$, normalized by $f(0) = 0$ and $|f'(0)| = 1$. (a) Show that there exists $\kappa > 0$ such that the image of the open unit disk $\mathbb{D}$ under any $f \in \mathcal{S}$ includes the open disk centered at 0 of radius κ. (b) Show that the function $f(z) = z/(1 - z)^2$ belongs to $\mathcal{S}$ and maps $\mathbb{D}$ onto the complex plane slit along the negative real axis from $-\frac{1}{4}$ to $-\infty$. Conclude that $\kappa \leq \frac{1}{4}$. *Remark.* The theorem in (a) was first proved by P. Koebe. L. Bieberbach showed that the estimate holds with $\kappa = \frac{1}{4}$, and this estimate is sharp. The theorem is known as **Koebe's one-quarter theorem**, and the function in (b) is referred to as the **Bieberbach function**.

8. Show that there is a constant $\beta > 0$ with the following property. If $f(z)$ is an analytic function on the open unit disk $\mathbb{D}$ such that $f(0) = 0$ and $f'(0) = 1$, there is a subdisk $D \subset \mathbb{D}$ such that $f(z)$ is one-to-one on D and $f(D)$ contains a disk of radius β. *Remark.* This is **Bloch's theorem**, and the optimal (largest) constant β is **Bloch's constant**.

9. Give a proof of Royden's theorem (Exercise 1.9) based on the Zalcman lemma.

10. A family $\mathcal{F}$ of meromorphic functions on a domain D is **normal at** $z_0 \in D$ if $\mathcal{F}$ is normal on some open disk centered at z_0. Show that $\mathcal{F}$ is normal at z_0 if and only if whenever $\{z_n\}$ is a sequence in D that converges to z_0 and $\rho_n \to 0$, then every sequence in $\mathcal{F}$ has a subsequence $\{f_n(z)\}$ for which the corresponding scaled functions $g_n(\zeta) = f_n(z_n + \rho_n \zeta)$ converge normally to a constant (possibly ∞).

11. Let $\mathcal{F}$ be a family of meromorphic functions on a domain D that is not normal at $z_0 \in D$, and suppose that $f_n \in \mathcal{F}$, $z_n \to z_0$, and $\rho_n \to 0$ are such that the scaled functions $g_n(\zeta) = f_n(z_n + \rho_n \zeta)$ converge normally to a nonconstant meromorphic function $g(\zeta)$. Let $\zeta_0 \in \mathbb{C}$, $w_0 = g(\zeta_0)$. (a) Show that there is a sequence $\xi_n \to z_0$ such that $f_n(\xi_n) = w_0$. (b) Show that if ζ_0 is not a critical point of $g(\zeta)$, then $f_n^\sharp(\xi_n) \to \infty$. (c) Show that if $\psi(w)$ is a meromorphic function defined near w_0 such that $\psi(w_0) = z_0$, then there is a sequence $\eta_n \to z_0$ such that η_n is a fixed point of $\psi \circ f_n$, that is, $\psi(f_n(\eta_n)) = \eta_n$. (d) Show that if ζ_0 is not a critical point of $g(\zeta)$, and if w_0 is not a critical point of $\psi(w)$, then $(\psi \circ f_n)^\sharp(\eta_n) \to \infty$.

3. Julia Sets

One of the early applications of Montel's theorem was to complex dynamics, the study of the behavior of the iterates of an analytic or meromorphic function. Julia and Fatou used Montel's theorem as a key tool for studying the iterates of a rational function. One of their main ideas was to understand the dynamical behavior of the iterates by splitting the extended complex plane $\mathbb{C}^*$ into two invariant subsets, on one of which (the Fatou set) the iterates are well behaved, and on the other of which (the Julia set) their behavior is chaotic. We will derive some basic facts about Fatou and Julia sets in this section and the next.

Let U be a domain in the extended complex plane $\mathbb{C}^*$, and let $f(z)$ be an analytic map from U to U. In other words, $f(z)$ is an analytic function on U (meromorphic if $\infty \in U$) whose range is contained in U. For the remainder of this chapter it will be convenient to denote the nth iterate

$f(f(\cdots(f(z))\cdots))$ (n times) of $f(z)$ by $f^n(z)$. *Danger!* This should not be confused with the nth power $f(z)^n$ of $f(z)$.

Example. If $f(z) = z + 1$, then $f^n(z) = z + n$.

Example. If $f(z) = \lambda z$, then $f^n(z) = \lambda^n z$.

Example. If $f(z) = z^d$, then $f^n(z) = z^{d^n}$.

Example. If $f(z)$ is a polynomial of degree d, then $f^n(z)$ is a polynomial of degree d^n.

In general it is not possible to express the iterates of $f(z)$ in a simple form. An important method for obtaining effective information on the behavior of iterates is to make a "change of variable" in order to express the function in a more tractable form, such as one of the forms above, with respect to the new variable. Such a change of variable is called an **analytic conjugation**. It is defined formally as follows.

Let $f(z)$ be an analytic map from U into U, and $g(\zeta)$ an analytic map of V into V. We say that $f(z)$ and $g(\zeta)$ are **conjugate** if there is a conformal map $\zeta = \varphi(z)$ of U onto V such that

$$\varphi(f(z)) \;=\; g(\varphi(z)), \qquad z \in U.$$

Thus $\varphi \circ f = g \circ \varphi$. We express this by saying that the following diagram commutes:

$$
\begin{array}{ccc}
U & \xrightarrow{\;f\;} & U \\
\varphi \downarrow & & \downarrow \varphi \\
V & \xrightarrow{\;g\;} & V.
\end{array}
$$

We can regard $f(z)$ and $g(\zeta)$ as the same analytic function, expressed in different coordinate systems. If φ conjugates $f(z)$ to $g(\zeta)$, then φ also conjugates each iterate of $f(z)$ to the corresponding iterate of $g(\zeta)$,

$$\varphi(f^n(z)) \;=\; g^n(\varphi(z)), \qquad z \in U, \, n \geq 1.$$

Further, the inverse of $\varphi(z)$ conjugates $g(\zeta)$ to $f(z)$, $\varphi^{-1} \circ g = f \circ \varphi^{-1}$.

Example. Every polynomial $P(z) = Az^d + \cdots$ of degree $d \geq 2$ is conjugate to a monic polynomial. Indeed, if we take $\zeta = \varphi(z) = cz$ and $Q = \varphi \circ P \circ \varphi^{-1}$, we obtain

$$
\begin{aligned}
Q(\zeta) \;&=\; \varphi(P(z)) \;=\; cP(z) \;=\; cAz^d + \cdots \\
&=\; c^{1-d} A\zeta^d + \text{lower order.}
\end{aligned}
$$

We take c such that $c^{1-d}A = 1$, and we have conjugated $P(z)$ to a monic polynomial $Q(\zeta)$ of degree d.

Example. Consider $P(z) = z^2 - 2$. Let $z = \psi(\zeta) = \zeta + 1/\zeta$ be the conformal map from $\{|\zeta| > 1\}$ to the slit z-plane $\mathbb{C}\backslash[-2, 2]$. We compute that

$$P(\psi(\zeta)) = \left(\zeta + \frac{1}{\zeta}\right)^2 - 2 = \zeta^2 + \frac{1}{\zeta^2} = \psi(\zeta^2).$$

Thus $\varphi = \psi^{-1}$ conjugates $P(z)$ on $\mathbb{C}\backslash[-2, 2]$ to the map $\zeta \mapsto \zeta^2$ on the exterior of the closed unit disk.

From the conjugation identity $\varphi(f(z)) = g(\varphi(z))$ we see that $f(z_0) = z_0$ if and only if $g(\varphi(z_0)) = \varphi(z_0)$, that is, z_0 is a fixed point of $f(z)$ if and only if $\varphi(z_0)$ is a fixed point of $g(\zeta)$. Differentiation of the conjugation identity yields $\varphi'(f(z))f'(z) = g'(\varphi(z))\varphi'(z)$, from which it follows that $f'(z_0) = 0$ if and only if $g'(\varphi(z_0)) = 0$. We say that fixed points and critical points are "conjugation invariants."

Example. Any fractional linear transformation $f(z)$ with exactly two fixed points can be conjugated to the multiplication $g(\zeta) = \lambda\zeta$ for some complex number $\lambda \neq 0$. Indeed, let $\varphi(z)$ be a fractional linear transformation that maps the fixed points z_0 and z_1 of $f(z)$ to 0 and ∞, respectively. Then $\zeta = \varphi(z)$ conjugates f to the fractional linear transformation $g = \varphi \circ f \circ \varphi^{-1}$. Since $g(\zeta)$ now has fixed points at 0 and ∞, $g(\zeta)$ must have the form $\lambda\zeta$. A similar argument shows that if $f(z)$ is a fractional linear transformation with only one fixed point, then $f(z)$ can be conjugated to the translation $g(\zeta) = \zeta + 1$. In this case we take $\varphi(z)$ to be the fractional linear transformation that maps the fixed point of $f(z)$ to ∞, some other specified point z_0 to 0, and $f(z_0)$ to 1. (See Exercise II.7.12.)

Now let $f(z)$ be a rational function, regarded as an analytic map from $\mathbb{C}^*$ to $\mathbb{C}^*$. The **Fatou set** of $f(z)$, denoted by $\mathcal{F} = \mathcal{F}(f)$, consists of all points $z_0 \in \mathbb{C}^*$ that have an open neighborhood W such that the restrictions of the iterates of $f(z)$ to W form a normal family of analytic functions on W. In this case, the Fatou set of f contains along with z all points in the open set W, so the Fatou set is open. The **Julia set** of $f(z)$, denoted by $\mathcal{J} = \mathcal{J}(f)$, is the complement of the Fatou set, $\mathcal{J} = \mathbb{C}^*\backslash\mathcal{F}$. Thus $\mathcal{J}$ is a closed subset of $\mathbb{C}^*$.

Example. For $P(z) = z^2$ we have $P^n(z) = z^{2^n}$. Thus the iterates $\{P^n(z)\}$ converge normally to ∞ on $\{|z| > 1\}$ and they converge normally to 0 on $\{|z| < 1\}$. If W is any disk containing a point z_0 with $|z_0| = 1$, then the iterates $P^n(z)$ converge to both 0 and ∞ on nonempty open subsets of W, so the iterates do not form a normal family on W. Thus the Julia set of $P(z)$ coincides with the unit circle, and the Fatou set of $P(z)$ is the complement of the unit circle.

Theorem. *The Fatou set and the Julia set of a rational function $f(z)$ are invariant, that is, $f(\mathcal{F}) \subseteq \mathcal{F}$ and $f(\mathcal{J}) \subseteq \mathcal{J}$.*

Since $\mathcal{F}$ and $\mathcal{J}$ are complementary sets, the theorem amounts to the assertion that $f(\mathcal{F}) \subseteq \mathcal{F}$ and $f^{-1}(\mathcal{F}) \subseteq \mathcal{F}$, that is, that $\mathcal{F}$ is **completely invariant**. Thus we must show that $z_0 \in \mathcal{F}$ if and only if $f(z_0) \in \mathcal{F}$. One direction is trivial. That $f(z_0) \in \mathcal{F}$ implies $z_0 \in \mathcal{F}$ follows from the observation that if a subsequence of iterates $\{f^{n_k-1}(z)\}$ converges normally on a disk V containing $f(z_0)$, then the compositions $\{f^{n_k}(z)\}$ converge normally on the open set $f^{-1}(V)$ containing z_0. The less obvious direction, that $z_0 \in \mathcal{F}$ implies $f(z_0) \in \mathcal{F}$, follows from the observation that if a subsequence of iterates $\{f^{n_k+1}(z)\}$ converges normally on a disk U containing z_0, then $\{f^{n_k}(z)\}$ converges normally on $f(U)$, which is an open set containing $f(z_0)$. (See Exercise 1.7.)

For simplicity we focus now on a monic polynomial $P(z)$ of degree $d \geq 2$. The main feature of $P(z)$ that we will exploit is its attracting fixed point at $z = \infty$. Since $P(z) \sim z^d$ as $z \to \infty$, there is $R > 0$ such that $|P(z)| > 2|z|$ for $|z| \geq R$. The iterates $P^n(z)$ then converge uniformly to ∞ on the exterior domain $U_R = \{|z| > R\}$. In particular, $U_R \subset \mathcal{F}$.

The **basin of attraction** of ∞, denoted by $A(\infty)$, is defined to be the set of all z such that $P^n(z) \to \infty$. This occurs if and only if $P^n(z) \in U_R$ for large n. We denote by $P^{-n}(U_R)$ the set of z such that $P^n(z) \in U_R$. This is the inverse image of U_R under $P^n(z)$, so it is open. Since $A(\infty)$ is the union of the open sets $P^{-n}(U_R)$, $A(\infty)$ is also open. Since $P^k(z)$ tends to ∞ as $k \to \infty$ uniformly on $P^{-n}(U_R)$, each $P^{-n}(U_R)$ is contained in $\mathcal{F}$, and $A(\infty) \subseteq \mathcal{F}$.

Theorem. *Let $P(z)$ be a polynomial of degree $d \geq 2$. The basin of attraction $A(\infty)$ of ∞ is an open connected subset of $\mathbb{C}^*$ containing ∞. The Julia set $\mathcal{J}$ of $P(z)$ coincides with the boundary of $A(\infty)$, and it is a nonempty compact subset of $\mathbb{C}$. Each bounded component of $\mathbb{C} \backslash \mathcal{J}$ is simply connected.*

The proof is straightforward. First note that $A(\infty)$ is completely invariant, that is, $P(A(\infty)) \subseteq A(\infty)$ and $P^{-1}(A(\infty)) \subseteq A(\infty)$. Thus $P(\partial A(\infty))$ is disjoint from $A(\infty)$, and $P(\partial A(\infty)) \subseteq \partial A(\infty)$. Then also $P^n(\partial A(\infty)) \subseteq \partial A(\infty)$ for all $n \geq 1$. In particular, the iterates $P^n(z)$ are uniformly bounded on $\partial A(\infty)$. By the maximum principle, the iterates $P^n(z)$ are uniformly bounded on the bounded components of the complement $\mathbb{C} \backslash \partial A(\infty)$ of $\partial A(\infty)$. Thus no bounded component of $\mathbb{C} \backslash \partial A(\infty)$ is iterated to ∞, and $A(\infty)$ consists of just one connected component, the unbounded component of $\mathbb{C} \backslash \partial A(\infty)$. By Montel's theorem, the iterates $P^n(z)$ form a normal family on each bounded component of the open set $\mathbb{C} \backslash \partial A(\infty)$, so that all these components belong to the Fatou set $\mathcal{F}$, and the Julia set $\mathcal{J}$ is contained in $\partial A(\infty)$. If $z_0 \in \partial A(\infty)$, and U is any open disk containing z_0, then $\{P^n(z)\}$ converges normally to ∞ on $U \cap A(\infty)$, while the sequence is bounded at z_0. Thus no subsequence of $\{P^n(z)\}$ can converge normally on U, and so $z_0 \in \mathcal{J}$. Thus $\mathcal{J} = \partial A(\infty)$. Since the polynomial $P(z)$ has degree $d \geq 2$, it has at least one fixed point $z_0 \in \mathbb{C}$. Evidently, $z_0 \notin A(\infty)$, so $A(\infty) \neq \mathbb{C}^*$, and $\mathcal{J} = \partial A(\infty)$ is not empty. Finally, since $A(\infty)$ is connected, also $A(\infty) \cup \partial A(\infty)$ is connected, and each bounded component of the Fatou set is simply connected. (See Section VIII.8.) This completes the proof.

Example. For $P(z) = z^2$, the basin of attraction of ∞ is the exterior of the unit circle, $A(\infty) = \{|z| > 1\}$. The other component of the Fatou set is the open unit disk, which is the basin of attraction of the fixed point 0.

Example. Let $P(z) = z^2 - 2$. We have seen that $P^n(z) \to \infty$ for all $z \in \mathbb{C}\backslash[-2, 2]$. On the other hand, if $-2 \leq x \leq 2$, then $-2 \leq P(x) \leq 2$, so the interval $[-2, 2]$ is invariant under $P(z)$ and disjoint from $A(\infty)$. Thus $A(\infty) = \mathbb{C}^*\backslash[-2, 2]$, and the Julia set is $\mathcal{J} = [-2, 2]$. In this case, the Fatou set coincides with $A(\infty)$.

These examples of "smooth" Julia sets are the exception rather than the rule. It turns out that the only values of c for which the Julia set of $z^2 + c$ lies on a smooth curve are the values $c = 0$ and $c = -2$ treated above. Several filled-in Julia sets for other values of the parameter c are depicted below. It can be shown that the Julia set of $z^2 - 0.6$ (next page) is a simple closed Jordan curve that is nowhere differentiable. This is true of all values of c in the "principal cardioid of the Mandelbrot set," except $c = 0$. We will show in the next section that the Julia set of $z^2 + 0.251$ (see below) is totally disconnected, that is, it contains no continuum.

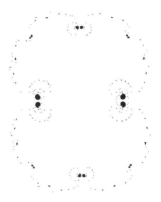

Cauliflower set $(c = \frac{1}{4})$ Totally disconnected $(c = 0.251)$

The theorem above forms the basis for a method for producing computer images of Julia sets, called the **boundary scanning method**. Fix R large, fix N large, and fix a grid of points in the z-plane. For each point z in the grid, we compute the iterates $P^k(z)$ until we reach an integer $n = n(z)$ such that $|P^n(z)| > R$, or until we reach $k = N$, in which case we set $n(z) = N$.

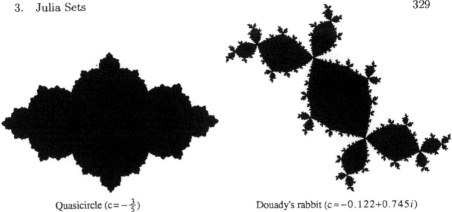

Quasicircle $(c = -\frac{3}{5})$ Douady's rabbit $(c \approx -0.122 + 0.745i)$

If we color blue those points z for which $n(z) < N$ and we color green those points for which $n(z) = N$, we obtain a picture of the **filled-in Julia set**, which is the union of $\mathcal{J}$ and the bounded components of the Fatou set. The integer $n(z)$ measures the escape rate of z to ∞. The escape rates can be captured visually by using different colors or shadings for different ranges of $n(z)$. An image of the Julia set can be obtained by recoloring red all those grid points for which $n(z) = N$ but that have a close neighbor satisfying $n(z) < N$.

Theorem. *Let $P(z)$ be a polynomial of degree $d \geq 2$. If z_0 belongs to the Julia set $\mathcal{J}$ of $P(z)$ and if U is any open neighborhood of z_0, then there exists $N \geq 1$ such that*

$$(3.1) \qquad \mathcal{J} \subseteq U \cup P(U) \cup \cdots \cup P^N(U).$$

Further, the inverse iterates $\bigcup_{k=1}^{\infty} P^{-k}(z_1)$ of any point $z_1 \in \mathcal{J}$ are dense in $\mathcal{J}$.

The proof of this theorem depends in an essential way on the general version of Montel's theorem. By our hypothesis, the restrictions of the sequence of iterates $\{P^n(z)\}$ to U do not form a normal family. By Montel's theorem, there is at most one value that is omitted by the sequence $\{P^n(z)\}$ on U. If $\{P^n(z)\}$ does not omit any value, then

$$(3.2) \qquad \bigcup_{n=1}^{\infty} P^n(U) \supseteq \mathcal{J}.$$

Since $\mathcal{J}$ is compact, and each set $P^n(U)$ is open, a finite number of these open sets already cover $\mathcal{J}$, and we obtain (3.1).

Suppose, on the other hand, there is an omitted value, call it w_0, so that $\bigcup P^n(U) = \mathbb{C} \backslash \{w_0\}$. Then there is no $z_0 \neq w_0$ such that $P(z_0) = w_0$, or else we would have $z_0 \in P^m(U)$ for some m, and $w_0 \in P^{m+1}(U)$. Since the only solution of $P(z) = w_0$ is w_0, $P(z) - w_0$ is a constant multiple of

$(z - w_0)^d$, and since $P(z)$ is assumed to be monic, $P(z) = w_0 + (z - w_0)^d$. Now, the iterates $P^n(z)$ tend uniformly to w_0 on any compact subset of $\{|z - w_0| < 1\}$. This can be seen directly, or by making the change of variable $\zeta = z - w_0$, which conjugates $P(z)$ to ζ^d. In any event, w_0 belongs to the Fatou set of $P(z)$, and we still obtain (3.2). Since $\mathcal{J}$ is compact, we then obtain (3.1) again. This proves the first statement of the theorem.

For the second statement, let U be an open nonempty subset of $\mathcal{J}$. Then $z_1 \in P^k(U)$ for some $k \geq 1$, and consequently, one of the points in $P^{-k}(z_1)$ belongs to U. This completes the proof.

This theorem provides another method for producing computer images of Julia sets, called the **inverse iteration method**. One selects a starting point $z_1 \in \mathcal{J}$, and one calculates the inverse iterates $P^{-k}(z_1)$ by solving the polynomial equations $P^k(z) = z_1$ for $1 \leq k \leq N$. For this method to be effective, the degree of $P(z)$ should be small (as $d = 2$). For a starting point in $\mathcal{J}$, one can look among the fixed points of $f(z)$. A rational function of degree $d \geq 2$ always has a fixed point in the Julia set (see Exercises 16 and 23).

The preceding theorem provides the basis for referring to Julia sets as "fractal sets," in the sense that shapes in the Julia set reappear infinitely often at different scales near any point of the Julia set. The idea is as follows. Let $z_0 \in \mathcal{J}$ be such that the forward orbit of z_0 does not contain any of the critical points of $P(z)$, and let z_1 be any point of $\mathcal{J}$. By the theorem, we can find images $P^n(z_0)$ arbitrarily close to z_1. If U is a small enough disk centered at z_0, then $P^n(z)$ maps U conformally onto an open set containing $P^n(z_0)$. Since $\mathcal{J}$ is completely invariant, $P^n(z)$ maps $\mathcal{J} \cap U$ onto $\mathcal{J} \cap P^n(U)$. Hence $\mathcal{J} \cap P^n(U)$ has the same "shape" as $\mathcal{J} \cap U$ though at a different scale.

Exercises for XII.3

1. Show that the Julia set $\mathcal{J}$ and the Fatou set $\mathcal{F}$ of a rational function $f(z)$ satisfy $f(\mathcal{J}) = \mathcal{J}$ and $f(\mathcal{F}) = \mathcal{F}$.

2. Show that the rational function $f(z) = z^2/(z^2 + 1)$ is conjugate to the quadratic polynomial $P(z) = z^2 + 2$.

3. For $\lambda, \mu \neq 0$, the two maps $f(z) = \lambda z$ and $g(z) = \mu z$ of $\mathbb{C}^*$ are conjugate if and only if $\lambda = \mu$ or $\lambda = 1/\mu$. *Hint.* A conjugation maps fixed points to fixed points.

4. Let z_0 be a fixed point of $f(z)$. Define the **multiplier** of the fixed point z_0 to be $\lambda = f'(z_0)$. Show that the multiplier at a fixed point is a conjugation invariant, that is, if $\zeta = \varphi(z)$ conjugates $f(z)$ to $g(\zeta)$, then the multiplier of $g(\zeta)$ at the fixed point $\varphi(z_0)$ is equal to the multiplier of $f(z)$ at z_0.

5. A fixed point z_0 of $f(z)$ is a **repelling fixed point** if $|f'(z_0)| > 1$. Show that the Julia set of a rational function $f(z)$ contains all its repelling fixed points.

6. Show that the Julia set of a fractional linear transformation is either empty or consists of one fixed point.

7. A fixed point z_0 of $f(z)$ is an **attracting fixed point** if $|f'(z_0)| < 1$. The **basin of attraction** of z_0, denoted by $A(z_0)$, is the set of z whose iterates $f^n(z)$ converge to z_0 as $n \to \infty$. Show that if $f(z)$ is a rational function, then $A(z_0)$ is an open subset of $\mathbb{C}^*$ containing z_0 whose boundary coincides with the Julia set.

8. Show that if $f(z)$ is a rational function of degree d, then the mth iterate $f^m(z)$ is a rational function of degree d^m.

9. Show that the Julia set of a rational function $f(z)$ coincides with the Julia set of its mth iterate $f^m(z)$.

10. A point z_0 is a **periodic point** of $f(z)$ if it is a fixed point of $f^m(z)$ for some $m \geq 1$. For such a z_0, set $z_1 = f(z_0)$, $z_2 = f(z_1)$, $\ldots$, $z_{m-1} = f(z_{m-2})$. Show that each z_j is a fixed point of $f^m(z)$ with the same multiplier $\lambda = f'(z_0) \cdots f'(z_{m-1})$. *Remark.* Assuming that the z_j's are distinct, we define the **multiplier of the cycle** $\{z_0, z_1, \ldots, z_{m-1}\}$ to be the multiplier of $f^m(z)$ at any of the points of the cycle. The cycle is an **attracting cycle** if its multiplier λ satisfies $|\lambda| < 1$, and it is a **repelling cycle** if $|\lambda| > 1$. The integer m is the **period** of the periodic point, or the **length** of the cycle.

11. Find all repelling cycles of the polynomial $f(z) = z^2$.

12. Show that all repelling cycles of a rational function are contained in its Julia set.

13. Find all attracting cycles of length two of the quadratic polynomial $z^2 + c$. Show that the values of the complex parameter c for which there is an attracting cycle of length two form an open disk.

14. Let $f(z)$ be a rational function. We define the **basin of attraction of an attracting cycle** of $f(z)$ to consist of the points $z \in \mathbb{C}^*$ whose iterates $f^n(z)$ accumulate on the cycle as $n \to \infty$. Show that the basin of attraction of an attracting cycle is an open subset of $\mathbb{C}^*$ whose boundary coincides with the Julia set. Show that different points of an attracting cycle lie in different components of the basin of attraction.

15. We define the **multiplicity of a fixed point** z_0 of $f(z)$ to be the order of the zero of $f(z) - z$ at $z = z_0$. Show that a fixed point has

multiplicity $m \geq 2$ if and only if its multiplier is 1. Show that the multiplicity of a fixed point is a conjugation invariant.

16. Show that a fixed point of a rational function $f(z)$ of multiplicity $m \geq 2$ belongs to the Julia set.

17. Show that a rational function of degree d has $d + 1$ fixed points, counting multiplicity.

18. Define the **analytic index** of a fixed point $z_0 \neq \infty$ of $f(z)$ to be the residue of $1/(z - f(z))$ at z_0. (a) Show that if z_0 is a fixed point of $f(z)$ with multiplier $\lambda \neq 1$, then the analytic index of $f(z)$ at z_0 is $1/(1 - \lambda)$. (b) Show that if $f(z)$ has a fixed point of multiplicity m at z_0, then for any small $\varepsilon > 0$, $f_\varepsilon(z) = f(z) - \varepsilon$ has m fixed points near z_0, each of multiplicity one, for which the sum of the analytic indices of $f_\varepsilon(z)$ tends to the analytic index of $f(z)$ at z_0 as $\varepsilon \to 0$. (c) Show that the analytic index of a fixed point is a conjugation invariant.

19. Find the fixed points and their analytic indices for the rational function $f(z) = (3z^2 + 1)/(z^2 + 3)$. Determine the Julia set and the Fatou set of $f(z)$.

20. Suppose that $f(z) = z - z^{m+1} + Az^{2m+1} + \mathcal{O}(z^{2m+2})$ for some integer $m \geq 1$. Show that the analytic index of $f(z)$ at the fixed point $z = 0$ is A.

21. Suppose that $g(z) = z - z^{m+1} + \mathcal{O}(z^{m+2})$ for some integer $m \geq 1$. Show that $g(z)$ can be conjugated near 0 to $f(z) = z - z^{m+1} + Az^{2m+1} + \mathcal{O}(z^{2m+2})$, where A is the analytic index of $f(z)$ at $z = 0$. *Hint.* Try conjugating by $\varphi(z) = z(1 + az^k)$.

22. If ∞ is a fixed point for $f(z)$, we define the **analytic index at ∞** of $f(z)$ to be the analytic index of $1/f(1/\zeta)$ at $\zeta = 0$. (*Danger!* This is not the residue of $z - f(z)$ at $z = \infty$.) Show that the sum of the analytic indices of a rational function $f(z)$ at its fixed points in $\mathbb{C}^*$ is $+1$.

23. Show that if $f(z)$ is a rational function of degree $d \geq 2$, and if all the fixed points of $f(z)$ have multiplicity one, then $f(z)$ has at least one repelling fixed point. *Hint.* Use Exercise 18a, and sum the real parts of the analytic indices at the $d + 1$ fixed points.

24. Let $f(z)$ be a rational function of degree $d \geq 2$. (a) Show that $\mathcal{J}$ is nonempty. (b) Show that either $\mathcal{F}$ is empty or $\mathcal{F}$ is dense in $\mathbb{C}^*$. (c) Show that $\mathcal{J}$ has no isolated points.

25. Let $f(z)$ be a rational function of degree $d \geq 2$. Show that the repelling periodic points of $f(z)$ are dense in the Julia set of $f(z)$. *Remark.* This theorem was proved independently by Fatou and Julia. It can be regarded as the first substantial theorem in rational iteration theory. For the proof, fill in the details of the following argument. Using the fact that $\mathcal{J}$ is a compact set with no isolated points, show that the points $z_0 \in \mathcal{J}$ that are not in the forward orbit of a critical point are dense in $\mathcal{J}$. For such a point z_0, refer to the entire function $g(\zeta)$ from Exercise 2.11. Show that there is a point $\zeta_0 \in \mathbb{C}$ such that ζ_0 is not a critical point of $g(\zeta)$ and $w_0 = g(\zeta_0)$ satisfies $f^m(w_0) = z_0$ for some m. Show that the points η_n from Exercise 2.11 are repelling periodic points of $f(z)$ that converge to z_0.

4. Connectedness of Julia Sets

The critical points and their forward orbits play a special role in complex dynamics. In this section we prove two results on connectedness of Julia sets, in which the critical points play the crucial role. While we focus on polynomials, the main ideas of the proofs are flexible and can be adapted to other situations (see the exercises).

We consider a monic polynomial $P(z) = z^d + \mathcal{O}\left(z^{d-1}\right)$ of degree $d \geq 2$. Let $R > 0$ be large, so that the circle $\Gamma_0 = \{|z| = R\}$ is contained in the basin of attraction $A(\infty)$ of ∞, and $|P(z)| > R$ for $|z| \geq R$. We may assume that $\Gamma_1 = P^{-1}(\Gamma_0)$ is "almost" a circle of radius $R^{1/d}$, which is mapped d-to-one by $P(z)$ onto Γ_0. Let U_0 be the "annular" domain between Γ_0 and Γ_1, and define $\Gamma_k = P^{-k}(\Gamma_0)$ and $U_k = P^{-k}(U_0)$. Then each U_k is an open set with boundary $\Gamma_k \cup \Gamma_{k+1}$, and $P(z)$ maps U_k d-to-one onto U_{k-1} (counting multiplicity at critical points).

If $z \in A(\infty)$, then either $|z| > R$, or there is a first integer $k \geq 0$ such that $|P^{k+1}(z)| > R$. In the latter case, $P^k(z)$ belongs to $U_0 \cup \Gamma_0$, and $z \in U_k \cup \Gamma_k$. Thus $A(\infty)$ is the disjoint union of the exterior disk $\{|z| > R\}$, the open sets U_k, and the curves Γ_k separating U_k and U_{k-1}. The effect of P is to map each Γ_k to Γ_{k-1} and each U_k to U_{k-1}.

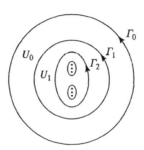

Since there are only finitely many critical points, we can arrange by adjusting R that no critical point of $P(z)$ lies on a curve Γ_k. Then $P(z)$ is conformal at each point of Γ_k and maps Γ_k d-to-one onto Γ_{k-1}. The composition $P^k(z)$ is then conformal at each point of Γ_k and maps Γ_k d^k-to-one onto the circle Γ_0. Thus each Γ_k is a finite union of disjoint simple closed analytic curves. The Γ_k's accumulate on $\partial A(\infty)$ as $k \to \infty$, and since $\mathcal{J} = \partial A(\infty)$, we have

$$\mathcal{J} \;=\; \lim_{k \to \infty} \Gamma_k \;=\; \lim_{k \to \infty} U_k\,.$$

Let $W_k = P^{-k}(\{|z| < R\})$. The boundary of W_k is Γ_k, and W_k is obtained from the disk $\{|z| < R\}$ by removing $U_0 \cup \Gamma_1 \cup \cdots \cup U_{k-1} \cup \Gamma_k$. Since the complement of W_k is connected, each component of W_k is simply connected. In view of the Riemann mapping theorem, we can think of W_k as the union of a finite number of "analytic disks" with analytic boundaries, one domain for each closed curve in Γ_k. Since $\mathbb{C} \backslash W_k \subset A(\infty)$, W_k contains all the bounded components of the Fatou set $\mathcal{F}$ together with the Julia set $\mathcal{J}$. Note also that $P^{-1}(W_k) = W_{k+1} \subset W_k$. Now the stage is set, and we state and prove our two theorems.

Theorem. *Let $P(z)$ be a polynomial of degree $d \geq 2$. The Julia set $\mathcal{J}$ of $P(z)$ is connected if and only if the iterates of each critical point of $P(z)$ are bounded.*

Suppose first that the iterates of critical points are all bounded, so that no critical point of $P(z)$ is in $A(\infty)$. Then at each point $w = P(z) \in A(\infty)$ we can define d analytic branches of the inverse function $P^{-1}(w)$, and these branches can be continued analytically along any curve in $A(\infty)$. Similarly, there are d^k analytic branches of the inverse $P^{-k}(w)$ of the kth iterate $P^k(z)$, and $P^{-k}(w)$ can be continued analytically along any path in $A(\infty)$. We deform Γ_0 to Γ_1 through a family of circular closed disjoint curves Γ_t in U_0, $0 \leq t \leq 1$. (In fact, since U_0 is conformally equivalent to an annulus, we can arrange for the Γ_t's to be simple closed analytic curves.) Suppose Γ_k consists of just one closed curve. Then as w traverses the closed curve Γ_0 d^k times, an analytic branch of $P^{-k}(w)$ traverses Γ_k once and returns to its initial point. We deform the curve Γ_0 through the Γ_t's to Γ_1, and we follow a branch of $P^{-k}(w)$ d^k times around each Γ_t. Since there are no critical points in $A(\infty)$, the branch of $P^{-k}(w)$ moves continuously with the curve Γ_t and cannot return to the initial value until w completes d^k full loops around Γ_t. Hence each $P^{-k}(\Gamma_t)$ consists of a single closed curve, and in particular, $\Gamma_{k+1} = P^{-k}(\Gamma_1)$ consists of a single closed curve. By induction on k we see, then, that each Γ_n consists of one closed curve, and in particular, each Γ_n is connected. Since a limit of compact connected sets is connected (Exercise 7), and since $\mathcal{J}$ is the limit of the connected sets Γ_n, $\mathcal{J}$ is connected.

To prove the converse, suppose next that there is a critical point in $A(\infty)$. Let k be the first integer such that there is a critical point in U_k, and for simplicity assume that there is only one critical point q in U_k. The argument in the preceding paragraph shows that each Γ_j consists of just one curve for $j \leq k$. As before, we consider the circular curves Γ_t and their inverse images $P^{-k}(\Gamma_t)$. Let r be the smallest parameter value such that $q \in P^{-k}(\Gamma_r)$. We can assume that Γ_t is a circle for t near r. Fix $\varepsilon > 0$ small. The argument in the preceding paragraph shows that $P^{-k}(\Gamma_{r-\varepsilon})$ consists of just one curve, which is the boundary of a bounded simply connected domain V containing q. Consider the inverse image $P^{-k}(\Gamma_r)$, which includes the critical point of q. If the critical point has order m, then the part of $P^{-k}(\Gamma_r)$ near q consists of $2m + 2$ analytic arcs that terminate at q and divide a disk centered at q into $2m + 2$ sector-like domains, as in the figure. In the sectors we alternately have $P^k(z)$ outside Γ_r and inside Γ_r. The curve $P^{-k}(\Gamma_{r-\varepsilon}) = \partial V$ enters and then exits every other sector, as in the figure. The curves in $P^{-k}(\Gamma_{r+\varepsilon})$ enter and then exit each of the remaining sectors. We can think of V as a disk, via the Riemann mapping theorem. Line segments drawn from q to the nearest point in ∂V in each of the $m+1$ sectors where ∂V approaches q then divide V into $m+1$ components. Since each of these components contains an arc of $P^{-k}(\Gamma_{r+\varepsilon})$, we see that $P^{-k}(\Gamma_{r+\varepsilon})$ consists of at least $m + 1$ analytic curves. Hence W_{k+1} has at least $m + 1$ components. Since each of these components contains curves in Γ_j for all $j > k + 1$, each of these components also contains points of $\mathcal{J}$, and the Julia set is not connected.

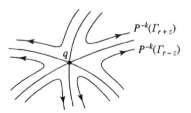

critical point of order $m = 2$

We say that a compact subset K of $\mathbb{C}$ is **totally disconnected** if for every $z_0 \in K$ and $\varepsilon > 0$ there is a subset E of K such that $z_0 \in E$, the diameter of E is less than ε, and E is at a positive distance from $K \backslash E$. Thus $K = E \cup (K \backslash E)$ is a decomposition of K into two disjoint compact subsets, one of which contains z_0 and has small diameter. At the opposite end of the spectrum from the preceding theorem, we now have the following.

Theorem. Let $P(z)$ be a polynomial of degree $d \geq 2$. If all the critical points of $P(z)$ are iterated to ∞, then the Julia set $\mathcal{J}$ of $P(z)$ is totally disconnected, and the Fatou set $\mathcal{F}$ of $P(z)$ coincides with the basin of attraction $A(\infty)$ of ∞.

In this case we choose N so large that none of the critical points of $P(z)$ are in W_N. Then since each component of W_N is simply connected, all d branches of $P^{-1}(w)$ can be defined analytically on each component of W_N. Each branch has image in W_N and maps points of $\mathcal{J}$ to points of $\mathcal{J}$. Fix $z_0 \in \mathcal{J}$. Let V be a component of W_N that contains $P^k(z_0)$ for infinitely many k's. For such a k, let $g_k(w)$ be the branch of $P^{-k}(w)$ on V satisfying $g_k\left(P^k(z_0)\right) = z_0$. The family $\{g_k(w)\}$ is uniformly bounded on V, hence a normal family of analytic functions on the open set V. Moreover, $g_k(w) \in W_{N+k} \cap A(\infty)$, so $g_k(w) \to \mathcal{J}$ for all $w \in W_N \cap A(\infty)$. Consequently, any limit $g(w)$ of the $g_k(w)$'s as $k \to \infty$ maps $W_N \cap A(\infty)$ into $\mathcal{J}$. Since $\mathcal{J}$ has no interior points, $g(w)$ is constant, and the constant value of $g(w)$ must be z_0. Thus there is a subsequence k_j for which $g_{k_j}(w)$ is defined and converges to z_0 uniformly on compact subsets of V. If we shrink V slightly, we obtain a compact subset E_0 of V such that $P^{k_j}(z_0) \in E_0$, $\partial E_0 \subset A(\infty)$, and the g_{k_j}'s converge uniformly on E_0 to z_0. Set $E_j = g_{k_j}(E_0)$. Then E_j is compact, $z_0 = g_{k_j}\left(P^{k_j}(z_0)\right) \in E_j$, and the diameter of E_j tends to 0. Since $\partial E_0 \subset A(\infty)$, also $\partial E_j \subset A(\infty)$. Thus $E_j \cap \mathcal{J}$ has positive distance from $\mathcal{J} \backslash E_j$, and the conditions in the definition of total disconnectedness are fulfilled at z_0. Since $z_0 \in \mathcal{J}$ is arbitrary, the Julia set is totally disconnected.

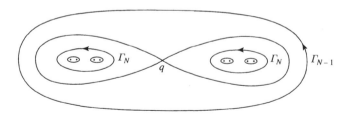

Exercises for XII.4

1. For each of the quadratic polynomials $P(z) = z^2$ and $P(z) = z^2 + 2$, sketch roughly the curves Γ_k for $0 \le k \le 3$. Take Γ_0 to be the circle $\{|z| = 9\}$.

2. Let $P(z)$ be a polynomial of degree $d \ge 2$ whose Julia set is connected. Let $G(z)$ be the Riemann map of $A(\infty)$ onto the exterior of the unit circle $\{|\zeta| > 1\}$ such that $G(\infty) = \infty$. Show that $\zeta = G(z)$ conjugates the polynomial $P(z)$ on $A(\infty)$ to the map ζ^d on the exterior $\{|\zeta| > 1\}$ of the unit circle. *Hint.* Apply the Schwarz reflection principle to $f = G \circ P \circ G^{-1}$.

3. Let $f(z)$ be a rational function with an attracting fixed point at z_0. Let U be the connected component of the basin of attraction of z_0 (see Exercise 3.7) containing z_0. Suppose there is no critical point of $f(z)$ in U except for possibly z_0. (a) Show that U is simply connected. *Hint.* Conjugate z_0 to ∞ by a fractional linear transformation and assume $z_0 = \infty$. Then modify the proof in the text. (b) Show that if $f(z)$ has degree $d \geq 2$, then U is conformally equivalent to the open unit disk, and $f(z)$ has a critical point at z_0. (c) Show that if z_0 is a critical point of $f(z)$ of order $m \geq 1$, and if $G(z)$ is the Riemann map of U onto the unit disk $\mathbb{D}$ such that $G(z_0) = 0$, then $\zeta = G(z)$ conjugates $f(z)$ on U to the map ζ^{m+1} on $\mathbb{D}$.

4. Let $f(z)$ be a rational function of degree $d \geq 2$, and suppose $\{z_0, f(z_0), \ldots, f^{n-1}(z_0)\}$ is an attracting cycle for $f(z)$. Show that $f(z)$ has a critical point in one of the connected components of the basin of attraction of the cycle that contains a point of the cycle. *Hint.* Apply the preceding exercise to $f^n(z)$. For definitions, see Exercises 3.10 and 3.14.

5. Show that a rational function of degree $d \geq 2$ has at most $2d - 2$ attracting cycles. *Hint.* Count critical points and apply the preceding exercise.

6. Show that a polynomial $P(z)$ of degree $d \geq 2$ has at most $d - 1$ attracting cycles in the (finite) complex plane.

7. Let $\{E_n\}$ be a sequence of compact subsets of $\mathbb{C}$ that converges to a compact set E. Thus E consists of those points z for which there are $z_n \in E_n$ satisfying $z_n \to z$. Show that if each E_n is connected, then E is connected.

8. Show that a compact subset K of $\mathbb{C}$ is totally disconnected if and only if for each $z_0 \in K$ and $\varepsilon > 0$, there is a simple closed curve γ in $\mathbb{C}\backslash K$ of diameter less than ε such that γ separates z_0 from ∞, that is, z_0 lies in the bounded component of $\mathbb{C}\backslash\gamma$.

9. Show that a compact subset K of $\mathbb{C}$ is totally disconnected if and only if K does not contain a continuum. *Strategy.* For fixed $z_0 \in K$, let $E(z_0)$ be the intersection of all closed subsets E of K such that $z_0 \in E$ and $K\backslash E$ is closed. Show that $E(z_0)$ is either a continuum or the singleton $\{z_0\}$. The set $E(z_0)$ is the "connected component of z_0 in K."

10. Let z_0 be a fixed point of a polynomial $P(z)$. Show that either $\{z_0\}$ is a connected component of the Julia set $\mathcal{J}$, or else the connected component of z_0 in the filled-in Julia set $\mathcal{K} = \mathbb{C}\backslash A(\infty)$ contains a critical point of $P(z)$.

5. The Mandelbrot Set

We now turn our attention to some special polynomials, the quadratic polynomials

$$P_c(z) = z^2 + c, \qquad z \in \mathbb{C},$$

where $c \in \mathbb{C}$ is a complex parameter. Let $\mathcal{J}_c$ denote the Julia set of $P_c(z)$. The only critical point of $P_c(z)$ is the origin $z = 0$. We consider the iterates $P_c(0) = c$, $P_c^2(0) = c^2 + c$, $P_c^3(c) = (c^2 + c)^2 + c$, $\ldots$. There are two cases that can occur, corresponding to the two theorems in the preceding section. If the iterates $P_c^k(0)$ are bounded, then the Julia set $\mathcal{J}_c$ is connected. Otherwise, the iterates $P_c^k(0)$ tend to ∞, and then $\mathcal{J}_c$ is totally disconnected.

We define the **Mandelbrot set** $\mathcal{M}$ to be the set of parameter values c such that the iterates $P_c^k(0)$ are bounded. Thus $c \in \mathcal{M}$ if and only if $\mathcal{J}_c$ is connected. Note that the Mandelbrot set is a subset of parameter space (c-space) and not of dynamic space (z-space).

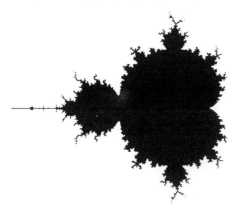

Theorem. *A complex number c belongs to the Mandelbrot set $\mathcal{M}$ if and only if $|P_c^n(0)| \le 2$ for all $n \ge 1$. The Mandelbrot set is a compact subset of the closed disk $\{|c| \le 2\}$. Further, $\mathbb{C}\backslash\mathcal{M}$ is connected.*

Suppose that $|c| > 2$. If $|z| = |c|$, then $|z^2 + c| \ge |c|^2 - |c| = (|c| - 1)|z|$. By the maximum principle, applied to $z/(z^2 + c)$ on the exterior of the circle, this estimate $|P_c(z)| \ge (|c| - 1)|z|$ persists for all z satisfying $|z| \ge |c|$. By iterating, starting with $P_c(0) = c$, we obtain successively $|P_c^2(0)| \ge (|c| - 1)|c|$, $|P_c^3(0)| \ge (|c| - 1)^2|c|$, and eventually $|P_c^k(0)| \ge (|c| - 1)^{k-1}|c|$. Thus $P_c^k(0) \to \infty$, so $c \notin \mathcal{M}$. It follows that $\mathcal{M}$ is a subset of the closed disk $\{|c| \le 2\}$.

Suppose next that $|c| \le 2$ and that $c \notin \mathcal{M}$. Let $n \ge 1$ be the first integer such that $|P_c^n(0)| > 2$. Then $|P_c^{n+1}(0)| = |P_c^n(0)^2 + c| \ge |P_c^n(0)|^2 - |c| > (|P_c^n(0)| - 1)|P_c^n(0)|$. We estimate $P_c^{n+2}(0)$ in the same way, and so on, and we obtain $|P_c^{n+k}(0)| > (|P_c^n(0)| - 1)^k |P_c^n(0)|$, which tends to $+\infty$ as $k \to \infty$. This proves the first statement in the theorem.

For fixed n, the set of $c \in \mathbb{C}$ such that $|P_c^n(0)| \leq 2$ is a closed set. Thus the description of $\mathcal{M}$ as the set of $c \in \mathbb{C}$ such that $|P_c^n(0)| \leq 2$ for all n shows that $\mathcal{M}$ is closed, and it is bounded; hence $\mathcal{M}$ is compact. Let U be any bounded domain with boundary contained in $\mathcal{M}$. Then $|P_c^n(0)| \leq 2$ for all $c \in \partial U$. Since $P_c(0)$ is a polynomial in c, we infer from the maximum principle that $|P_c^n(0)| \leq 2$ for all $c \in U$. Hence $U \subset \mathcal{M}$. It follows that $\mathbb{C} \backslash \mathcal{M}$ has no bounded components, and so $\mathbb{C} \backslash \mathcal{M}$ consists of a single unbounded component. The theorem is proved.

The constant 2 in the theorem is sharp. For $c = -2$, the origin is iterated in two steps to the repelling fixed point $z_0 = 2$, $0 \to -2 \to +2 \to +2 \to \cdots$. In particular, the iterates $P_{-2}^k(0)$ are bounded, so $-2 \in \mathcal{M}$. It is straightforward to show (Exercise 3) that $\mathcal{M}$ meets the real axis in the closed interval $\left[-2, \frac{1}{4}\right]$.

The theorem suggests an algorithm for obtaining computer pictures of $\mathcal{M}$. We fix a large integer N, and we lay a grid of points on the disk $\{|c| \leq 2\}$. For each point c in the grid we compute the iterates $P_c^k(0)$ of 0. If we reach $k < N$ for which $|P_c^k(0)| > 4$, we color the grid point sky blue; otherwise, we stop when we reach $k = N$ and we color the grid point black. The Mandelbrot set is approximated by the black grid points. By using different shades of blue to depict different escape rates, we obtain more attractive pictures. For an assortment of close-up pictures of various parts of the Mandelbrot set and a discussion of algorithms for obtaining the pictures, see the award-winning book of Peitgen and Richter [PR].

We describe now some results on the Mandelbrot set without giving proofs. Several of the proofs are laid out as exercises, while some are beyond our scope. We focus on the set W of those parameter values c such that $P_c(z)$ has an attracting cycle (other than ∞). Recall from the exercises for Sections 3 and 4 the definitions of attracting fixed points, attracting cycles, and their basins of attraction.

Suppose $P_c(z)$ has an attracting cycle, other than the fixed point at ∞. The basin of attraction of the cycle is contained in the Fatou set $\mathcal{F}_c$ and is disjoint from $A(\infty)$. Hence $\mathcal{J}_c$ cannot be totally disconnected, and $c \in \mathcal{M}$. Thus $W \subset \mathcal{M}$. It is straightforward to check that W is an open subset of the complex plane (Exercise 8). This depends on the observation that an attracting periodic point of period n is a zero of $P_c^n(z) - z$, and by Hurwitz's theorem the zeros of this polynomial vary continuously with the parameter c. With somewhat more effort it can be proved that $\partial W \subseteq \partial \mathcal{M}$, so that W is a union of connected components of the interior of $\mathcal{M}$ (Exercise 10). These components are called the **hyperbolic components** of the interior of $\mathcal{M}$. Each hyperbolic component corresponds to a single attracting cycle $\{z_0(c), z_1(c), \ldots, z_{n-1}(c)\}$ of length n, where each n-periodic point $z_j(c)$ is an analytic function of c in the hyperbolic component.

It turns out (see [CG]) that if U is a hyperbolic component of the interior of $\mathcal{M}$, and if $\lambda(c)$ is the multiplier of the attracting cycle of $P_c(z)$, then $c \mapsto \lambda(c)$ maps U conformally onto the open unit disk in the complex plane. In particular, each of the hyperbolic components contains exactly one parameter value c_0 such that $P_{c_0}(z)$ has a cycle with multiplier 0, that is, a "superattracting cycle." We refer to c_0 as the **center of the hyperbolic component**. The centers of the hyperbolic components corresponding to superattracting cycles of length n can be calculated by solving $P_c^n(0) = 0$, which is a polynomial equation in c, and discarding the solutions of $P_c^m(0) = 0$ for $m < n$.

Example. The c's for which $P_c(z)$ has an attracting fixed point form a cardioid, called the **principal cardioid**. It is a hyperbolic component of the interior of $\mathcal{M}$, with center at $c = 0$ corresponding to the superattracting fixed point $z = 0$ for $P_0(z) = z^2$. (See Exercise 4.)

Example. To find the superattracting 2-cycles, we solve $P_c^2(0) = c^2 + c = 0$. The solution $c = 0$ corresponds to the superattracting fixed point $z = 0$ for $P_0(z) = z^2$. The other solution $c = -1$ corresponds to the superattracting 2-cycle $\{0, -1\}$ for $P_{-1}(z) = z^2 - 1$. The corresponding hyperbolic component of the interior of $\mathcal{M}$ is the disk $\{|c + 1| < \frac{1}{4}\}$. (See Exercise 3.13.)

It was known for some time that the Mandelbrot set $\mathcal{M}$ is connected, that is, $\mathbb{C}^* \backslash \mathcal{M}$ is simply connected. Some intense interest has been focused on the following two questions. Does the conformal map from the open unit disk to $\mathbb{C}^* \backslash \mathcal{M}$ extend continuously to the boundary of the disk? Are all components of the interior of $\mathcal{M}$ hyperbolic? The answers remain a mystery. For more on the Mandelbrot set, see [CG].

Exercises for XII.5

1. Show that if $|P_c^n(0)| > 2$, then $|P_c^k(0)|$ is strictly increasing for $k \geq n$.

2. Show that if $|c| \leq 2$, then the Julia set $\mathcal{J}_c$ is a subset of the closed disk $\{|z| \leq 2\}$.

3. Show that $\mathcal{M} \cap \mathbb{R} = [-2, \frac{1}{4}]$. *Hint.* If $-2 \leq c \leq \frac{1}{4}$, let $r = \left(1 + \sqrt{1 - 4c}\right)/2$, the larger root of $x^2 + c = x$. Show that $P_c([c, r]) \subseteq [c, r]$.

4. Show that $P_c(z)$ has an attracting fixed point with multiplier λ, $|\lambda| < 1$, if and only if $c = \lambda/2 - \lambda^2/4$. Show that these values c form a cardioid with cusp at $c = \frac{1}{4}$. Sketch the cardioid. Show that the map $c \mapsto \lambda(c)$ is the Riemann map of the cardioid onto the open unit disk.

5. Find the values c for which there are superattracting cycles of length 3 and of length 4. Locate the values of c on the picture of the Mandelbrot set.

6. Show that any quadratic polynomial $P(z) = Az^2 + Bz + C$, $A \neq 0$, is conjugate to the polynomial $P_c(z)$ for a unique value of c.

7. Show that any quadratic polynomial $P(z) = Az^2 + Bz + C$ is conjugate to the polynomial $Q_\lambda(z) = \lambda z + z^2$ for some complex number λ. Show that $Q_\lambda(z)$ is conjugate to $Q_\mu(z)$ if and only if $\lambda = \mu$ or $\lambda = 2 - \mu$.

8. Show that the set of parameter values c such that $P_c(z)$ has an attracting cycle is an open subset of the complex plane that is contained in $\mathcal{M}$.

9. Use the result of Exercise 4.4 to show that if $P_c(z)$ has an attracting cycle of length n, then $P_c^{kn}(0)$ converges to a point $z_0(c)$ of the cycle as $k \to \infty$. Show further that $P_c^{kn+j}(0)$ converges to the point $z_j(c) = P_c(z_0(c))$ of the cycle as $k \to \infty$, and that the points $z_j(c)$ depend analytically on c, $0 \leq j < n - 1$.

10. Suppose $P_{c_0}(z)$ has an attracting cycle of length n. Let V be the connected component of the interior of $\mathcal{M}$ containing c_0. Show that $P_c(z)$ has an attracting cycle of length n for each $c \in V$. *Hint.* Use the fact that $P_c^k(0)$ is uniformly bounded for $c \in V$, hence a normal family. Use the preceding exercise, and show that any limit of a subsequence is an attracting periodic point.

XIII

Approximation Theorems

In this chapter we prove two fundamental theorems, one "additive" and the other "multiplicative," on prescribing zeros, and poles of meromorphic functions. The first is the Mittag-Leffler theorem, which asserts that we can prescribe the poles and principal parts of a meromorphic function. The second is the Weierstrass product theorem, which asserts that we can prescribe the zeros and poles, including orders, of a meromorphic function. The theorems are closely related. Both theorems are proved by the same type of approximation procedure, which depends on Runge's theorem on approximation by rational functions. We prove Runge's theorem in Section 1, followed by the Mittag-Leffler theorem in Section 2. In Section 3 we introduce infinite products, which can always be converted to infinite series by taking logarithms. We prove the Weierstrass product theorem in Section 4.

1. Runge's Theorem

Suppose $f(z)$ is a complex-valued function on a compact subset K of the complex plane. When can $f(z)$ be approximated uniformly on K by rational functions? When can $f(z)$ be approximated uniformly on K by polynomials (in z)? Any uniform limit of rational functions with poles off K must be continuous on K, and further, since a uniform limit of analytic functions is analytic, it must be analytic on the interior of K. These are necessary conditions for approximation. The first nontrivial sufficient condition for approximation is given by the following theorem.

Theorem (Runge's Theorem). *Let K be a compact subset of the complex plane. If $f(z)$ is analytic on an open set containing K, then $f(z)$ can be approximated uniformly on K by rational functions with poles off K.*

For the proof, let D be an open set with piecewise smooth boundary such that $K \subset D$ and such that $f(z)$ is analytic on $D \cup \partial D$. By the Cauchy integral formula,

$$f(z) = \frac{1}{2\pi i} \int_{\partial D} \frac{f(\zeta)}{\zeta - z} \, d\zeta, \qquad z \in D.$$

We chop ∂D up into a union of short curves γ_j such that each γ_j is contained in a disk $\{|z - c_j| < r_j\}$ that is at a positive distance from K. Then $f(z) = \sum f_j(z)$, where

$$f_j(z) = \frac{1}{2\pi i} \int_{\gamma_j} \frac{f(\zeta)}{\zeta - z} \, d\zeta, \qquad z \notin \gamma_j.$$

The function $f_j(z)$ is analytic off γ_j (Morera's theorem) and vanishes at ∞. Each $f_j(z)$ has a Laurent expansion in descending powers of $z - c_j$ that converges uniformly for $|z - c_j| > r_j$, hence uniformly on K. Thus $f_j(z)$ is uniformly approximable on K by polynomials in $1/(z - c_j)$. Adding these approximants, we see that $f(z)$ is uniformly approximable on K by rational functions with poles off K.

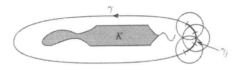

There is some flexibility with respect to the location of the poles of the approximating rational functions.

Lemma. *Let K be a compact subset of the complex plane, let U be a connected open subset of the extended complex plane $\mathbb{C}^*$ disjoint from K, and let $z_0 \in U$. Every rational function with poles in U can be approximated uniformly on K by rational functions with poles at z_0.*

Here we consider a polynomial in z to be a rational function with pole at ∞. The lemma is established by a "translation of poles" argument. We define the set V to consist of those points $\zeta \in U$ such that $1/(z - \zeta)$ is approximable uniformly on K by rational functions with pole at z_0. Thus if $\zeta \in V$, then each power $1/(z - \zeta)^k$ is also uniformly approximable on K by rational functions with pole at z_0. Since $z_0 \in V$, the set V is nonempty. Since U is connected, to show that $V = U$, it suffices to show that V is open and closed in U. The "closed" assertion is easy. If $\zeta \in U$ is a limit of a sequence $\zeta_j \in V$, then $1/(z - \zeta_j)$ converges uniformly on K to $1/(z - \zeta)$, so by the definition of V, also $\zeta \in V$. The crux of the proof then is to show that V is open, and this reduces to showing that if $\zeta_0 \in V$, and if ζ is near ζ_0, then $1/(z - \zeta)$ is uniformly approximable on K by polynomials in $1/(z - \zeta_0)$. We begin with the special case $\zeta_0 = \infty$. We must show that if ζ is near ∞ (that is, $|\zeta|$ is large), then $1/(z - \zeta)$ is uniformly approximable on K by polynomials in z. For this, we expand $1/(z - \zeta)$ in a geometric series,

$$\frac{1}{z - \zeta} = -\frac{1}{\zeta}\frac{1}{1 - z/\zeta} = -\frac{1}{\zeta}\sum_{k=0}^{\infty}\frac{z^k}{\zeta^k}.$$

If $|z| \le C$ for all $z \in K$, and if $|\zeta| > 2C$, then the kth term of the series is dominated by $1/2^k$. By the Weierstrass M-test, the series converges uniformly on K, and consequently, $\zeta \in V$ whenever $|\zeta| > 2C$. If ζ_0 is finite, the proof is essentially the same, with a change of variable $z \mapsto 1/(z - \zeta_0)$ to place ζ_0 at ∞. We choose $\varepsilon > 0$ less that the distance from ζ_0 to K, so that $|z - \zeta_0| \ge \varepsilon$ for $z \in K$. If $|\zeta - \zeta_0| < \varepsilon/2$, then $|(\zeta - \zeta_0)^k/(z - \zeta_0)^k| < 1/2^k$ for $z \in K$, so the geometric series

$$\frac{1}{z - \zeta} = \frac{1}{z - \zeta_0}\frac{1}{1 - (\zeta - \zeta_0)/(z - \zeta_0)} = \frac{1}{z - \zeta_0}\sum_{k=0}^{\infty}\frac{(\zeta - \zeta_0)^k}{(z - \zeta_0)^k}$$

is uniformly convergent for $z \in K$, by the Weierstrass M-test. Thus all ζ satisfying $|\zeta - \zeta_0| < \varepsilon/2$ belong to V, and V is an open set. This proves the lemma.

By approximating with rational functions and then using the lemma to translate the poles, we obtain immediately the following sharper version of Runge's theorem.

Theorem. *Let K be a compact subset of the complex plane, and suppose that $f(z)$ is analytic on an open set containing K. Let S be a subset of $\mathbb{C}^*\backslash K$ such that each connected component of $\mathbb{C}^*\backslash K$ contains a point of S. Then $f(z)$ can be approximated uniformly on K by rational functions with poles in S.*

In particular, if K is a compact subset of the complex plane, and if the complement of K is connected, then each function analytic in a neighborhood of K can be approximated uniformly on K by polynomials in z. There is a considerably more difficult theorem on polynomial approximation, **Mergelyan's theorem**, which asserts that any function that is continuous on K and analytic on the interior of K can be approximated uniformly on K by polynomials in z. The analogous statement for rational approximation is false.

Example. There is a compact set K with empty interior, and a continuous function $f(z)$ on K, such that $f(z)$ cannot be approximated uniformly on K by rational functions. We outline the construction of such an example, known as a **Swiss cheese set**. Let S be a sequence that is dense in the open unit disk $\mathbb{D}$. Select disks $D_k = \{|z - c_k| < r_k\}$ by induction, such that the center c_k of D_k is the first point in the sequence S that is not contained in any of the closed disks $E_j = \{|z - c_j| \le r_j\}$ for $j < k$, and then choose

r_k so small that E_k is disjoint from the sets E_j for $j < k$. We also assume that $\sum r_k$ converges and $\sum r_k^2 < 1$. Define $K = \{|z| \leq 1\} \backslash \cup_{j=1}^{\infty} D_j$, a compact set with no interior points. If $f(z)$ is a continuous function on K such that $|f(z)| \leq M$, then $|\int_{\partial D_k} f(z)dz| \leq 2\pi M r_k$, so that the series $\sum \int_{\partial D_k} f(z)dz$ converges absolutely. Further, $\sum \int_{\partial D_k} f(z)dz = \int_{\partial \mathbb{D}} f(z)dz$ if $f(z)$ is a rational function with poles off K, hence whenever $f(z)$ is any uniform limit on K of rational functions. However, by Green's theorem, or by a direct calculation, $\int_{\partial D_k} \bar{z}\, dz = 2\pi r_k^2 i$ and $\int_{\partial \mathbb{D}} \bar{z}\, dz = 2\pi i$. Since $\sum \int_{\partial D_k} \bar{z}\, dz \neq \int_{\partial \mathbb{D}} \bar{z}\, dz$, the function $\bar{z}$ cannot be approximated uniformly on K by rational functions.

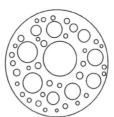

Swiss cheese set

There are many applications of Runge's theorem to construct functions that do one thing or another. We select one such theorem at random, in order to give an idea of how Runge's theorem can be applied. First we need a definition.

Let $f(z)$ be a function on the open unit disk $\mathbb{D}$, and let $\zeta \in \partial \mathbb{D}$. We define the **radial cluster set** of $f(z)$ at ζ to consist of all complex numbers w for which there is a sequence of radii r_j increasing to 1 such that $f(r_j\zeta) \to w$.

Theorem. *There is an analytic function $f(z)$ on the open unit disk $\mathbb{D}$ whose radial cluster set at any $\zeta \in \partial \mathbb{D}$ coincides with the entire complex plane.*

The proof construction goes as follows. Let $\{s_k\}$ be a sequence of positive numbers that increases to 1, let ε_k be a sequence of positive numbers that decreases to 0, and consider the circular arcs $\Gamma_k = \{s_k e^{i\theta} : \varepsilon_k \leq \theta \leq 2\pi\}$. Let S be a sequence of complex numbers that is dense in the complex plane, and let $\{w_k\}_{k=1}^{\infty}$ be an enumeration of S with repetitions, so that each number in S appears infinitely often in the sequence. Thus for any complex number w and for any $\zeta \in \partial \mathbb{D}$, there is a subsequence k_j of integers such that $s_{k_j}\zeta \in \Gamma_{k_j}$ and $w_{k_j} \to w$. Our strategy is to use Runge's theorem to approximate the function with constant value w_k on Γ_k by an analytic function $f(z)$. This is done by an induction argument, as follows. Let $f_1(z)$ be the constant function w_1. Having chosen $f_{m-1}(z)$, observe that $\Gamma_m \cup \{|z| \leq s_{m-1}\}$ is a compact subset of the complex plane with connected complement, so by Runge's theorem there is an analytic polynomial $f_m(z)$ such that $|f_m(z) - f_{m-1}(z)| < 1/2^m$ for $|z| \leq s_{m-1}$, while $|f_m(z) - w_m| < 1/2^m$ for $z \in \Gamma_m$. For $n \geq m$, $|f_n(z) - f_{n-1}(z)| < 1/2^n$ for $|z| \leq s_{m-1}$. Consequently, $\sum |f_n(z) - f_{n-1}(z)|$ converges uniformly for $|z| \leq s_{m-1}$, by the Weierstrass M-test. Hence

$$f(z) = \lim f_n(z) = f_1(z) + \sum_{n=2}^{\infty} [f_n(z) - f_{n-1}(z)]$$

is analytic on $\mathbb{D}$. From $f(z) = f_m(z) + \sum_{n=m+1}^{\infty} [f_n(z) - f_{n-1}(z)]$ we estimate $|f(z) - w_m|$ by

$$|f_m(z) - w_m| + \sum_{n=m+1}^{\infty} |f_n(z) - f_{n-1}(z)| \leq \frac{1}{2^m} + \sum_{n=m+1}^{\infty} \frac{1}{2^n} = \frac{1}{2^{m-1}},$$

for $z \in \Gamma_m$. Thus $\{f(s_{k_j}\zeta)\}$ has the same limit as $\{w_{k_j}\}$ as $j \to \infty$, which is w, and consequently, each complex number w is in the radial cluster set of $f(z)$ at ζ.

Exercises for XIII.1

1. Show that any analytic function $f(z)$ on a domain D can be approximated normally on D by a sequence of rational functions that are analytic on D.

2. Show that there is a sequence of polynomials $\{p_n(z)\}$ such that $p_n(z) \to 1$ if $\operatorname{Re} z > 0$, $p_n(z) \to 0$ if $\operatorname{Re} z = 0$, and $p_n(z) \to -1$ if $\operatorname{Re} z < 0$.

3. Let $\{z_j\}$ be a sequence of distinct points in a domain D that accumulates on ∂D, and let $\{w_j\}$ be a sequence of complex numbers. Show that there is an analytic function $f(z)$ on D such that $f(z_j) = w_j$ for all j. *Remark.* The sequence $\{z_j\}$ is called an **interpolating sequence** for analytic functions on D.

4. Let $\{z_k\}$ be a sequence of distinct points in a domain D that accumulates on ∂D. Let $\{m_k\}$ be a sequence of positive integers, and for each k, let $a_{k0}, \dots, a_{km_k}$ be complex numbers. Show that there is an analytic function $f(z)$ on D such that $f^{(j)}(z_k) = a_{kj}$ for $0 \leq j \leq m_k$ and for all k.

5. Let D be a domain in $\mathbb{C}$. Show that there is an analytic function $f(z)$ on D such that the cluster set of $f(z)$ at any point $\zeta \in \partial D$ includes the entire complex plane, that is, for any $\zeta \in \partial D$ and any complex number w, there is a sequence $\{z_j\}$ in D satisfying $z_j \to \zeta$ and $f(z_j) \to w$. *Hint.* Use Exercise 3.

6. Let $\{z_k\}$ be a sequence of distinct points in a domain D that accumulates at ∂D, and let E be a nonempty closed subset of the

extended complex plane $\mathbb{C}^*$. Show that there is an analytic function $f(z)$ on D such that E is the set of cluster values of $f(z)$ along the sequence $\{z_k\}$.

7. Show that there exist three analytic functions $f_1(z), f_2(z), f_3(z)$ on the open unit disk $\mathbb{D}$ such that the mapping $z \mapsto (f_1(z), f_2(z), f_3(z))$ embeds $\mathbb{D}$ as a closed submanifold of $\mathbb{C}^3$, that is, such that $|f_1(z)| + |f_2(z)| + |f_3(z)| \to +\infty$ as $z \to \partial\mathbb{D}$, and for each $z \in \mathbb{D}$ there is j such that $f_j'(z) \neq 0$. *Hint.* Take $f_3(z) = 1/(z-1)$. For $j \geq 1$, let E_j be the set of z satisfying $1 - 1/j \leq |z| \leq 1 - 1/(j+1)$ and $\varepsilon_j \leq \theta \leq 2\pi$, construct $f_1(z)$ to be approximately j on E_j for j odd, and construct $f_2(z)$ to be approximately j on E_j for j even.

8. Show that for any domain D in $\mathbb{C}$, there exist three analytic functions on D for which the mapping $z \mapsto (f_1(z), f_2(z), f_3(z))$ embeds D as a closed submanifold of $\mathbb{C}^3$. (See the preceding exercise.)

9. Suppose points $c_j \in \mathbb{D}$ and radii $r_j > 0$ are such that $\sum r_j < 1$ and such that the disks $D_j = \{|z - c_j| < r_j\}$ are disjoint subdisks of $\mathbb{D}$. Set $K = \{|z| \leq 1\} \setminus \cup D_j$, which is a compact set. By considering $\int_{|z|=1} (\bar{z} - f(z))\, dz - \sum \int_{|z-c_j|=r_j} (\bar{z} - f(z))\, dz$, show that

$$\sup_{z \in K} |\bar{z} - f(z)| \geq 1 - \sum r_j$$

for any rational function $f(z)$ with poles off K.

10. Let D be a bounded domain whose boundary Γ consists of a finite number of disjoint circles, and let $E = D \cup \Gamma$. Show that a function $f(z)$ on E can be approximated uniformly on E by rational functions with poles off E if and only if $f(z)$ is continuous on E and analytic on D. What functions on E can be uniformly approximated on E by polynomials in z?

11. Let K be a compact subset of the extended complex plane $\mathbb{C}^*$, and let S be a subset of $\mathbb{C}^* \setminus K$. Let J be the union of K and the components of $\mathbb{C}^* \setminus K$ that do not contain points of S. (a) Show that J is compact and that $\partial J \subset \partial K$. (b) Show that if $\zeta \in J \setminus K$, then the function $f(z) = 1/(z - \zeta)$ is not uniformly approximable on K by rational functions with poles in S. (c) Show that a function $f(z)$ analytic in a neighborhood of K is uniformly approximable on K by rational functions with poles in S if and only if $f(z)$ extends to be analytic in a neighborhood of J. (Polynomials are regarded as rational functions with poles at ∞.)

12. Let D be a bounded domain with smooth boundary Γ, and let K be a compact subset of D. Let $f(z)$ be a continuous function on $D \cup \Gamma$ that is analytic in D. Show that the Riemann sums $\sum [f(\zeta_j)/(\zeta_j -$

$z)](\zeta_j - \zeta_{j-1})$ for the Cauchy integral $\int_\Gamma f(\zeta)/(\zeta - z)\, d\zeta$ converge uniformly on K to $f(z)$ as the lengths of the arcs of Γ between successive points ζ_j of the partition tend to zero.

2. The Mittag-Leffler Theorem

Recall that if $f(z)$ is a meromorphic function with pole at z_0, and the Laurent expansion of $f(z)$ at z_0 is given by $\sum_{k=-m}^\infty a_k(z - z_0)^k$, then the principal part $P(z)$ of $f(z)$ at z_0 is the sum of the terms with the negative powers, $P(z) = \sum_{k=-m}^{k=-1} a_k(z - z_0)^k$. Thus $P(z)$ is a polynomial in $1/(z - z_0)$, and $f(z) - P(z)$ is analytic at z_0. The Mittag-Leffler theorem asserts that we can prescribe the poles and principal parts of a meromorphic function.

Theorem (Mittag-Leffler Theorem). *Let D be a domain in the complex plane. Let $\{z_k\}$ be a sequence of distinct points in D with no accumulation point in D, and let $P_k(z)$ be a polynomial in $1/(z - z_k)$. Then there is a meromorphic function $f(z)$ on D whose poles are the points z_k, such that $f(z) - P_k(z)$ is analytic at z_k.*

Let K_m be the set of $z \in D$ such that $|z| \leq m$ and the distance from z to ∂D is at least $1/m$. Then K_m is a compact subset of D, $K_m \subset K_{m+1}$, and each component of $\mathbb{C}^* \backslash K_m$ contains a point of $\mathbb{C}^* \backslash D$. Let $f_m(z)$ be the sum of the functions $P_k(z)$ for which $z_k \in K_{m+1} \backslash K_m$. There are only finitely many such functions $P_k(z)$. By Runge's theorem, there is a rational function $g_m(z)$ with poles in $\mathbb{C}^* \backslash D$ such that $|f_m(z) - g_m(z)| \leq 1/2^m$ for $z \in K_m$. Then $f(z) = \sum [f_m(z) - g_m(z)]$ converges uniformly on each compact subset of D, by the Weierstrass M-test. Since $\sum_{m=N+1}^\infty [f_m(z) - g_m(z)]$ is analytic on K_n for $N > n$, and $\sum_{m=1}^N [f_m(z) - g_m(z)]$ has poles at the points z_k that are in K_n, with the prescribed principal parts, $f(z)$ has the prescribed poles and principal parts in D.

In the case that D is the entire complex plane $\mathbb{C}$, we can take K_m to be the closed disk $\{|z| \leq m\}$. Then $f_m(z)$ is the sum of the functions $P_k(z)$ for which $m < |z_k| \leq m + 1$, and we can take $g_m(z)$ to be a partial sum of the power series expansion of $f_m(z)$ about $z = 0$.

Exercise. Find a meromorphic function on the complex plane whose poles are simple poles at the positive integers with residues all equal to 1.
Solution. This corresponds to the principal part $P_k(z) = 1/(z - k)$ at $z_k = k$. Our first guess is to try $f(z) = \sum 1/(z - k)$, but the sum does not converge. The constant term in the power series expansion of $1/(z - k)$ about $z = 0$ is $-1/k$, so our second guess is to set $g_k(z) = -1/k$ and try

$$f(z) = \sum_{k=1}^{\infty} \left(\frac{1}{z-k} + \frac{1}{k} \right) = z \sum_{k=1}^{\infty} \frac{1}{k} \frac{1}{z-k}.$$

This series *does* converge, uniformly on bounded sets, by comparison with the series $\sum 1/k^2$. Indeed, if $|z| \leq R$ and $k > 2R$, then $|z - k| > k/2$, so the kth summand is bounded by $2/k^2$.

The technique of subtracting off terms of a power series expansion to guarantee convergence can be used to construct doubly periodic functions with double poles. Let ω_1 and ω_2 be two complex numbers that do not lie on the same line through the origin. We sketch the construction of the **Weierstrass P-function** $\mathcal{P}(z)$ associated with ω_1 and ω_2. This is a doubly periodic meromorphic function on the complex plane, with periods ω_1 and ω_2, such that the poles of $\mathcal{P}(z)$ are double poles at the lattice points $m\omega_1 + n\omega_2$, $-\infty < m, n < \infty$, and the principal part of $\mathcal{P}(z)$ at $m\omega_1 + n\omega_2$ is $1/(z - m\omega_1 - n\omega_2)^2$. Again our first guess for constructing such a function might be to sum these principal parts over all integers m and n, but the sum does not converge. If we subtract the constant term of the power series expansion at $z = 0$, as in the preceding example, we are led to a series that *does* converge,

$$\mathcal{P}(z) = \frac{1}{z^2} + \sum_{(m,n) \neq (0,0)} \left[\frac{1}{(z - m\omega_1 - n\omega_2)^2} - \frac{1}{(m\omega_1 + n\omega_2)^2} \right].$$

To prove that the series converges normally, we require an estimate for the number $N(k)$ of lattice points $m\omega_1 + n\omega_2$ in a fixed annulus $\{k < |z| \leq k + 1\}$. We claim that $N(k) \leq ck$ for some constant c. To see this, choose $\delta > 0$ such that the distance between any two lattice points is at least 2δ. We consider the $N(k)$ disks of radius δ centered at lattice points in the annulus. Since none of the disks overlap, they cover an area of $\pi\delta^2 N(k)$. We may assume that $\delta < 1$, and then these disks are all contained in the enlarged annulus $\{k - 1 < |z| < k + 2\}$, which has area $6\pi k + 3\pi$. From $\pi\delta^2 N(k) \leq 6\pi k + 3\pi$ we obtain the asserted estimate $N(k) \leq ck$. If now $|z| \leq R$, and if $k > 2R$, then each summand with pole in the kth annulus is estimated easily by C/k^3, where the constant C depends on R but not on k. Since $\sum N(k)C/k^3$ converges, the Weierstrass M-test shows that the series defining $\mathcal{P}(z)$ converges uniformly for $|z| \leq R$. Hence the series converges normally on $\mathbb{C}$ to a meromorphic function. Further, the series converges absolutely at each z, so that we may arrange the terms of the series in any order. For each fixed n the series

$$\sum_{m=-\infty}^{\infty} \frac{1}{(z - m\omega_1 - n\omega_2)^2}$$

converges to a periodic function of z of period ω_1. Adding the appropriate constants and summing over n, we obtain $\mathcal{P}(z + \omega_1) = \mathcal{P}(z)$. Similarly, $\mathcal{P}(z + \omega_2) = \mathcal{P}(z)$, and hence $\mathcal{P}(z)$ is doubly periodic.

Let $f(z)$ be a meromorphic function on the complex plane, with poles at points z_k. A **partial fractions decomposition** for $f(z)$ is a normally converging series expansion $f(z) = \sum P_k(z)$, where $P_k(z)$ is a rational function whose only (finite) pole is at z_k. Thus $P_k(z)$ is the sum of the principal part of $f(z)$ at z_k and a polynomial. A partial fractions decomposition of $f(z)$ is never unique, as we can add any polynomial to one summand and subtract it from another.

Example. We show that $\pi^2 / \sin^2(\pi z)$ has the partial fractions decomposition

$$(2.1) \qquad \frac{\pi^2}{\sin^2(\pi z)} = \sum_{k=-\infty}^{\infty} \frac{1}{(z - k)^2}.$$

Set $f(z) = \sum 1/(z - k)^2$. If $|z| \leq R$, and if $|k| > 2R$, then $|z - k| > |k|/2$, and $1/|z - k|^2 < R/k^2$. By the Weierstrass M-test, the series converges uniformly for $|z| \leq R$. Thus $f(z)$ is a meromorphic function on the complex plane. The functions $f(z)$ and $\pi^2 / \sin^2(\pi z)$ have the same poles and principal parts, so that $f(z) - \pi^2 / \sin^2(\pi z)$ is entire. Evidently, $f(z + 1) = f(z)$, so that $f(z) - \pi^2 / \sin^2(\pi z)$ is periodic with period 1. We may focus on the period strip consisting of $z = x + iy$ for which $0 \leq x \leq 1$. In this strip we have

$$\left| \frac{1}{(z - k)^2} \right| = \frac{1}{y^2 + (k - x)^2} \leq \frac{1}{(k - 1)^2}, \qquad |k| \geq 2, \ |y| \geq 1.$$

By the Weierstrass M-test, the series for $f(z)$ converges uniformly and $f(z)$ is bounded for $|y| \geq 1$ and $0 \leq x \leq 1$. Since each summand tends to 0 as $|y| \to \infty$, the uniform convergence implies that $f(x + iy) \to 0$ as $|y| \to \infty$, $0 \leq x \leq 1$. The identity $|\sin z|^2 = |\sin x|^2 + |\sinh y|^2$ shows that also $\pi^2 / \sin^2(\pi z)$ is bounded for $|y| \geq 1$ and tends to 0 as $|y| \to \infty$. Thus $f(z) - \pi^2 / \sin^2(\pi z)$ is bounded for $|y| \geq 1$, hence bounded on the vertical strip $\{0 \leq x \leq 1\}$, hence by periodicity bounded on the entire complex plane. By Liouville's theorem it is constant. Since both $f(z)$ and $\pi^2 / \sin^2(\pi z)$ tend to 0 as $|y| \to \infty$, the constant is zero, and $f(z)$ coincides with $\pi^2 / \sin^2(\pi z)$.

Example. From the partial fractions decomposition of $\pi^2 / \sin^2(\pi z)$ we can obtain a partial fractions decomposition of $\pi \cot(\pi z)$ by integrating (2.1) term by term. This is justified on account of the normal convergence of the series. Since

$$\int_0^z \frac{d\zeta}{(\zeta - k)^2} = -\left(\frac{1}{z - k} + \frac{1}{k}\right), \qquad k \neq 0,$$

$$\int_0^z \left(\frac{\pi^2}{\sin^2(\pi\zeta)} - \frac{1}{\zeta^2}\right) d\zeta = -\pi \cot(\pi z) + \frac{1}{z},$$

we obtain from (2.1) the partial fractions decomposition

$$\pi \cot(\pi z) = \frac{1}{z} + \sum_{k \neq 0} \left(\frac{1}{z - k} + \frac{1}{k}\right).$$

If we combine the terms for $\pm k$, the constants cancel, and we obtain

$$\pi \cot(\pi z) = \frac{1}{z} + \sum_{k=1}^{\infty} \left(\frac{1}{z - k} + \frac{1}{z + k}\right) = \frac{1}{z} + 2z \sum_{k=1}^{\infty} \frac{1}{z^2 - k^2}.$$

Exercises for XIII.2

1. Use the partial fractions decomposition of $\pi^2/\sin^2(\pi z)$ to establish the formula

$$\sum_{n=1}^{\infty} \frac{1}{n^2} = \frac{\pi^2}{6}.$$

2. Establish the partial fractions decomposition

$$\pi \tan(\pi z) = -2z \sum_{n=0}^{\infty} \frac{1}{z^2 - \left(n + \frac{1}{2}\right)^2}.$$

   Hint. Use $\tan w = \cot w - 2\cot(2w)$.

3. Show that

$$\frac{\pi}{\sin(\pi z)} = \frac{1}{z} + 2z \sum_{n=1}^{\infty} \frac{(-1)^n}{z^2 - n^2}.$$

4. Show that

$$\frac{\pi}{\cos(\pi z)} = \sum_{n=1}^{\infty} \frac{(-1)^n(2n - 1)}{z^2 - \left(n - \frac{1}{2}\right)^2}.$$

5. Let $\{z_k\}$ be a sequence of distinct points such that $|z_k| \to \infty$ and $\sum |z_k|^{-m-1} < \infty$. Show that $z^m \sum 1/z_k^m(z - z_k)$ converges normally to a meromorphic function with principal part $1/(z - z_k)$ at z_k. (If $z_k = 0$, we replace the corresponding summand by $1/z$.)

6. Construct a meromorphic function on the complex plane whose poles are simple poles at the Gaussian integers $m + ni$ with residue 1.

7. Construct a meromorphic function on the complex plane whose poles are simple poles at the points $\log n$, $n \geq 1$, with principal parts $1/(z - \log n)$.

8. Construct a meromorphic function on the open unit disk $\mathbb{D}$ whose poles are simple poles at the points $(1 - 2^{-n})e^{2\pi i k/n}$, $1 \leq k \leq n$, $n \geq 1$, with residue 1.

9. Show that $\sum_{-\infty}^{\infty} 1/(z^3 - n^3)$ converges normally to a meromorphic function. Locate the poles and find the corresponding principal parts of the function. Express the function in terms of trigonometric functions (specifically, the cotangent function).

10. Show that the lattice points $m\omega_1 + n\omega_2$, $-\infty < m, n < \infty$, can be arranged in a sequence $\{z_k\}_{k=0}^{\infty}$ such that $|z_k| \geq c\sqrt{k}$.

11. Let $\{z_k\}$ be a sequence of distinct points such that $|z_k| > c\sqrt{k}$. Show that

$$\sum \left[\frac{1}{(z - z_k)^2} - \frac{1}{z_k^2} \right]$$

converges normally on $\mathbb{C}$ and absolutely at each $z \in \mathbb{C}$.

12. Let $f(z)$ be a doubly periodic meromorphic function on $\mathbb{C}$ with periods ω_1 and ω_2, and let $\mathcal{P}(z)$ be the Weierstrass P-function associated with the periods ω_1 and ω_2. (a) Show that if the only poles of $f(z)$ are double poles at the lattice points $m\omega_1 + n\omega_2$, $-\infty < m, n < \infty$, then there are constants a and b such that $f(z) = a\mathcal{P}(z) + b$. (b) Show that if the only poles of $f(z)$ are triple poles at the lattice points $m\omega_1 + n\omega_2$, $-\infty < m, n < \infty$, then there are constants a, b, c such that $f(z) = a\mathcal{P}(z) + b\mathcal{P}'(z) + c$. (c) Show that $\mathcal{P}'(z)^2 = 4\mathcal{P}(z)^3 + a\mathcal{P}(z)^2 + b\mathcal{P}(z) + c$ for some constants a, b, c.

13. Let D be a domain in $\mathbb{C}$, and let $E_k = \{|z - z_k| \leq r_k\}$, $k \geq 1$, be disjoint closed disks in D that accumulate only on the boundary of D. Suppose $Q_k(z)$ is analytic for $|z - z_k| > r_k$. Show that there is an analytic function $f(z)$ on $D \backslash \cup_{k=1}^{\infty} E_k$ such that for each k, $f(z) - Q_k(z)$ extends analytically to E_k.

3. Infinite Products

An **infinite product** is an expression of the form $\prod_{j=1}^{\infty} p_j$, where the p_j's are complex numbers. We say that the infinite product **converges** if $p_j \to 1$ and $\sum \text{Log } p_j$ converges, where we sum only over terms for which $p_j \neq 0$. If the infinite product converges, we define its value to be 0 if one of the p_j's is 0; otherwise, we define it to be

$$\prod_{j=1}^{\infty} p_j = \exp\left(\sum_{j=1}^{\infty} \operatorname{Log} p_j\right).$$

Thus any question we might ask about infinite products can be translated to a question about infinite series by taking logarithms.

To help clarify the definition of convergent infinite product, we make several simple observations. First, if $\prod p_j$ converges, then at most finitely many of the p_j's can be 0. This is because $p_j \to 1$. Second, if $\prod p_j$ converges, then

$$\prod_{j=1}^{\infty} p_j = \lim_{m \to \infty} \prod_{j=1}^{m} p_j = \lim_{m \to \infty} p_1 p_2 \cdots p_m.$$

This is because $\exp(\operatorname{Log} p_j) = p_j$. Third, we can always factor out a finite number of terms from a convergent infinite product,

$$\prod_{j=1}^{\infty} p_j = p_1 p_2 \cdots p_N \prod_{j=N+1}^{\infty} p_j.$$

Finally, if an infinite product converges, and none of the factors is 0, then the product cannot be 0.

Example. Consider

$$\prod_{k=1}^{\infty} \left(1 + \frac{(-1)^{k+1}}{k}\right) = (1+1)\left(1 - \frac{1}{2}\right)\left(1 + \frac{1}{3}\right)\left(1 - \frac{1}{4}\right)\cdots.$$

Since $(1 + 1/(2k-1))(1 - 1/(2k)) = 1$, the product of the first m terms is equal to 1 if m is even, and it is equal to the last factor $1 + 1/m$ if m is odd. Thus the product converges to 1.

It is often convenient to write $p_j = 1 + a_j$ and to express the product as $\prod(1 + a_j)$. If the product converges, then $a_j \to 0$, and only finitely many of the a_j's are equal to -1. For most purposes we can ignore the terms for which $1 + a_j = 0$ and work with the "tail" $\prod_{j=N}^{\infty}(1 + a_j)$ of the infinite product, for which a_j is near 0.

If $0 < t \le 1$, we have the estimate $t/2 \le \log(1+t) \le t$. From this estimate it follows that if $t_j \ge 0$, then $\sum t_j$ converges if and only if $\sum \operatorname{Log}(1 + t_j)$ converges. This leads immediately to the following test for convergence of infinite products.

Theorem. *If $t_j \ge 0$, then $\prod(1+t_j)$ converges if and only if $\sum t_j$ converges.*

Example. The infinite product

$$\prod_{k=1}^{\infty} \left(1 + \frac{1}{k^\alpha}\right) = (1+1)\left(1 + \frac{1}{2^\alpha}\right)\left(1 + \frac{1}{3^\alpha}\right)\cdots$$

converges for $\alpha > 1$ and diverges for $\alpha \le 1$.

The infinite product $\prod(1+a_j)$ is said to **converge absolutely** if $a_j \to 0$ and $\sum \text{Log}(1 + a_j)$ converges absolutely, where we sum over the terms for which $a_j \ne -1$. If $\prod(1 + a_j)$ converges absolutely, then $\sum \text{Log}(1 + a_j)$ converges, and $\prod(1 + a_j)$ converges.

Since $\text{Log}(1 + w)$ is analytic at $w = 0$ and has power series expansion $\text{Log}(1 + w) = w + \mathcal{O}(w^2)$, we see that $|\text{Log}(1 + a_j)|$ is comparable to $|a_j|$ when a_j is near 0. Consequently, $\sum |\text{Log}(1 + a_j)|$ converges if and only if $\sum |a_j|$ converges. This together with the preceding theorem yield the following.

Theorem. *The infinite product $\prod(1 + a_j)$ converges absolutely if and only if $\sum a_j$ converges absolutely. This occurs if and only if $\prod(1 + |a_j|)$ converges.*

Example. We have seen that $\prod \left(1 + (-1)^{k+1}/k\right)$ converges. However, it does not converge absolutely, on account of the divergence of the harmonic series, $\sum 1/k = \infty$.

Example. Consider the infinite product $\prod(1 + i/k)$. Since $\text{Log}(1 + i/k) = i/k + \mathcal{O}(1/k^2)$, the series $\sum \text{Log}(1 + i/k)$ does not converge, by comparison with the harmonic series. Consequently, $\prod(1 + i/k)$ does not converge. However, since $0 < \text{Log}|1 + i/k| = \frac{1}{2}\log(1 + 1/k^2) < 1/k^2$, the infinite product $\prod |1 + i/k|$ *does* converge. Thus absolute convergence of an infinite product is not equivalent to the convergence of the product of the absolute values of the factors.

Now we turn to infinite products of functions. The Weierstrass M-test for a sum of functions is converted easily to the following test for an infinite product of functions.

Theorem. *Suppose that $g_k(x) = 1 + h_k(x)$, $k \ge 1$, are functions on a set E. Suppose that there are constants $M_k > 0$ such that $\sum M_k < \infty$, and $|h_k(x)| \le M_k$ for $x \in E$. Then $\prod_{k=1}^{m} g_k(x)$ converges to $\prod_{k=1}^{\infty} g_k(x)$ uniformly on E as $m \to \infty$.*

Choose a constant C such that $|\log(1 + w)| \le C|w|$ for $|w| \le \frac{1}{2}$, and choose N such that $M_k \le \frac{1}{2}$ for $k \ge N$. The condition $|h_k(x)| \le M_k \le \frac{1}{2}$ implies that $|\text{Log}(1 + h_k(x))| \le CM_k$. By the Weierstrass M-test, the series $\sum_{k=N}^{\infty} \text{Log}(1 + h_k(x))$ converges uniformly on E. If we exponentiate we obtain uniform convergence of the partial products $\prod_{k=N}^{m} g_k(x)$ to

$\prod_{k=N}^{\infty} g_k(x)$ as $m \to \infty$. Since each $g_k(x)$ is bounded, when we multiply by the first $N - 1$ factors we obtain uniform convergence of $\prod_{k=1}^{m} g_k(x)$ to $\prod_{k=1}^{\infty} g_k(x)$ as $m \to \infty$.

If $G(z) = g_1(z) \cdots g_m(z)$ is a finite product of analytic functions, then by taking logarithms and differentiating, we obtain

(3.1) $$\frac{G'(z)}{G(z)} = \frac{g_1'(z)}{g_1(z)} + \cdots + \frac{g_m'(z)}{g_m(z)}.$$

This procedure is called **logarithmic differentiation**. The logarithmic differentiation formula also holds for uniformly convergent infinite products of analytic functions. It is proved by applying the formula above to finite subproducts and passing to the limit.

Theorem. Let $g_k(z)$, $k \geq 1$, be analytic functions on a domain D such that $\prod_{k=1}^{m} g_k(z)$ converges normally on D to $G(z) = \prod_{k=1}^{\infty} g_k(z)$. Then

(3.2) $$\frac{G'(z)}{G(z)} = \sum_{k=1}^{\infty} \frac{g_k'(z)}{g_k(z)}, \qquad z \in D,$$

where the sum converges normally on D.

Note that the function $G'(z)/G(z)$ has poles at the zeros of $G(z)$. However, the hypothesis implies that $g_k(z) \to 1$ uniformly on any compact subset of D, so the summands $g_k'(z)/g_k(z)$ are analytic on the compact subset for k large. Since the uniform convergence of a series is not affected by the first terms of the series, the poles do not affect the uniform convergence. Since normal convergence of a sequence of analytic functions implies uniform convergence of the derivatives of the functions in the sequence, we may apply (3.1) to the partial product $G_m(z) = g_1(z) \cdots g_m(z)$ and pass to the limit, to obtain (3.2).

Example. Consider the infinite product $f(z) = z \prod_{k=1}^{\infty} (1 - z^2/k^2)$. By the Weierstrass M-test, the series $\sum |z|^2/k^2$ converges uniformly on any bounded set. Thus the infinite product converges uniformly on any bounded set, and $f(z)$ is an entire function. Since the zeros of $f(z)$ are simple zeros at the integers, we suspect that $f(z)$ is related to $\sin(\pi z)$. To check this out, we differentiate logarithmically, to obtain

$$\frac{f'(z)}{f(z)} = \frac{1}{z} + 2z \sum_{k=1}^{\infty} \frac{1}{z^2 - k^2} = \frac{1}{z} + 2z \sum_{k=1}^{\infty} \left[\frac{1}{z - k} + \frac{1}{z + k} \right].$$

This expression we recognize from Section 2 as the partial fractions decomposition of $\pi \cot(\pi z) = \pi \cos(\pi z)/ \sin(\pi z)$. We integrate, and we see that $\log f(z)$ is $\log \sin(\pi z)$ up to adding a constant. Hence $f(z) = C \sin(\pi z)$. Since $f(z)/z \to 1$ as $z \to 0$, we have $C = 1/\pi$. Thus we obtain an infinite product expansion for the sine function,

$$(3.3) \qquad \sin(\pi z) = \pi z \prod_{k=1}^{\infty} \left(1 - \frac{z^2}{k^2}\right) = \pi z \left(1 - z^2\right) \left(1 - \frac{z^2}{4}\right) \cdots .$$

Exercises for XIII.3

1. Evaluate the following.

(a) $\displaystyle\prod_{n=1}^{\infty} \left(1 + \frac{1}{n(n+2)}\right)$ (b) $\displaystyle\prod_{n=2}^{\infty} \left(1 - \frac{1}{n^2}\right)$ (c) $\displaystyle\prod_{n=3}^{\infty} \left(\frac{n^2-1}{n^2-4}\right)$

2. Define $a_k = -\dfrac{1}{\sqrt{k}}$ if k is odd, and $a_k = \dfrac{1}{\sqrt{k}} + \dfrac{1}{k} + \dfrac{1}{k\sqrt{k}}$ if k is even. Show that $\prod(1 + a_k)$ converges, while $\sum a_k$ and $\sum a_k^2$ diverge.

3. Show that if $t_j \geq 0$, then $\prod(1 + t_j) \leq \exp(\sum t_j)$.

4. Show that if $0 < t_j < 1$, then $\prod(1 - t_j)$ converges if and only if $\sum t_j$ converges.

5. Show that the infinite product $\prod(1 + a_j)$ converges if and only if there is $N \geq 1$ such that $\lim_{m \to \infty} \prod_{j=N}^{m}(1 + a_j)$ exists and is nonzero.

6. Show that $\prod(1 + a_j)$ converges if and only if $\prod_{j=m}^{n}(1 + a_j) \to 1$ as $m, n \to \infty$. *Hint.* Take logarithms and invoke the Cauchy criterion for series.

7. Show that if $\prod(1 + a_k)$ converges, then $\prod |1 + a_k|$ converges.

8. Suppose $a_k \to 0$. Show that the series $\sum a_k$ converges absolutely if and only if both the series $\sum \text{Arg}(1 + a_k)$ and $\sum \text{Log} |1 + a_k|$ converge absolutely.

9. Show that $\displaystyle\prod_{n=1}^{\infty} \left(1 + \frac{1}{n^2}\right) = \frac{e^{\pi} - e^{-\pi}}{2\pi}$.

10. Show that $\displaystyle\prod_{n=0}^{\infty} \left(1 + z^{2^n}\right) = \frac{1}{1-z}$ for $|z| < 1$.

11. Show that if $p_k(z)$ is a polynomial of degree k such that $p_k(0) = 1$ and $p_k(z)$ has no zeros in the disk $\{|z| \leq k^3\}$, then $\prod p_k(z)$ converges normally.

12. Establish one of the following formulae, and deduce from it the other using logarithmic differentiation:

$$e^z - 1 = ze^{z/2} \prod_{k=1}^{\infty} \left(1 + \frac{z^2}{4\pi^2 k^2}\right),$$

$$\frac{1}{e^z - 1} = \frac{1}{z} - \frac{1}{2} + 2z \sum_{k=1}^{\infty} \frac{1}{z^2 + 4\pi^2 k^2}.$$

13. Use the infinite product expansion for $\sin(\pi z)$ to show that the **Wallis product**

$$\prod_{k=1}^{\infty} \frac{(2k)^2}{(2k-1)(2k+1)} = \lim_{n \to \infty} \frac{2 \cdot 2 \cdot 4 \cdot 4 \cdot 6 \cdot 6}{1 \cdot 3 \cdot 3 \cdot 5 \cdot 5 \cdot 7} \cdots \frac{(2n) \cdot (2n)}{(2n-1) \cdot (2n+1)}$$

converges to $\pi/2$. Use this to show that

$$\lim_{n \to \infty} \frac{[n!]^2}{(2n)!} \frac{2^{2n}}{\sqrt{n}} = \sqrt{\pi}.$$

14. Show that if $t > 0$, then $\displaystyle\prod_{-m \le k \le tm} \left(1 + \frac{z}{k}\right)$ converges to $\dfrac{\sin(\pi z)}{\pi z} t^z$ as $m \to \infty$.

15. Show that $\dfrac{1}{z} \displaystyle\prod_{n=1}^{\infty} \dfrac{n}{z+n} \left(\dfrac{n+1}{n}\right)^z$ converges to a meromorphic function $\Gamma(z)$ whose poles are simple poles at 0 and the negative integers. Show that

$$\Gamma(z) = \lim_{m \to \infty} \frac{(m-1)! \, m^z}{z(z+1) \cdots (z+m-1)}.$$

Show that $\Gamma(z+1) = z\Gamma(z)$. Show that $\Gamma(n+1) = n!$ for positive integers n. *Remark.* The function $\Gamma(z)$ is called the **gamma function**. It was first introduced by Euler, who defined it to be the limit above. We will give an equivalent definition in the next chapter.

16. Let α_k be a sequence of complex numbers, with possible repetitions, such that $|\alpha_k| < 1$ and $|\alpha_k| \to 1$, and consider the **infinite Blaschke product** defined by

$$B(z) = \prod \frac{\overline{\alpha_k}}{|\alpha_k|} \frac{\alpha_k - z}{1 - \overline{\alpha_k} z},$$

where the factors corresponding to $\alpha_k = 0$ are z.

(a) Suppose that $\sum (1 - |\alpha_k|) < \infty$. Let E be the set of accumulation points on the unit circle $\partial \mathbb{D}$ of the α_k's. Show that the infinite product converges normally on $\mathbb{C}^* \backslash E$ to a meromorphic

function $B(z)$, with the following properties: $|B(z)| < 1$ for
$z \in \mathbb{D}$, $|B(z)| = 1$ for $z \in \partial\mathbb{D}\backslash E$, and $B(z)$ has zeros precisely
at the points α_k.

(b) Show that if $\sum(1 - |\alpha_k|) = +\infty$, then the partial products
converge uniformly on compact subsets of $\mathbb{D}$ to 0.

(c) Suppose that $f(z)$ is a bounded analytic function on $\mathbb{D}$ that is
not identically zero. Show that $f(z)$ has a factorization $f(z) = B(z)g(z)$, where $B(z)$ is a (finite or infinite) Blaschke product,
and $g(z)$ is a bounded analytic function on $\mathbb{D}$ with no zeros. In
particular, the zeros $\alpha_1, \alpha_2, \ldots$ of $f(z)$, repeated according to
multiplicity, satisfy $\sum(1 - |\alpha_k|) < +\infty$.

4. The Weierstrass Product Theorem

The Weierstrass product theorem is a companion theorem to the Mittag-
Leffler theorem. The Mittag-Leffler theorem asserts that we can prescribe
the poles and principal parts of a meromorphic function. The Weierstrass
product theorem asserts that we can prescribe the zeros and poles of a
meromorphic function together with their orders.

Recall that the order of a meromorphic function $f(z)$ at a point z_0 is the
order of the zero if $f(z_0) = 0$, and it is minus the order of the pole if $f(z)$
has a pole at z_0. If z_0 is neither a pole nor a zero of $f(z)$, the order of $f(z)$
at z_0 is defined to be 0.

Theorem (Weierstrass Product Theorem). *Let D be a domain in
the complex plane. Let $\{z_k\}$ be a sequence of distinct points of D with no
accumulation point in D, and let $\{n_k\}$ be a sequence of integers (positive
or negative). Then there is a meromorphic function $f(z)$ on D whose only
zeros and poles are at the points z_k, such that the order of $f(z)$ at z_k is n_k.*

The proof runs parallel to the proof of the Mittag-Leffler theorem. Let
K_m be the set of $z \in D$ such that $|z| \leq m$ and the distance from z to ∂D is
at least $1/m$. Then K_m is a compact subset of D, $K_m \subset K_{m+1}$, and each
component of $\mathbb{C}^*\backslash K_m$ contains a point of $\mathbb{C}^*\backslash D$. Suppose $z_k \in K_{m+1}\backslash K_m$.
We connect z_k to a point $w_k \in C^*\backslash D$ by a simple curve γ_k in $\mathbb{C}^*\backslash K_m$. If
$w_k \neq \infty$ we define $f_k(z)$ to be an analytic branch of $\log((z - z_k)/(z - w_k))$
in the simply connected domain $\mathbb{C}^*\backslash\gamma_k$. If $w_k = \infty$, we take $f_k(z)$ to be an
analytic branch of $\log(1 - z/z_k)$. By Runge's theorem, there is a rational
function $g_k(z)$ with only pole at w_k such that $|f_k(z) - g_k(z)| \leq 2^{-k}/n_k$
on K_m. We consider the product

$$f(z) = \prod_{k=1}^{\infty} \left(\frac{z - z_k}{z - w_k}\right)^{n_k} e^{-n_k g_k(z)},$$

where we replace the factor in parentheses by $1 - z/z_k$ if $w_k = \infty$ and $z_k \neq 0$ and by z^{n_k} if both $w_k = \infty$ and $z_k = 0$. Now $n_k(f_k(z) - g_k(z))$ is not defined on γ_k. However, its exponential

$$\exp[n_k(f_k(z) - g_k(z))] = \left(\frac{z - z_k}{z - w_k}\right)^{n_k} e^{-n_k g_k(z)}$$

is meromorphic on D and has order n_k at z_k and no other poles or zeros. By the Weierstrass M-test, $\sum_{k=N}^{\infty} n_k(f_k(z) - g_k(z))$ converges uniformly on K_m, where N is chosen sufficiently large to exclude terms with $z_k \in K_m$. Hence the infinite product defining $f(z)$ converges normally on D to a meromorphic function. Clearly, $f(z)$ has the desired zeros and poles.

In the case that $D = \mathbb{C}$ is the entire complex plane, the points w_k are all ∞. Suppose $z_k \neq 0$ for $k \geq 1$, and let n_0 be the order (possibly zero) of $f(z)$ at $z = 0$. The product expansion then has the form

$$f(z) = z^{n_0} \prod_{k=1}^{\infty} \left(1 - \frac{z}{z_k}\right)^{n_k} e^{-n_k g_k(z)},$$

where $g_k(z)$ is an approximation to $\mathrm{Log}(1 - z/z_k)$ for $|z| < |z_k|$. We can take $g_k(z)$ to be a partial sum for the power series expansion of $\mathrm{Log}(1 - z/z_k)$ at $z = 0$. Since

$$\mathrm{Log}(1 - \zeta) = -\zeta - \frac{\zeta^2}{2} - \frac{\zeta^3}{3} - \cdots - \frac{\zeta^m}{m} + \mathcal{O}\left(\zeta^{m+1}\right),$$

we can take

$$g_k(z) = -\left(\frac{z}{z_k} + \frac{z^2}{2z_k^2} + \frac{z^3}{3z_k^3} + \cdots + \frac{z^{m_k}}{m_k z_k^{m_k}}\right),$$

where m_k is chosen large enough to guarantee convergence of the product.

Exercise. Find an entire function with simple zeros at the negative integers and no other zeros.
Solution. In this case the product

$$\prod_{k=1}^{\infty} \left(1 + \frac{z}{k}\right)$$

does not converge. However, from the estimate

$$\left|\mathrm{Log}\left(1 + \frac{z}{k}\right) - \frac{z}{k}\right| \leq C\frac{|z|^2}{k^2}, \qquad |z| \leq R, \, k \geq 2R,$$

we see that

(4.1)
$$\prod_{k=1}^{\infty} \left(1 + \frac{z}{k}\right) e^{-z/k}$$

converges normally on the entire complex plane to an entire function with the prescribed zeros.

Exercises for XIII.4

1. Let $G(z)$ be the entire function defined by the infinite product (4.1). Show that $\pi z G(z) G(-z) = \sin(\pi z)$.

2. Construct an entire function that has simple zeros at the points n^2, $n \geq 0$, and no other zeros.

3. Construct an entire function that has simple zeros on the real axis at the points $\pm n^{1/4}$, $n \geq 0$, and no other zeros.

4. Construct an entire function that has simple zeros on the positive real axis at the points $\sqrt{n}$, $n \geq 1$, and double zeros on the imaginary axis at the points $\pm i\sqrt{n}$, $n \geq 1$, and no other zeros.

5. Construct an entire function that has simple zeros at the Gaussian integers $m + ni$, $-\infty < m, n < \infty$, and no other zeros.

6. Find *all* entire functions $f(z)$ that satisfy the functional equation $f(2z) = (1 - 2z)f(z)$. Express the answer in terms of an infinite product.

7. Show that
$$\left| \mathrm{Log}(1 - \zeta) + \zeta + \frac{\zeta^2}{2} + \frac{\zeta^3}{3} + \cdots + \frac{\zeta^N}{N} \right| \leq \frac{1}{N} \frac{|\zeta|^{N+1}}{1 - |\zeta|},$$
for $|\zeta| < 1$.

8. Let $\{z_k\}$ be a sequence of distinct nonzero points such that $|z_k| \to \infty$, Let $N \geq 0$, and let $\{m_k\}$ be a sequence of positive integers such that $\sum m_k |z_k|^{-N-1} < \infty$. Show that
$$\prod_{k=1}^{\infty} \left(1 - \frac{z}{z_k}\right)^{m_k} \exp\left\{ m_k \left[\frac{z}{z_k} + \frac{z^2}{2z_k^2} + \cdots + \frac{z^N}{N z_k^N} \right] \right\}$$
converges normally to an entire function with zeros of order m_k at z_k and no other zeros. *Hint.* See the estimate in the preceding exercise.

9. Show that any meromorphic function $f(z)$ on a domain D is the quotient $f(z) = g(z)/h(z)$ of two analytic functions $g(z)$ and $h(z)$ on D.

10. Let $f(z)$ be a meromorphic function on a simply connected domain D. Show that the meromorphic functions with the same zeros and poles of the same orders as $f(z)$ are precisely the functions of the form $f(z)e^{h(z)}$, where $h(z)$ is analytic on D.

11. Give a brief solution of Exercise 1.5 on interpolating sequences based on the Mittag-Leffler theorem and the Weierstrass product theorem.

XIV

Some Special Functions

Our aim in this chapter is to illustrate the power of complex analysis by proving a deep theorem in number theory, the prime number theorem, which does not appear at first glance to be related to complex analysis. Along the way we introduce various functions that play an important role in complex analysis. In Section 1 we introduce the gamma function $\Gamma(z)$, which provides a meromorphic extension of the factorial function. We derive the asymptotic properties of the gamma function in Section 2 by viewing it as a Laplace transform. This yields Stirling's asymptotic formula for $n!$. In Section 3 we study the zeta function, which is a meromorphic function whose zeros are related to the asymptotic distribution of prime numbers. In Section 4 we study Dirichlet series associated with various number-theoretic functions, thereby giving a strong hint of the fecund relationship between complex analysis and number theory. The proof of the prime number theorem is given in Section 5.

1. The Gamma Function

The gamma function $\Gamma(z)$ is a meromorphic function that arises frequently in complex analysis. It extends the factorial function from the positive integers to the entire complex plane. It is defined in the right half-plane by

$$(1.1) \qquad \Gamma(z) = \int_0^\infty e^{-t} t^{z-1} dt, \qquad \operatorname{Re} z > 0.$$

The integral defining $\Gamma(z)$ is absolutely convergent, and

$$|\Gamma(x + iy)| \leq \Gamma(x) = \int_0^\infty e^{-t} t^{x-1} dt, \qquad x > 0.$$

Since the integrand depends analytically on the parameter z, the function $\Gamma(z)$ is analytic on the right half-plane.

To derive a functional equation for $\Gamma(z)$, we integrate by parts,

$$\Gamma(z+1) = \int_0^\infty e^{-t} t^z dt = -\int_0^\infty t^z d(e^{-t}) = -\left[t^z e^{-t}\right]_0^\infty + z \int_0^\infty e^{-t} t^{z-1} dt.$$

The term $t^z e^{-t}$ vanishes at 0 and at ∞ (technically, we integrate from 0 to R and take a limit as $R \to \infty$), and the identity becomes

$$(1.2) \qquad\qquad \Gamma(z+1) \ = \ z\Gamma(z), \qquad \operatorname{Re} z > 0.$$

To evaluate $\Gamma(z)$ at the positive integers, we begin with $\Gamma(1) = \int_0^\infty e^{-t} dt = 1$, and we use the functional equation to obtain successively $\Gamma(2) = 1$, $\Gamma(3) = 2 \cdot 1$, $\Gamma(4) = 3 \cdot 2 \cdot 1$, and by induction

$$(1.3) \qquad\qquad \Gamma(n+1) \ = \ n!, \qquad n \geq 1.$$

We declare that $0! = \Gamma(1) = 1$, and then (1.3) also holds for $n = 0$.

The functional equation (1.2) allows us to extend $\Gamma(z)$ to the left half-plane, as follows. We apply the functional equation m times to obtain $\Gamma(z+m) = (z+m-1)\cdots(z+1)z\Gamma(z)$, which we rewrite as

$$\Gamma(z) \ = \ \frac{\Gamma(z+m)}{(z+m-1)\cdots(z+1)z}.$$

The right-hand side is defined and meromorphic for $\operatorname{Re} z > -m$, with simple poles at $z = 0, -1, \cdots, -m+1$. By the uniqueness principle, the meromorphic extension is unique and it satisfies the functional equation (1.2). Passing to the limit as $m \to +\infty$, we obtain the following.

Theorem. *The function $\Gamma(z)$ extends to be meromorphic on the entire complex plane, where it satisfies the functional equation $\Gamma(z+1) = z\Gamma(z)$. Its poles are simple poles at $z = 0, -1, -2, \cdots$.*

Now we will express $\Gamma(z)$ in terms of an infinite product. Define

$$\Gamma_n(z) \ = \ \int_0^n t^{z-1}\left(1 - \frac{t}{n}\right)^n dt, \qquad \operatorname{Re} z > 0, \ n \geq 1.$$

Since $(1 - t/n)^n \leq e^{-t}$ and $(1 - t/n)^n \to e^{-t}$ for $t \geq 0$, we have

$$\lim_{n\to\infty} \Gamma_n(z) \ = \ \Gamma(z).$$

(The easiest way to justify passing to the limit is to appeal to Lebesgue's dominated convergence theorem.) The substitution $s = t/n$ yields

$$(1.4) \qquad \Gamma_n(z) \ = \ n^z \int_0^1 s^{z-1}(1 - s)^n ds, \qquad \operatorname{Re} z > 0, \ n \geq 1.$$

Thus

$$\Gamma_1(z) = \int_0^1 s^{z-1}(1 - s)\, ds = \left[\frac{s^z}{z} - \frac{s^{z+1}}{z+1}\right]_0^1 = \frac{1}{z} - \frac{1}{z+1} = \frac{1}{z(z+1)}.$$

If we integrate by parts in (1.4), we obtain

$$\Gamma_n(z) = \frac{n^z}{z} \int_0^1 (1-s)^n d(s^z) = -\frac{n^z}{z} \int_0^1 s^z d(1-s)^n$$

$$= \frac{n^{z+1}}{z} \int_0^1 s^z (1-s)^{n-1} ds = \left(\frac{n}{n-1}\right)^z \frac{n}{z} \Gamma_{n-1}(z+1).$$

We repeat this step $n-1$ times and use the formula for $\Gamma_1(z)$, and we obtain

$$\Gamma_n(z) = \frac{n^z n!}{z(z+1)\cdots(z+n-2)} \Gamma_1(z+n-1) = \frac{n^z n!}{z(z+1)\cdots(z+n)}.$$

We rewrite this as

$$\frac{1}{\Gamma_n(z)} = \frac{1}{n^z} z(1+z)\left(1+\frac{z}{2}\right)\cdots\left(1+\frac{z}{n}\right).$$

We compare this with the infinite product with zeros at the negative integers,

$$G(z) = \prod_{k=1}^{\infty} \left(1+\frac{z}{k}\right) e^{-z/k},$$

which was discussed in Section XIII.4. The exponential convergence factors guarantee that the product converges normally on the entire complex plane. In order to express $1/\Gamma(z)$ in terms of $G(z)$, we write

$$\frac{1}{\Gamma_n(z)} = z e^{\gamma_n z} \prod_{k=1}^{n} \left(1+\frac{z}{k}\right) e^{-z/k},$$

where

$$\gamma_n = 1 + \frac{1}{2} + \cdots + \frac{1}{n} - \log n.$$

We have seen (Section V.1) that the sequence $\{\gamma_n\}$ decreases to Euler's constant $\gamma = 0.5772\ldots$ as $n \to \infty$. Thus when we pass to the limit we obtain the following.

Theorem. *Let γ denote Euler's constant. Then*

(1.5) $$\frac{1}{\Gamma(z)} = z e^{\gamma z} \prod_{k=1}^{\infty} \left(1+\frac{z}{k}\right) e^{-z/k}, \qquad z \in \mathbb{C}.$$

Initially, the identity is established only in the right half-plane. However, both sides are meromorphic, so the identity persists for all $z \in \mathbb{C}$, by the uniqueness principle.

If we take logarithms in (1.5) and differentiate, we obtain

$$-\frac{\Gamma'(z)}{\Gamma(z)} = \frac{1}{z} + \gamma - \sum_{k=1}^{\infty} \frac{z/k^2}{1 + z/k} = \frac{1}{z} + \gamma - z \sum_{k=1}^{\infty} \frac{1}{k} \frac{1}{z + k}.$$

Another differentiation leads to the expression

(1.6) $$\frac{d}{dz} \frac{\Gamma'(z)}{\Gamma(z)} = \sum_{k=0}^{\infty} \frac{1}{(z + k)^2},$$

which we will return to in the next section.

There is another formula we require for later use, which follows from the infinite product expansion derived for $\sin(\pi z)$ in Section XIII.3.3. In terms of the function $G(z)$, it can be expressed (Exercise XIII.4.1) as $\sin(\pi z) = \pi z G(z) G(-z)$. Combining this with (1.5) we obtain

$$\frac{1}{\Gamma(z)\Gamma(-z)} = -z^2 G(z) G(-z) = -\frac{z \sin(\pi z)}{\pi}.$$

Since $\Gamma(1 - z) = -z\Gamma(-z)$, this yields

(1.7) $$\Gamma(z)\Gamma(1 - z) = \frac{\pi}{\sin(\pi z)}, \qquad z \in \mathbb{C}.$$

Exercises for XIV.1

1. Show that $\Gamma(z)$ has no zeros.

2. Show that for $m \geq 0$, the residue of $\Gamma(z)$ at $z = -m$ is $(-1)^m/m!$.

3. Show that $\Gamma(\frac{1}{2}) = \sqrt{\pi}$. Use this to show that

$$\Gamma\left(n + \frac{1}{2}\right) = \frac{1 \cdot 3 \cdot 5 \cdots (2n - 1)}{2^n}\sqrt{\pi} = 2^{1-2n}\frac{\Gamma(2n)}{\Gamma(n)}\sqrt{\pi}.$$

4. Use the Wallis product (Exercise XIII.3.13) to show that

$$\lim_{n \to \infty} 2^{2n-1}\sqrt{n}\,\frac{\Gamma(n)^2}{\Gamma(2n)} = \sqrt{\pi}.$$

5. Prove **Legendre's duplication formula,**

$$\Gamma(2z) = \frac{2^{2z-1}}{\sqrt{\pi}}\Gamma(z)\Gamma\left(z + \frac{1}{2}\right).$$

Hint. Show first that $\Gamma(2z)/\Gamma(z)\Gamma\left(z + \frac{1}{2}\right)$ has the form Ae^{Bz} by separating the product for $\Gamma(2z)$ into even and odd terms; then evaluate A and B.

6. Show that

$$\lim_{n \to \infty} \frac{\Gamma\left(n + \frac{1}{2}\right)}{\sqrt{n}\,\Gamma(n)} = 1.$$

7. Show that

$$\int_0^1 t^{p-1}(1-t)^{q-1}dt = \frac{\Gamma(p)\Gamma(q)}{\Gamma(p+q)}, \qquad p, q > 0.$$

Remark. This integral is called the **beta function**, denoted by $B(p,q)$. To establish the identity, begin by expressing $\Gamma(p)\Gamma(q)$ as a double integral and converting to polar coordinates. The r-integral yields $\Gamma(p+q)$ and the θ-integral leads to the integral above, with $t = \cos^2\theta$.

2. Laplace Transforms

Our aim in this section is to derive an asymptotic approximation to Laplace transforms of functions that do not grow too fast. This leads to an asymptotic approximation to $\Gamma(z)$, thence to Stirling's formula giving an asymptotic approximation to $n!$.

Let $h(s)$, $s \geq 0$, be a continuous or piecewise continuous function on the positive real axis. The **Laplace transform of** $h(s)$ is the function

$$(2.1) \qquad\qquad (\mathcal{L}h)(z) = \int_0^\infty e^{-sz}h(s)\, ds,$$

provided that the integral converges. We will always assume that $h(s)$ has at most exponential growth, that is, there are constants B, C such that

$$|h(s)| \leq Ce^{Bs} \qquad 0 \leq s < \infty.$$

Then the integral in (2.1) converges absolutely and defines an analytic function in the half-plane $\{\operatorname{Re} z > B\}$. The estimate

$$|(\mathcal{L}h)(x+iy)| \leq C\int_0^\infty e^{-xs}e^{Bs}ds = \frac{C}{x-B}, \qquad x > B,$$

shows that $(\mathcal{L}h)(z)$ is bounded on the half-plane $\{\operatorname{Re} z > B + \varepsilon\}$ for any $\varepsilon > 0$, and $(\mathcal{L}h)(z) \to 0$ as $\operatorname{Re} z \to \infty$.

Example. Suppose $h(s) = s^k$ for some integer $k \geq 0$. Since $s^k e^{-\varepsilon s} \to 0$ as $s \to \infty$ (polynomials grow more slowly than exponentials), there is for any $\varepsilon > 0$ an estimate of the form $|h(s)| \leq C_\varepsilon e^{\varepsilon s}$. Hence the Laplace transform $(\mathcal{L}s^k)(z)$ converges absolutely for all z satisfying $\operatorname{Re} z > 0$. We compute that

$$(\mathcal{L}s^k)(x) = \int_0^\infty e^{-xs}s^k ds = \frac{1}{x^{k+1}}\int_0^\infty e^{-t}t^k dt, \qquad x > 0.$$

The integral on the right is $\Gamma(k+1) = k!$, so $(\mathcal{L}s^k)(z)$ coincides with $k!/z^{k+1}$ when $z = x > 0$. By the uniqueness principle it coincides with $k!/z^{k+1}$ on the right half-plane,

$$(2.2) \qquad (\mathcal{L}s^k)(z) = \frac{k!}{z^{k+1}}, \qquad \operatorname{Re} z > 0, \ k \geq 0.$$

Lemma. *Suppose that* $|h(s)| \leq C s^N e^{Bs}$ *for some integer* $N \geq 0$ *and constants* B, C. *Then*

$$(2.3) \qquad |(\mathcal{L}h)(x + iy)| = \mathcal{O}\left(\frac{1}{x^{N+1}}\right), \qquad \text{as } x \to \infty.$$

This follows from the obvious estimates

$$|(\mathcal{L}h)(x+iy)| \leq C \int_0^\infty e^{-(x-B)s} s^N ds = C \frac{N!}{(x-B)^{N+1}} = \mathcal{O}\left(\frac{1}{x^{N+1}}\right).$$

Theorem. *Suppose that* $|h(s)| \leq C e^{Bs}$ *for* $s \geq 0$. *Fix* $N \geq 1$. *If* $h(s)$ *has the Taylor polynomial approximation*

$$(2.4) \qquad h(s) = b_0 + b_1 s + \frac{b_2}{2!} s^2 + \cdots + \frac{b_{N-1}}{(N-1)!} s^{N-1} + \mathcal{O}\left(s^N\right)$$

as $s \to 0+$, *then*

$$(2.5) \qquad (\mathcal{L}h)(z) = \frac{b_0}{z} + \frac{b_1}{z^2} + \frac{b_2}{z^3} + \cdots + \frac{b_{N-1}}{z^N} + \mathcal{O}\left(\frac{1}{x^{N+1}}\right)$$

as $x = \operatorname{Re} z \to \infty$.

To see this, we let $R(s)$ be the remainder term in the Taylor approximation, so that

$$h(s) = b_0 + b_1 s + \cdots + \frac{b_{N-1}}{(N-1)!} s^{N-1} + R(s), \qquad s \geq 0.$$

Since $|R(s)| \leq C_0 s^N$ for $s > 0$ near 0, there is $C_1 > 0$ such that

$$|R(s)| \leq C_1 s^N e^{(B+1)s}, \qquad 0 \leq s < \infty.$$

Since the Laplace transform operates linearly,

$$(\mathcal{L}h)(z) = b_0 (\mathcal{L}1)(z) + b_1 (\mathcal{L}s)(z) + \cdots + \frac{b_{N-1}}{(N-1)!} \left(\mathcal{L}s^{N-1}\right)(z) + (\mathcal{L}R)(z).$$

The result now follows from (2.2) and (2.3).

We wish to use the asymptotic approximation theorem to obtain asymptotic approximations to $\Gamma(z)$. We begin by expressing the derivative of $\Gamma'(z)/\Gamma(z)$ as a Laplace transform. From $(\mathcal{L}s)(z) = 1/z^2$ we obtain

$$\frac{1}{(z+k)^2} = (\mathcal{L}s)(z+k) = \int_0^\infty e^{-s(z+k)} s \, ds = \int_0^\infty e^{-sz} (e^{-s})^k s \, ds.$$

Summing over k from 0 to N, and recognizing the geometric series, we obtain

$$\sum_{k=0}^{N} \frac{1}{(z+k)^2} = \int_0^{\infty} e^{-sz} \sum_{k=0}^{N} (e^{-s})^k s\, ds = \int_0^{\infty} e^{-sz} \frac{1 - e^{-s(N+1)}}{1 - e^{-s}} s\, ds.$$

The exponentials $e^{-s(N+1)}$ decrease to 0, and passage to the limit in the right-hand side is justified. In view of (1.6), we obtain in the limit as $N \to \infty$ that

$$\frac{d}{dz} \frac{\Gamma'(z)}{\Gamma(z)} = \int_0^{\infty} e^{-sz} \frac{s}{1 - e^{-s}} ds, \qquad \operatorname{Re} z > 0.$$

The integral on the right is the Laplace transform of the function

$$g(s) = \frac{s}{1 - e^{-s}}.$$

This function $g(s)$ is a meromorphic function of the complex variable s, with poles at $s = 2\pi n i$ for $n = \pm 1, \pm 2, \ldots$. The function is analytic at $s = 0$, since $1 - e^{-s}$ has a simple zero there. The power series expansion of $g(s)$ about $s = 0$ is readily calculated to be

$$g(s) = 1 + \frac{1}{2}s + \frac{1}{12}s^2 + \frac{g^{(4)}(0)}{4!} s^4 + \cdots, \qquad |s| < 2\pi.$$

Since $g(s)/s \to 1$ as $s \to +\infty$, an estimate of the form $0 \le g(s) \le Ce^s$, $s \ge 0$, holds for some constant $C > 0$. The theorem then yields the asymptotic approximation

$$\frac{d}{dz} \frac{\Gamma'(z)}{\Gamma(z)} = \frac{1}{z} + \frac{1}{2z^2} + \frac{1}{6z^3} + \mathcal{O}\left(\frac{1}{x^5}\right), \qquad x = \operatorname{Re} z \to \infty.$$

To obtain an asymptotic approximation to $\Gamma(z)$, we must integrate twice. Except for a constant of integration, the antiderivative for any error term $\mathcal{O}(1/x^N)$, $N \ge 2$, is dominated by

$$\int_{x+iy}^{\infty} \frac{dt}{|t + iy|^N} \le \int_x^{\infty} \frac{dt}{t^N} = \frac{1}{(N-1)x^{N-1}} = \mathcal{O}\left(\frac{1}{x^{N-1}}\right), \qquad N \ge 2.$$

Hence

$$\frac{\Gamma'(z)}{\Gamma(z)} = \operatorname{Log} z + \alpha - \frac{1}{2z} - \frac{1}{12z^2} + \mathcal{O}\left(\frac{1}{x^4}\right),$$

where α is a constant of integration. If we integrate again, we obtain

$$\log \Gamma(z) = z \operatorname{Log} z - z + \alpha z - \frac{1}{2} \operatorname{Log} z + \beta + \frac{1}{12z} + \mathcal{O}\left(\frac{1}{x^3}\right),$$

where β is another constant of integration. If we substitute the asymptotic expansions for $\log \Gamma(z+1)$ and $\log \Gamma(z)$ into the relation $\log \Gamma(z+1) =$

$\log z + \log \Gamma(z)$, we obtain $\alpha = 0$. To evaluate β, we compare $\Gamma(2n)$ with $\Gamma(n)$, which can be done with the elementary relation in Exercise 1.3,

$$\Gamma(n)\Gamma\left(n + \frac{1}{2}\right) = 2^{1-2n}\Gamma(2n)\sqrt{\pi}.$$

If we take logarithms, substitute the approximations, and do some book-keeping, we obtain $\beta = \log \sqrt{2\pi}$. We have proved the following.

Theorem. *The gamma function $\Gamma(z)$ satisfies*

$$(2.6) \qquad \log \Gamma(z) = \left(z - \frac{1}{2}\right) \operatorname{Log} z - z + \log \sqrt{2\pi} + \frac{1}{12z} + \mathcal{O}\left(\frac{1}{x^3}\right)$$

as $x = \operatorname{Re} z \to \infty$.

The asymptotic approximation to $\Gamma(z)$ is obtained by exponentiating,

$$\Gamma(z) = z^z e^{-z} \sqrt{\frac{2\pi}{z}} \left(1 + \frac{1}{12z} + \mathcal{O}\left(\frac{1}{x^2}\right)\right), \qquad x = \operatorname{Re} z \to \infty.$$

Substituting $n! = n\Gamma(n)$, we obtain **Stirling's formula,**

$$n! = \left(\frac{n}{e}\right)^n \sqrt{2\pi n} \left(1 + \frac{1}{12n} + \mathcal{O}\left(\frac{1}{n^2}\right)\right).$$

It is natural to consider the asymptotic behavior of Laplace transforms in the right half-plane as $x = \operatorname{Re} z \to \infty$. However, for some classes of functions it is more appropriate to consider asymptotic behavior in a sector as $|z| \to \infty$ rather than in a half-plane. Let $f(z)$ be defined and analytic in a sector $S = \{-\theta_0 < \arg z < \theta_1\}$. We say that $\sum_{j=0}^{\infty} a_j/z^j$ is an **asymptotic series** for $f(z)$ in the sector S if for each $N \geq 0$,

$$f(z) - \sum_{j=0}^{N} \frac{a_j}{z^j} = \mathcal{O}\left(\frac{1}{z^{N+1}}\right)$$

as $z \to \infty$ in the sector. If $\sum a_j/z^j$ is an asymptotic series for $f(z)$, then

$$a_k = \lim_{z \to \infty} z^k \left[f(z) - a_0 - \frac{a_1}{z} - \cdots - \frac{a_{k-1}}{z^{k-1}}\right].$$

Hence the coefficients in the asymptotic series of $f(z)$ are uniquely determined. We refer to $\sum a_j/z^j$ as the **asymptotic expansion** of $f(z)$ in the sector.

If $f(z)$ is analytic at ∞, then its Laurent series is an asymptotic series for $f(z)$ in any sector. However, asymptotic series generally do not converge. (See the Exercises.) Further, it can happen that different functions have the same asymptotic expansion. For example, since $z^n e^{-z} \to 0$ as $z \to \infty$ in any sector $\{|\arg z| < (\pi/2) - \varepsilon\}$, the function e^{-z} has an asymptotic series in the sector that is identically zero.

The asymptotic approximation theorem shows that if $h(s)$ has at most exponential growth, and if $h(s)$ is infinitely differentiable at $s = 0$, then the Laplace transform $(\mathcal{L}h)(z)$ has the asymptotic expansion $\sum h^{(k)}(0)/z^{k+1}$ in any sector $S_\varepsilon = \{|\arg z| < (\pi/2) - \varepsilon\}$. In any such sector there is an estimate of the form $c_\varepsilon |z| \leq \operatorname{Re} z \leq |z|$; consequently, $|z| \to \infty$ if and only if $x = \operatorname{Re} z \to \infty$. In general, the series need not be an asymptotic expansion for $(\mathcal{L}h)(z)$ in the entire right half-plane, regarded as a sector. However, in the case of the function $\log \Gamma(z)$, it is. In fact, it can be shown that the series (2.6) is an asymptotic series for $\operatorname{Log} \Gamma(z)$ in any sector $\{|\arg z| < \pi - \varepsilon\}$, for any $\varepsilon > 0$.

Exercises for XIV.2

1. Show that if $\operatorname{Re} \lambda > -1$, then the Laplace transform of s^λ converges absolutely in the right half-plane, and

$$(\mathcal{L}s^\lambda)(z) = \frac{\Gamma(\lambda + 1)}{z^{\lambda+1}}, \qquad \operatorname{Re} z > 0.$$

2. Show that if $g(s) = e^{as} h(s)$, then $(\mathcal{L}g)(z) = (\mathcal{L}h)(z - a)$.

3. Suppose $h(s) = 0$ for $s < 0$. Show that if $a > 0$ and $g(s) = h(s - a)$, then $(\mathcal{L}g)(z) = e^{-az}(\mathcal{L}h)(z)$.

4. Show that if $h(s)$ is differentiable, then the Laplace transform of its derivative $h'(s)$ is $(\mathcal{L}h')(z) = z(\mathcal{L}h)(z) - h(0)$.

5. Compute $\left(\dfrac{n}{e}\right)^n \sqrt{2\pi n}$ and $\left(\dfrac{n}{e}\right)^n \sqrt{2\pi n}\left(1 + \dfrac{1}{12n}\right)$ for $n = 2, 3$, and 4, and compare the results with $n!$.

6. Define the **Bernoulli numbers** $B_1, B_2, B_3, \ldots$ by

$$1 - \frac{z}{2} \cot \frac{z}{2} = \frac{B_1}{2!} z^2 + \frac{B_2}{4!} z^4 + \frac{B_3}{6!} z^6 + \cdots .$$

(a) Show that $B_1 = 1/6$ and $B_2 = 1/30$. (b) Show that

$$\frac{s}{1 - e^{-s}} = 1 + \frac{1}{2}s + \frac{B_1}{2!} s^2 - \frac{B_2}{4!} s^4 + \frac{B_3}{6!} s^6 - \cdots .$$

(c) Show that $\log \Gamma(z)$ has asymptotic series

$$\log \Gamma(z) \sim (z - 1/2) \log z - z + \log \sqrt{2\pi} + \frac{B_1}{2} \frac{1}{z} - \frac{B_2}{3 \cdot 4} \frac{1}{z^3} + \frac{B_3}{5 \cdot 6} \frac{1}{z^5} - \cdots .$$

(d) Show that as $z \to \infty$ through the sector $|\arg z| \leq \pi/2 - \varepsilon$,

$$\Gamma(z) = z^z e^{-z} \sqrt{\frac{2\pi}{z}} \left(1 + \frac{1}{12z} + \frac{1}{288z^2} - \frac{139}{51840z^3} + \mathcal{O}\left(\frac{1}{|z|^4}\right)\right).$$

How many terms can be specified here using only B_1 and B_2?

7. Express the function

$$F(x) = \int_x^\infty \frac{e^{x-t}}{t}\,dt, \qquad x > 0,$$

as a Laplace transform, and show that it extends analytically to the right half-plane. Obtain the asymptotic expansion of the function, and show that the asymptotic expansion diverges at every point.

8. Let $h(s)$, $s \geq 0$, be a continuous function such that $|h(s)| \leq Ce^{Bs}$ for some constants $B, C > 0$. Suppose that $h(s)$ extends to be analytic in a neighborhood of the origin, and suppose that the power series expansion $h(s) = \sum_{k=0}^\infty h^{(k)}(0)s^k/k!$ has only a finite radius of convergence. Show that the asymptotic series $\sum_{k=0}^\infty h^{(k)}(0)/z^{k+1}$ of the Laplace transform $(\mathcal{L}h)(z)$ diverges at every point z.

9. Show that if $f(z)$ is analytic at ∞, then its power series expansion in descending powers of z is an asymptotic series for $f(z)$ in any sector.

10. Show that if $f(z)$ and $g(z)$ have asymptotic expansions in a sector, then so does their product $f(z)g(z)$.

11. Show that if $f(z)$ has an asymptotic series $\sum_{k=2}^\infty a_k/z^k$ in a sector S, beginning with the $1/z^2$ term, then any antiderivative $F(z)$ of $f(z)$ has an asymptotic series in S obtained by integrating term by term.

12. Show that if $f(z)$ has an asymptotic series $\sum_{k=0}^\infty a_k/z^k$ in a sector $S = \{\theta_0 < \arg z < \theta_1\}$, then $f'(z)$ has the asymptotic series $\sum(-ja_j)/z^{j+1}$, obtained by differentiating term by term, in any smaller sector $S_\varepsilon = \{\theta_0 + \varepsilon < \arg z < \theta_1 - \varepsilon\}$.

3. The Zeta Function

The zeta function was first introduced by Euler. He derived an infinite product representation for the zeta function that connected it to prime numbers. A century later Riemann discovered a close relationship between the zeta function and the asymptotic distribution of prime numbers. Except for certain "trivial" zeros on the negative axis, the zeros of the zeta function are easily seen to lie in the "critical" strip $\{0 \leq \operatorname{Re} s \leq 1\}$. Riemann related the zeros of the zeta function to the asymptotic distribution of prime numbers by an "explicit formula," and he conjectured that the zeros in the critical strip all lie on the line $\{\operatorname{Re} s = \frac{1}{2}\}$. At the beginning of the new millennium the most famous unsolved problem in complex analysis, if not in all of mathematics, is to determine whether the Riemann hypothesis holds. We will show in the next section that the zeta function has no zeros on the boundary lines of the critical strip, and this will lead to the prime number theorem.

In this context it is traditional to denote the complex variable by $s = \sigma + it$. With this notation, the **zeta function** is defined by

$$\zeta(s) = \sum_{n=1}^{\infty} \frac{1}{n^s}, \qquad \sigma = \mathrm{Re}\, s > 1.$$

The series converges absolutely for $\sigma > 1$, and it converges uniformly for $\sigma \geq 1 + \varepsilon$ for any $\varepsilon > 0$, by comparison with the numerical series $\sum n^{-(1+\varepsilon)}$. The function $\zeta(\sigma)$ is a decreasing function of σ for $1 < \sigma < \infty$, and $\zeta(\sigma) \to 1$ as $\sigma \to +\infty$. Since the harmonic series $\sum 1/n$ diverges, $\zeta(\sigma) \to +\infty$ as $\sigma \to 1+$. Note also that $|\zeta(\sigma + it)| \leq \zeta(\sigma)$ for all t. We have already met the series for $\zeta(2)$,

$$\zeta(2) = \sum_{n=1}^{\infty} \frac{1}{n^2} = \frac{\pi^2}{6}.$$

For the remainder of this chapter, we will always use p to denote a prime number, $p = 2, 3, 5, 7, 11, \cdots$. The following infinite product representation of the zeta function signals a connection between $\zeta(s)$ and prime numbers.

Theorem. *If* $\mathrm{Re}\, s > 1$, *then*

(3.1)
$$\frac{1}{\zeta(s)} = \prod_p \left(1 - \frac{1}{p^s}\right),$$

where the product is taken over all prime numbers p.

The series $\sum 1/p^{\sigma+it}$ converges absolutely for $\sigma > 1$, and it converges uniformly in any half-plane $\{\sigma \geq 1 + \varepsilon\}$. Hence the product (3.1) converges. Consider the geometric series

$$\frac{1}{1 - p^{-s}} = 1 + p^{-s} + p^{-2s} + \cdots, \qquad \mathrm{Re}\, s > 1.$$

If we multiply together the m series corresponding to primes $p_1, \cdots, p_m$, we obtain

$$\frac{1}{(1 - p_1^{-s}) \cdots (1 - p_m^{-s})} = \sum_{k_1, \dots, k_m = 0}^{\infty} (p_1^{k_1} \cdots p_m^{k_m})^{-s}.$$

Since every integer $n \geq 1$ has a unique representation as a product of powers of distinct primes, a summand $1/n^s$ appears at most once in this sum, so the sum is a subsum of the series $\sum n^{-s}$. As we incorporate more primes into the product, we eventually capture all terms n^{-s}, and in the limit we have

$$\prod_p \frac{1}{1 - p^{-s}} = \sum_{n=1}^{\infty} n^{-s} = \zeta(s).$$

Next we wish to extend $\zeta(s)$ to be a meromorphic function on the entire complex plane. To do this, we represent $\zeta(s)$ as a contour integral. We make a branch cut along the positive real axis and consider the function

$$(-z)^{s-1} = e^{(s-1)\,\mathrm{Log}(-z)}, \qquad z \in \mathbb{C}\backslash[0,\infty),$$

which depends analytically on both z and the complex parameter s. Let γ be the contour indicated below, following the top edge of the branch cut from $+\infty$ to ε, detouring along a circle of radius ε around $z = 0$, and returning from ε to $+\infty$ along the bottom edge of the branch cut. We consider the function

$$(3.2) \qquad\qquad \phi(s) = \frac{1}{2\pi i} \int_\gamma \frac{(-z)^{s-1}}{e^z - 1}\,dz.$$

The integral converges, and it represents an entire function of s. To evaluate it, we first assume that $\mathrm{Re}\ s > 1$, and we express

$$\phi(s) = \frac{1}{2\pi i} \int_\infty^\varepsilon \frac{e^{(s-1)(\log x - i\pi)}}{e^x - 1}\,dx + \frac{1}{2\pi i} \int_{|z|=\varepsilon} \frac{(-z)^{s-1}}{e^z - 1}\,dz$$
$$+ \frac{1}{2\pi i} \int_\varepsilon^\infty \frac{e^{(s-1)(\log x + i\pi)}}{e^x - 1}\,dx.$$

Since $e^z - 1$ has a simple zero at $z = 0$, the integrand is bounded on the circle $|z| = \varepsilon$ by $C\varepsilon^{\mathrm{Re}\ s-2}$, and the second integral is bounded by $C\varepsilon^{\mathrm{Re}\ s-1}$. Under our assumption that $\mathrm{Re}\ s > 1$, this integral tends to 0 with ε. Passing to the limit. we obtain

$$\phi(s) = \frac{1}{2\pi i} \left(e^{i\pi(s-1)} - e^{-i\pi(s-1)} \right) \int_0^\infty \frac{x^{s-1}}{e^x - 1}\,dx.$$

The term in parentheses is $2i\sin(\pi(s-1)) = -2i\sin(\pi s)$. Hence

$$(3.3) \qquad\qquad \phi(s) = -\frac{\sin(\pi s)}{\pi} \int_0^\infty \frac{x^{s-1}}{e^x - 1}\,dx.$$

We treat this integral by expanding $1/(e^x - 1) = e^{-x}/(1 - e^{-x})$ in a geometric series,

$$(3.4) \qquad \int_0^\infty \frac{x^{s-1}}{e^x - 1}\,dx = \int_0^\infty \left(\sum_{n=1}^\infty e^{-nx} \right) x^{s-1}\,dx.$$

The partial sums of the geometric series form an increasing sequence of functions that converges uniformly on each interval $[\varepsilon, \infty)$. From this it is straightforward to justify interchanging the summation and integration. Since

$$\int_0^\infty e^{-nx} x^{s-1}\,dx = \frac{1}{n^s} \int_0^\infty e^{-t} t^{s-1}\,dt = \frac{\Gamma(s)}{n^s},$$

(3.4) becomes

$$\int_0^\infty \frac{x^{s-1}}{e^x - 1} dx = \Gamma(s)\zeta(s), \qquad \text{Re}\, s > 1.$$

Substituting this into (3.3) and using $\Gamma(s)\Gamma(1-s) = \pi/\sin(\pi s)$ from (1.7), we obtain

$$\phi(s) = -\frac{\sin(\pi s)}{\pi}\Gamma(s)\zeta(s) = -\frac{\zeta(s)}{\Gamma(1-s)}.$$

We have established this identity for $\text{Re}\, s > 1$. However, $\phi(s)$ is an entire function, so we can use this identity to extend $\zeta(s)$ to be a meromorphic function in the entire complex plane. The extended function then satisfies the following.

Theorem. For the branch of $(-z)^{s-1}$ and the contour γ described above, we have

$$(3.5) \qquad \zeta(s) = -\Gamma(1-s)\phi(s) = -\frac{\Gamma(1-s)}{2\pi i}\int_\gamma \frac{(-z)^{s-1}}{e^z - 1} dz.$$

Since $\phi(s)$ is an entire function, and the poles of $\Gamma(1-s)$ are simple poles at $s = 1, 2, 3, \ldots$, $\zeta(s)$ is analytic except for possibly a simple pole at $s = 1$. Now, $\Gamma(s)$ has a simple pole at $s = 0$ with residue 1, so $\Gamma(1-s) \sim 1/(1-s)$ as $s \to 1$. Since

$$\phi(1) = \frac{1}{2\pi i}\int_\gamma \frac{dz}{e^z - 1} = \text{Res}\left[\frac{1}{e^z - 1}, 0\right] = 1,$$

we see that $\zeta(s) \sim 1/(s-1)$ as $s \to 1$. Thus $\zeta(s)$ has a simple pole with residue 1 at $s = 1$. We collect these observations.

Theorem. The zeta function $\zeta(s)$ is a meromorphic function on the complex plane with only one pole, a simple pole at $s = 1$ with residue 1.

Next we derive a functional equation for the zeta function.

Theorem. The function $\zeta(s)$ satisfies the functional equation

$$(3.6) \qquad \zeta(s) = 2^s \pi^{s-1} \sin\left(\frac{\pi s}{2}\right)\Gamma(1-s)\zeta(1-s).$$

For the proof, we modify the contour γ used to define $\phi(s)$. Fix s to be real and negative. Let γ_n be the contour indicated in the figure below, obtained from γ by replacing the segment starting at n on the top edge of the slit and going around 0 and back along the bottom edge of the slit to n

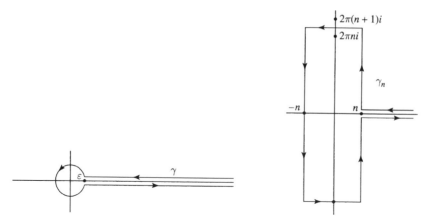

by straight line segments running counterclockwise around the boundary of the rectangle R_n with vertices at $\pm n \pm (2n+1)\pi i$. Define

$$\phi_n(s) \;=\; \frac{1}{2\pi i} \int_{\gamma_n} \frac{(-z)^{s-1}}{e^z - 1} dz.$$

Since $|e^z - 1| > \frac{1}{2}$ on the edges of the rectangle, the integrand is bounded on the edges by $2n^{s-1}$. By the ML-estimate, the integrals over these edges are bounded by Cn^s. Since $s < 0$, this tends to 0 as $n \to \infty$, and consequently, $\phi_n(s) \to 0$ as $n \to \infty$. Now, the difference $\int_{\gamma_n} - \int_{\gamma}$ is the integral around the closed contour consisting of the boundary of the rectangle R_n with a detour along an elongated indentation that does not cross over the branch cut. The contour encloses only the poles of the integrand at the points $\pm 2\pi k i$, $0 < k \leq n$. Each pole is a simple pole, with residue

$$\left. \frac{(-z)^{s-1}}{e^z} \right|_{z=\pm 2\pi k i} \;=\; (2\pi)^{s-1} |k|^{s-1} e^{(s-1)\,\mathrm{Log}(\pm i)}.$$

If we combine the residues for $\pm k$ and sum, we obtain from the residue theorem

$$\phi_n(s) - \phi(s) \;=\; (2\pi)^{s-1} \left(e^{(s-1)i\pi/2} + e^{-(s-1)i\pi/2} \right) \sum_{k=1}^{n} k^{s-1}.$$

We substitute

$$e^{(s-1)i\pi/2} + e^{-(s-1)i\pi/2} \;=\; 2\cos\left((s-1)\frac{\pi}{2} \right) \;=\; 2\sin\left(\frac{\pi s}{2} \right)$$

and let $n \to \infty$. Since $\phi_n(s) \to 0$, we obtain in the limit

$$-\phi(s) \;=\; 2^s \pi^{s-1} \sin\left(\frac{\pi s}{2} \right) \zeta(1-s).$$

If we combine this with (3.5), we obtain the functional equation (3.6), at least for $s < 0$. Since both sides of the identity (3.6) are meromorphic, (3.6) holds for all s, by the uniqueness principle.

The functional equation of the zeta function yields information on its zeros. On account of the product representation (3.1), the zeta function has no zeros in the half-plane $\{\text{Re}\,s > 1\}$. Since $1/\Gamma(z)$ has a product representation expressing it as an entire function, $\Gamma(z)$ has no zeros. The functional equation then shows that the only zeros of $\zeta(s)$ in the left half-plane $\{\text{Re}\,s < 0\}$ are the zeros of $\sin(\pi s/2)$, which are simple zeros at $s = -2, -4, -6, \ldots$. As mentioned above, these zeros are called the **trivial zeros** of the zeta function. The nontrivial zeros of the zeta function lie in the strip $\{0 \le \text{Re}\,s \le 1\}$, which is called the **critical strip**. Many outstanding mathematicians have worked on the **Riemann hypothesis**, that the nontrivial zeros of the zeta function lie on the line $\{\text{Re}\,s = \frac{1}{2}\}$. It is known that $\zeta(s)$ has infinitely many zeros in the critical strip. Their asymptotic distribution is known, and it is known that at least a third of them lie on the critical line $\{\text{Re}\,s = \frac{1}{2}\}$. Further, the first zillion or so zeros are known to lie on the critical line. Meanwhile, whether the Riemann hypothesis is true remains a famous unresolved problem.

Exercises for XIV.3

1. Show that if a is a zero of the zeta function in the critical strip, then so are $\bar{a}$, $1 - a$, and $1 - \bar{a}$.

2. Show that the function $\xi(s) = \frac{1}{2}s(1-s)\pi^{-s/2}\Gamma(s/2)\zeta(s)$ is an entire function that satisfies $\xi(1 - s) = \xi(s)$. Show that $\xi(s)$ is real when s is real and also when $\text{Re}\,s = \frac{1}{2}$. Where are the zeros of $\xi(s)$?

3. By comparing $\zeta(\sigma)$ with $\int_1^\infty x^{-\sigma}dx$, show directly from the definition of $\zeta(s)$ that $(\sigma - 1)\zeta(\sigma) \to 1$ as $\sigma \to 1+$.

4. Show that $(1 - 2^{1-s})\zeta(s)$ is an entire function, which is represented by the series

$$(1 - 2^{1-s})\zeta(s) = 1 - \frac{1}{2^s} + \frac{1}{3^s} - \frac{1}{4^s} + \cdots, \qquad \text{Re}\,s > 1.$$

Remark. We will see in the next section that the series converges and represents the function for $\text{Re}\,s > 0$.

5. By comparing the partial fractions decomposition

$$\pi \cot(\pi z) = \frac{1}{z} + \sum_{n \ne 0}\left(\frac{1}{z - n} + \frac{1}{n}\right)$$

with the series defining the Bernoulli numbers, and differentiating, derive the power series expansion

$$\sum_{n \ne 0}\frac{1}{(z - n)^2} = \sum_{k=0}^{\infty}\frac{2^{2k+1}\pi^{2k+2}B_{k+1}}{(2k)!(k + 1)}z^{2k}, \qquad |z| < 1.$$

Use this to show that

$$\zeta(2m) = \frac{2^{2m-1}\pi^{2m}B_m}{(2m)!}, \qquad m \geq 1.$$

4. Dirichlet Series

A **Dirichlet series** is a series of the form

(4.1)
$$\sum_{n=1}^{\infty} \frac{a_n}{n^s} = a_1 + \frac{a_2}{2^s} + \frac{a_3}{3^s} + \frac{a_4}{4^s} + \cdots,$$

which we will often abbreviate to $\sum a_n n^{-s}$. The prototypical Dirichlet series is the series $\sum n^{-s}$ representing the zeta function. Dirichlet series are important in number theory. They were used by Dirichlet to study the distribution of prime numbers in arithmetic progressions.

The modulus $|a_n n^{-(\sigma+it)}| = |a_n| n^{-\sigma}$ of the nth term of the series depends only on the real part σ of $s = \sigma + it$. The modulus decreases as σ increases. Thus if the series (4.1) converges absolutely for $s = s_0$, then it converges absolutely for all s satisfying $\operatorname{Re} s \geq \operatorname{Re} s_0$. By the Weierstrass M-test, it also converges uniformly in the half-plane $\{\operatorname{Re} s \geq \operatorname{Re} s_0\}$, by comparison with the numerical series $\sum |a_n| n^{-\operatorname{Re} s_0}$. Proceeding in analogy with the definition of radius of convergence of a power series, we define σ_a to be the infimum of σ for which $\sum |a_n| n^{-\sigma}$ converges, and we have the following.

Theorem. *For each Dirichlet series* (4.1), *there is a unique extended real number* σ_a, $-\infty \leq \sigma_a \leq +\infty$, *such that the series converges absolutely if* $\operatorname{Re} s > \sigma_a$, *and the series does not converge absolutely if* $\operatorname{Re} s < \sigma_a$. *For any* $\sigma_0 > \sigma_a$, *the series converges uniformly on the half-plane* $\{\operatorname{Re} s \geq \sigma_0\}$. *The sum* $f(s) = \sum a_n n^{-s}$ *is analytic for* $\operatorname{Re} s > \sigma_a$.

The extended real number σ_a is called the **abscissa of absolute convergence** of the series (4.1).

Example. The series $\sum n^{-s}$ for $\zeta(s)$ has abscissa of absolute convergence $\sigma_a = 1$. It does not converge at $s = 1$.

Suppose the function $f(s)$ is represented by a Dirichlet series, say $f(s) = \sum a_n n^{-s}$ for $\operatorname{Re} s > \sigma_a$. Since the series converges uniformly in some half-plane, and each of the terms $a_n n^{-s} \to 0$ as $\sigma \to \infty$ for $n \geq 2$, we have

$$\sum_{n=2}^{\infty} \frac{a_n}{n^s} \to 0 \quad \text{as} \quad \sigma = \operatorname{Re} s \to \infty.$$

Consequently, $f(s) \to a_1$ as $\sigma \to \infty$. Similarly, we can capture the other coefficients of the series from $f(s)$. We write

$$(4.2) \qquad n^s \left[f(s) - a_1 - \frac{a_2}{2^s} - \cdots - \frac{a_{n-1}}{(n-1)^s} \right] = a_n + \sum_{k=n+1}^{\infty} a_k \frac{n^s}{k^s}.$$

To estimate the series on the right, choose σ_0 large so that $\sum |a_k| k^{-\sigma_0} = M < \infty$. Then

$$\left| \sum_{k=n+1}^{\infty} a_k \frac{n^s}{k^s} \right| \leq \sum_{k=n+1}^{\infty} \frac{n^\sigma}{k^{\sigma-\sigma_0}} \frac{|a_k|}{k^{\sigma_0}} \leq \frac{n^\sigma}{(n+1)^{\sigma-\sigma_0}} M.$$

Since $n^\sigma / (n+1)^{\sigma-\sigma_0} \to 0$ as $\sigma \to \infty$, the series on the right-hand side of (4.2) tends to 0 as $\sigma \to \infty$, and the left-hand side of (4.2) tends to a_n as $\sigma \to \infty$. Thus starting with a_1, we can determine in succession the coefficients a_n from $f(s)$. In particular, the coefficients a_n are determined uniquely by the function $f(s)$, and we have proved the first statement of the following theorem.

Theorem. If an analytic function $f(s)$ can be represented by a Dirichlet series in some half-plane $\{\operatorname{Re} s > \sigma_0\}$, then that Dirichlet series is unique. If $f(s) = \sum a_n n^{-s}$ and $g(s) = \sum b_n n^{-s}$ are represented by Dirichlet series in a half-plane, then so are their sum $f(s) + g(s)$ and their product $f(s)g(s)$. Further, the Dirichlet series of the product $f(s)g(s)$ is $\sum c_n n^{-s}$, where

$$c_n = \sum_{d|n} a_d b_{n/d}.$$

Here the notation "$d|n$" means "d divides n." The theorem follows by multiplying the series for $f(s)$ and $g(s)$ term by term and regrouping the terms, which we can do for absolutely convergent series.

The relation between convergence and absolute convergence of Dirichlet series is more delicate than the corresponding relationship for power series. It may occur that a Dirichlet series converges not only for $\sigma > \sigma_a$, but also for $\sigma > \sigma_a - 1$.

Example. The Dirichlet series $\sum (-1)^n n^{-s}$ has $\sigma_a = 1$. However, it converges whenever $s = \sigma$ is real and positive, by the alternating series test. The following theorem shows that it converges at every point of the right half-plane $\{\operatorname{Re} s > 0\}$.

Theorem. If the Dirichlet series $\sum a_n n^{-s}$ converges at a point $s = s_0$, then it converges for all s such that $\operatorname{Re} s > \operatorname{Re} s_0$. Further, it converges uniformly in any sector $\{|\arg(s - s_0)| \leq (\pi/2) - \varepsilon\}$ with vertex s_0 and aperture strictly less than π.

This is proved using Abel's formula for summation by parts. Let $\sum a_k$ be a series, with partial sums

$$A_n = \sum_{k=1}^{n} a_k.$$

The formula for summation by parts is analogous to the formula for integration by parts,

$$(4.3) \qquad \sum_{k=m}^{n} a_k b_k = A_n b_n - A_{m-1} b_m + \sum_{k=m}^{n-1} A_k (b_k - b_{k+1}).$$

This can be checked by substituting $a_k = A_k - A_{k-1}$ and adjusting the indices.

Before applying the formula, we make two preliminary reductions in order to simplify the proof. First, observe that $\sum a_n n^{-s}$ converges at $s = s_1$ if and only if $\sum (a_n n^{-s_0}) n^{-s}$ converges at $s = s_1 - s_0$. Thus by replacing $\sum a_n n^{-s}$ by $\sum b_n n^{-s}$, where $b_n = a_n n^{-s_0}$, we can assume that $s_0 = 0$ and that $\sum a_n$ converges. The second reduction is to replace a_1 by $a_1 - \sum a_n$, so that we can assume that $\sum a_n = 0$. We must show that for any $C > 0$, the series $\sum a_n n^{-s}$ converges uniformly for $|s| \leq C\sigma$, where $s = \sigma + it$ and $\sigma > 0$. We do this by invoking the summation by parts formula, which in this case becomes

$$(4.4) \qquad \sum_{k=m}^{n} \frac{a_k}{k^s} = \frac{A_n}{n^s} - \frac{A_{m-1}}{m^s} + \sum_{k=m}^{n-1} A_k \left(\frac{1}{k^s} - \frac{1}{(k+1)^s} \right).$$

Let ε_m be the maximum of $|A_k|$ for $k \geq m - 1$. Then $\varepsilon_m \to 0$ as $m \to \infty$. From (4.4) we obtain

$$(4.5) \qquad \left| \sum_{k=m}^{n} \frac{a_k}{k^s} \right| \leq \varepsilon_m \left(\frac{1}{n^\sigma} + \frac{1}{m^\sigma} + \sum_{k=m}^{n-1} \left| \frac{1}{k^s} - \frac{1}{(k+1)^s} \right| \right).$$

The first two summands are each bounded by 1, and

$$\left| \frac{1}{k^s} - \frac{1}{(k+1)^s} \right| = |s| \left| \int_k^{k+1} r^{-s-1} dr \right| \leq C\sigma \int_k^{k+1} r^{-\sigma-1} dr.$$

If we substitute these estimates in (4.5) and sum, we obtain

$$\left| \sum_{k=m}^{n} \frac{a_k}{k^s} \right| \leq \varepsilon_m \left(2 + C\sigma \int_m^n r^{-\sigma-1} dr \right) \leq \varepsilon_m (2 + C).$$

Since this tends to 0 as $m, n \to \infty$, the series converges uniformly in the sector, and consequently it converges pointwise in the open half-plane. This completes the proof of the theorem.

The preceding theorem allows us to define the **abscissa of convergence** of a Dirichlet series $\sum a_n n^{-s}$ to be the extended real number σ_c, $-\infty \leq$

$\sigma_c \leq +\infty$, such that the series converges if $\mathrm{Re}\, s > \sigma_c$ and does not converge if $\mathrm{Re}\, s < \sigma_c$. Clearly, $\sigma_c \leq \sigma_a$. It is easy to see (Exercise 3) that $\sigma_a \leq \sigma_c + 1$, and further, this estimate is sharp.

Example. Consider the series representation from Exercise 3.4,

$$(4.6) \qquad \left(1 - 2^{1-s}\right)\zeta(s) \;=\; 1 - \frac{1}{2^s} + \frac{1}{3^s} - \frac{1}{4^s} + \cdots, \qquad \mathrm{Re}\, s > 1.$$

The abscissa of absolute convergence of the series on the right is $\sigma_a = 1$. By the alternating series test, the series converges for any positive real number $s > 0$. Consequently, the abscissa of convergence is $\sigma_c = 0$, and the series converges in the right half-plane to an analytic function. By the uniqueness principle, the identity (4.6) then holds in the entire right half-plane.

Exercises for XIV.4

1. Show using (4.6) that $\zeta(\sigma) < 0$ for $0 < \sigma < 1$. Conclude that the only zeros of the zeta function on the real line are the trivial zeros at the points $s = -2, -4, -6, \dots$.

2. Show that if the Dirichlet series $f(s) = \sum a_n n^{-s}$ converges in some half-plane, and if the a_n's are not all zero, then there is a half-plane $\{\mathrm{Re}\, s > \sigma_0\}$ on which $f(s)$ has no zeros.

3. Show that if the terms $a_n n^{-s}$ are bounded for $s = s_0$, then the Dirichlet series $\sum a_n n^{-s}$ converges absolutely for $\mathrm{Re}\, s > 1 + \mathrm{Re}\, s_0$. Deduce that $\sigma_a \leq \sigma_c + 1$.

4. A coefficient sequence $\{a_n\}$ is **multiplicative** if $a_1 = 1$ and $a_{mn} = a_m a_n$ whenever m and n are relatively prime. It is **strongly multiplicative** if $a_1 = 1$ and $a_{mn} = a_m a_n$ for all m and n. Show that if $\{a_n\}$ is multiplicative, then

$$\sum_{n=1}^{\infty} \frac{a_n}{n^s} \;=\; \prod_p \left(1 + a_p p^{-s} + a_{p^2} p^{-2s} + \cdots\right), \qquad \mathrm{Re}\, s > \sigma_a,$$

while if $\{a_n\}$ is strongly multiplicative, then

$$\sum_{n=1}^{\infty} \frac{a_n}{n^s} \;=\; \prod_p \left(1 - a_p p^{-s}\right)^{-1}, \qquad \mathrm{Re}\, s > \sigma_a.$$

5. The **Möbius μ-function** is defined on positive integers n by $\mu(1) = 1$, $\mu(n) = (-1)^r$ if n is a product of r distinct primes, and $\mu(n) = 0$ otherwise. (a) Show that μ is multiplicative but not strongly multiplicative. (b) Show that

$$\frac{1}{\zeta(s)} \;=\; \sum_{n=1}^{\infty} \frac{\mu(n)}{n^s}, \qquad \mathrm{Re}\, s > 1.$$

(c) Show directly, or from the uniqueness of Dirichlet series, that

$$\sum_{d|n} \mu(d) = 0, \qquad n \geq 2.$$

(d) Show that if $f(s) = \sum a_n n^{-s}$ where $\{a_n\}$ is strongly multiplicative, then $1/f(s) = \sum a_n \mu(n) n^{-s}$.

6. The **Dirichlet convolution** of the sequences $\{a_n\}_{n=1}^{\infty}$ and $\{b_n\}_{n=1}^{\infty}$ is the sequence $\{c_n\}_{n=1}^{\infty}$ defined by

$$c_n = \sum_{d|n} a_d b_{n/d}.$$

Show that if the Dirichlet series $\sum a_n n^{-s}$ and $\sum b_n n^{-s}$ converge in some half-plane to $f(s)$ and $g(s)$ respectively, then the Dirichlet series $\sum c_n n^{-s}$ converges in some half-plane to $f(s)g(s)$.

7. Let $\varphi(n)$ be **Euler's totient function**, defined to be the number of positive integers not exceeding n that are relatively prime to n. Show that

$$\frac{\zeta(s-1)}{\zeta(s)} = \sum_{n=1}^{\infty} \frac{\varphi(n)}{n^s}, \qquad \mathrm{Re}\, s > 2.$$

Hint. Use the fact that $\sum_{d|n} \varphi(d) = n$.

8. If $\sigma_a(n)$ denotes the sum of the ath powers of the divisors of n, then

$$\zeta(s)\zeta(s-a) = \sum_{n=1}^{\infty} \frac{\sigma_a(n)}{n^s}, \qquad \mathrm{Re}\, s > 1,\ \mathrm{Re}\, s > a+1.$$

9. Let $d(n)$ denote the number of divisors of n. Establish the following formulae:

$$\zeta(s)^2 = \sum_{n=1}^{\infty} \frac{d(n)}{n^s}, \qquad \frac{\zeta(s)^3}{\zeta(2s)} = \sum_{n=1}^{\infty} \frac{d(n^2)}{n^s}, \qquad \frac{\zeta(s)^4}{\zeta(2s)} = \sum_{n=1}^{\infty} \frac{d(n)^2}{n^s}.$$

Hint. Use Exercise 4.

10. Show that

$$\frac{\zeta'(s)}{\zeta(s)} = -\sum_{n=2}^{\infty} \frac{\Lambda(n)}{n^s}, \qquad \mathrm{Re}\, s > 1,$$

where $\Lambda(n) = \log p$ if n is a power of a prime number p, and $\Lambda(n) = 0$ otherwise.

11. Define $a_n = 1$ and $b_n = \pm 1$ if $n \equiv \pm 1 \mod 6$; otherwise, set $a_n = b_n = 0$. Define two "Dirichlet L-series" by

$$L_1(s) = \sum \frac{a_n}{n^s} = 1 + \frac{1}{5^s} + \frac{1}{7^s} + \frac{1}{11^s} + \cdots,$$

$$L_{-1}(s) = \sum \frac{b_n}{n^s} = 1 - \frac{1}{5^s} + \frac{1}{7^s} - \frac{1}{11^s} + \cdots.$$

(a) Show that $\{a_n\}$ and $\{b_n\}$ are strongly multiplicative. (See Exercise 4.)

(b) Show that $L_1(s)$ has $\sigma_a = \sigma_c = 1$, and that $L_1(s) \to +\infty$ as $s \to 1+$.

(c) Show that $L_{-1}(s)$ has $\sigma_a = 1$ and $\sigma_c = 0$.

(d) Show that $\log L_1(s) - \log L_{-1}(s)$ can be expressed as

$$\sum_{p \equiv -1 \mod 6} \log(1 + p^{-s}) - \sum_{p \equiv -1 \mod 6} \log(1 - p^{-s}),$$

and deduce from (b) that there are infinitely many primes congruent to -1 modulo 6.

(e) Show that $\log L_1(s) + \log L_{-1}(s)$ can be expressed as

$$-2 \sum_{p \equiv 1 \mod 6} \log(1 - p^{-s}) - \sum_{p \equiv -1 \mod 6} \log(1 - p^{-2s}),$$

and deduce that there are infinitely many primes congruent to 1 modulo 6.

Remark. Dirichlet showed using this technique that if $1 \le k < m$, and if k is relatively prime to m, then the arithmetic progression $\{k, k+m, k+2m, \dots\}$ contains infinitely many primes. Further, for fixed m the primes are distributed asymptotically evenly among such arithmetic progressions.

12. Suppose that the partial sums of the Dirichlet series $\sum a_n n^{-s}$ are bounded at $s = s_0$. Show that the series converges for $\operatorname{Re} s > \operatorname{Re} s_0$.

13. Suppose $a_n \ge 0$ for all n, and the abscissa of absolute convergence σ_a for the Dirichlet series $f(s) = \sum a_n n^{-s}$ is finite. Show that $f(s)$ cannot be extended analytically to any neighborhood of σ_a.

14. Denote by σ_c and σ_a the abscissas of convergence and of absolute convergence, respectively, for the Dirichlet series $\sum a_n n^{-s}$. Show that

$$\sigma_a - 1 \le \limsup_{n \to \infty} \frac{\log |a_n|}{\log n} \le \sigma_c.$$

15. Denote by σ_c the abscissa of convergence of the Dirichlet series $\sum a_n n^{-s}$. (a) Show that if $\sum a_n$ converges, and $R_n = a_n + a_{n+1} +$

$\cdots$, then

$$\sigma_c = \limsup_{n \to \infty} \frac{\log |R_n|}{\log n}.$$

(b) Show that if $\sum a_n$ does not converge, and $A_n = a_1 + a_2 + \cdots + a_n$, then

$$\sigma_c = \limsup_{n \to \infty} \frac{\log |A_n|}{\log n}.$$

5. The Prime Number Theorem

We say that two functions $f(x)$ and $g(x)$ are **asymptotic** as $x \to \infty$, written $f \sim g$, if $f(x)/g(x) \to 1$ as $x \to \infty$. Let $\pi(x)$ be the number of prime numbers p such that $p \le x$. Thus $\pi(2) = 1$, $\pi(6) = 3$, $\pi(10) = 4$, and so on. The prime number theorem gives the asymptotic behavior of $\pi(x)$ as $x \to \infty$.

Theorem (Prime Number Theorem). *The number $\pi(x)$ of prime numbers not exceeding x satisfies $\pi(x) \sim x/\log x$ as $x \to \infty$.*

This formula for the asymptotic distribution of primes was surmised by several outstanding mathematicians, including Gauss. Tables of factorizations of the first several hundred thousand positive integers were already published before 1700. Gauss (or, more accurately, his assistant) calculated all primes up to about 3,000,000, and Gauss used the results to aid his researches into number theory.

After much effort by many mathematicians, the prime number theorem was proved in 1896 independently by Hadamard and de la Vallée Poussin. Their proofs were very difficult. Subsequently, several new proofs of the prime number theorem have been given, and the various proofs have been gradually simplified. We will follow D. Zagier's exposition of a proof due to D.J. Newman (1972). As with most proofs of the prime number theorem, the proof is divided into two parts. The first is to show that the zeta function has no zeros on the line $\{\operatorname{Re} s = 1\}$. The second is to establish a "Tauberian theorem" relating the zeta function and the distribution of primes.

An important role in the proof is played by the function $\Phi(s)$ defined by the Dirichlet series

$$(5.1) \qquad\qquad \Phi(s) = \sum_p \frac{\log p}{p^s}, \qquad \operatorname{Re} s > 1,$$

where as usual we use p for prime numbers. The series converges absolutely, and $\Phi(s)$ is analytic for $\operatorname{Re} s > 1$. If we differentiate logarithmically the product formula (3.1) for the zeta function, we obtain

$$(5.2) \qquad -\frac{\zeta'(s)}{\zeta(s)} \;=\; \sum_p \frac{\log p}{p^s - 1} \;=\; \Phi(s) + \sum \frac{\log p}{p^s(p^s - 1)}.$$

The series on the right converges to an analytic function for $\operatorname{Re} s > \frac{1}{2}$. Since $\zeta'(s)/\zeta(s)$ is meromorphic on the entire complex plane, (5.2) shows that $\Phi(s)$ extends to be meromorphic for $\{\operatorname{Re} s > \frac{1}{2}\}$, with simple poles at the poles and zeros of the zeta function. Thus $\Phi(s)$ has a simple pole at $s = 1$ with residue 1, and $\Phi(s)$ has a simple pole at $s = s_0$ with residue $-m$ if $\zeta(s)$ has a zero at $s = s_0$ of order m.

Theorem. *The meromorphic function $\Phi(s)$ has no poles on the vertical line $\{\operatorname{Re} s = 1\}$ except at $s = 1$. The zeta function $\zeta(s)$ has no zeros on the line $\{\operatorname{Re} s = 1\}$.*

The proof is by magic. Fix $t > 0$, and suppose that $\zeta(s)$ has order q at $1 + it$ and order q' at $1 + 2it$, where $q, q' \geq 0$. Then it has orders q and q' at $1 - it$, and $1 - 2it$, respectively. Consider

$$0 \leq \left(p^{it/2} + p^{-it/2}\right)^4 = p^{2it} + 4p^{it} + 6 + 4p^{-it} + p^{-2it}.$$

Multiplying by $\varepsilon(\log p)/p^{1+\varepsilon}$ and summing over the primes, we obtain

$$0 \leq \varepsilon[\Phi(1 + \varepsilon + 2it) + 4\,\Phi(1 + \varepsilon + it) + 6\Phi(1 + \varepsilon)$$
$$+ 4\Phi(1 + \varepsilon - it) + \Phi(1 + \varepsilon - 2it)].$$

Since the residues of $\Phi(s)$ at the points $1, 1\pm it, 1\pm 2it$ are respectively $1, -q, -q'$, we have $\varepsilon\Phi(1+\varepsilon) \to 1$, $\varepsilon\Phi(1+\varepsilon\pm it) \to -q$, and $\varepsilon\Phi(1+\varepsilon\pm 2it) \to -q'$ as $\varepsilon \to 0$. In the limit we then obtain

$$0 \leq -q' - 4q + 6 - 4q - q' \leq 6 - 8q - 2q'.$$

Since q and q' are nonnegative integers, this can occur only when $q = 0$. Thus $1 + it$ is not a zero of $\zeta(s)$ nor a pole of $\Phi(s)$. It follows that $\Phi(s)$ has no poles on the line $1 + i\mathbb{R}$ except $s = 1$, and $\zeta(s)$ has no zeros on the line.

None of the proofs of the prime number theorem treats the function $\pi(x)$ directly. They treat the function

$$(5.3) \qquad\qquad\qquad \vartheta(x) \;=\; \sum_{p \leq x} \log p.$$

The function $\vartheta(x)$ is an increasing, piecewise continuous function of x. Since there are at most $\pi(x)$ summands in (5.3),

$$(5.4) \qquad\qquad 0 \leq \vartheta(x) \leq \pi(x)\log x, \qquad x \geq 1.$$

For fixed $\varepsilon > 0$, we have

$$\vartheta(x) \geq \sum\{\log p : x^{1-\varepsilon} < p \leq x\} \geq (1 - \varepsilon)(\log x)(\pi(x) - \pi(x^{1-\varepsilon}))$$
$$\geq (1 - \varepsilon)(\log x)(\pi(x) - x^{1-\varepsilon}).$$

Combining this with (5.4), we obtain

$$\frac{\vartheta(x)}{x} \leq \pi(x)\frac{\log x}{x} \leq \frac{1}{1 - \varepsilon}\frac{\vartheta(x)}{x} + \frac{\log x}{x^{\varepsilon}}.$$

Since $(\log x)/x^{\varepsilon} \to 0$ as $x \to \infty$, we see from this that $\vartheta(x)/x \to 1$ if and only if $\pi(x)(\log x)/x \to 1$. We have proved the following.

Lemma. *The prime number theorem holds if and only if $\vartheta(x) \sim x$.*

It is clear that $\vartheta(x) \leq x \log x$. We need the following stronger estimate for the growth of $\vartheta(x)$, which goes back to Chebyshev.

Lemma. *$\vartheta(x) \leq Cx$ for some constant C and $x \geq 0$.*

To see this, we consider the binomial coefficient

$$\binom{2n}{n} = \frac{(2n)!}{(n!)^2} < 2^{2n}.$$

Each prime number p between n and $2n$ divides the binomial coefficient. Consequently, their product does, and their product is bounded by 2^{2n}, so

$$\sum\{\log p : n < p < 2n\} \leq 2n \log 2.$$

Applying this estimate to $n = 2^k$ and summing, we obtain

$$\vartheta(2^m) = \sum_{k=1}^{m}\sum\{\log p : 2^{k-1} < p < 2^k\}$$

$$\leq 2(1 + 2^2 + \cdots + 2^{m-1})\log 2 < 2^{m+1}\log 2.$$

Given x, we choose m such that $2^{m-1} < x \leq 2^m$, and then $\vartheta(x) \leq \vartheta(2^m) \leq 2^{m+1}\log 2 \leq (4\log 2)x$.

Next we connect $\Phi(s)$ and $\vartheta(t)$ via the Laplace transform. This requires a logarithmic change of variable, to convert the multiplication on $(0, \infty)$ to addition on $(-\infty, \infty)$. We consider the function $t \mapsto \vartheta(e^t)$, which is a piecewise continuous function of t.

Lemma. *The Laplace transform $(\mathcal{L}\vartheta(e^t))(s)$ of the function $\vartheta(e^t)$ converges absolutely for $\operatorname{Re} s > 1$, and*

$$(\mathcal{L}\vartheta(e^t))(s) = \frac{\Phi(s)}{s}, \qquad \operatorname{Re} s > 1.$$

Since $\vartheta(e^t) \leq Ce^t$, the Laplace transform of $\vartheta(e^t)$ converges absolutely for $\operatorname{Re} s > 1$. Let p_n denote the nth prime. Then $\vartheta(e^t)$ is constant for $\log p_n < t < \log p_{n+1}$, so

$$\int_{\log p_n}^{\log p_{n+1}} e^{-st}\vartheta(e^t)\,dt \;=\; \vartheta(p_n)\frac{e^{-st}}{-s}\Big|_{\log p_n}^{\log p_{n+1}} \;=\; \frac{1}{s}\vartheta(p_n)\left(p_n^{-s} - p_{n+1}^{-s}\right).$$

We sum over the primes and use $\vartheta(p_n) - \vartheta(p_{n-1}) = \log p_n$ to obtain

$$\int_0^\infty e^{-st}\vartheta(e^t)\,dt = \frac{1}{s}\sum \vartheta(p_n)\left(p_n^{-s} - p_{n+1}^{-s}\right) = \frac{1}{s}\sum(\vartheta(p_n) - \vartheta(p_{n-1}))p_n^{-s}$$

$$= \frac{1}{s}\sum \frac{\log p_n}{p_n^s} = \frac{1}{s}\Phi(s), \qquad \operatorname{Re} s > 1.$$

Instead of dealing with the Laplace transform of $\vartheta(e^t)$, it will be convenient to treat the Laplace transform of the function

$$(5.5) \qquad\qquad\qquad h(t) \;=\; \vartheta(e^t)e^{-t} - 1.$$

The effect of multiplying by e^{-t} is to translate the variable s of the Laplace transform by 1, and the effect of subtracting 1 is to kill the pole of $\Phi(s)$ at $s = 1$.

Lemma. The function $h(t)$ defined by (5.5) is a bounded piecewise continuous function on $[0,\infty)$, whose Laplace transform converges absolutely in the right half-plane and satisfies

$$(5.6) \qquad\qquad (\mathcal{L}h)(s) \;=\; \frac{\Phi(s+1)}{s+1} - \frac{1}{s}, \qquad \operatorname{Re} s > 0.$$

The Laplace transform $(\mathcal{L}h)(s)$ extends analytically across the imaginary axis.

The link between the analyticity of $\Phi(s)$ and the asymptotic behavior of $h(t)$ is contained in the following "Tauberian theorem."

Theorem. Let $h(t)$, $t \geq 0$, be a bounded piecewise continuous function whose Laplace transform $(\mathcal{L}h)(z)$ extends analytically across the imaginary axis. Then

$$\lim_{T\to\infty} \int_0^T h(t)\,dt \;=\; \lim_{\varepsilon\to 0+} (\mathcal{L}h)(\varepsilon).$$

Note that the Laplace transform $(\mathcal{L}h)(z)$ is defined only for $\operatorname{Re} z > 0$, where the defining integral converges absolutely. Let $g(z)$ denote the analytic continuation of $(\mathcal{L}h)(z)$ across the imaginary axis. The theorem asserts that the Laplace transform exists, possibly as an improper integral, at $z = 0$, and is equal to $g(0)$.

For the proof, let $\varepsilon > 0$ be small. For fixed $T > 0$, define

$$g_T(z) = \int_0^T e^{-zt} h(t)\, dt, \qquad z \in \mathbb{C}.$$

Then $g_T(z)$ is an entire function. We will show that $|g(0) - g_T(0)| < \varepsilon$ for large T.

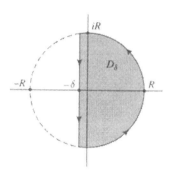

Suppose $|h(t)| \le M$ for $0 \le t < \infty$. Choose $R > 0$ so large that $M/R < \varepsilon/4$, and then choose $\delta > 0$ very small, so that $g(z)$ is defined and analytic in and across the boundary of the domain D_δ consisting of z such that $|z| < R$ and $\operatorname{Re} z > -\delta$. From Cauchy's theorem we have

$$(5.7) \qquad g(0) - g_T(0) = \frac{1}{2\pi i} \int_{\partial D_\delta} (g(z) - g_T(z)) e^{zT} \left(1 + \frac{z^2}{R^2}\right) \frac{dz}{z}.$$

We break ∂D_δ into pieces. Let γ_+ be the semicircle $\{|z| = R,\ \operatorname{Re} z > 0\}$, let α_δ be the vertical interval along ∂D_δ where $\operatorname{Re} z = -\delta$, and let β_δ be the two short arcs of the circle $\{|z| = R\}$ connecting α_δ to $\pm iR$, all oriented according to the orientation of ∂D_δ. For $z = x + iy \in \gamma_+$ we have

$$|g(z) - g_T(z)| = \left| \int_T^\infty h(t) e^{-zt} dt \right| \le M \int_T^\infty e^{-xt} dt = \frac{M e^{-xT}}{x}.$$

Since also $|1 + z^2/R^2| = 2|x|/R$ for $|z| = R$, we obtain from the ML-estimate that the right-hand side of (5.7), integrated over γ_+, is bounded by

$$\left| \frac{1}{2\pi i} \int_{\gamma_+} (g(z) - g_T(z)) e^{zT} \left(1 + \frac{z^2}{R^2}\right) \frac{dz}{z} \right|$$

$$\le \frac{1}{2\pi} \frac{M e^{-xT}}{x} \cdot e^{xT} \cdot \frac{2x}{R} \cdot \frac{1}{R} \cdot \pi R = \frac{M}{R} < \frac{\varepsilon}{4}.$$

For the integral of $g(z) - g_T(z)$ over the remainder of ∂D_δ, we treat the functions separately. Let γ_- be the semicircle $\{|z| = R,\ \operatorname{Re} z < 0\}$. Since

$g_T(z)$ is entire,

$$\frac{1}{2\pi i}\int_{\alpha_\delta \cup \beta_\delta} g_T(z)e^{zT}\left(1+\frac{z^2}{R^2}\right)\frac{dz}{z} = \frac{1}{2\pi i}\int_{\gamma_-} g_T(z)e^{zT}\left(1+\frac{z^2}{R^2}\right)\frac{dz}{z},$$

and we estimate the left-hand side just as for the first integral. If $z = x + iy \in \gamma_-$, then

$$|g_T(z)| = \left|\int_0^T h(t)e^{-zt}\,dt\right| \le M\int_0^T e^{-xt}\,dt < \frac{Me^{-xT}}{|x|},$$

so

$$\left|\frac{1}{2\pi i}\int_{\gamma_-} g_T(z)e^{zT}\left(1+\frac{z^2}{R^2}\right)\frac{dz}{z}\right| \le \frac{M}{R} < \frac{\varepsilon}{4}.$$

Now $|e^{zT}| \le 1$ on β_δ and the length of β_δ tends to 0 with δ, so we can choose $\delta > 0$ so small that for any $T > 0$,

$$\left|\frac{1}{2\pi i}\int_{\beta_\delta} g(z)e^{zT}\left(1+\frac{z^2}{R^2}\right)\frac{dz}{z}\right| < \frac{\varepsilon}{4}.$$

Finally, $|e^{zT}| = e^{-\delta T}$ on α_δ, so for T large the integral over α_δ is also bounded by $\varepsilon/4$. Adding the four estimates, we obtain $|g(0) - g_T(0)| < \varepsilon$ for T large, as required.

If we apply the Tauberian theorem to the function $h(t)$ defined by (5.5), we conclude that

$$\lim_{T\to\infty}\int_0^T \left[\vartheta(e^t)e^{-t} - 1\right]\,dt$$

exists. After a change of variable $x = e^t$, we find that

$$(5.8) \qquad\qquad \lim_{R\to\infty}\int_1^R \left(\frac{\vartheta(x)}{x} - 1\right)\frac{dx}{x}$$

exists. We complete the proof of the prime number theorem by showing that this implies $\vartheta(x) \sim x$.

Lemma. *If $\vartheta(x)$ is an increasing function of x such that the limit (5.8) exists, then $\vartheta(x) \sim x$.*

Suppose that $\vartheta(x) > (1 + \varepsilon)x$ for some $\varepsilon > 0$ and for arbitrarily large values of x. For any such x we have

$$\int_x^{(1+\varepsilon)x} \left(\frac{\vartheta(t)}{t} - 1 \right) \frac{dt}{t} \geq \int_x^{(1+\varepsilon)x} \left(\frac{\vartheta(x)}{t} - 1 \right) \frac{dt}{t}$$

$$\geq \int_x^{(1+\varepsilon)x} \left[(1 + \varepsilon)\frac{x}{t} - 1 \right] \frac{dt}{t}$$

$$= \int_1^{1+\varepsilon} \left[\frac{1 + \varepsilon}{r} - 1 \right] \frac{dr}{r} = c > 0.$$

Thus

$$\int_1^R \left(\frac{\vartheta(t)}{t} - 1 \right) \frac{dt}{t}$$

increases by c over infinitely many disjoint intervals of the form $(x, (1+\varepsilon)x)$, so the integral cannot have a limit as $R \to \infty$. A similar calculation shows that we cannot have $\vartheta(x) < (1 - \varepsilon)x$ for arbitrarily large values of x. Consequently, $\vartheta(x)/x \to 1$ as $x \to \infty$.

Exercises for XIV.5

1. Why is the binomial coefficient "m choose k" less than 2^m?

2. Show that $|1 + z^2/R^2| = 2|\operatorname{Re} z|/R$ for $|z| = R$.

3. Compute $\pi(x)$ for $x = 10$ and $x = 100$, and compare the result with $x/\log x$.

4. Show that $\pi(x) \sim \int_2^x (1/\log t)dt$. *Remark.* This asymptotic formula (with equivalent notation) was found scribbled on the back page of a table of logarithms that Gauss was given at age fourteen.

5. Show that if p_n is the nth prime, then $p_n \sim n \log n$ as $n \to \infty$.

6. Show that $\sum 1/p = +\infty$, where the sum is over the primes, by filling in the details of the following elementary argument. If the sum converges, choose N such that the tail $S = \sum_{k>N} 1/p_k$ satisfies $S < 1$. Show that $S + S^2 + S^3 + \cdots$ dominates the sum of $1/n$ over positive integers not divisible by $p_1, p_2, \ldots, p_N$; hence it dominates $\sum_{m=1}^{\infty} 1/(1 + mp_1 p_2 \cdots p_N)$, which diverges.

7. Exactly where in the proof of the prime number theorem is the Chebyshev estimate $\vartheta(x) \leq Cx$ absolutely required? Can you modify the proof so that it is not required?

8. Let $h(t)$, $0 \leq t < \infty$, be a piecewise continuous function such that

$$H(T) = \int_0^T e^{-z_0 t} h(t) \, dt, \qquad T \geq 0,$$

is bounded. Show that $(\mathcal{L}h)(z) = \int_0^\infty e^{-zt} h(t)\, dt$ exists as an improper integral for all z satisfying $\operatorname{Re} z > \operatorname{Re} z_0$. *Hint.* Show that

$$\int_0^T e^{-zt} h(t)\, dt = e^{-(z-z_0)T} H(T) + (z - z_0) \int_0^T e^{-(z-z_0)t} H(t)\, dt.$$

XV

The Dirichlet Problem

In Chapter X we used the Poisson kernel to solve the Dirichlet problem for the unit disk. In this chapter we study the Dirichlet problem for more general domains in the plane. The basic method, due to O. Perron, is to look for the solution of the Dirichlet problem as the upper envelope of a family of subsolutions. In Section 2 we introduce subharmonic functions, which play the role of the subsolutions. In Section 3 we derive Harnack's inequality, which provides a compactness criterion for families of harmonic functions. Perron's procedure for solving the Dirichlet problem is developed in Section 4. We apply the method to sketch another proof of the Riemann mapping theorem in Section 5. In Sections 6 and 7 we introduce Green's function.

1. Green's Formulae

In this section we will establish three fundamental formulae for dealing with harmonic and subharmonic functions. All are consequences of Green's theorem. We assume that the boundary curves and functions we treat are smooth enough so that Green's theorem is valid.

Let γ be a smooth curve in the plane. As before, we denote by $\mathbf{n}$ the unit normal vector to the curve. If γ is parameterized by arc length, $\gamma(s) = (x(s), y(s))$, then $\mathbf{n} = (dy/ds, -dx/ds)$. We denote the derivative of a smooth function $v(x, y)$ in the normal direction by $\partial v/\partial n$. Thus

$$\frac{\partial v}{\partial n} = \nabla v \cdot \mathbf{n} = \frac{\partial v}{\partial x}\frac{dy}{ds} - \frac{\partial v}{\partial y}\frac{dx}{ds}, \qquad \frac{\partial v}{\partial n}\, ds = \frac{\partial v}{\partial x}\, dy - \frac{\partial v}{\partial y}\, dx.$$

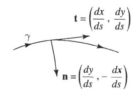

Let D be a bounded domain with piecewise smooth boundary, and let u and v be smooth functions on $D \cup \partial D$. We consider integrals of the form

$$\int_{\partial D} u \frac{\partial v}{\partial n} ds = \int_{\partial D} -u \frac{\partial v}{\partial y} dx + u \frac{\partial v}{\partial x} dy.$$

By Green's theorem, this is equal to

$$\iint_D \left(\frac{\partial u}{\partial x} \frac{\partial v}{\partial x} + u \frac{\partial^2 v}{\partial x^2} + \frac{\partial u}{\partial y} \frac{\partial v}{\partial y} + u \frac{\partial^2 v}{\partial y^2} \right) dx \, dy.$$

Rewriting this in terms of the gradient and Laplacian operators, we obtain **Green's first formula**,

$$(1.1) \qquad \int_{\partial D} u \frac{\partial v}{\partial n} ds = \iint_D \nabla u \cdot \nabla v \, dx \, dy + \iint_D u \Delta v \, dx \, dy.$$

If we interchange u and v and subtract, the gradient terms cancel, and we obtain **Green's second formula**,

$$(1.2) \qquad \int_{\partial D} \left(u \frac{\partial v}{\partial n} - v \frac{\partial u}{\partial n} \right) ds = \iint_D (u \Delta v - v \Delta u) \, dx \, dy.$$

We will be particularly interested in this identity when one or both of u and v are harmonic. Then the corresponding Laplacian term is zero, and the formula simplifies. In the very special case that u is harmonic and $v \equiv 1$, the formula reduces to $\int_{\partial D} \partial u / \partial n \, ds = 0$, which was used in Chapter III.

Suppose $v(z)$ is harmonic in a punctured neighborhood of a point ζ. We say that $v(z)$ has a **logarithmic pole** at ζ if $v(z) - \log(1/|z - \zeta|)$ is harmonic at ζ. The following theorem provides a version of Green's second formula for functions with a logarithmic pole, which is analogous to Pompeiu's formula (Section IV.8) for the operator $\partial/\partial z$. It is sometimes referred to as **Green's third formula**.

Theorem. *Let D be a bounded domain with piecewise smooth boundary, and let u be a smooth function on $D \cup \partial D$. Let $\zeta \in D$, and let $v(z)$ be a harmonic function on $D \backslash \{\zeta\}$ such that v extends smoothly to ∂D and such that $v(z) - \log 1/|z - \zeta|$ is harmonic at ζ. Then*

$$(1.3) \qquad u(\zeta) = -\frac{1}{2\pi} \iint_D v \Delta u \, dx \, dy - \frac{1}{2\pi} \int_{\partial D} \left(u \frac{\partial v}{\partial n} - v \frac{\partial u}{\partial n} \right) ds.$$

Note that by Green's second formula, the right-hand side of (1.3) is zero if we replace $v(z)$ by a function that is harmonic on all of D. Thus we can add to $v(z)$ any harmonic function without affecting the right-hand side. We add the harmonic function $-v(z) + \log(1/|z - \zeta|)$ to $v(z)$, and we see that it suffices to establish (1.3) when $v(z)$ is replaced by $\log(1/|z - \zeta|)$.

So suppose $v(z) = \log(1/|z - \zeta|)$. We follow the idea of the proof of Pompeiu's theorem. (The picture is the same as the picture for the proof of

Cauchy's integral formula in Section IV.4.) We excise a small disk centered at ζ from D and set $D_\varepsilon = D\backslash\{|z - \zeta| \leq \varepsilon\}$, and we apply Green's second formula to D_ε. Taking into account $\Delta v = 0$ on D_ε, we obtain

$$\int_{\partial D} \left(u\frac{\partial v}{\partial n} - v\frac{\partial u}{\partial n} \right) ds - \int_{|z|=\varepsilon} \left(u\frac{\partial v}{\partial n} - v\frac{\partial u}{\partial n} \right) ds = -\iint_{D_\varepsilon} v\,\Delta u\,dx\,dy.$$

Here we have reversed the orientation of the circle $\{|z-\zeta| = \varepsilon\}$ to the usual counterclockwise orientation, so now for $z = \zeta + re^{i\theta}$ we have $\partial/\partial n = \partial/\partial r$. The logarithmic singularity at ζ is absolutely integrable,

$$\iint_{|z-\zeta|<1} \log\frac{1}{|z - \zeta|}\,dx\,dy = \int_0^{2\pi}\int_0^1 \left(\log\frac{1}{r}\right) r\,dr\,d\theta < \infty.$$

Consequently, the area integral over D_ε tends to the area integral over D as $\varepsilon \to 0$. To establish (1.3), it remains to show that

$$(1.4) \qquad\qquad -\int_{|z-\zeta|=\varepsilon} \left(u\frac{\partial v}{\partial n} - v\frac{\partial u}{\partial n} \right) ds \;\to\; 2\pi u(\zeta).$$

If $|(\nabla u)(z)| \leq M$ for z near ζ, then

$$\left| \int_{|z-\zeta|=\varepsilon} v\frac{\partial u}{\partial n}\,ds \right| \leq \left(\log\frac{1}{\varepsilon} \right) M \cdot 2\pi\varepsilon,$$

which tends to 0 with ε. Since $\dfrac{\partial v}{\partial n} = \dfrac{\partial}{\partial r}\log\dfrac{1}{r} = -\dfrac{1}{r}$ and $ds = r\,d\theta$, the other summand in (1.4) is

$$-\int_{|z-\zeta|=\varepsilon} u\frac{\partial v}{\partial n}\,ds = \int_0^{2\pi} u(\zeta + \varepsilon e^{i\theta})\,d\theta.$$

Since the averages of $u(z)$ around small circles centered at ζ tend to $u(\zeta)$, we see that this tends to $2\pi u(\zeta)$ as $\varepsilon \to 0$. This establishes (1.4) and with it the formula (1.3).

Exercises for XV.1

1. Show that $v(z)$ has a logarithmic pole at ζ if and only if $v(z) = \log|f(z)|$ for some function $f(z)$ that is meromorphic near ζ and has a simple pole at ζ.

2. Let D be a bounded domain with smooth boundary, and let $u(z)$ be a smooth function on $D \cup \partial D$ such that $u(z) = 0$ for $z \in \partial D$. Show that $\iint_D u(z)(\Delta u)(z)\,dx\,dy \leq 0$, with strict inequality unless $u(z) = 0$ for all $z \in D$.

3. Let D be a bounded domain with smooth boundary, and let λ be a real number. Suppose there is a nonzero smooth function $u(z)$ on

$D \cup \partial D$ such that $\Delta u = \lambda u$ on D and $u = 0$ on ∂D. Show that $\lambda < 0$.

4. Let D be a bounded domain with smooth boundary, and let $u(z)$ and $v(z)$ be smooth functions on $D \cup \partial D$ such that $u(z)$ is harmonic on D and $v(z) = 0$ on ∂D. Show that $\iint_D \nabla u \cdot \nabla v \, dx \, dy = 0$.

5. Let D be a bounded domain with smooth boundary, and let $u(z)$ and $v(z)$ be smooth functions on $D \cup \partial D$ such that $u(z)$ is harmonic on D and $v(z) = u(z)$ on ∂D. Show that

$$\iint_D |\nabla v|^2 dx \, dy \;=\; \iint_D |\nabla u|^2 dx \, dy + \iint_D |\nabla(v - u)|^2 dx \, dy.$$

Remark. The quadratic form $\iint |\nabla v|^2 dx \, dy$ is called the **Dirichlet integral**. The identity shows that the solution to the Dirichlet problem minimizes the Dirichlet integral. This is an instance of **Dirichlet's principle**.

6. Let D be a bounded domain with smooth boundary, and let h be a continuous function on ∂D. The **Neumann problem** is to find a harmonic function u on D such that $\partial u / \partial n = h$ on ∂D. Show that the Neumann problem is not solvable unless $\int_{\partial D} h \, ds = 0$.

7. Let $h\left(e^{i\theta}\right)$ be a smooth (say twice continuously differentiable) real-valued function on the unit circle, with Fourier series $h\left(e^{i\theta}\right) = \sum_{-\infty}^{\infty} b_n e^{in\theta}$. Show that the Neumann problem is solvable on the unit disk with boundary function $h\left(e^{i\theta}\right)$ if and only if $b_0 = 0$. Express the solution explicitly as a Fourier series, and comment on the uniqueness of the solution.

8. Let $h\left(e^{i\theta}\right)$ be a smooth real-valued function on the boundary of the annulus $\{\rho < |z| < \sigma\}$. Show that the Neumann problem is solvable with boundary function $h(z)$ if and only if $\rho \int_0^{2\pi} h\left(\rho e^{i\theta}\right) d\theta = \sigma \int_0^{2\pi} h\left(\sigma e^{i\theta}\right) d\theta$. Express the solution explicitly as a Fourier series. (See Problem VI.1.6.)

9. Let $u(z)$ be a smooth function on the complex plane that is zero outside of some compact set. Show that

$$u(\zeta) \;=\; -\frac{1}{2\pi} \iint_{\mathbb{C}} (\Delta u)(z) \log \frac{1}{|z - \zeta|} \, dx \, dy, \qquad \zeta \in \mathbb{C}.$$

Remark. If we integrate this formally by parts twice, we obtain

$$u(\zeta) \;=\; -\frac{1}{2\pi} \iint_{\mathbb{C}} u(z) \, \Delta \left(\log \frac{1}{|z - \zeta|} \right) dx \, dy, \qquad \zeta \in \mathbb{C}.$$

Thus the "distribution" Laplacian of $-\dfrac{1}{2\pi} \log \dfrac{1}{|z - \zeta|}$ with respect to z is the point mass at ζ (the "Dirac delta-function"), in the sense that it is equal to 0 away from ζ, and it is infinite at ζ in such a way that its integral (total mass) is equal to 1.

2. Subharmonic Functions

Let D be a domain, and let $u(z)$ be a continuous function from D to the extended real line $[-\infty, \infty)$. We say that $u(z)$ is **subharmonic** if for each $z_0 \in D$, there is $\varepsilon > 0$ such that $u(z)$ satisfies the mean value inequality

$$(2.1) \qquad u(z_0) \le \int_0^{2\pi} u\left(z_0 + re^{i\theta}\right) \frac{d\theta}{2\pi}, \qquad 0 < r < \varepsilon.$$

If $u(z_0) = -\infty$, this inequality is automatically satisfied. Note also that ε may depend on z_0. This definition of subharmonicity is the same as the definition of the mean value property (Section III.4), except that "$=$" is replaced by "$\le$" and the function $u(z)$ is allowed to assume the value $-\infty$.

Example. Any harmonic function is subharmonic. If $u(z)$ is harmonic, then equality holds in (2.1) for all r such that the closed disk $\{|z - z_0| \le r\}$ is contained in D.

Example. If $f(z)$ is analytic, then $\log |f(z)|$ is subharmonic. More generally, if $u(z)$ is harmonic, or even just subharmonic, on the open set where $u(z) > -\infty$, then $u(z)$ is subharmonic.

Example. If $f(z)$ is analytic, then $|f(z)|$ is subharmonic. In this case the integral representation $f(z_0) = \int_0^{2\pi} f\left(z_0 + re^{i\theta}\right) d\theta/2\pi$ and an obvious estimate yield (2.1) for $|f(z)|$.

Subharmonicity is analogous in many respects to convexity of functions of one real variable. A smooth function $\chi(t)$ is convex if and only if $\chi''(t) \ge 0$. (See Exercise 9.) There is a corresponding characterization of subharmonic functions in which the operator d^2/dt^2 is replaced by the Laplacian Δ.

Theorem. *A smooth real-valued function $u(z)$ is subharmonic on D if and only if $\Delta u \ge 0$ on D.*

This follows from Green's third formula (1.3), applied to the functions $u(z)$ and $v(z) = \log(\rho/|z - z_0|)$ on a disk $\{|z - z_0| \le \rho\}$, which we assume is contained in D. In this case, formula (1.3) becomes

$$u(z_0) = -\frac{1}{2\pi} \iint_{|z - z_0| \le \rho} (\Delta u)(z) \log\left(\frac{\rho}{|z - z_0|}\right) dx\, dy + \int_0^{2\pi} u\left(z_0 + \rho e^{i\theta}\right) \frac{d\theta}{2\pi}.$$

This shows that the mean value inequality holds for all $z_0 \in D$ and all sufficiently small $\rho > 0$ if and only if $\Delta u \geq 0$.

We list several simple consequences of the definition of subharmonicity.

Lemma. *If $u(z)$ and $v(z)$ are subharmonic, then*

    (i) $u(z) + v(z)$ *is subharmonic,*
    (ii) $c\,u(z)$ *is subharmonic for any constant $c \geq 0$,*
    (iii) $w(z) = \max(u(z), v(z))$ *is subharmonic.*

To check (iii), for instance, let $z_0 \in D$, and suppose the mean value estimate (2.1) holds for $u(z_0)$ and $v(z_0)$. Then since $u(z) \leq w(z)$ and $v(z) \leq w(z)$, both $u(z_0)$ and $v(z_0)$ are bounded by $\int_0^{2\pi} w(z_0 + re^{i\theta})d\theta/2\pi$. Hence so is their maximum $w(z_0)$.

Both the maximum principle and the strict maximum principle hold for subharmonic functions. The proofs are virtually the same as those for harmonic functions.

Theorem (Strict Maximum Principle). *Let $u(z)$ be a subharmonic function on a domain D. If $u(z)$ attains its maximum at some point of D, then $u(z)$ is constant.*

To prove the strict maximum principle, one shows first that the set where $u(z)$ attains its maximum M is open. The proof of this given in Section III.5 carries over directly. It requires only the mean value inequality (2.1) in place of the mean value equality. The rest of the proof is easy. Since $u(z)$ is continuous, the set $\{u(z) < M\}$ is also open. Since D is connected, the set $\{u(z) = M\}$ must then be either empty or all of D.

Theorem (Maximum Principle). *Let D be a bounded domain, and let $u(z)$ be a subharmonic function on D. If $\limsup u(z) \leq 0$ as $z \to \partial D$, then $u(z) \leq 0$ for all $z \in D$.*

This follows from the strict maximum principle. We can argue as follows. The "lim sup" condition implies that the subharmonic function $w(z) = \max(u(z), 0)$ tends to 0 as z tends to ∂D. If we define $w(z)$ to be 0 on ∂D, then $w(z)$ is continuous on the compact set $D \cup \partial D$, so it attains its maximum at some point z_0. If $w(z_0) > 0$, then $z_0 \in D$. By the strict maximum principle, $w(z) \equiv w(z_0)$ is constant, contradicting $w(z) = 0$ on ∂D. It follows that the maximum of $w(z)$ is 0, and consequently, $u(z) \leq 0$ on D.

The following corollary to the maximum principle is obtained by applying the maximum principle to the function $u(z) - v(z)$. It justifies the nomenclature "subharmonic."

Theorem. *Let D be a bounded domain, and let $u(z)$ be a subharmonic function on D. Let $v(z)$ be a continuous finite real-valued function on $D \cup \partial D$ that is harmonic on D. If $\limsup_{z \to \zeta} u(z) \le v(\zeta)$ for all $\zeta \in \partial D$, then $u(z) \le v(z)$ for all $z \in D$.*

We conclude with a technique for constructing subharmonic functions that will play a key role in Perron's procedure for solving the Dirichlet problem.

Lemma. *Let $u(z)$ be subharmonic on a domain D. Let $D_0 = \{|z - z_0| < \rho\}$ be a disk such that $D_0 \cup \partial D_0$ is contained in D and such that $u(z) \ne -\infty$ on ∂D_0. Define the function $v(z)$ on D to be equal to $u(z)$ on $D \backslash D_0$ and to be the harmonic extension of $u|_{\partial D_0}$ on D_0. Then $v(z)$ is subharmonic on D, and $u(z) \le v(z)$ on D.*

The proof is a simple application of the maximum principle. The function $v(z)$ on D_0 is the Poisson integral of the restriction of u to ∂D_0, as in Section X.1. It is harmonic on D_0 and attains the boundary values $u(z)$ continuously on ∂D_0. By the maximum principle, $u(z) \le v(z)$ on D_0. Consequently, $v(z)$ is continuous and satisfies $u(z) \le v(z)$ on D. Evidently, $v(z)$ satisfies the mean value inequality at any point of $D \backslash D_0$, because $u(z)$ does and $u(z) = v(z)$ there. On D_0, $v(z)$ has the mean value property, since it is harmonic there. Thus $v(z)$ is subharmonic on D.

Exercises for XV.2

1. Show that $u(z)$ is harmonic if and only if both $u(z)$ and $-u(z)$ are subharmonic.

2. Suppose $u(z)$ is subharmonic on a domain D. Show that the mean value inequality (2.1) holds for all r such that the disk $\{|z - z_0| \le r\}$ is contained in D.

3. Let $u(z)$ be a continuous function from D to $[-\infty, \infty)$. Suppose $u_n(z)$ is an increasing sequence of subharmonic functions on D such that $u_n(z) \to u(z)$ for $z \in D$. Show that $u(z)$ is subharmonic. *Hint.* The difficulty is that the ε in (2.1), in the definition of subharmonicity, is allowed to depend on n. To circumvent this difficulty, use the result of Exercise 2.

4. Let $u(z)$ be a continuous function from D to $[-\infty, \infty)$. Suppose $u_n(z)$ is a decreasing sequence of subharmonic functions on D such that $u_n(z) \to u(z)$ for $z \in D$. Show that $u(z)$ is subharmonic.

5. Let $u(x, y)$ be a real-valued function on D with continuous first- and second-order partial derivatives. (a) For each $z_0 \in D$, show

that there is a quadratic polynomial $P(z)$ such that

$$u(z) = \text{Re}(P(z)) + \frac{(\Delta u)(z_0)}{4}|z - z_0|^2 + R(z),$$

where $R(z) = o(|z - z_0|^2)$, that is, $R(z)/|z - z_0|^2 \to 0$ as $z \to z_0$.
(b) Use (a) to show that if $(\Delta u)(z_0) > 0$, then $u(z)$ satisfies the mean value inequality at z_0. (c) Use (a) and (b) to show that $u(x, y)$ is subharmonic on D if and only if $(\Delta u)(z) \geq 0$ on D. *Hint.* If $\Delta u \geq 0$, show first that $u + \varepsilon|z|^2$ is subharmonic for any $\varepsilon > 0$.

6. Show that if $u(x, y)$ is harmonic and $p > 0$, then $|\nabla u|^p$ is subharmonic. *Hint.* Use Exercise 5c.

7. Let $h(z)$ be a smooth real-valued function on a neighborhood of the closure of a bounded domain D. Show that $h(z)$ can be expressed as the difference of two smooth subharmonic functions. *Hint.* Consider $u(z) = h(z) + C|z|^2$.

8. Let E be a subset of ∂D for which there is a subharmonic function $v(z)$ on D such that $v(z) < 0$ for $z \in D$, and $v(z)$ tends to $-\infty$ as $z \in D$ tends to any point of E. Show that if $u(z)$ is a subharmonic function on D that is bounded above, and if $\limsup u(z) \leq 0$ as $z \in D$ tends to any point of ∂D that is not in E, then $u(z) \leq 0$ in D.

9. A real-valued function $\chi(t)$ on an interval $I \subseteq \mathbb{R}$ is **convex** if $\chi(st_0 + (1 - s)t_1) \leq s\chi(t_0) + (1 - s)\chi(t_1)$ whenever $t_0, t_1 \in I$ and $0 \leq s \leq 1$.
(a) Show that a convex function on an open interval is continuous.
(b) Show that a smooth function $\chi(t)$ is convex if and only if $\chi''(t) \geq 0$. (c) Show that if $\chi(t)$ is convex, and if $s_1, \ldots, s_n \geq 0$ satisfy $\sum s_j = 1$, then

$$\chi\left(\sum_{j=1}^n s_j t_j\right) \leq \sum_{j=1}^n s_j\chi(t_j), \qquad t_1, \ldots, t_n \in I.$$

(d) Suppose $w(t)$ is a continuous function on an interval J such that $w(t) \geq 0$ and $\int_J w(t)dt = 1$. Show that if $\chi(t)$ is a convex function on I, and if $h(t)$ is any continuous function from J to I, then

$$\chi\left(\int_J h(t)w(t)dt\right) \leq \int_J \chi(h(t))w(t)dt.$$

Remark. This is **Jensen's inequality**. The relevant property of $w(t)$ is that $w(t)dt$ is a probability measure. To prove Jensen's inequality, approximate the integral by sums and use (c).

10. Show that if $u(z)$ is a subharmonic function on a domain D with values in an interval I, and if $\chi(t)$ is an increasing convex function on I, then $\chi(u(z))$ is subharmonic. *Hint.* Use Jensen's inequality.

11. Show that if $f(z)$ is analytic and $p > 0$, then $|f(z)|^p$ is subharmonic. *Hint.* Use the convexity of e^{pt} and Exercise 10, or use Exercise 6.

3. Compactness of Families of Harmonic Functions

The Arzelà-Ascoli theorem asserts that a uniformly bounded family of equicontinuous functions is compact. For a family of analytic functions, the Cauchy integral representation formula shows that uniform boundedness already implies equicontinuity. Hence a uniformly bounded family of analytic functions is compact. We can draw the same conclusion for a uniformly bounded family of harmonic functions using the Poisson integral representation formula. The Poisson integral formula for the unit disk is

$$(3.1)\quad u\left(re^{i\theta}\right) = \int_0^{2\pi} u\left(e^{i\varphi}\right) P_r(\theta - \varphi)\frac{d\varphi}{2\pi} = \int_0^{2\pi} u\left(e^{i(\theta-\varphi)}\right) P_r(\varphi)\frac{d\varphi}{2\pi},$$

where

$$(3.2)\qquad P_r(\theta) = \frac{1 - r^2}{1 + r^2 - 2r\cos\theta}, \qquad 0 \le r < 1,\ 0 \le \theta \le 2\pi.$$

The Poisson kernel is an infinitely differentiable function of x and y. We can differentiate (3.1) under the integral sign with respect to x and y, and when we do, we obtain integral representations of the derivatives of $u(x,y)$ in terms of the function $u\left(e^{i\theta}\right)$ and the derivatives of the Poisson kernel,

$$\frac{\partial}{\partial x}u(z) = \int_0^{2\pi} u\left(e^{i\varphi}\right)\frac{\partial}{\partial x}P_r(\theta - \varphi)\frac{d\varphi}{2\pi}, \qquad z = x + iy = re^{i\theta} \in \mathbb{D},$$

with similar formulae for the other derivatives. These integral formulae allow us to estimate the derivatives of $u(x,y)$ in terms of the maximum of M of $\left|u\left(e^{i\theta}\right)\right|$ on the unit circle,

$$\left|\frac{\partial}{\partial x}u(z)\right| \le M\int_0^{2\pi}\left|\frac{\partial}{\partial x}P_r(\theta - \varphi)\right|\frac{d\varphi}{2\pi}, \qquad z = x + iy = re^{i\theta} \in \mathbb{D},$$

and similarly for the other derivatives. Since each derivative of the Poisson kernel is uniformly bounded on each proper subdisk of the unit disk, we obtain uniform bounds for each derivative of $u(z)$ in terms of the maximum of $|u(z)|$ on the unit circle. (See also Section X.1, Exercises 8 and 9.)

   The same procedure works if we replace the Poisson integral representation for the unit disk by the Poisson integral representation for an arbitrary closed disk. We obtain thus the following.

Lemma. *If $\mathcal{F}$ is a family of harmonic functions on a domain D that is uniformly bounded on each closed disk in D, then the partial derivatives of the functions in $\mathcal{F}$ of any fixed order are uniformly bounded on each closed disk in D.*

In particular, the gradients of the functions in $\mathcal{F}$ are uniformly bounded on any closed disk in D, so the functions in $\mathcal{F}$ are equicontinuous. (See Section XI.2.) From the Arzelà-Ascoli theorem and a diagonalization argument, we obtain the following analogue for harmonic functions of Montel's theorem.

Theorem. *Let $\mathcal{F}$ be a family of harmonic functions on a domain D that is uniformly bounded on each compact subset of D. Then every sequence in $\mathcal{F}$ has a subsequence that converges uniformly on each compact subset of D.*

Families of positive harmonic functions are also compact, provided that we allow $+\infty$ as a possible limit for a sequence of positive harmonic functions. This follows again from the Poisson integral representation. This time it is the positivity of the Poisson kernel that is crucial.

Theorem (Harnack's Inequality). *If $u(z)$ is a positive harmonic function on the open unit disk $\mathbb{D}$, then*

$$(3.3) \qquad \frac{1-r}{1+r}\, u(0) \;\leq\; u\!\left(re^{i\theta}\right) \;\leq\; \frac{1+r}{1-r}\, u(0), \qquad re^{i\theta} \in \mathbb{D}.$$

The proof depends on the estimate

$$\frac{1-r}{1+r} \;=\; \frac{1-r^2}{1+r^2+2r} \;\leq\; P_r(\theta) \;\leq\; \frac{1-r^2}{1+r^2-2r} \;=\; \frac{1+r}{1-r}.$$

We can always approximate $u(z)$ by a dilate $u(\rho z)$, $\rho < 1$, and assume that $u(z)$ extends harmonically across the unit circle. Then $u(z)$ is represented by the Poisson integral formula (3.1). Substituting the estimates for $P_r(\theta)$ into (3.1) and using the positivity of $u(z)$, we obtain

$$\frac{1-r}{1+r} \int_0^{2\pi} u\!\left(e^{i(\theta-\varphi)}\right) \frac{d\varphi}{2\pi} \;\leq\; \int_0^{2\pi} u\!\left(e^{i(\theta-\varphi)}\right) P_r(\varphi)\frac{d\varphi}{2\pi}$$

$$\leq\; \frac{1+r}{1-r} \int_0^{2\pi} u\!\left(e^{i(\theta-\varphi)}\right) \frac{d\varphi}{2\pi}.$$

Since the average value of $u\!\left(e^{i(\theta-\varphi)}\right)$ is $u(0)$, this becomes (3.3).

The estimate (3.3) is sharp. It becomes an equality for the function $u(z) = \mathrm{Re}((1+z)/(1-z))$ at $z = \pm r$.

If we scale and translate Harnack's inequality to an arbitrary disk, of radius $R > 0$ and center z_0, we obtain

$$(3.4) \qquad \frac{R-r}{R+r}\, u(z_0) \;\leq\; u(z) \;\leq\; \frac{R+r}{R-r}\, u(z_0), \qquad |z - z_0| \leq r,\ r < R,$$

for all positive harmonic functions $u(z)$ on the disk $|z - z_0| < R$. The estimate (3.4) is also referred to as **Harnack's inequality**. It allows us to draw some striking conclusions. From the left-hand inequality it follows that if $\{u_n(z)\}$ is a sequence of positive harmonic functions on the disk $\{|z - z_0| < R\}$ such that $u_n(z_0) \to +\infty$, then $u_n(z) \to +\infty$ uniformly on any subdisk $\{|z - z_0| \leq \rho\}$, $\rho < R$. From the right-hand inequality it follows that if $\{u_n(z_0)\}$ is bounded, then $u_n(z)$ is uniformly bounded on any subdisk $\{|z - z_0| \leq \rho\}$, $\rho < R$. If we combine these observations with a connectedness argument, we obtain the following.

Lemma. *Suppose $\{u_n(z)\}$ is a sequence of positive harmonic functions on a domain D. If there is $z_0 \in D$ such that $u_n(z_0) \to +\infty$, then $u_n(z) \to +\infty$ uniformly on each compact subset of D. If there is $z_0 \in D$ such that $\{u_n(z_0)\}$ is bounded, then $\{u_n(z)\}$ is uniformly bounded on each compact subset of D.*

To see this, let U be the set of $z \in D$ such that $u_n(z) \to +\infty$. The remarks preceding the lemma show that both U and $D\backslash U$ are open. Since D is a domain, either $U = D$ or U is empty. In the case $U = D$, we can cover any compact set by a finite number of disks on which $u_n(z) \to +\infty$ uniformly, and we see that $u_n(z) \to +\infty$ uniformly on compact subsets of D. If U is empty, we cover any compact set by a finite number of disks, and we see that $\{u_n(z)\}$ is uniformly bounded on each compact subset of D.

If we combine the preceding lemma and the compactness theorem for families of harmonic functions, we obtain immediately the following compactness theorem for families of positive harmonic functions.

Theorem. *Let $\mathcal{F}$ be a family of positive harmonic functions on a domain D. Every sequence in $\mathcal{F}$ has a subsequence that either converges uniformly on compact subsets of D to a harmonic function or converges uniformly on compact subsets of D to $+\infty$.*

The compactness theorem for positive harmonic functions can be applied to monotone sequences of harmonic functions.

Theorem. *Let $\{u_n(z)\}$ be an increasing sequence of harmonic functions on a domain D. Then $\{u_n(z)\}$ converges uniformly on compact subsets of D, either to a harmonic function or to $+\infty$.*

To see this, consider the functions $v_n(z) = u_n(z) - u_1(z)$. For each fixed z the sequence $\{v_n(z)\}$ is increasing, hence convergent either to a finite value or to $+\infty$. The preceding theorem shows that $\{v_n(z)\}$ has a subsequence that converges uniformly on compact subsets of D, either to a harmonic function or to $+\infty$. It follows that $\{u_n(z)\}$ also converges uniformly on compact subsets of D.

Exercises for XV.3

1. Let $\{u_n(z)\}$ be a sequence of positive harmonic functions on a domain D. Show that either $\sum u_n(z)$ converges to a harmonic function uniformly on each compact subset of D, or the partial sums of the series converge uniformly to $+\infty$ on each compact subset of D.

2. Let $u(z)$ be a positive harmonic function on the horizontal strip $\{-1 < \operatorname{Im} z < 1\}$. Show that if $u(n) \to +\infty$ as $n \to +\infty$, then $u(x + iy) \to +\infty$ as $x \to +\infty$ for all y, $-1 < y < 1$.

3. Let $u(z)$ be a positive harmonic function on the horizontal strip $\{-1 < \operatorname{Im} z < 1\}$. Show that if $u(n) \to 0$ as $n \to +\infty$, then $u(x + iy) \to 0$ as $x \to +\infty$ for all y, $-1 < y < 1$.

4. Let K be a compact subset of a domain D. Show that there are constants $c > 0$ and $C > 1$ such that $c \le u(z)/u(w) \le C$ for all positive harmonic functions u on D and all points $z, w \in K$.

5. Let D be a bounded domain. For $z, w \in D$, define $d(z, w)$ to be the infimum of $\log C$, taken over all constants $C > 1$ such that $(1/C) \le u(z)/u(w) \le C$ for all positive harmonic functions u on D. (a) Show that $d(z, w) \le d(z, \zeta) + d(\zeta, w)$ for all $z, \zeta, w \in D$. (b) Show that $d(z_n, z_0) \to 0$ if and only if $|z_n - z_0| \to 0$. (c) Find $d(z, w)$ in the case that $D = \mathbb{D}$ is the open unit disk.

6. Prove the compactness theorem for families of positive harmonic functions by applying Montel's theorem to functions of the form $f = e^{-(u+iv)}$, where v is a (locally defined) harmonic conjugate for the positive harmonic function u.

7. Let $u(z)$ be a harmonic function on a domain D. Show that if $|u(z)| \le M$ for $z \in D$, then $|(\nabla u)(z)| \le 2M/\operatorname{dist}(z, \partial D)$. *Hint.* Use "scaling." Check the estimate for the unit disk and $z = 0$ from the Poisson integral representation, and transfer the estimate to a disk centered at z of appropriate radius. The gradient scales according to one over the radius of the disk.

4. The Perron Method

Let D be a bounded domain. Recall that the Dirichlet problem for a given continuous function $h(z)$ on the boundary of D is to find a harmonic function $u(z)$ on D that has boundary values $h(z)$. Any such function $u(z)$ is unique. However, the Dirichlet problem need not be solvable.

Example. Let $D = \mathbb{D}\backslash\{0\}$ be the punctured unit disk, and let $h(z)$ be the continuous function on ∂D defined by $h(z) = 0$ for $|z| = 1$, while $h(0) = 1$. If $u(z)$ is a harmonic function on D such that $\limsup u(z) \le h(\zeta)$ for $\zeta \in \partial D$, then for any $\varepsilon > 0$, $u(z) \le \varepsilon \log(1/|z|)$ for $|z| = 1$ and also for z near 0. By the maximum principle, the estimate holds on all of D. Letting $\varepsilon \to 0$, we obtain $u(z) \le 0$ on D. Consequently, $u(z)$ cannot assume the boundary value 1 at $z = 0$.

Let $h(\zeta)$ be a continuous real-valued function on ∂D. We define the **Perron family of subsolutions** corresponding to $h(\zeta)$, denoted by $\mathcal{F}_h$, to be the family of all subharmonic functions $u(z)$ on D such that

$$\limsup_{D \ni z \to \zeta} u(z) \le h(\zeta), \qquad \zeta \in \partial D.$$

We define the **Perron solution** to the Dirichlet problem corresponding to the boundary function $h(\zeta)$ to be the upper envelope $\tilde{h}(z)$ of the family $\mathcal{F}_h$,

$$\tilde{h}(z) = \sup\{u(z) : u \in \mathcal{F}_h\}, \qquad z \in D.$$

In the case that there is a harmonic function $u(z)$ on D that attains the boundary values $h(\zeta)$ continuously on ∂D, then happily $u(z)$ coincides with the Perron solution. Indeed, since $u \in \mathcal{F}_h$, we have $u(z) \le \tilde{h}(z)$ on D. The reverse inequality follows from the maximum principle. If $v \in \mathcal{F}_h$, then by the maximum principle, $v(z) \le u(z)$ on D, and consequently, $\tilde{h}(z) \le u(z)$ on D.

Several elementary properties of the Perron solution are listed in the next lemma.

Lemma. *Let D be a bounded domain, and let $h(\zeta)$ be a continuous real-valued function on ∂D.*

(i) *If $\alpha \le h(\zeta) \le \beta$ on ∂D, then $\alpha \le \tilde{h}(z) \le \beta$ on D.*

(ii) *If $h_1(\zeta) \le h_2(\zeta)$ on ∂D, then $\widetilde{h_1}(z) \le \widetilde{h_2}(z)$ on D.*

(iii) *$\widetilde{ah}(z) = a\tilde{h}(z)$ for all $a > 0$.*

(iv) *$\widetilde{h_1}(z) + \widetilde{h_2}(z) \le \widetilde{(h_1 + h_2)}(z), \qquad z \in D$.*

(v) *$\widetilde{(h + v)}(z) = \tilde{h}(z) + v(z)$ on D whenever $v(z)$ is continuous on $D \cup \partial D$ and harmonic on D.*

In (i), the constant function α belongs to $\mathcal{F}_h$, so $\alpha \le \tilde{h}$. The estimate $\tilde{h} \le \beta$ follows from the maximum principle, as above. Properties (ii), (iii), and (v) follow virtually immediately from the definitions. For (ii) note that $\mathcal{F}_{h_1} \subseteq \mathcal{F}_{h_2}$, for (iii) that $\mathcal{F}_{ah} = a\mathcal{F}_h$, and for (v) that $\mathcal{F}_{h+v} = \mathcal{F}_h + v$. To prove (iv), suppose $u_1 \in \mathcal{F}_{h_1}$ and $u_2 \in \mathcal{F}_{h_2}$. Then $u_1 + u_2 \in \mathcal{F}_{h_1+h_2}$, and consequently, $u_1(z) + u_2(z) \leq (\widehat{h_1 + h_2})(z)$. If we take the supremum over $u_1 \in \mathcal{F}_{h_1}$ and $u_2 \in \mathcal{F}_{h_2}$, we obtain (iv).

Theorem. *The Perron solution $\tilde{h}(z)$ defined above is harmonic on D.*

For the proof, let D_0 be a fixed disk such that $D_0 \cup \partial D_0 \subset D$, and let $\{z_j\}_{j=1}^\infty$ be a sequence of points in D_0 that is dense in D_0. Let $\{u_{jk}(z)\}_{k=1}^\infty$ be a sequence of functions in $\mathcal{F}_h$ such that $u_{jk}(z_j) \to \tilde{h}(z_j)$ for $j \geq 1$. Define $v_m(z) = \max\{u_{jk}(z) : 1 \leq j, k \leq m\}$. Then $v_m \in \mathcal{F}_h$, $\{v_m(z)\}$ is an increasing sequence, and $v_m(z_j) \to \tilde{h}(z_j)$ for all j. Define $w_m(z)$ so that $w_m(z) = v_m(z)$ for $z \in D \backslash D_0$, and $w_m(z)$ is the harmonic extension of $v_m|_{\partial D_0}$ on D_0. Then $w_m \in \mathcal{F}_h$, $\{w_m(z)\}$ is an increasing sequence, and $w_m(z_j) \to \tilde{h}(z_j)$ for all j. Further, the functions $w_m(z)$ are all harmonic on D_0. By Harnack's theorem, the sequence $\{w_m(z)\}$ converges uniformly on compact subsets of D_0 to a harmonic function $w(z)$, which evidently satisfies $w(z_j) = \tilde{h}(z_j)$ for all j. Now let z_0 be any other point of D_0. We go through this procedure for the sequence $\{z_j\}_{j=0}^\infty$ obtained by adjoining z_0 to the original sequence, and we obtain a harmonic function $w_0(z)$ on D_0 such that $w_0(z_j) = \tilde{h}(z_j)$ for all $j \geq 0$. Since $w(z) = w_0(z)$ on a dense subset of D_0, in fact $w(z) = w_0(z)$ for all $z \in D_0$, and in particular $w(z_0) = w_0(z_0) = \tilde{h}(z_0)$. Since z_0 is arbitrary, $w(z) = \tilde{h}(z)$ for all $z \in D_0$, and $\tilde{h}(z)$ is harmonic on D_0, hence on D.

Now we address the problem of whether $\tilde{h}(z)$ has boundary values $h(\zeta)$ on ∂D. We say that a point $\zeta_0 \in \partial D$ is a **regular boundary point of D** if

$$\lim_{D \ni z \to \zeta_0} \tilde{h}(z) = h(\zeta_0)$$

for any continuous function $h(\zeta)$ on ∂D.

Example. The origin $\zeta_0 = 0$ is *not* a regular boundary point of the punctured unit disk $D = \mathbb{D}\backslash\{0\}$. We will see shortly that each boundary point on the unit circle is regular.

Suppose that $\zeta_0 \in \partial D$ is a regular boundary point. We solve the Dirichlet problem for the function $h(\zeta) = |\zeta - \zeta_0|$, $\zeta \in \partial D$. Since the function $|z - \zeta_0|$ belongs to $\mathcal{F}_h$, the Perron solution $\tilde{h}(z)$ satisfies $\tilde{h}(z) \geq |z - \zeta_0|$. Since ζ_0 is regular, $\tilde{h}(z) \to 0$ as $z \to \zeta_0$. Thus the function $w(z) = -\tilde{h}(z)$ is harmonic on D, $w(z) \leq -|z - \zeta_0|$ on D, and $w(z) \to 0$ as $z \to \zeta_0$. The function $w(z)$ "peaks" at ζ_0, in the sense that $w(z)$ has a limit at ζ_0 that is strictly greater than $\limsup w(z)$ as z tends to any other boundary point of D. This motivates the following definition.

A **subharmonic barrier** at $\zeta_0 \in \partial D$ is a subharmonic function $w(z)$ defined on $D \cap \{|z - \zeta_0| < \delta\}$ for some $\delta > 0$, such that

   (i) $w(z) < 0$ for $z \in D$, $|z - \zeta_0| < \delta$,
   (ii) $w(z) \to 0$ as $z \in D$ tends to ζ_0,
   (iii) $\limsup w(z) < 0$ as $z \in D$ tends to ζ for any $\zeta \in \partial D, 0 < |\zeta - \zeta_0| < \delta$.

While only defined on the part of D near ζ_0, the subharmonic barrier can be adjusted so that it is defined on all of D and satisfies (iii) at any $\zeta \in \partial D \backslash \{\zeta_0\}$. The conditions (i) and (iii) guarantee that if $\varepsilon > 0$ is sufficiently small, then $w(z) \leq -2\varepsilon$ for all $z \in D$ that satisfy $|z - \zeta_0| = \delta/2$. If we set $u(z) = \max(w(z), -\varepsilon)$ for $z \in D \cap \{|z - \zeta_0| < \delta/2\}$, and $u(z) = -\varepsilon$ for other points of D, then $u(z)$ is a subharmonic barrier at ζ_0 that is now defined on all of D.

Example. In the case $D = \mathbb{D}$ is the open unit disk, the function $w(z) = \operatorname{Re}\left(\bar{\zeta}_0 z\right) - 1$ is a subharmonic barrier at the boundary point $\zeta_0 \in \partial \mathbb{D}$. In this case, the barrier is harmonic.

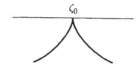

graph of subharmonic barrier at ζ_0

Example. Suppose there is a straight line segment I in $\mathbb{C} \backslash D$ that has one endpoint at $\zeta_0 \in \partial D$. We map the complement $\mathbb{C}^* \backslash I$ of the segment onto the open unit disk $\mathbb{D}$ by a conformal map $\varphi(z)$ such that $\varphi(\zeta_0) = 1$. Then $w(z) = \operatorname{Re} \varphi(z) - 1$ is a subharmonic barrier at ζ_0.

Example. If D is bounded by a finite number of smooth boundary curves, then every point of ∂D has a subharmonic barrier. In this case, each $\zeta_0 \in \partial D$ is the endpoint of a line segment contained in $\mathbb{C} \backslash D$, in the direction of the outer normal to D at ζ_0.

We showed above that if ζ_0 is a regular boundary point of D, then there is a subharmonic barrier at ζ_0. The following theorem shows that the existence of subharmonic barriers characterizes regular boundary points.

Theorem. *Let D be a bounded domain, and let $\zeta_0 \in \partial D$. If there is a subharmonic barrier at ζ_0, then ζ_0 is a regular boundary point of D.*

To prove this, let $h(\zeta)$ be a continuous real-valued function on ∂D. We assume for convenience that $h(\zeta_0) = 0$ and that $|h(\zeta)| \leq 1$ for $\zeta \in \partial D$. Let $\varepsilon > 0$, and choose $\delta > 0$ such that $|h(\zeta)| < \varepsilon$ for $\zeta \in \partial D, |\zeta - \zeta_0| \leq \delta$. Let $w(z)$ be a subharmonic barrier at ζ_0. We can assume that the subharmonic barrier $w(z)$ is defined on all of D. Choose $\rho > 0$ such that $w(z) \leq -\rho$ for $|z| \geq \delta$. Then $u(z) = (w(z)/\rho) - \varepsilon$ is subharmonic on D, $u(z) \leq -\varepsilon$ on D, and $\limsup_{z \to \zeta} u(z) \leq -1$ for any point $\zeta \in \partial D$ satisfying $|\zeta| \geq \delta$. Thus $\limsup_{z \to \zeta} u(z) \leq h(\zeta)$ for all $\zeta \in \partial D$, and $u \in \mathcal{F}_h$. Hence $u(z) \leq \tilde{h}(z)$. Since $u(z) \to -\varepsilon$ as $z \to \zeta_0$, we obtain $\tilde{h}(z) \geq -2\varepsilon$ for ζ near ζ_0. The same argument, applied to $-h(\zeta)$, shows that $\widetilde{(-h)}(z) \geq -2\varepsilon$ for ζ near ζ_0. By part (iv) of the lemma, $\tilde{h}(z) + \widetilde{(-h)}(z) \leq 0$. Hence $\tilde{h}(z) \leq -\widetilde{(-h)}(z) \leq 2\varepsilon$ for ζ near ζ_0. It follows that $|\tilde{h}(z)| \leq 2\varepsilon$ for ζ near ζ_0. Since this is true for any $\varepsilon > 0$, $\tilde{h}(z) \to 0$ as $z \to \zeta_0$.

Since there is a subharmonic barrier at each smooth boundary point, we obtain the following corollary.

Corollary. *Let D be a bounded domain whose boundary ∂D consists of a finite number of piecewise smooth closed curves. If $h(\zeta)$ is a continuous function on ∂D, then $\tilde{h}(z)$ is a harmonic function on D that tends to $h(\zeta)$ as $z \in D$ tends to each boundary point ζ of ∂D.*

Exercises for XV.4

1. Let D be the annulus $\{a < |z| < b\}$. Find $\tilde{h}(z)$ for the function $h(\zeta)$ defined by $h(\zeta) = \alpha$ for $|\zeta| = a$ and $h(\zeta) = \beta$ for $|\zeta| = b$.

2. Let $D = \mathbb{D}\backslash\{0\}$ be the punctured unit disk, and let $h(\zeta)$ be a continuous function on ∂D. Show that $\tilde{h}(z)$ is the Poisson integral of the restriction of $h(\zeta)$ to the unit circle $\partial \mathbb{D}$, irrespective of the value $h(0)$.

3. Let D be a bounded domain, and let $h(\zeta)$ be continuous on ∂D. Let $z_0 \in D$, and for $\varepsilon > 0$, set $D_\varepsilon = D \cap \{|z - z_0| > \varepsilon\}$. Define h_ε on ∂D_ε so that $h_\varepsilon = h$ on ∂D, and $h_\varepsilon(\zeta) = 0$ for $|\zeta - z_0| = \varepsilon$. Show that $\tilde{h}_\varepsilon(z)$ converges to $\tilde{h}(z)$ as $\varepsilon \to 0$, uniformly on any subset of D at a positive distance from z_0.

4. Let D be a bounded domain, and let $h(z)$ be a continuous function on ∂D. Let $\{D_m\}$ be a sequence of smoothly bounded domains that increase to D. Let $u(z)$ be a continuous extension of $h(z)$ to $D \cup \partial D$, and let $u_m(z)$ be the harmonic extension of $u|_{\partial D_m}$ to D_m. (a) Show that u_m converges uniformly on compact subsets of D to a harmonic function Wh on D. *Hint.* If $u(z)$ is smooth, represent $u(z)$ as the difference of two subharmonic functions, as in Exercise 2.6. (b) Show that Wh depends only on h, and not on the particular extension u of h to D nor on the sequence $\{D_m\}$. (c) Show that W is linear, that is, $W(ah_1 + bh_2) = aW(h_1) + bW(h_2)$. (d) Show that $\tilde{h} \leq Wh$, where $\tilde{h}$ is the Perron solution to the Dirichlet problem. *Remark.* The harmonic function Wh is the **Wiener solution** to

the Dirichlet problem with boundary function h. It can be shown
that the Wiener solution coincides with the Perron solution.

5. The Riemann Mapping Theorem Revisited

The Riemann mapping theory can be proved by solving a Dirichlet problem. This was Riemann's original strategy, though he could not solve the
Dirichlet problem in the full generality of simply connected domains. We
will sketch a proof along the lines envisioned by Riemann.

We say that a subset E of the extended complex plane $\mathbb{C}^*$ is a **continuum** if it is compact and connected and has more than one point. If E is a
continuum, then $\mathbb{C}^*\backslash E$ is simply connected. Conversely, if E is a compact
set such that $\mathbb{C}^*\backslash E$ is simply connected, and if E has more than one point,
then E is a continuum. (See Section VIII.8.) The following criterion allows
us to solve the Dirichlet problem for a simply connected domain.

Lemma. *Let D be a domain, and let $\zeta_0 \in \partial D$. If ζ_0 lies on a continuum
in $\mathbb{C}\backslash D$, then ζ_0 is a regular boundary point of D.*

Proofsketch. Let E be the continuum in $\mathbb{C}\backslash D$ containing ζ_0, and let ζ_1 be
any other point of E. Since $\mathbb{C}^*\backslash E$ is simply connected, $\log((z-\zeta_0)/(z-\zeta_1))$
has an analytic branch on $\mathbb{C}\backslash E$; call it $f(z)$. The set $D_0 = \{|z-\zeta_0|/|z-\zeta_1| <
1/e\}$ is a disk containing ζ_0, and $\operatorname{Re} f(z) < -1$ for $z \in D \cap D_0$. Define
$h(z) = \operatorname{Re}(1/f(z))$ for $z \in D \cap D_0$. Then $h(z) < 0$. Since $\operatorname{Re} f(z) \to -\infty$
as $z \to \zeta_0$, $h(z) \to 0$ as $z \to \zeta_0$, and $h(z)$ is "almost" a barrier at ζ_0.
The difficulty is that $h(z)$ might tend to 0 at other boundary points of
$D \cap D_0$. The proof is completed by following the procedure in the proof of
Bouligand's lemma (pp. 8–9 of Tsuji's book; see the references), in which
a genuine barrier at ζ_0 is constructed from $h(z)$.

Corollary. *Let D be a bounded domain. If $\mathbb{C}^*\backslash D$ consists of finitely many
continua, then every point of ∂D is a regular boundary point.*

Now we can prove the Riemann mapping theorem. Suppose D is a
simply connected domain and that $\mathbb{C}^*\backslash D$ has at least two points. As in
Section XI.3, we can use the logarithm function (or the square root function) and a fractional linear transformation to map D conformally onto
a bounded domain. Indeed, let $\zeta_0, \zeta_1 \in \partial D$, and let $f(z)$ be an analytic
branch of $\log((z - \zeta_0)/(z - \zeta_1))$ in D. If $w_0 = f(z_0)$ for some $z_0 \in D$, then
the image $f(D)$ covers a disk D_0 centered at w_0. Consequently, $f(z)$ cannot assume any of the values in $D_0 + 2\pi i$ on D, so that $1/(f(z) - w_0 - 2\pi i)$
maps D conformally onto a bounded domain.

So we assume that D is bounded, and we also assume for convenience
that $0 \in D$. We solve the Dirichlet problem for the function $\log|\zeta|$ on ∂D,

and we obtain a harmonic function $u(z)$ on D such that $u(z)$ tends to $\log|\zeta|$ as $z \in D$ tends to any point $\zeta \in \partial D$. Since D is simply connected, the function $u(z)$ has a harmonic conjugate $v(z)$ in D. Define

$$\varphi(z) = ze^{-(u(z)+iv(z))}, \qquad z \in D.$$

Then $\varphi(z)$ is analytic on D, and $|\varphi(z)| \to 1$ as $z \to \partial D$. By the maximum principle, $|\varphi(z)| < 1$ for $z \in D$. Further, $\varphi(z)$ has only one zero in D, a simple zero at $z = 0$. Fix w such that $|w| < 1$. We apply the argument principle to the domain $\{|\varphi(z)| < 1 - \varepsilon\}$, where $|w| < 1 - \varepsilon$, and we see that $\varphi(z) - w$ has exactly one zero on D. Hence $\varphi(z)$ maps D conformally onto the open unit disk.

If we analyze this proof and compare it to the proof given in Chapter XI, we see that both proofs depend on the idea of compactness of families of functions, but in other respects they are quite different. In this proof, Harnack's theorem on equicontinuity of uniformly bounded families of harmonic functions replaces the thesis grade version of Montel's theorem. These two compactness theorems are essentially equivalent. The use of regular boundary points and criteria for regularity replace the use of the hyperbolically expanding property of the square root function, and these techniques are quite different.

Exercises for XV.5

1. Let D be a bounded domain such that ∂D consists of two disjoint continua E_0 and E_1. Let $u(z)$ be the harmonic function on D with boundary values $u(z) = 0$ on E_0 and $u(z) = 1$ on E_1, and let $v(z)$ be a (locally defined) conjugate harmonic function of $u(z)$. Show that there is a constant $\alpha > 0$ such that $\varphi(z) = e^{\alpha(u(z)+iv(z))}$ is a single-valued analytic function on D, and $w = \varphi(z)$ maps D conformally onto the annulus $\{1 < |w| < e^\alpha\}$.

2. Suppose D is a doubly connected domain in the extended complex plane $\mathbb{C}^*$, that is, $\mathbb{C}^* \backslash D$ consists of two disjoint connected sets. Show that D can be mapped conformally onto exactly one of (a) the punctured complex plane $\mathbb{C} \backslash \{0\}$, (b) the punctured unit disk $\mathbb{D} \backslash \{0\}$, or (c) an annulus $\{1 < |z| < R\}$. You may use the result of Exercise 1.

6. Green's Function for Domains with Analytic Boundary

We say that the domain D has **analytic boundary** if the boundary of D consists of a finite number of disjoint simple closed analytic boundary curves. We will focus in this section on bounded domains with analytic boundary. In this case, every point of ∂D is a regular boundary point, and we can solve the Dirichlet problem for any continuous boundary function. Further, any harmonic function on D that is constant on a boundary curve extends harmonically across the boundary curve, by the Schwarz reflection principle.

So let D be a bounded domain with analytic boundary. Fix $\zeta \in D$, and let $h(z) = \log|z - \zeta|$ for $z \in \partial D$. Then $h(z)$ is a continuous function on ∂D, and the Perron solution $\bar{h}(z)$ is a harmonic function on D that assumes the boundary values $\log|z - \zeta|$ continuously on ∂D. We define **Green's function** for D with pole at ζ to be the function $g(z)$ on $D\backslash\{\zeta\}$ given by

$$g(z) = \bar{h}(z) + \log\left(\frac{1}{|z - \zeta|}\right).$$

It satisfies

(6.1) $g(z)$ is harmonic on $D\backslash\{\zeta\}$,

(6.2) $g(z) - \log\dfrac{1}{|z - \zeta|}$ is harmonic at ζ,

(6.3) $g(z) \to 0$ as z tends to the boundary of D.

Occasionally, it will be convenient to denote Green's function by $g(z, \zeta)$, to emphasize the dependence on ζ. If $h(z)$ is another function with the properties (6.1), (6.2), and (6.3), then $h(z) - g(z)$ is a harmonic function on D that tends to zero on the boundary of D. By the maximum principle, $h(z) - g(z) = 0$ on D, and $h(z) = g(z)$. Thus Green's function with pole at ζ is uniquely determined by the properties (6.1), (6.2), and (6.3).

Example. Green's function for the unit disk $\mathbb{D}$ with pole at $z = 0$ is $g(z) = \log(1/|z|)$.

Example. Suppose D is a simply connected domain, $D \neq \mathbb{C}$, and let $\zeta \in D$. Let $\varphi(z)$ be the Riemann map of D onto the open unit disk $\mathbb{D}$ satisfying $\varphi(\zeta) = 0$, and set $g(z) = -\log|\varphi(z)|$. Since $|\varphi(z)| \to 1$ as $z \to \partial D$, $g(z) \to 0$ as $z \to \partial D$. Since $\varphi(z)/(z - \zeta)$ is analytic and nonzero at $z = \zeta$, we see by taking logarithms that $g(z) + \log|z - \zeta|$ is harmonic at $z = \zeta$, and so $g(z)$ has a logarithmic pole at $z = \zeta$. Thus $g(z)$ satisfies (6.1), (6.2), and (6.3) above, and $g(z)$ is Green's function for D with pole at ζ.

Lemma. *Let D be a bounded domain with analytic boundary, and let $g(z)$ be Green's function for D with pole at $\zeta \in D$. Then $g(z) > 0$ on D, $g(z) = 0$ on ∂D, and $g(z)$ reflects harmonically across ∂D so that $g(z) < 0$ on the side of ∂D outside of D. Further, the directional derivative of $g(z)$ in the direction of the exterior normal satisfies $\partial g/\partial n < 0$ on ∂D.*

The positivity follows from the maximum principle. Since $g(z) \to +\infty$ as $z \to \zeta$, we can choose $\varepsilon > 0$ so small that $g(z) > 0$ for $|z - \zeta| \le \varepsilon$. The maximum principle, applied to the domain $D\backslash\{|z - \zeta| \le \varepsilon\}$, shows that $g(z) \ge 0$ for all $z \in D$. The strict maximum principle guarantees that $g(z)$ does not assume a minimum value 0 at any point of D, so that

$g > 0$ on D. By the Schwarz reflection principle (Section X.2), g reflects harmonically across the analytic boundary curves in ∂D and $g < 0$ on the side of ∂D outside D. Thus $\partial g/\partial n \leq 0$ on ∂D. Let $z_0 \in \partial D$, and let $f(z)$ be an analytic function defined on a disk centered at z_0 whose real part is $g(z)$. Then $|f'(z_0)| = |\nabla g(z_0)|$. Since the set $\{\operatorname{Re} f(z) = \operatorname{Re} f(z_0)\}$ consists of only one analytic arc passing through z_0, $f'(z_0) \neq 0$, and consequently, $(\nabla g)(z_0) \neq 0$. Since $g = 0$ on ∂D, $(\nabla g)(z_0)$ is a nonzero multiple of the normal vector at z_0. Thus $\partial g/\partial n \neq 0$, and we conclude that there is strict inequality $\partial g/\partial n < 0$ on ∂D. The lemma is proved.

Again let D be a bounded domain with analytic boundary, and let $g(z)$ be Green's function for D with pole at $\zeta \in D$. Let $u(z)$ be any harmonic function on D that extends continuously to ∂D. Green's third formula (1.3), applied to $u(z)$ and to Green's function $g(z)$ with pole at ζ, becomes

$$(6.4) \qquad u(\zeta) = -\frac{1}{2\pi} \int_{\partial D} u \frac{\partial g}{\partial n} \, ds.$$

We have established Green's formulae only for functions that are smooth across the boundary, but in the case of an analytic boundary it is easy to argue (Exercise 8) that the formula holds whenever u is only continuous up to the boundary.

We define **harmonic measure** on ∂D for $\zeta \in D$ to be the differential

$$(6.5) \qquad d\mu_\zeta(z) = -\frac{1}{2\pi} \frac{\partial g}{\partial n} \, ds.$$

Since $\partial g/\partial n < 0$ on ∂D, harmonic measure is positive, in the sense that $\int_{\partial D} v(z) d\mu_\zeta(z) \geq 0$ whenever $v \geq 0$ on ∂D. If we apply (6.4) to the function $u(z) = 1$, we obtain

$$\int_{\partial D} d\mu_\zeta(z) = 1.$$

We say that $d\mu_\zeta$ is a probability measure on ∂D in the sense that it is positive and has unit total mass. In terms of harmonic measure, (6.4) reads as follows.

Theorem. *Let D be a bounded domain with analytic boundary, and let $g(z)$ be Green's function with pole at $\zeta \in D$. Define harmonic measure $d\mu_\zeta(z)$ by (6.5). Then*

$$u(\zeta) = \int_{\partial D} u(z) \, d\mu_\zeta(z), \qquad \zeta \in D,$$

for every harmonic function $u(z)$ on D that extends continuously to ∂D.

Example. In the case of the unit disk, Green's function with pole at 0 is $\log(1/|z|)$. and

$$\frac{\partial}{\partial n} g(z) \;=\; \frac{\partial}{\partial r} \log \frac{1}{r} \Big|_{r=1} \;=\; -1, \qquad ds = d\theta.$$

Hence harmonic measure on $\partial \mathbb{D}$ for $0 \in \mathbb{D}$ is the normalized arc length measure

$$d\mu_0 \;=\; \frac{d\theta}{2\pi}.$$

It is straightforward to check (Exercise 2) that the harmonic measure for an arbitrary point $\zeta \in \mathbb{D}$ is given by the Poisson kernel.

We say that z_0 is a **critical point of Green's function** if $(\nabla g)(z_0) = 0$. The value $g(z_0)$ is called a **critical value** of $g(z)$. If $g(z)$ is the real part of the analytic function $f(z)$ near z_0, then $|f'(z_0)| = |\nabla g(z_0)|$, so that z_0 is a critical point of $g(z)$ if and only if z_0 is a zero of $f'(z)$. We define the **order of the critical point** z_0 of $g(z)$ to be the order of zero of $f'(z)$ at $z = z_0$. According to our earlier definition, this is the order of the critical point z_0 of $f(z)$. (See Section VIII.5.) Further, if z_0 is a critical point of $g(z)$ of order m, then the set $\{g(z) > g(z_0)\}$ consists of $m + 1$ sector-shaped domains with vertex z_0, each of aperture $\pi/(m + 1)$. This makes it clear that no point of ∂D can be a critical point of $g(z)$, because near each $z_0 \in \partial D$ the set $\{g(z) > 0\}$ consists of just a half-disk, which can be regarded as a sector of aperture π, so that $m = 0$.

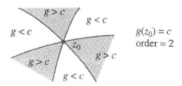

Theorem. *Let D be a bounded domain with analytic boundary consisting of N closed curves. Let $g(z)$ be Green's function for D with pole at $\zeta \in D$. Then $g(z)$ has $N - 1$ critical points in D, counting multiplicities.*

The critical points of $g(z)$ are isolated, since they are the zeros of an analytic function. There are only finitely many critical points of $g(z)$ in D, and there are no critical points of $g(z)$ on ∂D.

For $0 \le r < \infty$, let D_r be the set of $z \in D$ such that $g(z) > r$ and let $\Gamma_r = \partial D_r$ be its boundary. For r large, Γ_r is a circular curve surrounding ζ, and D_r is approximately a disk centered at ζ. As r decreases, from $+\infty$ to 0, the domains D_r increase until at $r = 0$ we reach $D_0 = D$ and $\Gamma_0 = \partial D$. Each D_r is a domain. Indeed, Green's function is constant on the boundary of each component of D_r, and Green's function cannot be constant on the boundary of any component of D_r on which $g(z)$ is harmonic, so there can be only one component of D_r, the component containing the pole ζ.

Except at a critical point, each curve in Γ_r is locally the level set of the real part of an analytic function with nonvanishing derivative, hence an analytic curve. Since the curves in Γ_r cannot approach the boundary of D, each curve in Γ_r is either a simple closed analytic curve, or it is an analytic curve that begins and terminates at (possibly different) critical points. Thus if r is not a critical value of $g(z)$, then Γ_r consists of a finite number of simple closed analytic curves.

Denote by U_r the complement of $D_r \cup \Gamma_r$ in the extended complex plane. Each component of U_r is a simply connected domain bounded by a curve in Γ_r, and there are as many components of U_r as there are analytic curves in Γ_r. We will count the number of components of U_r and see how this number changes as r decreases from $+\infty$ to 0. The analytic curves in Γ_r move continuously with r when r is not a critical value of $g(z)$, so the number of components of U_r remains constant as r decreases between critical values. We focus on what happens to a component of U_r when r crosses over a critical value.

Let z_0 be a critical point of $g(z)$ of order m, with critical value $g(z_0) = c$, and let $\varepsilon > 0$ be small. Let V be the component of $U_{c+\varepsilon}$ containing z_0, so that ∂V is one of the curves in $\Gamma_{c+\varepsilon}$. Then $\{g(z) > c\}$ consists of $m+1$ sector-like domains near z_0 with vertices at z_0. The curve ∂V approaches z_0 through each of these sectors and then retreats, as in the figure. In each sector we connect z_0 to the nearest point of ∂V by a straight line segment. We can think of V as a disk, via the Riemann mapping theorem, and then it is apparent that the $m+1$ segments divide V into $m+1$ components, as in the figure. Each of these $m+1$ components contains one of the $m+1$ sector-like domains with vertex at z_0 where $g(z) < c$. We conclude that there are exactly $m+1$ components of $\{g(z) < c\}$ that have z_0 in their boundary. Thus as r decreases from $c+\varepsilon$ to $c-\varepsilon$, the critical point z_0 of order m accounts for an increase in the number of components of U_r by m. The same is true for all critical points, even though some might correspond to the same critical value. Since U_r has one component when r is large, it eventually consists of $1 + L$ components when r is small, where L is the total number of critical points counting multiplicity. Thus $1 + L = N$, and the theorem is proved.

The idea of using the sublevel sets of a function such as $g(z)$ to describe the topology of D is the starting point for "Morse theory." One way to visualize the procedure is to place the graph of $g(z)$ in a water reservoir and gradually to drain the water. The curves Γ_r become the shoreline and D_r an island. As the level of water recedes, the islands D_r evolve to D and the shoreline to ∂D. Dramatic changes occur only when r passes through a critical value. The figure on the right depicts the level sets of Green's function of an annulus. Note that there is one critical point of order one, which corresponds to a saddle point of the graph of $g(z)$.

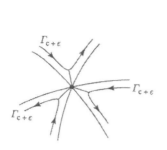

Exercises for XV.6

1. Show that for the unit disk $\mathbb{D}$, Green's function is given by

$$g(z, \zeta) = -\log \left| \frac{z - \zeta}{1 - \bar{\zeta}z} \right|, \qquad z, \zeta \in \mathbb{D}.$$

2. Show by computing the normal derivative of Green's function that harmonic measure for the unit disk $\mathbb{D}$ is given by the Poisson kernel,

$$d\mu_\zeta \left(e^{i\theta} \right) = \frac{1 - \rho^2}{1 + \rho^2 - 2\rho\cos(\theta - \varphi)} \frac{d\theta}{2\pi}, \qquad \zeta = \rho e^{i\varphi} \in \mathbb{D}.$$

3. Suppose that $g(z)$ is Green's function for D with pole at ζ, and suppose that $\varphi(z)$ is a conformal map of a domain W onto D such that $\varphi(\xi) = \zeta$. Show that $h(w) = g(\varphi(w))$ is Green's function for W with pole at ξ.

4. Suppose that D is a simply connected domain, mapped by $\varphi(z)$ conformally onto the open unit disk. Show that Green's function for D is given by

$$g(z, \zeta) = -\log \left| \frac{\varphi(z) - \varphi(\zeta)}{1 - \overline{\varphi(\zeta)}\varphi(z)} \right|, \qquad z, \zeta \in D.$$

5. Let D be a bounded domain with analytic boundary, and let $g(z, \zeta)$ be Green's function for D with pole at ζ. Show that if $u(z)$ is a smooth function on $D \cup \partial D$, then

$$u(\zeta) = \int_{\partial D} u(z) \, d\mu_\zeta(z) - \frac{1}{2\pi} \iint_D g(z, \zeta)(\Delta u)(z) \, dx \, dy, \qquad \zeta \in D.$$

6. Let D be a bounded domain with analytic boundary, let $g(z)$ be Green's function for D with pole at ζ, and let $v(z)$ be the (locally defined) harmonic conjugate function of $g(z)$. Fix $r \geq 0$. Show that

if we orient the curves of the level set $\{g(z) = r\}$ as boundary curves of the domain $\{g(z) > r\}$, then $v(z)$ is strictly decreasing along each curve in the level set. Show further that the total increase of $v(z)$ along the curves in the level set $\{g(z) = r\}$ is -2π.

7. Let D be a bounded domain with analytic boundary, let $g(z)$ be Green's function for D with pole at ζ, and let $v(z)$ be the (locally defined) harmonic conjugate function of $g(z)$. Fix one of the boundary curves γ of ∂D. Show that there is $\alpha > 0$ such that $w = e^{\alpha(g+iv)}$ is single-valued in a neighborhood of γ and maps an open set on the side of γ in D conformally onto an annulus $\{1 < |w| < 1 + \delta\}$.

8. Establish formula (6.4) in the case that $u(z)$ is only continuous on ∂D by applying the smooth case to the domain $\{g(z) > \varepsilon\}$ and letting $\varepsilon \to 0$.

9. A point (x_0, y_0) is a **critical point** of order m for a smooth function $u(x, y)$ if all partial derivatives of $u(x, y)$ with respect to x and y of order $k \le m$ vanish at (x_0, y_0), while at least one partial derivative of order $m + 1$ does not vanish there. In the case that $u(x, y)$ is the real part of the analytic function $f(z)$, show that (x_0, y_0) is a critical point of $u(x, y)$ of order m if and only if $z_0 = x_0 + iy_0$ is a critical point of $f(z)$ of order m.

7. Green's Function for General Domains

Green's function plays a useful role for domains without smooth boundaries and for unbounded domains. In this section we discuss briefly Green's function for general domains.

Let D be an arbitrary domain, and fix $\zeta \in D$. Let $\{D_n\}$ be an increasing sequence of bounded domains with analytic boundaries such that $\zeta \in D_1$ and $\cup D_n = D$. Let $g_n(z) = g_n(z, \zeta)$ be Green's function for D_n with pole at ζ. Since $D_n \subset D_{n+1}$ and $g_n(z) \le g_{n+1}(z)$ on ∂D_n, the maximum principle shows that $g_n(z) \le g_{n+1}(z)$ on D_n. Thus $\{g_n(z)\}$ is an increasing sequence, and $g_n(z) + \log |z - \zeta|$ is harmonic on D_n. According to Harnack's theorem, there are two cases that can occur. Either $g_n(z)$ tends to $+\infty$ uniformly on compact subsets of D, in which case we say that "Green's function for D does not exist." Or $g_n(z)$ converges uniformly on compact subsets of $D\backslash\{\zeta\}$, in which case we define the limit $g(z) = g(z, \zeta)$ to be **Green's function** of D with pole at ζ.

Lemma. *Green's function $g(z, \zeta)$ with pole at ζ satisfies the following:*

(7.1) $g(z, \zeta)$ *is a positive harmonic function on* $D\backslash\{\zeta\}$;

(7.2) $g(z, \zeta) - \log \dfrac{1}{|z - \zeta|}$ *is harmonic at* ζ,

(7.3) *if $h(z)$ is a positive harmonic function on $D\backslash\{\zeta\}$ that has a loga-*
     rithmic pole at ζ, then $h(z) \geq g(z, \zeta)$.

Properties (7.1), (7.2), and (7.3) characterize Green's function uniquely.

Properties (7.1) and (7.2) follow from the corresponding properties of
$g_n(z)$, since $g_n(z) + \log|z - \zeta|$ converges to $g(z) + \log|z - \zeta|$ uniformly on
a neighborhood of ζ. Property (7.3) follows from the maximum principle.
Indeed, $h(z) + \log|z - \zeta|$ dominates $g_n(z) + \log|z - \zeta|$ on ∂D_n hence on D_n,
and in the limit we obtain $h(z) \geq g(z)$ on D. If $g_0(z)$ is any other function
with the properties (7.1), (7.2), and (7.3), then (7.3) applies to each of $g(z)$
and $g_0(z)$, so $g(z) \geq g_0(z) \geq g(z)$, and equality holds.
    The uniqueness assertion shows that the definition of $g(z)$ is independent
of the sequence of domains $\{D_n\}$ increasing to D. Further, this definition
of $g(z)$ coincides with the definition given in the preceding section where
D is a bounded domain with analytic boundary. In this case we can take
$D_n = D$ for all n.

Example. Let $D = \mathbb{C}$ be the complex plane, and let $\zeta = 0$. We take
$D_n = \{|z| < n\}$, which has Green's function $g_n(z) = \log n - \log|z|$. Since
$g_n(z)$ tends to $+\infty$ as $n \to \infty$, Green's function does not exist for $\mathbb{C}$.

Lemma. *If D is a bounded domain, and $\zeta \in D$, then Green's function*
$g(z, \zeta)$ for D with pole at ζ exists. Further,

$$g(z, \zeta) = \tilde{h}(z) + \log \frac{1}{|z - \zeta|}, \qquad z \in D,$$

where $\tilde{h}(z)$ is the Perron solution of the Dirichlet problem with boundary
function $h(z) = \log|z - \zeta|$, $z \in \partial D$.

This follows from two applications of the maximum principle. Let $\{D_n\}$
and $g_n(z, \zeta)$ be as above. Define u_n by $u_n(z) = g_n(z, \zeta)$ for $z \in D_n$, and
$u_n(z) = 0$ for $z \in D\backslash D_n$. Then $u_n(z) + \log|z - \zeta|$ is subharmonic on D,
and it coincides with $\log|z - \zeta|$ on ∂D. Hence $u_n(z) + \log|z - \zeta| \in \mathcal{F}_h$,
and $u_n(z) + \log|z - \zeta| \leq \tilde{h}(z)$. Passing to the limit on D, we obtain
$g(z, \zeta) + \log|z - \zeta| \leq \tilde{h}(z)$.
    For the reverse inequality, let $\varepsilon > 0$, and let $u \in \mathcal{F}_h$. Then $u(z) - \varepsilon \leq$
$\log|z - \zeta|$ for n large. From the maximum principle for the harmonic
function $g_n(z, \zeta) + \log|z - \zeta|$ on D_n, we have $u(z) - \varepsilon \leq g_n(z, \zeta) + \log|z - \zeta|$
on D_n. Passing to the limit, first as $n \to \infty$ then as $\varepsilon \to 0$, we obtain
$u(z) \leq g(z, \zeta) + \log|z - \zeta|$ on D. Taking the supremum over such u, we
obtain $\tilde{h}(z) \leq g(z, \zeta) + \log|z - \zeta|$ on D. Consequently, equality holds, and
the lemma is established.

Example. Let $D = \mathbb{D}\backslash\{0\}$ be the punctured unit disk, and let $\zeta \in D$. We
have seen that for any continuous function $h(z)$ on ∂D, the Perron solu-

tion $\tilde{h}(z)$ to the Dirichlet problem is the Poisson integral of $h\left(e^{i\theta}\right)$, which coincides with the solution to the Dirichlet problem on $\mathbb{D}$ with boundary function $h|_{\partial\mathbb{D}}$. The preceding theorem then shows that Green's function for $\mathbb{D}\backslash\{0\}$ coincides with Green's function for $\mathbb{D}$. The isolated point $\{0\}$ of ∂D is a removable singularity for Green's function.

Now we consider the dependence of $g(z,\zeta)$ on the parameter ζ. We show that the existence of Green's function for a domain D does not depend on the pole ζ. One way to see this is as follows. Let $\zeta_0 \in D$, and let D_n increase to D as before, with $\zeta_0 \in D_1$. Choose $r > 0$ such that $\{|z - \zeta_0| \leq r\} \subset D_1$. Then

$$u_n(z,\zeta) = g_n(z,\zeta) + \log|z - \zeta| - g_n(z,\zeta_0) - \log|z - \zeta_0|$$

is harmonic on D_n. Further, since Green's function is zero on ∂D_n, $u_n(z,\zeta)$ is bounded in modulus by a constant C on ∂D_n, independent of n, provided that $|\zeta - \zeta_0| < r/2$. By the maximum principle,

$$-C \leq g_n(z,\zeta) + \log|z - \zeta| - g_n(z,\zeta_0) - \log|z - \zeta_0| \leq C$$

for all $z \in D$ and for $|\zeta - \zeta_0| < r/2$. This estimate shows that if $g_n(z,\zeta_0) \to +\infty$ as $n \to \infty$, then $g_n(z,\zeta) \to +\infty$ as $n \to \infty$ for $|\zeta - \zeta_0| < r/2$, while if $g_n(z,\zeta_0)$ is bounded, then $g_n(z,\zeta)$ is bounded for $|\zeta - \zeta_0| < r/2$. Thus the sets of ζ for which $g_n(z,\zeta) \to +\infty$ and for which $g_n(z,\zeta)$ is bounded form a decomposition of D into two disjoint open subsets. Since D is connected, one of these subsets is empty and the other is all of D. Thus the existence of Green's function does not depend on the pole ζ.

In Section XVI.3 we will treat Green's function in the setting of Riemann surfaces, and we will show that the existence of Green's function does not depend on the pole. In Section XVI.4 we will show that Green's function is symmetric, that is, it satisfies $g(z,\zeta) = g(\zeta,z)$. In particular, $g(z,\zeta)$ depends harmonically on the parameter ζ.

Exercises for XV.7

1. Suppose that $\zeta \in D_0 \subset D$, and suppose that Green's function $g(z)$ for D with pole at ζ exists. Show that Green's function $g_0(z)$ for D_0 with pole at ζ exists and satisfies $g_0(z) \leq g(z)$.

2. Let D be a domain for which Green's function $g(z)$ with pole at ζ exists. Show that for each $r > 0$, the set of $z \in D$ satisfying $g(z) > r$ is a domain, which has Green's function $g(z) - r$ with pole at ζ.

3. Let D be a domain for which Green's function $g(z)$ with pole at ζ exists. Suppose that $\varphi(z)$ is a conformal map of a domain W onto D such that $\varphi(\xi) = \zeta$. Show that $h(w) = g(\varphi(w))$ is Green's function for W with pole at ξ.

4. Show that Green's function for the upper half-plane $\mathbb{H}$ is

$$g(z, \zeta) = \log \left| \frac{z - \bar{\zeta}}{z - \zeta} \right|, \qquad z, \zeta \in \mathbb{H}.$$

5. Show that Green's function for the horizontal strip $\{|\operatorname{Im} z| < \pi/2\}$ is

$$g(z, \zeta) = \log \left| \frac{e^z + e^{\bar{\zeta}}}{e^z - e^{\zeta}} \right|, \qquad -\frac{\pi}{2} < \operatorname{Im} z, \operatorname{Im} \zeta < \frac{\pi}{2}.$$

6. Let $g(z, \zeta)$ be Green's function for the strip $\{|\operatorname{Im} z| < \pi/2\}$. (See the preceding exercise.) Show that for ζ fixed and $\beta > 0$, the series

$$\sum_{n=-\infty}^{\infty} g(z + \beta n, \zeta)$$

converges uniformly on compact subsets of the strip to a function $h(z)$ harmonic on the strip except for logarithmic poles at $\zeta + \beta n$, $-\infty < n < \infty$. Show that $h(z)$ is periodic with period β. Show that $h(\pi i (\log w)/2\alpha)$ is Green's function for the annulus $\{e^{-\alpha} < |w| < e^{\alpha}\}$ with pole at $e^{-2\alpha i \zeta/\pi}$, for $\alpha = \pi^2/\beta$.

7. Show that Green's function of D with pole at ζ is the upper envelope of the family of functions $u(z)$ on D such that $u(z) + \log |z - \zeta|$ is subharmonic on D, and $u(z) = 0$ off a compact subset of D.

8. Suppose that Green's function $g(z)$ for D exists and that the level set $\{g(z) = r\}$ does not contain a critical point of $g(z)$. Show that each connected component of the level set is either a simple closed analytic curve in D or an analytic arc, each end of which terminates at an irregular boundary point of D.

9. Let D be a domain, let $\zeta_0 \in \partial D$, and suppose there is a subharmonic barrier at ζ_0. Show that Green's function $g(z)$ for D exists, and $g(z) \to 0$ as $z \to \zeta_0$. *Remark.* In particular, if Green's function for D does not exist, then $\mathbb{C} \backslash D$ does not contain a continuum, so $\mathbb{C} \backslash D$ is totally disconnected (see Section XII.4).

10. Show that every domain D in $\mathbb{C}$ is the increasing union of bounded domains D_m with analytic boundaries.

11. Let D be an exterior domain, that is, a domain containing the exterior of a disk. If there is a harmonic function $g(z)$ on D such that $g(z) \to 0$ as $z \to \partial D$, and $g(z) - \log |z|$ is bounded as $z \to \infty$, we say that $g(z)$ is **Green's function for D with pole at ∞.** (a) Show that Green's function with pole at ∞ is unique (if it exists). (b) Show that $g(z) > 0$ for $z \in D$. (c) Show that $g(z) - \log |z|$ has a

limit as $z \to \infty$. *Remark.* The limit is called **Robin's constant** of D. (d) Show that if D is simply connected, then Robin's constant is $-\log |\varphi'(\infty)|$, where $\varphi(z)$ is the Riemann map of D onto the open unit disk satisfying $\varphi(\infty) = 0$, and $\varphi'(\infty) = \lim_{z \to \infty} z\varphi(z)$.

12. Show that the extended complex plane $\mathbb{C}^*$ has no Green's function with pole at ∞, in the sense that there is no harmonic function on the complex plane such that $g(z) - \log |z|$ is bounded as $z \to \infty$.

XVI

Riemann Surfaces

Our goal in this chapter is to prove the uniformization theorem for Riemann surfaces and to indicate its usefulness as a tool in complex analysis. We begin in Sections 1 and 2 by defining Riemann surfaces, providing examples, and showing how various local notions as analytic function, meromorphic function, and harmonic function carry over to Riemann surfaces. In Sections 3 and 4 we define Green's function for Riemann surfaces and show that Green's function is symmetric. In Section 5 we show that every Riemann surface has bipolar Green's functions. We prove the uniformization theorem in Section 6. The proof depends on Green's function when it exists and on bipolar Green's function otherwise. In Section 7 we define covering spaces and covering maps, and we state several results that indicate the power of the uniformization theorem.

1. Abstract Riemann Surfaces

Roughly speaking, a Riemann surface is a space that is locally the same as the complex plane. A nearsighted person standing on the surface could not distinguish it from the complex plane. We will use the following definition.

A **Riemann surface** is a set R with a collection of subsets $\{U_\alpha\}$ and complex-valued functions $z_\alpha(p)$, $p \in U_\alpha$, such that

(1.1) each $z_\alpha(p)$ is a one-to-one function from U_α onto a domain $z_\alpha(U_\alpha)$ in the complex plane;

(1.2) each composition $z_\beta \circ z_\alpha^{-1}$ is analytic wherever it is defined, that is, from $z_\alpha(U_\alpha \cap U_\beta)$ to $z_\beta(U_\alpha \cap U_\beta)$;

(1.3) R is connected, that is, for any two points p and q in R, there is a finite collection of indices $\alpha_1, \ldots, \alpha_m$ such that $p \in U_{\alpha_1}$, $q \in U_{\alpha_m}$, and $U_{\alpha_j} \cap U_{\alpha_{j+1}}$ is nonempty for $1 \le j < m$;

(1.4) R is a "Hausdorff space," that is, if p_0 and p_1 are distinct points of R, and if $p_0 \in U_\alpha$ and $p_1 \in U_\beta$, then there are (small) disks D_0 and D_1 centered at $z_\alpha(p_0)$ and $z_\beta(p_1)$, respectively, such that $z_\alpha^{-1}(D_0)$ and $z_\beta^{-1}(D_1)$ are disjoint subsets of R.

We refer to U_α as a **coordinate patch** and to $z_\alpha(p)$ as the **coordinate map** on U_α, mapping U_α onto $z_\alpha(U_\alpha)$. The composition $z_\beta \circ z_\alpha^{-1}$ corresponds to a change of variable. If D_0 is a disk in $z_\alpha(U_\alpha)$, we refer to $W_0 = z_\alpha^{-1}(D_0)$ as a **coordinate disk** in R. The coordinate function z_α maps W_0 one-to-one onto D_0.

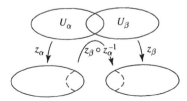

Example. Each domain D in the complex plane can be regarded as a Riemann surface. We require only one coordinate patch $U_1 = D$, with coordinate map the identity function $z_1(\zeta) = \zeta$.

Example. The simplest nontrivial Riemann surface is the extended complex plane $\mathbb{C}^* = \mathbb{C} \cup \{\infty\}$. This can be coordinatized by two coordinate patches $U_0 = \mathbb{C}$ and $U_1 = \mathbb{C}^* \backslash \{0\}$, with coordinate maps $z_0(\zeta) = \zeta$ and $z_1(\zeta) = 1/\zeta$. In this case, $z_0(U_0 \cap U_1) = \mathbb{C} \backslash \{0\}$, and the change of coordinate variable is given by

$$(z_1 \circ z_0^{-1})(\zeta) = 1/\zeta, \qquad \zeta \in \mathbb{C} \backslash \{0\},$$

which is analytic.

Example. Let ω_1 and ω_2 be two complex numbers that do not lie on the same line through the origin, and let L be the lattice they generate, consisting of all linear combinations $m\omega_1 + n\omega_2$ where m and n are integers. We denote by $T = \mathbb{C}/L$ the torus determined by L. Points of T can be regarded as congruence classes $\zeta + L$, where $\zeta \in \mathbb{C}$. Choose $\varepsilon > 0$ so small that all nonzero lattice points satisfy $|m\omega_1 + n\omega_2| > \varepsilon$. For each complex number $\lambda \in \mathbb{C}$, let U_λ be the subset of T defined by

$$U_\lambda = \{\zeta + L : |\zeta - \lambda| < \varepsilon\}.$$

The coordinate map $z_\lambda : U_\lambda \to \mathbb{C}$ is defined by $z_\lambda(\zeta + L) = \zeta$ for $|\zeta - \lambda| < \varepsilon$. It maps U_λ onto the disk $\{|z - \lambda| < \varepsilon\}$. The condition on ε guarantees that the coordinate map z_λ is one-to-one. A change of coordinate $z_\alpha \circ z_\lambda^{-1}$ is the identity, wherever defined, hence analytic.

Example. Consider the Riemann surface R of $w = \sqrt{z}$ as constructed in Section I.4. It is obtained from two copies S_1, S_2 of the slit z-plane $\mathbb{C} \backslash (-\infty, 0]$ by identifying edges appropriately. We can make R into an abstract Riemann surface with four coordinate patches U_1, U_2, U_3, U_4, where U_1 is S_1 (without the edges) and z_1 is the natural map of S_1 onto $\mathbb{C} \backslash (-\infty, 0]$; similarly for U_2 and z_2; U_3 consists of the top half of S_1, the bottom half of S_2, and the slit along which they are joined, and z_3 is the natural map of U_3 onto $\mathbb{C}\backslash[0, +\infty)$; and similarly for U_4. In this case, the change of variable corresponding to two coordinate maps is the identity function wherever defined, hence analytic.

Example. The Riemann surface R of $w = \log z$, as constructed in Section I.5, can be made into an abstract Riemann surface in the same way. In this case, R is obtained from an infinite sequence of copies of the slit plane $\mathbb{C}\backslash(-\infty, 0]$ by identifying edges of slits appropriately. This time we use infinitely many coordinate patches, one for each sheet and one to cover each line where two sheets are attached.

Topological notions and notions of convergence and continuity are carried over to a Riemann surface R by referring them to the complex plane via the coordinate maps. We define a subset W of R to be **open** if for all α the image $z_\alpha(W \cap U_\alpha)$ is an open subset of the complex plane. Since a change of variable $z_\beta \circ z_\alpha^{-1}$ maps open sets to open sets, a subset of a coordinate patch U_α is open in R if and only if its image under the single coordinate map z_α is open.

We say that a sequence $\{p_n\}$ of points of R **converges** to $p \in R$ if for any coordinate patch U_α containing p, the p_n's eventually belong to U_α, and $z_\alpha(p_n)$ converges to $z_\alpha(p)$. The "Hausdorff" condition (1.4) asserts simply that the limit of a convergent sequence is unique. It is possible to construct pathological spaces that satisfy the other three axioms (1.1)-(1.3) but for which limits of sequences are not unique.

A subset K of R is **compact** if K is contained in a finite union of coordinate patches, and every sequence in K has a subsequence that converges to a point in K. This occurs if and only if K can be expressed as a finite union $K = K_1 \cup \cdots \cup K_m$, where each K_j is a compact subset of a single coordinate patch.

A complex-valued function $f(p)$ defined on R is **continuous at** q if $f \circ z_\alpha^{-1}$ is continuous at $z_\alpha(q)$ for any coordinate patch U_α containing q. This occurs if and only if $f(p_n) \to f(q)$ whenever p_n converges to q. Similarly, a function g from a subset E of $\mathbb{R}^n$ to R is **continuous** if for any coordinate patch U_α, the composition $z_\alpha \circ g$ is continuous on $g^{-1}(U_\alpha)$.

A **path** in R from p_0 to p_1 is a continuous function $\gamma(t)$, $0 \le t \le 1$, from the unit interval $[0, 1]$ into R such that $\gamma(0) = p_0$ and $\gamma(1) = p_1$. The path is **smooth** if the compositions $z_\alpha \circ \gamma$ are smooth wherever defined. Piecewise smooth paths and analytic curves are defined similarly.

An open subset D of a Riemann surface is **connected** if any two points of D can be joined by a path in D. The connectivity hypothesis (1.3) in the definition of Riemann surface is equivalent to requiring that the surface be connected. Indeed, since each $z_\alpha(U_\alpha)$ is a domain, any two points of U_α can be joined by a path, and by concatenating a finite number of paths we can join any two points of R.

Example. A connected open subset of a Riemann surface can be regarded as a Riemann surface, by restricting the coordinate maps to the subset. We refer to the connected open subset as a **subsurface** of the Riemann surface, or as a **domain** on the surface.

In addition to properties involving convergence and continuity, any "local" property that is invariant under an analytic change of variable can be carried over to a Riemann surface by means of the coordinate maps. We list several such properties.

A complex-valued function $f(p)$ defined in an open subset V of R is **analytic** if $f \circ z_\alpha^{-1}$ is analytic on $z_\alpha(V \cap U_\alpha)$ for each coordinate patch U_α. A function $f : V \to \mathbb{C}^*$ is **meromorphic** if $f \circ z_\alpha^{-1}$ is meromorphic on each $z_\alpha(V \cap U_\alpha)$. The function $f(p)$ is **analytic at** $p_0 \in R$ (respectively **meromorphic at** $p_0 \in R$) if it is analytic (respectively meromorphic) in some coordinate disk containing p_0.

The order of a zero of an analytic function at a point is invariant under an analytic change of variable. Thus we can talk about the zeros and their orders of an analytic function on a Riemann surface. Similarly, the orders of poles of meromorphic functions on a Riemann surface are well-defined.

Example. For a domain in the complex plane, regarded as a Riemann surface, the definitions of analytic and meromorphic functions reduce to the usual definitions.

Example. Since the coordinate map for $\mathbb{C}^* \backslash \{0\}$ is $z_1(\zeta) = 1/\zeta$, the analytic functions on a domain R on the Riemann sphere $\mathbb{C}^*$ are the functions $f(\zeta)$ that are analytic on $R \cap \mathbb{C}$, such that $g(\zeta) = f(1/\zeta)$ is analytic at $\zeta = 0$. Thus the definition of a function being analytic at ∞ when we regard R as an abstract Riemann surface is the same as the definition given in Section V.5 for analyticity at ∞. Similarly, the definition of a function being meromorphic at ∞ when we regard R as an abstract Riemann surface is the same as the definition of being meromorphic at ∞ given in Section VI.3.

Example. We saw in Section I.4 that the function $\sqrt{z}$ can be defined continuously on its Riemann surface R, by defining it to be the principal branch on one sheet and the negative of the principal branch on the second sheet. Thus defined, the function $\sqrt{z}$ is actually analytic on R. On each of the coordinate patches U_1, U_2, U_3, U_4, the function corresponds to an analytic branch of $\sqrt{z}$ on a slit plane, with respect to the local coordinate map.

Example. Each doubly periodic meromorphic function $f(z)$ on the complex plane with periods ω_1, ω_2 determines a meromorphic function $g(p)$ on the torus $T = \mathbb{C}/L$ by $g(z + L) = f(z)$. Every meromorphic function on the torus arises in this manner from a doubly periodic function on the complex plane.

We have defined analyticity for complex-valued functions on a Riemann surface. We can also define a notion of analyticity for a function from one Riemann surface to another.

Let R and S be two Riemann surfaces. A function $f : R \rightarrow S$ is **analytic at** $p \in R$ if for any coordinate patch U_α in R containing p with coordinate map z_α, and any coordinate patch V_β in S containing $f(p)$ with coordinate map w_β, the composition $w_\beta \circ f \circ z_\alpha^{-1}$ is analytic at $z_\alpha(p)$. We refer to f also as an **analytic map** from R to S.

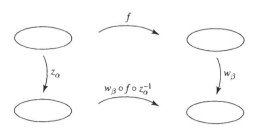

Example. The analytic functions from a Riemann surface R to the Riemann sphere $\mathbb{C}^*$ are the meromorphic functions on R, together with the constant function ∞ on R (Exercise 4).

Technically, our definition of Riemann surface depends not only on the set R but the specific coordinate patches $\{U_\alpha\}$ and coordinate maps z_α. We will regard two Riemann surfaces as being the "same" if the underlying sets are the same and the identity map between them is analytic and has an analytic inverse. This occurs if and only if they have the same analytic functions, and in this case we can use as a coordinate map any one-to-one analytic function from an open subset U of R onto a domain in the complex plane. This allows us to focus on coordinate disks when convenient.

We say that two Riemann surfaces R and S are **conformally equivalent** if there is an analytic map $f : R \rightarrow S$ that is one-to-one and onto. In this case, the inverse map $f^{-1} : S \rightarrow R$ is also analytic.

Example. The Riemann surface of $\sqrt{z}$ is conformally equivalent to the punctured complex plane $\mathbb{C}\backslash\{0\}$. The analytic map giving the equivalence is the continuous extension of the function $f(z) = \sqrt{z}$ to the Riemann surface.

Let S be a Riemann surface. A subsurface R of S with boundary ∂R is a **finite bordered subsurface** if $R \cup \partial R$ is a compact subset of S, and ∂R consists of a finite number of disjoint simple closed analytic curves in S. Thus for each $p_0 \in \partial R$ there is a coordinate disk U in S with coordinate map $z(p)$ that maps $U \cap R$ onto a semidisk in the upper half-plane and maps $U \cap \partial R$ onto an interval on the real axis. The analytic structure on $R \cup \partial R$ is actually independent of the ambient surface S, and is determined by the coordinate disks for points of R and coordinate semidisks for points of ∂R. We refer to $R \cup \partial R$ as a **finite bordered Riemann surface**.

Example. Let S be a compact Riemann surface, and let $D_1, \ldots, D_N$ be N disjoint closed coordinate disks in S. If we punch out the closed disks, we obtain a finite bordered Riemann surface $R = S \backslash \cup D_j$ with border ∂R consisting of the N analytic boundary curves ∂D_j, $1 \leq j \leq N$. It turns out (Exercise 2.5) that every finite bordered Riemann surface can be obtained in this way, from a compact surface by excising a finite number of coordinate disks.

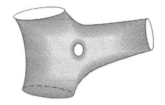

Exercises for XVI.1

1. Define the Riemann surface R of $\log z$ in terms of explicit coordinate patches. Define explicitly the function on R determined by $\log z$. Show that it is a one-to-one analytic map of R onto the complex plane $\mathbb{C}$.

2. What is the smallest number of coordinate patches necessary to make a torus into a Riemann surface?

3. Show (a) that the residue of a meromorphic function $f(p)$ at a pole $p_0 \in R$ depends on the choice of the coordinate at p_0, and (b) if the pole is not simple, a coordinate can be chosen with respect to which $f(p)$ has residue zero at p_0.

4. Show that the analytic maps from a Riemann surface R to the Riemann sphere $\mathbb{C}^*$ are the meromorphic functions on R and the constant function ∞.

5. Let $\omega \neq 0$, and let $\mathbb{Z}\omega$ be the integral multiples of ω. Let R be the set of congruence classes $z + \mathbb{Z}\omega$, $z \in \mathbb{C}$. Show that R is a Riemann surface that is conformally equivalent to the punctured plane $\mathbb{C} \backslash \{0\}$.

6. Consider the Riemann surface R of $\sqrt{z}$ as described in the example in this section. Let $S = R \cup \{0, \infty\}$. Define a coordinate disk V_0 containing 0 to be the point 0 together with the set of $z \in S_1 \cup S_2$ such that $|z| < 1$, and define the coordinate map $\pi_0 : V_0 \rightarrow \mathbb{D}$ by $\pi_0(z) = \sqrt{z}$, $\pi_0(0) = 0$. Similarly, define a coordinate disk V_1 containing ∞ to be the point ∞ together with the set of $z \in S_1 \cup S_2$

such that $|z| > 1$, with coordinate map $\pi_1 : V_1 \to \mathbb{D}$, $\pi_1(z) = 1/\sqrt{z}$, $\pi_1(\infty) = 0$. Show that S is a Riemann surface. Show that the function $\sqrt{z}$ on R extends to a one-to-one analytic map of S onto the Riemann sphere $\mathbb{C}^*$.

7. Suppose that $f : R \to S$ is a nonconstant analytic map. (a) Show that $f(R)$ is an open subset of S. (b) Show that if R is compact, then $f(R) = S$. (c) Use (b) to prove the fundamental theorem of algebra, that every nonconstant polynomial $P(z)$ has a zero.

8. Show that if U and V are disjoint open subsets of a Riemann surface R such that $R = U \cup V$, then either $U = R$ or $V = R$. Use this to show that the uniqueness principle for meromorphic functions holds on a Riemann surface R, that is, if $f(p)$ and $g(p)$ are two meromorphic functions on R such that $f(p) = g(p)$ for p in some nonempty open subset of R, then $f(p) = g(p)$ for all $p \in R$.

9. Let $f(\zeta)$ be analytic at $\zeta = \zeta_0$, and let f_{ζ_0} be its power series expansion at ζ_0. Let R be the set of all pairs (ζ, f_ζ), where $\zeta \in \mathbb{C}$, f_ζ is a power series expansion centered at ζ with a positive radius of convergence, and f_ζ is the analytic continuation of f_{ζ_0} along some path from ζ_0 to ζ. Define a coordinate patch containing (ζ, f_ζ) to consist of all pairs (ξ, f_ξ) such that $|\xi - \zeta| < \varepsilon$ and f_ξ is the power series expansion of $f_\zeta(z)$ about $z = \xi$. Define the coordinate map on this coordinate patch by $\pi(\xi, f_\xi) = \xi$. (a) Show that R is a Riemann surface. (b) Show that the projection function $\pi(\xi, f_\xi) = \xi$ is analytic on R. (c) Show that $F(\xi, f_\xi) = f_\xi(\xi)$ is analytic on R and that in terms of the above coordinate map at ζ_0 it coincides with $f(\zeta)$ in a coordinate disk centered at ζ_0. (d) Show that if $P(\zeta, w)$ is a polynomial in ζ and w such that $P(\zeta, f(\zeta)) = 0$ for $|\zeta - \zeta_0| < \varepsilon$, then $P(\xi, f_\xi(\xi)) = 0$ for all $(\xi, f_\xi) \in R$.

10. Let $x_1 < x_2 < \cdots < x_n$ be n consecutive points on the real axis. Let ζ_0 be distinct from the x_j's, and let $f(\zeta)$ be a branch of $\sqrt{(\zeta - x_1) \cdots (\zeta - x_n)}$ defined at ζ_0. Describe the Riemann surface R of $f(\zeta)$ constructed in the preceding exercise, and show that it is the same as the Riemann surface described in Exercise I.7.7. Show that $\pi(\xi, f_\xi) = \xi$ is a two-to-one map of R onto $\mathbb{C} \backslash \{x_1, \ldots, x_n\}$. Show that the surface R is equivalent to a compact surface S with a finite number of punctures. *Remark.* The Riemann surface S is a **hyperelliptic** Riemann surface.

11. Let R be a finite bordered Riemann surface, with border ∂R. Let $\tilde{R}$ be a duplicate copy of R, and denote by $\tilde{p}$ the point in $\tilde{R}$ corresponding to $p \in R$. Let $S = R \cup \tilde{R} \cup \partial R$. Define $\tau : S \to S$ by $\tau(p) = p$ if $p \in \partial R$ and by $\tau(p) = \tilde{p}$, $\tau(\tilde{p}) = p$ if $p \in R$.

(a) Show that S can be made into a compact Riemann surface with R as a subsurface so that τ is anticonformal, that is, $f(p)$ is analytic on an open set U if and only if $\overline{f(\tau(p))}$ is analytic on $\tau(U)$. *Remark.* The surface S is the **doubled surface** of R, and τ is the reflection in ∂R.

(b) Show that the doubled surface of the unit disk is the Riemann sphere $\mathbb{C}^*$.

(c) What is the doubled surface of an annulus?

(d) Show that each meromorphic function f on S has the form $f = g + ih$, where g and h are meromorphic functions on S that are real-valued on ∂R.

(e) Show that if f is meromorphic on R, and if $\operatorname{Im} f(p) \to 0$ as $p \to \partial R$, then f extends to be meromorphic on S, and the extension is analytic and real-valued on ∂R.

12. Show that a hyperelliptic Riemann surface (Exercise 10) is the double of a finite bordered subsurface. Specify the anticonformal reflection τ.

13. For τ in the open upper half-plane $\mathbb{H}$, denote by L_τ the lattice $\mathbb{Z} + \mathbb{Z}\tau$ generated by 1 and τ, and denote the Riemann surface $\mathbb{C}/L_\tau$ by T_τ.

(a) Show that the Riemann surface $T = \mathbb{C}/L$ in the example in this section is conformally equivalent to the Riemann surface T_τ for some $\tau \in \mathbb{H}$. *Hint.* Take $\tau = \pm\omega_1/\omega_2$, where the sign is chosen so that $\operatorname{Im}\tau > 0$.

(b) Show that T_τ is conformally equivalent to $T_{\tau'}$ if and only if there is a fractional linear transformation of the form $f(z) = (az + b)/(cz + d)$ where a, b, c, d are integers satisfying $ad - bc = 1$, such that $f(\tau) = \tau'$. *Remark.* The matrix with entries a, b, c, d is called a **unimodular matrix**. The unimodular matrices form a group, which is the special linear group $\mathrm{SL}(2, \mathbb{R})$.

14. Let $\mathcal{U}$ denote the fractional linear transformations $f(z) = (az + b)/(cz + d)$ corresponding to unimodular matrices. Let U be the domain in the upper half-plane consisting of points $z \in \mathbb{H}$ satisfying $|z| > 1$ and $-\frac{1}{2} < \operatorname{Re} z < \frac{1}{2}$.

(a) Show that $g(z) = -1/z$ belongs to $\mathcal{U}$, and $g(z)$ has a fixed a point at i. Sketch the image of U under $g(z)$.

(b) Show that $h(z) = 1 - 1/z$ belongs to $\mathcal{U}$, and $h(z)$ has a fixed point at the vertex $(1 + \sqrt{3}i)/2$ of U. Sketch the images of U under $h(z)$ and $(h \circ h)(z)$.

(c) Show that the images of U and its boundary curves under the fractional linear transformations in $\mathcal{U}$ fill out the upper half-plane. *Hint.* Show that the images fill out an open set containing $U \cup \partial U$.

(d) Show that if $f \in \mathcal{U}$, then either $f(z) = z + m$ for some integer m, or $f(z)$ maps the outside of the circle $|z + d/c| = 1/|c|$ onto the inside of a circle centered on the real axis of the same radius $1/|c|$. *Remark.* The circle $|z + d/c| = 1/|c|$ is the **isometric circle** of $f(z)$.

(e) Show that if $z, w \in U \cup \partial U$ satisfy $w = f(z)$ for some $f \in \mathcal{U}$ other than the identity transformation, then either $w = z \pm 1$ and both points lie on opposite vertical sides of $\mathcal{U}$, or $w = -1/z$ and both points lie on the unit circle. *Remark.* Thus the set E consisting of U and the points $z \in \partial U$ satisfying $\operatorname{Re} z \leq 0$ is a "fundamental domain" for $\mathcal{U}$, in the sense that for each $w \in \mathbb{H}$ there is a unique $z \in E$ such that $w = f(z)$ for some $f \in \mathcal{U}$. The images of U under the functions in $\mathcal{U}$ form a "tiling" of $\mathbb{H}$. *Remark.* With composition as the operation, $\mathcal{U}$ is a group, called the **modular group**. If A is a unimodular matrix, then both $\pm A$ give rise to the same $f \in \mathcal{U}$, and every $f \in \mathcal{U}$ arises from exactly two unimodular matrices. Thus the modular group is isomorphic to the quotient $\mathrm{SL}(2, \mathbb{R})/\{\pm I\}$.

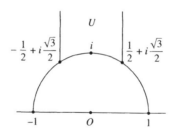

2. Harmonic Functions on a Riemann Surface

In this section we adapt the Perron procedure to a Riemann surface to solve the Dirichlet problem. The development proceeds very much along the same lines as for planar domains. The success of the procedure depends on the fact that the compactness theorems for harmonic functions are local, as are the notions of subharmonic function and subharmonic barrier.

A real-valued function $u(p)$ on a Riemann surface R is **harmonic** at $p_0 \in R$ if there is a coordinate disk containing p_0 such that $u(p)$ is a harmonic function of the coordinate $z = z(p)$ on the coordinate disk. This occurs if and only if $u(p)$ is the real part of an analytic function on the coordinate disk. Thus the definition of harmonicity is independent of the coordinate map.

The local arguments involving harmonic functions on parameter disks, together with compactness and connectedness arguments, allow us to carry over to Riemann surfaces the compactness theorems for families of harmonic functions from Section XV.3. One specific result we will use is the following version of Harnack's estimate.

Lemma. *For each compact subset K of a Riemann surface R, there is a constant $C > 0$ such that*

$$\frac{1}{C} \leq \frac{u(p)}{u(q)} \leq C, \qquad p, q \in K,$$

for every positive harmonic function on R.

If p belongs to a closed subdisk of a coordinate disk that has q as its center, this is simply Harnack's inequality. If we replace C by C^2, we obtain an estimate of the same form for arbitrary points p and q in a fixed closed subdisk of a coordinate disk. For the general case we cover K by a connected set that is a finite union of closed subdisks of coordinate disks, and we apply the estimate to successive points in a chain $p = p_0, p_1, \dots, p_n = q$, where each pair p_j, p_{j+1} belong to the same disk. (See Exercises XV.3.4-5.)

Now let R be a Riemann surface, and let W be an open subset of R. A continuous function $u : W \to [-\infty, +\infty)$ is **subharmonic** on W if each $p \in W$ belongs to a coordinate patch on which $u(p)$ is a subharmonic function of the coordinate $z = z(p)$. (We could allow subharmonic functions to be only "upper semicontinuous," but for our goals this leads to no gain.)

Theorem (Strict Maximum Principle). *If u is a subharmonic function on a Riemann surface R, and u attains its maximum at some point of R, then u is constant on R.*

The proof of the strict maximum principle for subharmonic functions (Section XV.2) carries over to Riemann surfaces. Let E be the set of $p \in R$ where u attains its maximum. The local argument used in the plane shows that E is open, while the continuity of u implies that $R \backslash E$ is also open. The connectedness of R then implies that either $E = R$ or E is empty. In the case at hand, we must have $E = R$, and then u is constant.

Since continuous functions on a compact set attain their maximum, we obtain as an immediate consequence the following version of the maximum principle.

Theorem (Maximum Principle). *Let u be a subharmonic function on a Riemann surface R. If $u(p) \leq c$ for $p \in R \backslash K$, where K is a compact subset of R, then $u(p) \leq c$ for all $p \in R$.*

Now we may solve the Dirichlet problem on a Riemann surface by the Perron process of taking an upper envelope of subsolutions. To make this explicit, it is convenient to formalize the Perron procedure.

Let W be an open subset of the Riemann surface R. We say that a nonempty family $\mathcal{F}$ of subharmonic functions on W is a **Perron family of subharmonic functions** if the following two conditions are satisfied:

(2.1) if $u, v \in \mathcal{F}$, then $\max(u, v) \in \mathcal{F}$, where $\max(u, v)$ is the maximum of u and v,

(2.2) if $u \in \mathcal{F}$, and D_0 is a coordinate disk in W such that u is finite on ∂D_0, then the function v defined to be u on $W \backslash D_0$ and the harmonic extension of $u|_{\partial D_0}$ on D_0 is in $\mathcal{F}$.

The proof given in Section XV.4 carries over to show the following.

Theorem. *Let $\mathcal{F}$ be a Perron family of subharmonic functions on a domain W on a Riemann surface R, and let u be the upper envelope of the family $\mathcal{F}$,*

$$u(p) \;=\; \sup\{v(p) : v \in \mathcal{F}\}.$$

Then either u is harmonic on W, or $u(p) = +\infty$ for all $p \in W$.

The notion of a subharmonic barrier (Section XV.4) is also local. The proof given in Section XV.4 shows that if $W \cup \partial W$ is compact, and there is a subharmonic barrier at every point of ∂W, then the Dirichlet problem is solvable for W. Since there is a subharmonic barrier at every point on a boundary arc of W, we obtain the following.

Theorem. *Let W be a domain on a Riemann surface such that $W \cup \partial W$ is compact and ∂W consists of a finite number of simple closed piecewise smooth boundary curves. Then each continuous function $h(q)$ on ∂W is the boundary values of a (unique) harmonic function $\tilde{h}(p)$ on W.*

As an application, we show that any Riemann surface can be exhausted by finite bordered subsurfaces.

Theorem. *For each compact subset K of a Riemann surface R, there is a domain W in R such that $K \subset W$, $W \cup \partial W$ is compact, and ∂W consists of a finite number of disjoint simple closed analytic curves.*

For the proof, we cover K by a finite number of coordinate disks, and then we connect these by finitely many more coordinate disks, to obtain a domain $U_0 \subset R$ such that $U_0 \supset K$ and U_0 is a finite union of coordinate disks. It is straightforward then to construct a domain $U \subset U_0$ such that $U \supset K$ and ∂U consists of a finite number of smooth arcs. Let D_0 be a small coordinate disk in U, with boundary Γ_0 contained in U. We solve the Dirichlet problem on the domain $U \backslash (D_0 \cup \Gamma_0)$ and obtain a harmonic function u with boundary values $u(p) = 1$ on Γ_0 and $u(p) = 0$ on ∂U. Since the critical points of u are isolated points that can accumulate only on $\partial U \cup \Gamma_0$, we can choose $\varepsilon > 0$ small such that the level set $\{u = \varepsilon\}$ does not contain a critical point of u, and further such that $u > \varepsilon$ on $K \cap U$. Then the domain $W = \{p \in U : u(p) > \varepsilon\} \cup \Gamma_0 \cup D_0$ has the desired properties.

Exercises for XVI.2

1. Let $w = w(z)$ be analytic on a domain D in the complex plane. Show that

$$\frac{\partial^2}{\partial z \partial \bar{z}} = \left|\frac{dw}{dz}\right|^2 \frac{\partial^2}{\partial w \partial \bar{w}}.$$

Deduce that a smooth function $h(w)$ is harmonic on $w(D)$ if and only if $h(w(z))$ is harmonic on D.

2. Let W be an open subset of a Riemann surface such that $W \cup \partial W$ is compact, and let $u(p)$ be a subharmonic function on W. Show that if $\limsup u(p) \leq c$ as $p \to \partial W$, then $u(p) \leq c$ for all $p \in W$.

3. Let $\{u_n\}$ be a sequence of positive harmonic functions on a Riemann surface R that converges uniformly on some coordinate disk. Show that the sequence converges uniformly on each compact subset of R.

4. Let $\{u_n\}$ be a sequence of harmonic functions on a Riemann surface R that is uniformly bounded on each compact subset of R and that converges uniformly on some coordinate disk. Show that the sequence converges uniformly on each compact subset of R.

5. Show that any finite bordered Riemann surface can be obtained from a compact Riemann surface by excising a finite number of closed coordinate disks. *Hint.* Assume that there is one boundary curve in ∂R. By punching out a small disk D_0 and solving a Dirichlet problem, find a positive harmonic function u on $R \backslash D_0$ such that $u = 0$ on ∂R. Use u and its harmonic conjugate to map $\{0 < u < \varepsilon\}$ onto an annulus $\{1 < |w| < \sigma\}$. Take $S = R \cup \{|w| \leq 1\}$, and use $\{|w| < \sigma\}$ as a coordinate disk.

3. Green's Function of a Surface

To define Green's function for a Riemann surface R, we adapt the definition given for an arbitrary planar domain. Fix a point $q \in R$, and let $z(p)$ be a coordinate map for a coordinate disk at q satisfying $z(q) = 0$. Let $\mathcal{F}_q$ be the family of subharmonic functions $u(p)$ on $R \backslash \{q\}$ such that $u = 0$ off some compact subset of R, and $u(p) + \log|z(p)|$ is subharmonic on the coordinate disk at q. Then $\mathcal{F}_q$ is a Perron family of subharmonic functions on $R \backslash \{q\}$. The family is nonempty, since the constant function $u = 0$ belongs to $\mathcal{F}_q$.

There are now two cases that can occur. If the upper envelope of $\mathcal{F}_q$ is finite on $R\backslash\{q\}$, we say that **Green's function for R with pole at q exists**, and we denote it by

$$g(p,q) \;=\; \sup\{u(p) : u \in \mathcal{F}_q\}, \qquad p \in R\backslash\{q\}.$$

Otherwise, the upper envelope of $\mathcal{F}_q$ is $+\infty$ at all points $p \in R\backslash\{q\}$, and in this case we say that **Green's function does not exist**.

Theorem. *Suppose Green's function $g(p,q)$ with pole at q exists, and let $z(p)$ be a coordinate map for a coordinate disk at q with $z(q) = 0$. Then $g(p,q)$ is harmonic and (strictly) positive for $p \in R\backslash\{q\}$, and $g(p,q) + \log|z(p)|$ is harmonic at q. If $h(p)$ is a positive harmonic function on $R\backslash\{q\}$ such that $h(p) + \log|z(p)|$ is harmonic at q, then $h(p) \geq g(p,q)$ for $p \in R\backslash\{q\}$.*

As the upper envelope of a Perron family, $g(p,q)$ is harmonic, and further $g(p,q) \geq 0$. Let $\rho > 0$ be small, so that the coordinate disk $\{|z(p)| \leq \rho\}$ is defined. The function defined to be $-\log|z(p)| + \log\rho$ on this coordinate disk and 0 outside it belongs to $\mathcal{F}_q$. Consequently,

$$(3.1) \qquad g(p,q) \geq -\log|z(p)| + \log\rho, \qquad 0 < |z(p)| \leq \rho,$$

and $g(p,q) \to +\infty$ as $p \to q$. Since $g(p,q)$ cannot attain its minimum unless it is constant, $g(p,q) > 0$ on $R\backslash\{q\}$. Let M be the maximum of $g(p,q)$ on the coordinate circle $\{|z(p)| = \rho\}$. Any $u \in \mathcal{F}_q$ satisfies $u(p) \leq M$ for $|z(p)| = \rho$, and consequently,

$$u(p) + \log|z(p)| \;\leq\; M + \log\rho$$

for $|z(p)| = \rho$. Since $u(p) + \log|z(p)|$ is subharmonic on the coordinate disk, this estimate persists for $|z(p)| \leq \rho$. Taking the supremum over $u \in \mathcal{F}_q$, we obtain

$$(3.2) \qquad g(p,q) + \log|z(p)| \;\leq\; M + \log\rho, \qquad |z(p)| \leq \rho.$$

Combining this with (3.1), we see that $g(p,q) + \log|z(p)|$ is a bounded harmonic function on the punctured coordinate disk $\{0 < |z(p)| < \rho\}$. Since isolated singularities of bounded harmonic functions are removable, $g(p,q) + \log|z(p)|$ is harmonic at q.

For the final statement of the theorem, suppose $h(p)$ is a positive harmonic function on $R\backslash\{q\}$ with a logarithmic pole at q. If $u \in \mathcal{F}_q$, then $u - h$ is subharmonic on R, and $u - h < 0$ off a compact subset of R. By the maximum principle, $u - h < 0$ on R. Taking the upper envelope of such u, we obtain $g(p,q) - h(p) \leq 0$. This completes the proof.

Our next goal is to show that if $g(p,q)$ exists for one point $q \in R$, then it exists for all $q \in R$. The proof is based on an auxiliary harmonic function and a clever estimation technique that merits special attention.

Fix $q_0 \in R$, and let $z(p)$ be a coordinate map for a coordinate disk at q_0 with $z(q_0) = 0$. Fix $r > 0$ small so that the closed coordinate disk $B_r = \{|z(p)| \leq r\}$ is defined. Let $\mathcal{F}$ be the family of subharmonic functions u on $R \backslash B_r$ such that $u \leq 1$, and $u = 0$ off a compact subset of R. Since $\mathcal{F}$ is a Perron family, the upper envelope

$$\omega(p) \;=\; \sup\{u(p) : u \in \mathcal{F}\}, \qquad p \in R \backslash B_r,$$

is a harmonic function, which evidently satisfies $0 \leq \omega \leq 1$. The function $\omega(p)$ can be viewed as the Perron solution to the Dirichlet problem on $R \backslash B_r$ with boundary function 1 on ∂B_r and 0 at ∞. Since there is a subharmonic barrier at each point of ∂B_r, $\omega(p) \to 1$ as $p \to \partial B_r$. Now, there are two cases that can occur: Either $0 < \omega < 1$ on $R \backslash B_r$, or $\omega \equiv 1$ on $R \backslash B_r$.

Lemma. If $g(p,q)$ exists for some point q in the open coordinate disk $B_r \backslash \partial B_r = \{|z(p)| < r\}$, then $0 < \omega < 1$ on $R \backslash B_r$. Conversely, if $0 < \omega < 1$ on $R \backslash B_r$, then $g(p,q)$ exists for all $q \in B_r \backslash \partial B_r$.

The first statement is the easier to prove. Suppose $g(p,q)$ exists, and choose $c > 0$ such that $g(p,q) \geq c$ for all $p \in \partial B_r$. By the maximum principle, $u(p) \leq g(p,q)/c$ for all u in the Perron family $\mathcal{F}$ defining ω. Consequently, $\omega(p) \leq g(p,q)/c$. Now, the infimum of $g(p,q)$ for $p \in R$ is 0; if it were $a > 0$, then $g(p,q) - a$ would be a positive harmonic function on $R \backslash \{q\}$ with a logarithmic pole at q, contradicting the minimality property of $g(p,q)$ in the preceding theorem. Hence the infimum of $\omega(p)$ is also 0, and the case $0 < \omega < 1$ obtains.

For the converse, suppose that $0 < \omega < 1$ on $R \backslash B_r$. Choose $s > r$ such that the closed coordinate disk $B_s = \{|z(p)| \leq s\}$ is defined. Choose $\kappa < 1$ such that $\omega(p) \leq \kappa$ for $p \in \partial B_s$. Choose $C > 0$ such that $|\log |z(p) - z(q)|| \leq C$ for $p \in \partial B_r$. Let $u \in \mathcal{F}_q$, and let M be the maximum of $u(p)$ for $p \in \partial B_s$. Since $u(p) + \log |z(p) - z(q)|$ is subharmonic on B_s and bounded by $M + C$ on ∂B_s, we have

$$u(p) + \log |z(p) - z(q)| \;\leq\; M + C, \qquad p \in B_s.$$

In particular, this holds for $p \in \partial B_r$, and consequently,

$$u(p) \;\leq\; M + 2C, \qquad p \in \partial B_r.$$

Since $\omega(p) = 1$ on ∂B_r, we have

$$u(p) \;\leq\; (M + 2C)\omega(p)$$

for $p \in \partial B_r$, hence by the maximum principle for all $p \in R \backslash B_r$. If we take the maximum over $p \in \partial B_s$, we then obtain

$$M \;\leq\; (M + 2C)\kappa.$$

This can be solved for M to yield $M \leq 2C\kappa/(1 - \kappa)$. Thus we have shown that any $u \in \mathcal{F}_q$ satisfies $u(p) \leq 2C\kappa/(1 - \kappa)$ for $p \in \partial B_s$. Consequently, Green's function with pole at q exists.

The lemma shows that the set of $q \in R$ for which Green's function $g(p,q)$ exists is open, and the set of $q \in R$ for which $g(p,q)$ does not exist is also open. Since R is connected, one of these sets coincides with R and the other is empty. We have proved the following.

Theorem. *If Green's function with pole at q exists for some $q \in R$, then it exists for all $q \in R$.*

Another consequence of the lemma is the following theorem, which implies in particular that Green's function exists for any finite bordered Riemann surface with nonempty border.

Theorem. *Suppose the Riemann surface R is an open subset of a Riemann surface S, and suppose there is a boundary arc for R in S. Then Green's function for R exists.*

In this case, there is a subharmonic barrier at each point of the arc, so the auxiliary function $\omega(p)$ tends to 0 as $p \in R$ tends to the boundary arc. Hence we are in the case where $0 < \omega < 1$, and Green's function exists.

Next we prove a theorem of T. Radó to the effect that any Riemann surface is the union of an increasing sequence (!) of compact subsets. This property is useful, but it is not essential, in the sense that if we could not derive it from the definition of a Riemann surface, we would simply incorporate it into the definition as property (1.5). (In fact, this property is incorporated into the definition of complex manifolds of higher dimension, and it is incorporated into the definition of differentiable manifolds of all dimensions.)

Theorem. *Let R be a Riemann surface. There is an increasing sequence $\{R_n\}$ of finite bordered subsurfaces of R such that $\cup R_n = R$.*

Suppose first that Green's function $g(p,q)$ exists. Fix $p_0 \neq q$, and choose a sequence $u_n \in \mathcal{F}_q$ such that $u_n(p_0) \to g(p_0, q)$. Let R_n be a finite bordered subsurface of R containing p_0 and q such that $u_n = 0$ off a compact subset of R_n. We can arrange also that R_n includes the compact set $R_{n-1} \cup \partial R_{n-1}$. We adjoin to R_n any component of $R \backslash R_n$ that is compact, and we can assume that there are no compact components of $R \backslash R_n$. Now, $g_n(p,q)$ is an increasing sequence, and $u_n(p_0) \leq g_n(p_0, q) \leq g(p_0, q)$. By Harnack's theorem, $g_n(p,q)$ converges uniformly on compact subsets of $R \backslash \{q\}$ to a harmonic function $u(p)$ on $R \backslash \{q\}$. Since $u(p) - g(p,q) \leq 0$ on $R \backslash \{q\}$, and $u(p_0) - g(p_0, q) = 0$, the strict maximum principle implies that $u(p) - g(p,q) \equiv 0$. Consequently, $g_n(p,q)$ converges to $g(p,q)$ uniformly on compact subsets of $\cup R_n$, and in particular it converges uniformly on ∂R_n for all n.

We claim that $\cup R_n = R$. We argue by contradiction. Suppose there is a point $p_0 \in R \backslash \cup R_n$. Let D_0 be a coordinate disk at p_0 with coordinate map $z(p)$ satisfying $z(p_0) = 0$. No boundary curve of ∂R_n lies completely in D_0, or else it would bound a compact component of $R \backslash R_n$. Hence by following a straight line to the nearest curve in ∂R_n and then following this to ∂D_0, we obtain a curve γ_n in $D_0 \backslash R_n$ connecting p_0 to ∂D_0. Let E be the lim sup of the γ_n's, that is, $p \in E$ if and only if there is a sequence $p_n \in \gamma_n$ such that $p_n \to p$. Then E is a compact connected subset of R, and E is disjoint from R_k for each k, so $E \subset R \backslash \cup R_n$. Note that $p_0 \in E$, and since each γ_k meets ∂D_0, also E meets ∂D_0. Thus E has more than one point. By the condition given in Section XV.4, there is a subharmonic barrier at each point of E for the open set $R \backslash E$. Hence Green's function $g_E(p, q)$ of the component of $R \backslash E$ containing q tends to 0 as $p \to E$. Since $g_n(p, q) \le g_E(p, q)$ for all n, we then obtain $g(p, q) \to 0$ as $p \to E$. However, $g(p, q) > 0$ for all p, so that E is empty. This contradiction establishes the theorem in the case that Green's function exists.

In the case that Green's function does not exist, we excise a coordinate disk D_0 and consider the surface $R \backslash D_0$. Green's function does exist for $R \backslash D_0$. We apply the result already proved to $R \backslash D_0$, and the theorem for R follows easily.

Exercises for XVI.3

1. Show that if Green's function exists for S, and R is a subsurface of S, then Green's function exists for R, and $g_R \le g_S$.

2. Show that if $f : R \to S$ is an analytic map of R to S, and if Green's function exists for S, then Green's function exists for R, and $g_R(p, q) \le g_S(f(p), f(q))$.

3. Suppose Green's function $g(p, q)$ exists for R. Let $z(p)$ be a coordinate map at q with $z(q) = 0$. Show that if $u(p)$ is a positive harmonic function on $R \backslash \{q\}$ such that $u(p) + \log |z(p)|$ is bounded below near q, then $u(p) \ge g(p, q)$.

4. Suppose Green's function $g(p, q)$ exists for R. Let $z(p)$ be a coordinate map at q with $z(q) = 0$. Show that if $u(p)$ is a subharmonic function on $R \backslash \{q\}$ such that $u(p) = 0$ off some compact subset of R, and $u(p) + \log |z(p)|$ is bounded above near q, then $u(p) \le g(p, q)$.

5. Let R be a Riemann surface for which Green's function does not exist. Let K be a compact subset of R. Show that if $u(p)$ is a subharmonic function on $R \backslash K$ such that $u(p)$ is bounded above and $\limsup u(p) \le 0$ as $p \to \partial K$, then $u(p) \le 0$ for all $p \in R \backslash K$.

Remark. If Green's function for R exists, and the function ω is as in this section, then this statement fails for $-\omega$.

6. Show that Green's function for R exists if and only if there is a nonconstant subharmonic function on R that is bounded above. *Hint.* If Green's function $g(p,q)$ exists, consider $-g(p,q)$. For the reverse implication, use the preceding exercise.

4. Symmetry of Green's Function

We are interested now in how Green's function $g(p,q)$ depends on the parameter q. Our goal is to prove the following.

Theorem. *Suppose R is a Riemann surface for which Green's function $g(p,q)$ exists. Then Green's function is symmetric,*

$$g(p,q) \;=\; g(q,p), \qquad p,q \in R, \quad p \neq q.$$

It suffices to prove the theorem in the case that R is a finite bordered Riemann surface. The general case can then be obtained by taking limits.

To separate the analytic difficulties from the topological difficulties, we first prove the theorem in the case that D is a bounded domain in the complex plane with analytic boundary. In this case, the proof is reminiscent of the proof of Pompeiu's formula. We excise two closed disks U_0, U_1 of small radius $\varepsilon > 0$ centered at points $\zeta_0, \zeta_1 \in D$. Let $D_\varepsilon = D \backslash (U_0 \cup U_1)$. The functions $u_0(z) = g(z, \zeta_0)$ and $u_1(z) = g(z, \zeta_1)$ are harmonic on $D_\varepsilon \cup \partial D_\varepsilon$. From Green's second formula we obtain

$$\int_{\partial D_\varepsilon} \left(u_0 \frac{\partial u_1}{\partial n} - u_1 \frac{\partial u_0}{\partial n} \right) ds \;=\; 0.$$

Since $u_0 = u_1 = 0$ on ∂D, this becomes

$$\int_{\partial U_0} \left(u_0 \frac{\partial u_1}{\partial n} - u_1 \frac{\partial u_0}{\partial n} \right) ds + \int_{\partial U_1} \left(u_0 \frac{\partial u_1}{\partial n} - u_1 \frac{\partial u_0}{\partial n} \right) ds \;=\; 0.$$

We may reverse the direction of the unit normal vector, and take it to be the usual outer normal direction to the disks U_0 and U_1. With this switch in sign, the integral over ∂U_0 is $2\pi u_1(\zeta_0)$ and the integral over ∂U_1 is $-2\pi u_0(\zeta_1)$, by Green's third formula (1.3). Hence $u_1(\zeta_0) = u_0(\zeta_1)$, and $g(\zeta_0, \zeta_1) = g(\zeta_1, \zeta_0)$.

If we had available the machinery of differential forms and Green's theorem for manifolds, we could carry over the proof for planar domains immediately to a finite bordered Riemann surface. However, we can also proceed directly by "triangulating" R and by applying Green's formula on each triangle.

Let R be a finite bordered Riemann surface, with Green's function $g(p,q)$, and fix points $q_0, q_1 \in R$. Our first step is to cut R into curvilinear "triangles" $T_0, T_1, \ldots, T_n$ such that each T_j is contained in some coordinate patch, say with coordinate map $z_j(p)$, and ∂T_j consists of three smooth arcs joining the three "vertices" of T_j. Thus if $j \neq k$, then either T_j is disjoint from T_k, or they share only a common vertex, or they share a common side consisting of a smooth arc that has opposite orientations with respect to the two triangles. We can carry out this triangulation procedure for any planar domain with piecewise smooth boundary, and since R is covered by a finite number of coordinate patches, we can triangulate R by focusing on the coordinate patches one by one. We assume that q_0 lies inside T_0 and q_1 inside T_1.

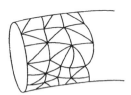

Define $u_0(p) = g(p, q_0)$ and $u_1(p) = g(p, q_1)$. From Green's third formula for $j = 0, 1$, and from Green's second formula for $j \geq 2$, we obtain

(4.1) $$\int_{\partial T_j} \left(u_1 \frac{\partial u_2}{\partial n} - u_2 \frac{\partial u_1}{\partial n} \right) ds = \begin{cases} 2\pi u_1(q_0), & j = 0, \\ -2\pi u_0(q_1), & j = 1, \\ 0, & j \geq 2, \end{cases}$$

where the integral over ∂T_j is taken with respect to the coordinate map $z_j(p)$. Suppose γ is a common side of two triangles. Let $v_0(p)$ and $v_1(p)$ be conjugate harmonic functions for $u_0(p)$ and $u_1(p)$, defined on γ. Then we can express the normal derivatives of the u_j's in terms of the tangential derivatives of the v_j's by the Cauchy-Riemann equations, and obtain

$$\int_\gamma \left(u_0 \frac{\partial u_1}{\partial n} - u_1 \frac{\partial u_0}{\partial n} \right) ds = -\int_\gamma \left(u_0 \frac{\partial v_1}{\partial s} - u_1 \frac{\partial v_0}{\partial s} \right) ds$$

$$= \int_\gamma u_1 dv_0 - u_0 dv_1.$$

Now, the right-hand integral is independent of the coordinate map, and it changes sign if we reverse the direction of γ. Thus if we add up all the integrals in (4.1), the integrals over the common sides of adjacent triangles cancel, and we are left with the integrals over the sides of triangles on ∂R. Since both u_1 and u_2 are zero on ∂R, the integrals over these sides vanish, and we obtain

$$2\pi(u_0(q_1) - u_1(q_0)) = \sum \int_{\partial T_j} \left(u_1 \frac{\partial u_2}{\partial n} - u_2 \frac{\partial u_1}{\partial n} \right) ds = 0.$$

Thus $u_0(q_1) = u_1(q_0)$, and the symmetry is established.

Example. Green's function $g(z, \zeta) = \log|(z - \zeta)/(1 - \bar{\zeta}z)|$ for the open unit disk $\mathbb{D}$ is symmetric in z and ζ.

Exercises for XVI.4

1. Let R be a finite bordered Riemann surface with border ∂R, and fix $q \in R$. Let $h(p)$ be the (locally defined) conjugate harmonic function for Green's function $g(p, q)$ for R. Show that if $u(p)$ is a harmonic function on R that is smooth up to ∂R, then

$$u(q) = \frac{1}{2\pi} \int_{\partial R} u(p)\, dh(p).$$

Show further that $h(p)$ is strictly increasing on ∂R, and the total increase of $h(p)$ around the curves in ∂R is 2π. *Remark.* The integral can be defined by breaking ∂R into arcs contained in coordinate disks. It does not depend on the coordinate maps, but it does depend on the orientation of the curves in ∂R, which must be oriented positively with respect to R.

5. Bipolar Green's Function

Not every Riemann surface has a Green's function. However, every surface *does* have a bipolar Green's function. In this section we will prove the existence of bipolar Green's function, and in the next we will use it to prove the celebrated uniformization theorem.

Let q_1, q_2 be distinct points of the Riemann surface R, and let Δ_1 and Δ_2 be disjoint coordinate disks containing q_1 and q_2, respectively, with coordinate maps $z_1(p)$ and $z_2(p)$ satisfying $z_1(q_1) = 0$ and $z_2(q_2) = 0$. We define **bipolar Green's function** with poles at q_1 and q_2 to be any harmonic function $G(p, q_1, q_2)$ on $R\backslash\{q_1, q_2\}$ such that

(5.1) $G(p, q_1, q_2) + \log|z_1(p)|$ is harmonic at q_1,
(5.2) $G(p, q_1, q_2) - \log|z_2(p)|$ is harmonic at q_2,
(5.3) $G(p, q_1, q_2)$ is bounded on $R\backslash(\Delta_1 \cup \Delta_2)$.

Thus $G(p, q_1, q_2)$ has a logarithmic pole at q_1, and $-G(p, q_1, q_2)$ has a logarithmic pole at q_2. Note that $G(p, q_1, q_2)$ is not uniquely determined. It is unique up to adding a bounded harmonic function.

Example. Bipolar Green's function for the Riemann sphere $\mathbb{C}^*$ with poles at $q_1 = 0$ and $q_2 = \infty$ is given by $G(z, 0, \infty) = -\log|z|$. In this case, $G(z, 0, \infty)$ is uniquely determined up to adding a constant, since a bounded harmonic function on $\mathbb{C}^*$ is constant.

Example. If Green's function $g(p, q)$ exists for the Riemann surface R, then

$$G(p, q_1, q_2) = g(p, q_1) - g(p, q_2)$$

is a bipolar Green's function for R.

Theorem. *For each pair of distinct points q_1, q_2 on a Riemann surface, there is a bipolar Green's function $G(p, q_1, q_2)$.*

The technical part of the proof is contained in the following lemma.

Lemma. *Let S be a finite bordered Riemann surface, and let q_1, q_2 be distinct points of S. Let $B_1 = \{|z_1(p)| \leq \sigma\}$ and $B_2 = \{|z_2(p)| \leq \sigma\}$ be disjoint closed coordinate disks, where $z_1(q_1) = 0$, $z_2(q_2) = 0$. There is a constant $C > 0$ such that*

$$|g_R(p, q_1) - g_R(p, q_2)| \leq C, \qquad p \in R \backslash (B_1 \cup B_2),$$

for all Riemann surfaces R containing $S \cup \partial S$ for which Green's function $g_R(p, q)$ exists.

It suffices to establish the lemma for finite bordered Riemann surfaces R containing $S \cup \partial S$. Let $\rho > 0$ satisfy $\rho < \sigma$. For $j = 1, 2$, let A_j be the closed coordinate disk $\{|z_j(p)| \leq \rho\}$, and let $M_j = M_j(R)$ be the maximum of $g_R(p, q_j)$ on ∂A_j. For $p \in \partial B_j$ we have

$$g_R(p, q_j) + \log |z_j(p)| \leq \max\{g_R(q, q_j) : q \in \partial B_j\} + \log \sigma,$$

and $g_R(p, q_j) + \log |z_j(p)|$ is harmonic on B_j. By the maximum principle, the estimate persists for all $p \in B_j$. Taking the supremum over $p \in \partial A_j$, we obtain

$$M_j + \log \rho \leq \max\{g_R(q, q_j) : q \in \partial B_j\} + \log \sigma.$$

Thus there is $p_j \in \partial B_j$ such that $M_j + \log \rho \leq g_R(p_j, q_j) + \log \sigma$, or

$$M_j - g_R(p_j, q_j) \leq \log(\sigma / \rho).$$

Now, $M_j - g_R(p, q_j)$ is a positive harmonic function on $S \backslash (A_1 \cup A_2)$. If we apply the Harnack estimate from Section 2 to the Riemann surface $S \backslash (A_1 \cup A_2)$ and the compact subset $\partial B_1 \cup \partial B_2$, we obtain a constant C_0, independent of R, such that $M_j - g_R(p, q_j) \leq C_0$ for $p \in \partial B_1 \cup \partial B_2$. Thus

$$(5.4) \qquad M_j - C_0 \leq g_R(p, q_j) \leq M_j, \qquad p \in \partial B_1 \cup \partial B_2.$$

Since $g_R(p, q_1)$ is harmonic for $p \in B_2$ and satisfies (5.4) on ∂B_2, it must satisfy (5.4) for all $p \in B_2$. In particular,

$$M_1 - C_0 \leq g_R(q_2, q_1) \leq M_1.$$

Similarly,

$$M_2 - C_0 \ \leq \ g_R(q_1, q_2) \leq M_2.$$

Since $g_R(q_2, q_1) = g_R(q_1, q_2)$, these estimates imply that $|M_1 - M_2| \leq C_0$. From (5.4) we then obtain

$$|g_R(p, q_1) - g_R(p, q_2)| \ \leq \ 2C_0, \qquad p \in \partial B_1 \cup \partial B_2.$$

Since Green's function vanishes on ∂R, this estimate persists on $R \backslash (B_1 \cup B_2)$, by the maximum principle. This establishes the lemma, with $C = 2C_0$.

Now we complete the proof by appealing to compactness results for uniformly bounded families of harmonic functions. Conceptually, the easiest way to proceed is to use Radó's theorem and exhaust R by a sequence of finite bordered Riemann surfaces with Green's functions $g_n(p, q)$. The differences $G_n(p, q_1, q_2) = g_n(p, q_1) - g_n(p, q_2)$ are then uniformly bounded for p outside of coordinate disks containing q_1 and q_2. We pass to a convergent subsequence, and the limit is bipolar Green's function $G(p, q_1, q_2)$.

There is a useful trick that allows us to bypass Radó's theorem. It is to approximate R by surfaces $R \backslash B_\varepsilon$, where B_ε is a closed coordinate disk $\{|z_0(p)| \leq \varepsilon\}$ centered at some arbitrary point $p_0 \in R$. Green's function $g_\varepsilon(p, q)$ exists for $R \backslash B_\varepsilon$, and we can form bipolar Green's function $g_\varepsilon(p, q_1) - g_\varepsilon(p, q_2)$. The lemma and compactness results allow us to pass to a normal limit through some sequence $\varepsilon_j \to 0$, to obtain a bipolar Green's function $G(p, q_1, q_2)$ for $R \backslash \{p_0\}$. Since $G(p, q_1, q_2)$ is bounded near the isolated singularity p_0, it extends harmonically to p_0 and is a bipolar Green's function for R.

Exercises for XVI.5

1. Determine all bipolar Green's functions for the complex plane $\mathbb{C}$.

2. Determine all bipolar Green's functions for the Riemann sphere $\mathbb{C}^*$.

3. Write out the details for the proof that bypasses Radó's theorem. *Hint.* See Exercise 2.3.

6. The Uniformization Theorem

A Riemann surface R is **simply connected** if any closed path in R can be deformed continuously to a point. This occurs if and only if any two paths in R from p_0 to p_1 can be deformed to each other through a continuous family of paths from p_0 to p_1. If R is simply connected, and if S is a Riemann surface conformally equivalent to R, then S is simply connected. Our aim in this section is to prove the following fundamental theorem of Koebe and Poincaré.

Theorem (Uniformization Theorem for Riemann Surfaces). *Each simply connected Riemann surface is conformally equivalent to either the open unit disk $\mathbb{D}$, the complex plane $\mathbb{C}$, or the Riemann sphere $\mathbb{C}^*$.*

One of the main ingredients of the proof is the idea of analytic continuation along a path and the monodromy theorem. Since analytic continuation along paths is a local notion, it can be extended to paths on Riemann surfaces. The proof of the monodromy theorem (Section V.8) is local, and so the monodromy theorem holds on Riemann surfaces.

Our strategy is to split the proof into two cases. First we show that if Green's function for R exists, it can be used to map R conformally to the open unit disk. Then if Green's function for R does not exist, we use bipolar Green's function to map R to the sphere or the punctured plane.

So suppose R is a simply connected Riemann surface for which Green's function exists. Fix a point $q_0 \in R$. Since $g(p, q_0)$ has a logarithmic pole at q_0, there is an analytic function φ defined near q_0 with a simple zero at q_0 such that $|\varphi(p)| = e^{-g(p,q_0)}$. The function φ can be continued analytically along any path in R from q_0 to any other point p, by continuing the harmonic conjugate function of $g(p, q_0)$ along the path and taking an exponential. Since R is simply connected, the monodromy theorem asserts that the analytic continuation does not depend on the path from q_0 to p, and consequently, it defines an analytic function φ on R such that

$$|\varphi(p)| = e^{-g(p,q_0)}, \qquad p \in R.$$

In particular, $|\varphi(p)| < 1$ for $p \in R$, and φ has only one zero, a simple zero at q_0. Fix $q_1 \in R$, and define

$$\psi(p) = \frac{\varphi(p) - \varphi(q_1)}{1 - \overline{\varphi(q_1)}\varphi(p)}, \qquad p \in R.$$

Then ψ is analytic on R, $|\psi(p)| < 1$ for $p \in R$, and $\psi(q_1) = 0$. If $u \in \mathcal{F}_{q_1}$, then $u(p) + \log|\psi(p)|$ is subharmonic on R, including at q_1, since $\psi(q_1) = 0$, and $u(p) + \log|\psi(p)| < 0$ off a compact subset of R. By the maximum principle, $u(p) + \log|\psi(p)| < 0$ on R. Taking the supremum over $u \in \mathcal{F}_{q_1}$, we obtain

(6.1) $$g(p, q_1) + \log|\psi(p)| \leq 0, \qquad p \in R.$$

Since $\psi(q_0) = -\varphi(q_1)$, we obtain using the symmetry of Green's function that

$$g(q_0, q_1) + \log|\psi(q_0)| = g(q_0, q_1) - g(q_1, q_0) = 0.$$

By the strict maximum principle, equality holds in (6.1). Thus $\log|\psi(p)| = -g(p, q_1)$ for all $p \in R$, and consequently, ψ has no zeros on $R \backslash \{q_1\}$. It follows that φ assumes the value $\varphi(q_1)$ only at $p = q_1$. Since q_1 is arbitrary, φ is one-to-one. Thus φ maps R conformally onto a domain $\varphi(R)$ in the unit disk, and further $\varphi(R)$ is simply connected. By the Riemann mapping theorem, $\varphi(R)$ is conformally equivalent to the open unit disk, and then so is R.

For the remaining case of the proof, we require the following lemma. (For a more general statement, see the exercises for Section 3.)

Lemma. *If there is a nonconstant bounded analytic function on a Riemann surface R, then Green's function for R exists.*

In this case, for a prescribed point $q \in R$ we can find a nonconstant analytic function f on R such that $f(q) = 0$ and $|f(p)| < 1$ for $p \in R$. If $u \in \mathcal{F}_q$, then $u(p) + \log|f(p)|$ is subharmonic on R, and $u(p) + \log|f(p)| < 0$ off a compact subset of R. By the maximum principle, $u(p) + \log|f(p)| < 0$ on R. This provides an upper bound $-\log|f(p)|$ for the functions in $\mathcal{F}_q$, and consequently, Green's function with pole at q exists.

Now suppose that Green's function for R does not exist. Let $G(p, q_1, q_2)$ be a bipolar Green's function for R. Since R is simply connected, we can then find by analytic continuation a meromorphic function φ on R such that

$$|\varphi(p)| = e^{-G(p,q_1,q_2)}, \qquad p \in R.$$

In particular, φ has only one zero on R, a simple zero at q_1, and only one pole on R, a simple pole at q_2. Further, there are coordinate disks B_1 at q_1 and B_2 at q_2 such that

$$\frac{1}{C} \leq |\varphi(p)| \leq C, \qquad p \in R \backslash (B_1 \cup B_2).$$

We claim that φ is one-to-one. To see this, let q_0 be any point of R distinct from q_1 and q_2, select a bipolar Green's function $G(p, q_0, q_2)$, and let φ_0 be a meromorphic function such that

$$|\varphi_0(p)| = e^{-G(p,q_0,q_2)}, \qquad p \in R.$$

Consider the quotient

$$\psi(p) = \frac{\varphi(p) - \varphi(q_0)}{\varphi_0(p)}, \qquad p \in R.$$

The poles at q_2 cancel, so ψ is a bounded analytic function on R. Since Green's function does not exist, the lemma shows that ψ is constant. Since the poles at q_2 are both simple, the quotient ψ is not zero at q_2, and consequently, $\psi(p) \equiv c \neq 0$. It follows that φ assumes the value $\varphi(q_0)$ at only one point, at $p = q_0$, and consequently, φ is one-to-one. Now, φ maps R conformally onto a simply connected domain $\varphi(R)$ on the Riemann sphere $\mathbb{C}^*$. If $\mathbb{C}^* \backslash \varphi(R)$ has more than one point, then by the Riemann mapping theorem $\varphi(R)$ can be mapped conformally onto the unit disk, and consequently, Green's functions for $\varphi(R)$ and hence for R exist. We conclude that $\mathbb{C}^* \backslash \varphi(R)$ has at most one point. Thus either $\varphi(R) = \mathbb{C}^*$, or else $\varphi(R)$ is the punctured sphere, which is conformally equivalent to $\mathbb{C}$.

Exercises for XVI.6

1. Show that the symmetry of Green's function used in the proof can be derived directly by obtaining an inequality from (6.1) and then interchanging the roles of q_0 and q_1.

2. Consider the fundamental domain U for the modular group $\mathcal{U}$ defined in Exercise 1.14.
   (a) Show that with appropriate identification of points of ∂U the fundamental domain can be made into a Riemann surface.
   (b) Show that the Riemann surface is conformally equivalent to the complex plane.
   (c) Show that there is an analytic function $J(\tau)$ on the upper half-plane $\mathbb{H}$ such that $J(\tau) = J(\tau')$ if and only if there is $f \in \mathcal{U}$ such that $f(\tau) = \tau'$. This occurs if and only if the Riemann surface T_τ (defined in Exercises 1.13) is conformally equivalent to $T_{\tau'}$.
   (d) Show that $J(\tau)$ has critical points at $(\sqrt{3}\,i \pm 1)/2$ of order 2 and at i of order 1.

7. Covering Surfaces

We have seen how a torus is obtained from the complex plane by identifying points that are congruent modulo a lattice L. In this section we indicate how any Riemann surface can be obtained from a simply connected Riemann surface in roughly the same manner, by identifying congruent points. The main new ideas are "covering maps" and "covering transformations."

Let R and S be two Riemann surfaces. An analytic map $\pi : S \to R$ is a **covering map** if each $p \in R$ belongs to an open subset U of R whose inverse image $\pi^{-1}(U) = \cup V_\alpha$ is a union of disjoint open subsets V_α of S, each of which is mapped one-to-one by π onto U. If U is a coordinate disk in R with coordinate map $z(p)$, then each V_α is a coordinate disk in S with coordinate map $(z \circ \pi)(q)$.

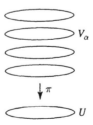

Example. Let S be the Riemann surface of $\sqrt{z}$, constructed from two copies of the slit plane $\mathbb{C}\backslash(-\infty, 0]$ by identifying edges appropriately. The natural map π of S onto the punctured plane $\mathbb{C}\backslash\{0\}$ is a covering map.

If D_0 is a disk in $\mathbb{C}\backslash(-\infty, 0]$, then $\pi^{-1}(D_0)$ consists of two disjoint disks, one on each sheet. If D_0 is a disk in $\mathbb{C}\backslash\{0\}$ overlapping the negative axis $(-\infty, 0)$, then $\pi^{-1}(D_0)$ still consists of two disjoint coordinate disks, one in each of the coordinate patches U_3 and U_4 described in Section 1.

Example. There is a natural projection π of the Riemann surface of $\log z$ onto the punctured complex plane $\mathbb{C}\backslash\{0\}$. In this case, π is a covering map, and each $\pi^{-1}(D_0)$ consists of infinitely many coordinate disks on the surface, one for each sheet used in constructing the surface.

Example. Let T be the torus determined by the lattice L generated by ω_1 and ω_2. The natural projection $\pi(z) = z + L$, $z \in \mathbb{C}$, is then a covering map of $\mathbb{C}$ onto T. If D_ε is a small open disk in $\mathbb{C}$, then $D_\varepsilon + L$ is a coordinate disk in the torus, and $\pi^{-1}(D_\varepsilon + L)$ consists of all translates $D_\varepsilon + m\omega_1 + n\omega_2$, where m and n are integers. Each of these translates is an open disk that is mapped one-to-one onto the coordinate disk in T.

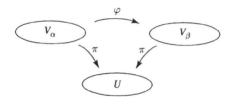

Suppose $\pi : S \to R$ is a covering map. A **covering transformation** is a one-to-one analytic map φ of S onto S such that $\pi(\varphi(p)) = \pi(p)$ for all $p \in S$. The composition of two covering transformations is a covering transformation and the inverse of a covering transformation is a covering transformation. A covering transformation φ permutes the points of each set $\pi^{-1}(\{q\})$, $q \in R$. Further, if U is a coordinate disk on R such that $\pi^{-1}(U) = \cup V_\alpha$ is a disjoint union of coordinate disks on S as above, then φ permutes the V_α's, and φ maps each V_α one-to-one onto some V_β by $\varphi = \pi_\beta^{-1} \circ \pi_\alpha$ on V_α, where π_α is the restriction of π to V_α. In particular, if $\varphi(p) = p$ for some point $p \in V_\alpha$, then $\varphi = \pi_\alpha^{-1} \circ \pi_\alpha$ is the identity map on the entire coordinate disk V_α. Since S is connected, the uniqueness principle holds for analytic maps of S (see the exercises in Section 1), and consequently, φ is the identity map of S. We state this result formally.

Theorem. Let $\pi : S \to R$ be a covering map, and let $\varphi : S \to S$ be a covering transformation. Then either φ is the identity transformation $\varphi(p) \equiv p$, or φ has no fixed points.

Example. Consider the covering map $\pi : \mathbb{C} \to T$ of the complex plane onto the torus determined by the lattice L generated by ω_1 and ω_2. Any translation of the form

$$\psi(z) \ = \ z + m\omega_1 + n\omega_2, \qquad z \in \mathbb{C},$$

is a covering transformation. Let φ be an arbitrary covering transformation. Since $\varphi(0) \in \pi^{-1}(\{0\}) = L$, there are integers m and n such that $\varphi(0) = m\omega_1 + n\omega_2$. Then the covering transformation $\psi^{-1} \circ \varphi$ satisfies $(\psi^{-1} \circ \varphi)(0) = 0$. By the theorem, $\psi^{-1} \circ \varphi$ is the identity, and consequently, $\varphi = \psi$. Thus the covering transformations are precisely the translations by the lattice points in L.

Example. Let $\varphi : \mathbb{C}^* \to \mathbb{C}^*$ be a covering transformation for some covering map. Then φ is a fractional linear transformation, so φ has a fixed point. From the theorem we conclude that φ is the identity.

It turns out that every Riemann surface can be obtained from a simply connected surface by identifying points. We state the relevant theorem without proof, though the proof will be outlined in the exercises.

Theorem. *For each Riemann surface R, there is a covering map $\pi : S \to R$ of a simply connected Riemann surface S onto R. If $p_0 \in R$ and $q_0, q_1 \in \pi^{-1}(\{p_0\})$, there is a unique covering transformation φ of S such that $\varphi(q_0) = q_1$.*

The surface S is called the **universal covering surface** of R. The surface R can be regarded as a quotient surface of S, obtained by identifying two points q_0 and q_1 of S if there is a covering transformation $\varphi : S \to S$ such that $\varphi(q_0) = q_1$.

According to the uniformization theorem, there are only three possibilities for the universal covering surface: the open unit disk $\mathbb{D}$, the complex plane $\mathbb{C}$, and the Riemann sphere $\mathbb{C}^*$. Most surfaces have the unit disk as their universal covering surface.

Theorem. *The only Riemann surface having the Riemann sphere $\mathbb{C}^*$ as its universal covering surface is the sphere itself. The only Riemann surfaces having the complex plane $\mathbb{C}$ as universal covering surface are the complex plane itself, the punctured complex plane $\mathbb{C}\backslash\{0\}$, and tori. All other Riemann surfaces have the open unit disk as universal covering surface.*

The first statement is a consequence of the fact that the only covering transformation $\varphi : \mathbb{C}^* \to \mathbb{C}^*$ is the identity. For the second statement, one shows that the only analytic functions $\varphi : \mathbb{C} \to \mathbb{C}$ that can be covering transformations are translations.

Example. Consider the domain consisting of the points z in the upper half-plane $\mathbb{H}$ such that $0 < \operatorname{Re} z < 1$ and $|z - \frac{1}{2}| > \frac{1}{2}$. The domain D is a geodesic triangle with respect to the hyperbolic metric of $\mathbb{H}$, with vertices

at 0, 1, and $i\infty$. We reflect D in the circular boundary arc $\left|z - \frac{1}{2}\right| = \frac{1}{2}$, and
we obtain a geodesic triangle with vertices at 0, $\frac{1}{2}$, and 1, and with sides
the semicircles determined by $\left|z - \frac{1}{2}\right| = \frac{1}{2}$, $\left|z - \frac{1}{4}\right| = \frac{1}{4}$, and $\left|z - \frac{3}{4}\right| = \frac{1}{4}$.
We continue reflecting in the semicircular sides. The reflected images of D
are disjoint geodesic triangles that, together with their sides, fill out the
semistrip $\{0 < \operatorname{Re} z < 1, \operatorname{Im} z > 0\}$. If we reflect D also in its vertical sides,
and continue reflecting, we fill out the entire upper half-plane with the
various reflections of D and its sides. The result is a "tiling" of the upper
half-plane by geodesic triangles. Consider the Riemann map π of D onto
the upper half-plane, normalized by $\pi(0) = 0$, $\pi(1) = 1$, and $\pi(i\infty) = i\infty$.
By the Schwarz reflection principle, the map π reflects across the circle
$\left|z - \frac{1}{2}\right| = \frac{1}{2}$ and maps the reflected geodesic triangle onto the lower half-
plane, so that the semicircle where $\left|z - \frac{1}{4}\right| = \frac{1}{4}$ is mapped to $(-\infty, 0)$
and $\left|z - \frac{3}{4}\right| = \frac{1}{4}$ to $(1, +\infty)$. We continue reflecting, and we obtain an
analytic function π on the upper half-plane that maps each reflection of D
onto either the upper or the lower half-plane, and that maps a common
side between two such reflections to one of the three intervals $(-\infty, 0)$,
$(0, 1)$, and $(1, +\infty)$. If D_0 is an open disk disjoint from the real axis,
then $\pi^{-1}(D_0)$ is a disjoint union of domains, one in "every other" of the
reflected images of D, each mapped one-to-one onto D_0. Something similar
happens for disks in $\mathbb{C}\backslash\{0, 1\}$ meeting $\mathbb{R}$. Thus π is a covering map of the
upper half-plane onto the twice-punctured plane $\mathbb{C}\backslash\{0, 1\}$. It represents
the upper half-plane as the universal covering surface of $\mathbb{C}\backslash\{0, 1\}$. The
composition of two successive reflections is a conformal self-map of the
upper half-plane, hence a fractional linear transformation. Each of these
is a covering transformation, and in fact, all covering transformations are
fractional linear transformations that are compositions of an even number
of reflections.

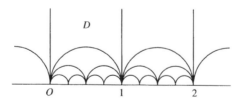

Exercises for XVI.7

1. Show that for $m \geq 1$, the map $\pi(z) = z^m$ is a covering map of
   $\mathbb{C}\backslash\{0\}$ onto itself.

2. Show that $\pi(z) = e^z$ is a covering map of $\mathbb{C}$ onto $\mathbb{C}\backslash\{0\}$.

3. Let $\pi : S \to R$ be a covering map. Show that if $\varphi : S \to S$ is
   analytic and satisfies $\pi \circ \varphi = \pi$, then φ is onto.

4. Show that if $\{\varphi_n\}$ is a sequence of covering transformations that converges normally to φ, then $\varphi_n = \varphi$ for large n. *Remark.* The covering transformations form a "discrete group."

5. Consider the covering map $\pi : \mathbb{H} \to \mathbb{C}\backslash\{0,1\}$ discussed above. Let V be the domain consisting of $z \in \mathbb{H}$ satisfying $|\operatorname{Re} z| < 1$, $|z - \frac{1}{2}| > \frac{1}{2}$, and $|z + \frac{1}{2}| > \frac{1}{2}$.

   (a) Show that the images of $V \cup \partial V$ under the covering transformations fill out $\mathbb{H}$. Sketch how the twice-punctured plane $\mathbb{C}\backslash\{0,1\}$ can be obtained from V by identifying points of ∂V.

   (b) Identify the covering transformations corresponding to reflection in the imaginary axis followed by reflection in one of the semicircles in ∂V. *Ans.* $w = z/(1 \pm 2z)$.

   (c) Show that each covering transformation is a finite composition of the four covering transformations $w = z \pm 2$ and $w = z/(1 \pm 2z)$.

   (d) Show that the covering transformations form a subgroup of the modular group of index 6. *Strategy.* Show that the fundamental domain V is filled out by six congruent copies of the fundamental domain U for the modular group. (See Exercise 1.14.)

   (e) Show that a unimodular matrix corresponds to a covering transformation for π if and only if the matrix is congruent to the identity matrix modulo 2. *Remark.* The covering transformations form the **principal congruence subgroup** of the modular group of level 2.

6. Let $\pi : S \to R$ be a covering map, let $q_0 \in S$, and let $p_0 = \pi(q_0) \in R$.
   (a) Show that if $\gamma(t)$, $0 \le t \le 1$, is a path in R from $\gamma(0) = p_0$ to $\gamma(1) = p$, then there is a unique path $\sigma(t)$, $0 \le t \le 1$, in S starting at $\sigma(0) = q_0$, such that $\gamma = \pi \circ \sigma$. *Remark.* We say that σ is the **lift** of γ from R to S starting at q_0. (b) Let γ_0 and γ_1 be two paths in R from p_0 to p_1, and suppose there is a continuous deformation γ_s, $0 \le s \le 1$, of γ_0 to γ_1 with endpoints fixed. Let σ_s be the lift of γ_s to S starting at q_0. Show that the paths σ_s all terminate at the same point of S, and that σ_s, $0 \le s \le 1$, is a continuous deformation of σ_0 to σ_1.

7. Let γ be the closed path in $\mathbb{C}\backslash\{0,1\}$ starting at $\frac{1}{2}$ and looping once around the circle $|z| = \frac{1}{2}$ counterclockwise. Let α be the closed path in $\mathbb{C}\backslash\{0,1\}$ starting at $\frac{1}{2}$ and looping once around the circle $|z - 1| = \frac{1}{2}$ counterclockwise. Let $\alpha\gamma$ denote the path α followed by γ, and similarly for $\gamma\alpha$. Sketch the lifts of each of the four paths α, γ, $\alpha\gamma$, $\gamma\alpha$ to paths in $\mathbb{H}$ starting at the lift $\frac{1}{2} + \frac{1}{2}i$ of $\frac{1}{2}$. Use the result to show that $\alpha\gamma$ cannot be continuously deformed to $\gamma\alpha$.

8. Let $\pi : S \to R$ be a covering map, let $q_0 \in S$, and let $p_0 = \pi(q_0) \in R$. Show that if $f : \mathbb{D} \to R$ is an analytic map such that $f(0) = p_0$, then there is a unique analytic map $g : \mathbb{D} \to S$ such that $g(0) = q_0$ and $f = \pi \circ g$. *Remark.* We say that g is a **lift** of f.

9. Use the covering map $\pi : \mathbb{H} \to \mathbb{C} \backslash \{0, 1\}$ and the preceding exercise to prove that if $\mathcal{F}$ is a family of analytic functions on the open unit disk $\mathbb{D}$ that omits the points 0 and 1, then $\mathcal{F}$ is a normal family (Montel's theorem). *Strategy.* Let $\{f_n\}$ be a sequence in $\mathcal{F}$, and assume that it has no subsequence that converges normally to 0, 1, or ∞. Find $z_n \in \mathbb{D}$ with $|z_n| \leq r < 1$ such that the $f(z_n)$'s have lifts ζ_n, $\pi(\zeta_n) = f(z_n)$, that belong to a compact subset of $\mathbb{H}$. Consider the lifts $g_n : \mathbb{D} \to \mathbb{H}$ of f_n that satisfy $g_n(z_n) = \zeta_n$.

10. Let R be a Riemann surface, and fix a point $p_0 \in R$. Let S be the set of pairs $(p, [\gamma])$, where $p \in R$ and $[\gamma]$ is a homotopy class of paths in R from p_0 to p. If $p \in R$, U_α is a coordinate disk at p, and γ is a path in R from p_0 to p, define $V_{\alpha,\gamma}$ to be the pairs $(q, [\sigma]) \in S$ such that $q \in U_\alpha$ and σ is the path in R that follows γ from p_0 to p and then any path in U_α from p to q. Define $\pi : S \to R$ to be the coordinate projection $\pi(q, [\gamma]) = q$. (a) Show that S can be made into a Riemann surface in which the sets $V_{\alpha,\gamma}$ are coordinate disks and π is analytic. (b) Show that π is a covering map of S onto R. (c) Show that S is simply connected. *Remark.* Thus S is the universal covering surface of R.

Hints and Solutions for Selected Exercises

Chapter I: The Complex Plane and Elementary Functions

Section I.1

1. (a) circle, (b) annulus, (c) disk, (d) $[-1, 1]$, (e) half-plane, (f) horizontal strip, (g) vertical strip, (h) $\mathbb{C}\backslash\mathbb{R}$, (i) half-plane, (j) empty set.

2. (a) Substitute $z = x + iy$, $w = u + iv$, and use the definitions.

3. Set $z = x + iy$, $a = \alpha + i\beta$, then $\mathrm{Re}(\bar{a}z) = \alpha x + \beta y$, and the equation becomes $(x - \alpha)^2 + (y - \beta)^2 = \rho^2$.

4. Apply triangle inequality to $z = \mathrm{Re}\, z + i\,\mathrm{Im}\, z$, to obtain $|z| \le |\mathrm{Re}\, z| + |\mathrm{Im}\, z|$. Equality holds only when z is real or pure imaginary.

6. If $|z| = 1$, then $|z - a| = |z - a||\bar{z}| = |z\bar{z} - a\bar{z}| = |1 - a\bar{z}| = |1 - \bar{a}z|$.

8. Write $p(z) = a_n z^n + \cdots + a_0$ and $h(z) = b_{n-1}z^{n-1} + \cdots + b_0$. Equate coefficients in the polynomial identity $p(z) = (z - z_0)h(z) + p(z_0)$, and solve for the b_j's in terms of the a_j's.

11. If n is even, $h(z) = z^{n-2} + z^{n-4} + \cdots + 1$, $r(z) = 1$. If n is odd, $h(z) = z^{n-2} + z^{n-4} + \cdots + z$, $r(z) = z$.

Section I.2

1. (a) $e^{i\pi/4} = (1 + i)/\sqrt{2}$, $e^{5\pi i/4} = -(1 + i)/\sqrt{2}$, (c) $e^{\pm i\pi/4} = (1 \pm i)/\sqrt{2}$, $e^{\pm 3\pi i/4} = (-1 \pm i)/\sqrt{2}$, (e) 2, $2e^{\pm 2\pi i/3} = -1 \pm i\sqrt{3}$, (g) 16.

3. For $0 < b < 1$, an ellipse $x^2/(1+b)^2 + y^2/(1-b)^2 = 1$, traversed in positive direction with increasing θ. For $b = 1$, an interval $[-2, 2]$. For $1 < b < +\infty$, an ellipse traversed in negative direction. For $b = \rho e^{i\varphi}$, express equation as $e^{i\varphi/2}(e^{i(\theta-\varphi/2)} + \rho e^{-i(\theta-\varphi/2)})$ to see that curve is rotate of ellipse or interval by $\varphi/2$.

4. $n = 4, 8, 12, \cdots$.

5. (b) Apply (a) to $z = e^{i\theta}$ and to $z = e^{-i\theta}$, add the identities, and use the definitions of sine and cosine.

6. (a) Apply the fundamental theorem of algebra. (b) Expand the product in (a) and find the coefficient of z^{n-1}. (c) Evaluate the identity in (a) at 0. (d) Apply 5(a) to $z = \omega_1^k$.

7. $|z^n + 1| \ge |z^n| - 1 = R^n - 1$, so $1/|z^n + 1| \le 1/(R^n - 1)$. Now multiply

447

by $|z^m| = R^m$.

8. $\cos 4\theta = \cos^4 \theta - 6\cos^2 \theta \sin^2 \theta + \sin^4 \theta$, $\sin 4\theta = 4\cos^3 \theta \sin \theta - 4\cos \theta \sin^3 \theta$

Section I.3

2. If $P = (X, Y, Z)$ corresponds to $z = (X + iY)/(1 - Z)$, then $-P$ corresponds to $-(X + iY)/(1 + Z) = -(X^2 + Y^2)/[(1 + Z)(X - iY)] = -(1 - Z^2)/[(1 + Z)(X - iY)] = -(1 - Z)/(X - iY) = -1/\bar{z}$.

4. Show that if z corresponds to (X, Y, Z), then $1/z$ corresponds to its rotate $(X, -Y, -Z)$.

6. (a) Follows from the fact that the usual euclidean distance in $\mathbb{R}^3$ is a metric. (b) Start with $d(z_1, z_2)^2 = (X_1 - X_2)^2 + (Y_1 - Y_2)^2 + (Z_1 - Z_2)^2$, expand, substitute, and do the algebra. (c) $d(z, \infty) = 2/\sqrt{1 + |z|^2}$.

Section I.4

4. (a) Use four sheets, can make branch cuts along real axis from $-\infty$ to 0. (b) Use two sheets, can make branch cuts along horizontal line from $i - \infty$ to i. (c) Use five sheets, can make branch cuts along real axis from $-\infty$ to 1.

Section I.5

3. Check from definitions that $e^{\bar{z}} = e^x(\cos y - i\sin y) = \overline{e^z}$.

4. Substitute $z = 0$, obtain $e^\lambda = 1$, $\lambda = 2\pi mi$.

Section I.6

3. $f(re^{i\theta}) = \log r + i\theta$, $-\pi/2 < \theta < 3\pi/2$.

4. Any straight line cut from z_0 to ∞, in any direction, will do.

Section I.7

1. (a) $e^{2\pi n}e^{-\pi/4}e^{i\log\sqrt{2}}$, $-\infty < n < \infty$, (b) $-ie^{2\pi n}e^{\pi/2}$, $-\infty < n < \infty$, (c) $\pm 1/\sqrt{2}$, (d) $e^{2\pi n}e^{\log 2 + \pi/3}e^{i(-\log 2 + \pi/3)}$, $-\infty < n < \infty$,

2. $-\pi/2 + i\log 2 + 4\pi m + 2\pi in$, $-\infty < m, n < \infty$,

5. i^i has values $e^{2\pi n}e^{-\pi/2}$, $-\infty < n < \infty$, so that $(i^i)^i$ has values $e^{i(2\pi n - \pi/2 + 2\pi ki)}$, which coincides with $-ie^{2\pi m}$, $-\infty < m < \infty$.

6. Phase factors $e^{2\pi ia}$ at 0 and $e^{2\pi ib}$ at 1. Require $e^{2\pi ia}e^{2\pi ib} = 1$, or $a + b =$ an integer, to have a continuous single-valued determination of $z^a(1 - z)^b$.

7. Use two sheets with slits $[x_1, x_2]$, $[x_3, x_4]$, $\cdots$. If n is odd, also need slit $[x_n, +\infty)$. Identify top edge of slit on one sheet with bottom edge of slit on other sheet. Topologically the surface is a sphere with one handle for each slit after the first one, and punctures corresponding to ∞. If $n = 3$ or $n = 4$, surface is a torus.

8. The function is $z\sqrt{1 - 1/z^3}$. If $|z| > 1$, can use the principal value of the square root to define a branch of the function. There are branch points at $z = 0$ and $z^3 = 1$, that is, at $0, 1, \pm e^{2\pi i/3}$. Make two branch cuts by connecting any two pairs of points by curves; for instance, connect 0 to 1 by a straight line, and the other cube roots of unity by a straight line or arc of unit circle. The resulting two-sheeted surface with identification of cuts and with points at infinity is a torus.

9. Function is $(z+1)\sqrt{z}\sqrt{(1 - 1/z^3)(1 + 1/z)}$. If $|z| > 1$, the second square root can be defined to be single-valued. The values then return to the neg-

ative of their initial values when z traverses the circle, because $\sqrt{z}$ does.

10. Values return to their initial values.

11. Use three sheets. Make two cuts, from 1 to $\pm e^{2\pi i/3}$, on each sheet. In this case the cuts share a common endpoint.

Section I.8

2. Use $\cos(x+iy) = \cos x \cos(iy) - \sin x \sin(iy) = \cos x \cosh y - i \sin x \sinh y$, take the modulus squared, and use $\cosh^2 y = 1 + \sinh^2 y$. The identity for $|\cos z|^2$ shows that the only zeros of $\cos z$ are the zeros of $\cos x$ on the real axis, that is, $\pi/2 + m\pi$, $m = 0, \pm 1, \pm 2, \cdots$. Translation by any period λ of $\cos z$ sends zeros to zeros. Thus any period is an integral multiple of π, and since odd integral multiples are not periods, the only periods of $\cos z$ are $2\pi n$, $-\infty < n < \infty$.

3. Since $\cosh z = \cos(iz)$, the zeros of $\cosh z$ are at $i\pi/2 + im\pi$, $m = 0, \pm 1, \pm 2, \cdots$, and the periods of $\cosh z$ are $2\pi mi$, $m = 0, \pm 1, \pm 2, \cdots$. The function $\sinh z = -i \sin(iz)$ is treated similarly.

4. If $z = \tan w$, then $iz = (e^{iw} - e^{-iw})/(e^{iw} + e^{-iw}) = (e^{2iw} - 1)/(e^{2iw} + 1)$. Solve for e^{2iw} and take logarithms.

6. Use one copy of the doubly slit plane S for each integer n, and define $f_n(z) = \mathrm{Tan}^{-1} z + n\pi$ on the nth sheet. Attach the nth sheet to the $(n+1)$th sheet along one of the cuts, so that $f_n(z)$ and $f_{n+1}(z)$ have the same values at the junction.

Chapter II: Analytic Functions

Section II.1

1. (a) 1, (b) 0, (c) 2, (d) 0.

2. Bounded for $|z| \le 1$, $\to 0$ for $|z| < 1$.

4. For fixed k, $(1 - 1/N)(1 - 2/N) \cdots (1 - (k-1)/N) \to 1$ as $N \to \infty$.

5. Use $b_n - b_{n-1} = (1/n) - \int_{n-1}^{n} (1/t)dt < 0$.

9. (a) $\limsup = 2$, $\liminf = 0$, (b) $\limsup = +\infty$, $\liminf = -\infty$, (c) $\limsup = 1$, $\liminf = -1$, (d) $\limsup = +\infty$ if $|x| > 1$, 1 if $|x| = 1$, 0 if $|x| < 1$; $\liminf = -\infty$ if $-\infty < x < -1$, -1 if $x = -1$, 0 if $|x| < 1$, 1 if $x = 1$, $+\infty$ if $x > 1$.

10. (a) continuous on $\mathbb{C}$, (b) continuous for $z \ne 0$, (c) continuous for $z \ne 0$, (d) continuous on $\mathbb{C}$.

11. No limit at points of $(-\infty, 0]$, continuous on $\mathbb{C} \setminus (-\infty, 0]$.

12. No limit at 0, has limit at all other points of closed lower half-plane, limit $= -\pi$ at points of $(-\infty, 0)$.

13. $\to 0$ if $\mathrm{Re}\, \alpha > 0$, $\to 1$ if $\alpha = 0$, otherwise no limit at 0.

15. (a) open, (b) closed, (c) open, (d) neither, (e) open, (f) neither, (g) both open and closed.

Section II.2

1. (a) $2z$, (e) $-2z/(z^2 + 3)^2$, (g) $(ad - bc)/(cz + d)^2$.

2. Differentiate $1 + z + z^2 + \cdots + z^n = (1 - z^{n+1})/(1 - z)$.

4. Use $(f(z + \Delta z) - f(z))/\Delta z \approx 2az + b\bar{z} + (bz + 2c\bar{z})\overline{\Delta z}/\Delta z$.

5. Use $(g(z+\Delta z) - g(z))/\Delta z = \overline{(f(\bar{z} + \overline{\Delta z}) - f(\bar{z}))/\overline{\Delta z}}$.

6. Use twice the theorem that if $g_n(t)$ converges uniformly to $g(t)$ for $0 \le t \le 1$, then $\int_0^1 g_n(t)dt$ has limit $\int_0^1 g(t)dt$, once to show that $H(z)$ is differentiable, and again to show that the derivative is continuous.

Section II.3

1. (a) $1/\cos^2 z$, (b) $1/\cosh^2 z$, (c) $\tan z \sec z$.

2. $i \cos z$.

5. Use $f' = (\partial u)/(\partial x) - i(\partial u)/(\partial y) = (\partial v)/(\partial y) + i(\partial v)/(\partial x)$.

8. Since ∇v is the rotate of ∇u by $\pi/2$, the directional derivative of v in the θ-direction is equal to the directional derivative of u in the r-direction. This is the first polar Cauchy-Riemann equation. The second follows from the same argument.

Section II.4

2. Write $f(z) = ce^{a\,\mathrm{Log}\,z}$ and differentiate.

3. $f'(z) = (1 - 2z)/(2f(z))$.

4. Derivative $= 1/(1+z^2)$. Any two branches of $\tan^{-1} z$ differ by a constant, so derivatives are same.

5. Derivative $= \pm 1/\sqrt{1 - z^2}$. Derivatives of branches of $\cos^{-1} z$ are not always the same.

7. This is the change of variable formula for a double integral, since the Jacobian is $|f'(z)|^2$.

8. 6π.

9. Use the area formula separately on the top half and the bottom half of the unit disk, where the function is one-to-one. Integral $= 2\pi$.

Section II.5

1. (a) $2xy + C$, (c) $-\cosh x \cos y + C$, (e) $-(1/2)\log(x^2 + y^2) + C$.

4. Use $\partial^2 (zh)/\partial x^2 = z\partial^2 h/\partial x^2 + 2\partial h/\partial x$ and similar identity for y-derivatives, add, equate to 0, set $h = u + iv$, and take real and imaginary parts, to obtain Cauchy-Riemann equations for u and v.

7. $\mathrm{Arg}\, z$ is a harmonic conjugate on $\mathbb{C}\backslash(-\infty, 0]$. Any other harmonic conjugate has form $\mathrm{Arg}\, z + C$, which does not extend continuously to $\mathbb{C}\backslash\{0\}$.

8. $-(i/2)(\log z)^2$.

Section II.6

1. To aid sketching, express z in polar coordinates.

4. $w = e^{\pi z/(2A)}$.

5. $w = z^{\pi/(2B)}$.

6. Not defined at $z = 0$, not conformal at $z = \pm 1$. If $f(z) = \lambda$, then $z = (\lambda \pm \sqrt{\lambda^2 - 4})/2$. If $z = \rho e^{i\theta}$, then $w = u + iv = \rho\cos\theta + i\rho\sin\theta + (\cos\theta)/\rho - i(\sin\theta)/\rho$, and $u^2/(\rho+1/\rho)^2 + v^2/(\rho-1/\rho)^2 = 1$. Since $f(D) = f(\mathbb{D})$, and f is at most two-to-one, f is one-to-one on D. Since f maps $\partial\mathbb{D}$ onto $[-2,2]$, and f maps onto $\mathbb{C}$, the image of D is $\mathbb{C}\backslash[-2,2]$.

8. Not defined at $z = 0$, not conformal at $z = \pm e^{i\alpha/2}$. The expression $f(z) = e^{i\alpha/2}(e^{-i\alpha/2}z + 1/e^{-i\alpha/2}z)$ shows that $f(z)$ is a composition of the rotation by $-a/2$, the function of Exercises 6 and 7, and a rotation by $\alpha/2$.

Thus f maps $\{|z| > 1\}$ one-to-one onto the complement rotate of $[-2, 2]$ by $\alpha/2$.

9. Suppose $f(0) = 0$ and $u_x \neq 0$ at 0. The tangents in the orthogonal directions $(1, t)$ and $(-t, 1)$ are mapped to the tangents in the directions $(u_x, v_x) + t(u_y, v_y)$ and $-t(u_x, v_x) + (u_y, v_y)$. The orthogonality of these directions for all t yields $u_x u_y + v_x v_y = 0$ and $u_y^2 + v_y^2 - u_x^2 - v_x^2 = 0$. These equations and some algebra lead to $u_x = \pm v_y$ and $u_y = \mp v_x$.

Section II.7

1. (a) $w = 2i(z - (1+i))/(z - 2)$, (c) $w = (i - 1)/(z - 2)$, (f) $w = z/(z - i)$, (h) $w = i(z - i)/(z + i)$.

2. Circle $\to$ straight line through 0 and $i - 1$, disk $\to$ half-plane on lower-left of line, real axis $\to$ straight line through $i - 1$ orthogonal to image of circle, that is, with slope $= 1$.

4. Unit circle $\to$ imaginary axis, unit disk $\to$ right half-plane, $[-1, 1] \to$ arc of circle $|w - i/2| = 3/2$ in right half-plane from $-i$ to $2i$.

8. If $w/(\alpha z + \beta)/(\gamma z + \delta)$, divide each coefficient by the square root of $\alpha\delta - \beta\gamma$ to obtain representation with $ad - bc = 1$. Representation is not unique, as can multiply all coefficients by -1.

9. If $f(z)$ maps the three real numbers x_1, x_2, x_3 to 0, 1, ∞, then $f(z)$ is represented explicitly as $A(z - x_1)/(z - x_2)$, where A is real. Any representation with $ad - bc = 1$ is obtained from this by multiplying each coefficient by the same (real) constant.

Chapter III: Line Integrals and Harmonic Functions

Section III.1

1. (a) 72/5, (b) 16, (c) 32.

3. 2/3.

4. $-\pi R^2/2$.

6. Differentiate by hand, use uniform convergence of $1/[(z - w)(z - (w + \Delta w))]$ to $1/(z - w)^2$ for $z \in \gamma$.

Section III.2

1. (a) $h = (x^2 + y^2)/2$, (b) $h = (2x^3 + y^6)/6$, (c) $h = xy$, (d) not independent of path, $\int_{|z|=1} = -2\pi$.

2. $\int_{|z|=r} = 2\pi/r^2 \neq 0$.

6. Suppose $|\gamma(t)| = 1$ for $0 \leq t \leq 1$, $\gamma(0) = \gamma(1) = 1$. Follow hint, write $\gamma(t) = e^{i\theta_j(t)}$, $t_{j-1} \leq t \leq t_j$. Note $\theta_j(t_j) - \theta_{j+1}(t_j)$ is an integral multiple of 2π. Add multiples of 2π to the θ_j's, obtain $\theta(t)$ continuous for $0 \leq t \leq 1$ such that $\theta(0) = 0$ and $\gamma(t) = e^{i\theta(t)}$. Note $\theta(1) = 2\pi m$ for some integer m. Deform by $\gamma_s(t) = e^{i[(1-s)\theta(t)+2\pi sm]}$.

Section III.3

1. (a) $du = dx - dy$, $dv = dx + dy$, $v = x + y$, (b) $v = 3x^2 y - y^3$, (c) $v = \cosh x \sin y$, (d) $v = x/(x^2 + y^2)$.

4. u has a harmonic conjugate v_1 on the annulus slit along $(-b, -a)$, and also a harmonic conjugate v_2 on the annulus slit along (a, b). Since $v_1 - v_2$

is constant above the slit $(-b, -a)$, and also constant below the slit, v_1 jumps by a constant across the slit. Arg z also jumps by a constant across the slit. By appropriate choice of C, $v_1 - C \operatorname{Arg} z$ is continuous across the slit $(-b, -a)$, and $u - C \log |z|$ has a harmonic conjugate $v_1 - C \operatorname{Arg} z$ on the annulus. For the identity, use the polar form of the Cauchy-Riemann equations to convert the r-derivative of u to a θ-derivative of v.

Section III.4
1. Express $dxdy$ in polar coordinates centered at z_0 and integrate first with respect to θ.

Section III.5
1. Note that if u attains its maximum or minimum on D, it is constant.
2. $|z^n + \lambda| \le r^n + \rho$, with equality at $re^{i\varphi/n}e^{2\pi ik/n}$.
6. Set $g = (z+1)^{-\varepsilon}f$. Then $|g| \le |f|$. Take R large so that $|g(z)| \le M$ for $|z| \ge R$. The lim sup condition implies that there is $\delta > 0$ such that $|g(z)| \le M + \varepsilon$ for $|z| < R$, $0 < \operatorname{Re} z \le \delta$. Apply the maximum principle to the domain $\{|z| < R, \operatorname{Re} z > \delta\}$, to obtain $|g(z)| \le M + \varepsilon$. Then let $\varepsilon \to 0$.
9. Let $\varepsilon > 0$. Take $\delta > 0$ such that $u(z) \ge -\varepsilon$ for $z \in D$, $0 < |z| < \delta$. Let $\rho > 0$. Take $R > 0$ so large that $u(z) + \rho \log |z| > 0$ for $|z| > R$. By the maximum principle, $u(z) + \rho \log |z| \ge -\varepsilon + \rho \log \delta$ for $z \in D$, $\delta < |z| < R$, hence for all $z \in D$ such that $|z| > \delta$. Let $\rho \to 0$, then let $\varepsilon \to 0$, to obtain $u \ge 0$ on D.

Section III.6
1. $\phi = 2x + y$, $\psi = 2y - x$, $f(z) = (2 - i)z$. Streamlines are straight lines with slope $1/2$. Flux across $[0, 1]$ is -1, across $[0, i]$ is 2.
2. (a) $\phi = \alpha \log r + \beta\theta$, which is defined only locally, or on a slit plane. (b) $\psi = \alpha\theta - \beta \log r$, $f(z) = (\alpha - i\beta) \log z$. (c) Flux is $2\pi\alpha$, the increase of ψ around circle centered at 0. Origin is source if $\alpha > 0$, sink if $\alpha < 0$.
5. $\psi(z) = A[2 \arg(e^z - 1) - \pi - 1]$, where $A > 0$.
6. Stream function of conjugate flow is $-\phi$, complex velocity potential is $-i(\phi + i\psi)$.
7. Stream function of conjugate flow is $- \arg z$, complex velocity potential is $-i \log z$, speed is $1/|z|$, particles travel faster near 0.

Section III.7
1. $(T_1 - T_2)(2/\pi)[\operatorname{Arg}(1 + z) - \operatorname{Arg}(1 - z)] + T_2$
2. $\phi = V_1(2/\pi)[\operatorname{Arg}(1 + z^2) - \operatorname{Arg}(1 - z^2)]$
4. $\phi = 50[\operatorname{Arg}(\sin z + 1) - \operatorname{Arg}(\sin z - 1)]$
5. $u = (2/\pi)(T_1 - T_0)x + (T_0 - T_1)/2$
7. $\phi(x, y, z) = cz$
8. $\phi = -1/r$
9. $\mathbf{F} = (2 - n)\mathbf{u_r}/r^{n-1}$

Chapter IV: Complex Integration and Analyticity

Section IV.1
1. (a) $i/2$, (b) $-1/2$, (c) 0.
2. (a) $2\pi i$ for $m = -1$, otherwise 0, (b) $2\pi i$ for $m = 1$, otherwise 0, (c) 2π for $m = 0$, otherwise 0.
3. (a) 0, (b) $2\pi R^{m+1}$, (c) $2\pi i R^2$ for $m = 1$, otherwise 0.
4. Substitute $\bar{z} = x - iy$, $dz = dx + idy$, and apply Green's theorem.
6. $|\text{Log } z| = ((\log R)^2 + \theta^2)^{1/2} \leq \sqrt{2}\log R$ for $R > e^\pi$. Apply the ML-estimate, with $M = \sqrt{2}(\log R)/R^2$, $L = 2\pi R$.

Section IV.2
1. The integrals are all independent of path and can be evaluated by finding a primitive. (a) $2\pi^5 i/5$, (b) 0, (c) $i(e^\pi - e^{-\pi})$, (d) 0.
2. In the right half-plane, use $\text{Log } z$ as a primitive. Integral $= \text{Log}(i\pi) - \text{Log}(-i\pi) = i\pi/2 - i(-\pi/2) = \pi i$. In the left half-plane, use $\log r + i\theta$, $0 < \theta < 2\pi$, as a primitive. Integral $= i\pi/2 - i(3\pi/2) = -\pi i$.
3. A primitive is $z^{m+1}/(m + 1)$, $m \neq -1$.

Section IV.3
1. Apply Cauchy's theorem and pass to limit. Integrals over vertical sides of rectangles $\to 0$ as $R \to \infty$, by the ML-estimate. $\int_{-R}^{R} \to \sqrt{2\pi}$, and the integral over other horizontal side $\to e^{t^2/2} \int_{\infty}^{-\infty} e^{-x^2/2} e^{-ixt} dx$.
3. (a) Follow hint to estimate $\int_{-1}^{1}$ by $\int_{0}^{\pi} |f(e^{i\theta})|^2 d\theta$ and by $\int_{\pi}^{2\pi} |f(e^{i\theta})|^2 d\theta$. Use $\int_{0}^{2\pi} |f(e^{i\theta})|^2 d\theta = \int_{0}^{2\pi} f(e^{i\theta})\overline{f(e^{i\theta})} d\theta = 2\pi \sum c_k^2$. (b) Write $c_k = a_k + ib_k$ and apply estimate in (a) twice. (c) The double sum is $\int_{0}^{1} f(x)^2 dx$. Estimate this by $\int_{-1}^{1} |f(x)|^2 dx$.

Section IV.4
1. (a) $2\pi i$, (b) 0, (c) 0, (d) πi, (e) $2\pi i/[(m-1)!]$ for $m \geq 1$, 0 for $m \leq 0$, (f) $2\pi i$, (g) $\pi i/2$, (h) $-\pi i/2 + \pi i/4e^2$.
2. A harmonic function is locally the real part of an analytic function, and any derivative of a harmonic function can be expressed in terms of the real part of a some complex derivative of that analytic function.

Section IV.5
1. If $u \leq C$, and $u = \text{Re } f$, then e^f is bounded, hence constant by Liouville's theorem, and f is constant.
2. If f does not attain values in the disk $|w - c| < \varepsilon$, then $1/(f - c)$ is bounded, hence constant by Liouville's theorem, and f is constant.
4. Apply the Cauchy estimates for $f^{(m+1)}(z)$ to a disk $|z - z_0| < R$, and let $R \to \infty$, to obtain $f^{(m+1)}(z_0) = 0$. If $f(z)$ is a polynomial of degree $\leq m$ such that $f(z)/z^m$ is bounded near 0, then $f(z) = cz^m$.

Section IV.6
1. Rotate and apply result in text.
2. The analyticity of $H(z)$ follows from the theorem in the text. If $|h| \leq M$, take $C = M(b - a)$ and $A = \max(|a|, |b|)$.

Section IV.8

2. By the Leibniz rule and Exercise 1, the $\bar{z}$-derivative is $bz + 2c\bar{z}$. The function is complex-differentiable at z if and only if $bz + 2c\bar{z} = 0$. If $b = c = 0$, the function is entire. Otherwise this locus is either $\{0\}$ or a straight line through 0, and there is no open set on which the function is analytic.

6. Using the Leibniz rule, obtain $(\partial/\partial\bar{z})(f\bar{g}) = f(\partial/\partial\bar{z})\bar{g} + \bar{g}(\partial/\partial\bar{z})f = f(\partial/\partial\bar{z})\bar{g} = f\overline{g'}$, then apply (8.4).

7. Since the coefficients of the Taylor expansion depend linearly on f, and since any f is a linear combination of the functions $1, z, \bar{z}, z^2, \bar{z}^2, |z|^2$ and an error term $\mathcal{O}(|z|^3)$, it suffices to check the formula for these six functions.

Chapter V: Power Series

Section V.1

6. Note that $x \log x$ and $x(\log x)^2$ are increasing for $x > 1$, so the terms of both series are decreasing. For $\sum 1/(k \log k)$, there are 2^n terms between $k = 2^n + 1$ and $k = 2^{n+1}$, each $\geq 1/(2^{n+1}(n+1) \log 2)$, so these terms have sum $\geq 1/(2(n + 1) \log 2)$. Thus the series diverges, by comparison with the harmonic series. The other series is treated similarly, using an upper estimate. These series can be also treated using the integral test.

7. Observe that $\sum_m^n a_k = S_n - S_{m-1}$, where S_n is the nth partial sum of the series.

Section V.2

1. Set $f_k'(x) = 0$, check that $f_k(x)$ attains its maximum $\varepsilon_k = 1/2\sqrt{k}$ when $x^k = \sqrt{k}$. Since $\varepsilon_k \to 0$, $f_k \to 0$ uniformly.

2. $g_k(x) \to g(x)$, where $g(x) = 0$ for $0 \leq x < 1$, $g(1) = 1/2$, and $g(x) = 1$ for $x > 1$. Convergence is uniform on $[0, 1 - \varepsilon]$ for any $\varepsilon > 0$. Not uniform on $[0, 1]$, because limit function is not continuous at 1. Convergence is uniform on $[1 + \varepsilon, +\infty)$ for any $\varepsilon > 0$.

3. For any $\varepsilon > 0$, $f_k'(z) = z^{k-1}$ converges uniformly for $|z| \leq 1 - \varepsilon$.

4. Apply the Weierstrass M-test.

5. Converges for $x \neq 1$.

6. Check that $x^k/(1 + x^{2k})$ is decreasing for $x \geq 1$. Apply Weierstrass M-test, with M_k the value of the summand at $1 + \varepsilon$.

8. Apply the Weierstrass M-test.

9. If it converges uniformly for $|z| < 1$, then it also converges uniformly for $|z| \leq 1$.

Section V.3

1. (a) 1/2, (b) 6, (c) 1, (d) 5/3, (e) $1/\sqrt{2}$, (f) $+\infty$, (g) 2, (h) 0, (i) e.

2. (a) $|z - 1| < 1$, (b) all z, (c) $|z - 2| < 1/2$, (d) $|z + i| \leq 1$, (e) $z = 3$, (f) $|z - 2 - i| \leq 1/2$.

3. Neither series converges at $z = 1$, so $R = 1$.

5. Differentiate the geometric series twice, obtain (a) $z/(1 - z)^2$, (b) $-z + z/(1 - z)^2 + 2z^2/(1 - z)^3$.

Section V.4

1. (a) $\sqrt{2}$, (b) $\pi/2$, (c) $\pi/2$, (d) $\sqrt{5}$, (e) 3, (f) 2.
2. Rewrite $f(z)$ as $(z+1)/(z - e^{2\pi i/3})(z + e^{2\pi i/3})$. Singularities of $f(z)$ are at $\pm e^{2\pi i/3}$, and distance from 2 to nearest singularity is $\sqrt{7}$.
3. Log z extends to be analytic for $|z - (i-2)| < \sqrt{5}$, though the extension does not coincide with Log z in the part of the disk in the lower half-plane.
4. Near 0 the function coincides with one of the branches of $(1\pm\sqrt{1 - 4z})/2$. The radius of convergence of the power series of either branch is $1/4$, which is the distance to the singularity at $1/4$.
6. $\cosh z = \sum_{n=0}^{\infty} z^{2n}/(2n)!$, $\sinh z = \sum_{n=0}^{\infty} z^{2n+1}/(2n + 1)!$, $R = \infty$.
7. $\mathrm{Tan}^{-1}(z) = z - z^3/3 + z^5/5 - z^7/7 + \cdots$, converges for $|z| < 1$.
10. Use $f^{(n)}(z) = \alpha(\alpha - 1)\cdots(\alpha - n + 1)(1 + z)^{\alpha-n}$ and the formula for the coefficient of z^n. The series reduces to a polynomial for $\alpha = 0, 1, 2, \cdots$. Otherwise radius of convergence is 1, which is distance to the singularity at -1. Can obtain the radius of convergence also from the ratio test.
13. Use power series.

Section V.5

1. (a) $\sum_{n=0}^{\infty} (-1)^n/z^{2n+2}$, (c) $\sum_{n=0}^{\infty} 1/n! z^{2n}$, (d) $\sum_{n=0}^{\infty} 1/(2n + 1)! z^{2n}$.
4. If $|z| \leq M$ for $z \in E$, and $R > M$, then $1/(w - z) = \sum z^n/w^{n+1}$ converges uniformly for $z \in E$ and $|w| > R$. Integrate term by term, obtain $f(w) = \sum_{n=0}^{\infty} b_n/w^{n+1}$, $|w| > R$, where $b_n = \iint_E z^n dxdy$.
5. $f(w) = \pi/w$ for $|w| \geq 1$, $f(w) = \pi\bar{w}$ for $|w| \leq 1$. To find the formula for $|w| < 1$, break the integral into two pieces corresponding to $|z| > |w|$ and to $|z| < |w|$, and use geometric series.

Section V.6

1. $1/\cos z = 1 + (1/2)z^2 + (5/24)z^4 + (1/12)z^6 + \mathcal{O}(z^8)$.
2. $z/\sin z = 1 + (1/6)z^2 + (7/360)z^4 + \mathcal{O}(z^6)$.
3. $R = 1 =$ distance to singularity at -1.
4. $B_1 = 1/6$, $B_2 = B_4 = 1/30$, $B_3 = 1/42$, $B_5 = 5/66$.
5. $E_0 = 1$, $E_2 = -1$, $E_4 = 5$, $E_6 = -61$.
6. $f_m^{(k)}(z) \to f^{(k)}(z)$ uniformly for $|z| \leq \rho - \varepsilon$, so $a_{k,m} = f_m^{(k)}(0)/k! \to f^{(k)}(0)/k! = a_k$.

Section V.7

1. (a) simple zeros at $\pm i$, (b) simple zeros at $\pm e^{\pi i/4}, \pm e^{3\pi i/4}$, (c) triple zero at 0, simple zeros at $n\pi$, $n = \pm1, \pm2, \cdots$, (d) double zeros at $n\pi$, $n = 0, \pm1, \pm2, \cdots$, (h) simple zeros at $-\pi i/4 + n\pi i/2$, $n = 0, \pm1, \pm2, \cdots$, (i) no zeros.
2. (a) analytic at ∞, (b) analytic at ∞, simple zero, (c)-(i) not analytic at ∞.
6. Apply argument in text to $f(z) - f(z_0)$.
9. Write $f(z) = (z - z_0)^N h(z)$, where $h(z)$ has a convergent power series and $h(z_0) \neq 0$. Take $g(z) = (z - z_0)e^{(\log h(z))/N}$ for an appropriate branch of the logarithm.
10. Show that the zeros of $f(z)$ are isolated. At a zero of $f(z)$, write

$f(z)^N = (z - z_0)^m h(z)$ where $h(z_0) \neq 0$, and show that N divides m.

13. $f(D) \subset D \cup \partial D$. If $f(z)$ is not constant, then $f(D)$ is open, and $f(D)$ cannot contain any point of ∂D.

Section V.8

1. $\sqrt[3]{z^3 - 1}$ returns to $e^{2\pi i/3}$ times initial value, other functions return to initial values.

2. $f_t(z) = it + \sum_{m=1}^{\infty}((-1)^{m-1}e^{-itm}/m)(z - e^{it})^m$, $f_{2\pi}(z) = f_0(z) + 2\pi i$.

9. Let t_1 be the infimum of $t > a$ such that the power series expansion of $P(z, f_t(z))$ at $\gamma(t)$ is not identically zero, and apply the uniqueness principle for $t < t_1$ near t_1.

Chapter VI: Laurent Series and Isolated Singularities

Section VI.1

1. (a) Laurent expansions $-\sum_{n=-1}^{\infty} z^n$ for $0 < |z| < 1$, and $\sum_{n=-\infty}^{-2} z^n$ for $|z| > 1$.

2. (a) $\sum a_n(z + 1)^n$, where $a_n = -1$ for $n \leq -1$ and $a_n = -1/2^{n+1}$ for $n \geq 0$. Converges for $1 < |z + 1| < 2$.

4. If $f(z) = \sum a_n z^n$ is even, then $f(z) = f(-z) = \sum a_n(-1)^n z^n$. By the uniqueness of the expansion, $a_n = (-1)^n a_n$, and $a_n = 0$ if n is odd.

5. Let $f(z) = \sum a_n z^n$. Then $a_{-1} = \int_{|z|=r} f(z)dz$, and $f(z) - a_{-1}/z$ has the primitive $\sum_{n \neq -1}(a_n/(n + 1))z^{n+1}$.

Section VI.2

1. (a) double poles at ± 1, principal parts $\mp(1/4)/(z \pm 1)^2$, (c) removable singularity at 0, (e) essential singularity at 0, (g) analytic on $\mathbb{C} \setminus [0, 1]$, no isolated singularities, (h) double pole at 1, principal part $1/(z - 1)^2 - (1/2)/(z - 1)$.

2. (a) 3, (b) ∞, (c) $\pi\sqrt{2}$, (d) π.

3. (a) $\tan z$ has two poles in the disk $|z| < 4$, simple poles at $\pm\pi/2$, principal parts $-1/(z \pm \pi/2)$. If $f_1(z) = -1/(z - \pi/2) - 1/(z + \pi/2)$, then $f_0(z) = f(z) - f_1(z)$ is analytic for $|z| < 4$, and $f_1(z)$ is analytic for $|z| > 3$ and $\to 0$ as $z \to \infty$. By uniqueness, $f(z) = f_0(z) + f_1(z)$ is the Laurent decomposition. (b) Use geometric series. Converges for $|z| > \pi/2$. (c) $a_0 = a_2 = 0$, $a_1 = 1 + 8/\pi^2$. (d) $3\pi/2$.

4. Use uniqueness of the Laurent decomposition.

7. By Riemann's theorem, $g(z) = (z - z_0)^N f(z)$ has a removable singularity at z_0, hence $f(z) = g(z)/(z - z_0)^N$ is meromorphic at z_0.

11. For starters, check the function $f(z) = 1/(z - z_0)^N$.

13. Suppose values of $f(z)$ do not cluster at L as $z \to z_0$. Then $g(z) = 1/(f(z) - L)$ is bounded for $|z - z_0| < \varepsilon$, $z \neq z_j$. Apply Riemann's theorem first to the z_j's for j large, then to z_0, to see that $g(z)$ extends to be analytic for $|z - z_0| < \varepsilon$, and $f(z)$ is meromorphic there.

Section VI.3

1. (a) removable singularity at ∞, (c) essential singularity at ∞, (e) simple pole at ∞, (g) removable singularity at ∞, (h) ∞ not an isolated singularity.

2. Laurent expansion has infinitely many negative powers of $1/z$, so singularity at ∞ is essential.

3. If e^f has a removable singularity at ∞, then it is bounded, hence constant by Liouville's theorem, and f is constant. If e^f has a pole at ∞, then e^{-f} is bounded, hence constant by Liouville's theorem, and f is constant.

4. (c) Use binomial series to expand each branch of $f(z) = z(1 - 1/z^3)^{1/2}$ in a Laurent series at ∞. Each branch is analytic for $|z| > 1$ and has a simple pole at ∞.

Section VI.4

1. (b) $-1/z + 1/(z-1) + 1/(z+1)$ (d) $(-1/4)/(z-i)^2 + (-i/4)/(z-i) + (-1/4)/(z+i)^2 + (i/4)/(z+i)$ (f) $1 - 3/(z+2)$.

2. (a) $z - [(1+i)/2]/(z-i) + [(-1+i)/2]/(z+i)$ (b) $z^3 - (1/3)/(z-1) + (\omega/3)/(z-\omega) - (\omega/3)/(z+\omega)$, where $\omega = e^{2\pi i/3}$.

3. Dimension $= 5$, with basis $1, 1/z, 1/z^2, 1/(z-i), 1/(z-i)^2$.

Section VI.5

2. $1/\cos(2\pi z) = 2e^{2\pi i z}/(1 + e^{4\pi i z}) = 2e^{2\pi i z}\sum_{m=0}^{\infty}(-1)^m e^{4m\pi i z}$, converges absolutely for $\operatorname{Im} z > 0$. For any $\varepsilon > 0$, converges uniformly for $\operatorname{Im} z \geq \varepsilon$.

5. If ω is a period $\neq 0$ or ± 1, then $|\omega \pm 1| \geq 1$. If moreover $\omega \neq \pm i$, then $|\omega^2 + 1| \geq 1$. The only possibilities for periods on unit circle are then $\{\pm 1\}$, $\{\pm 1, \pm i\}$, and $\{\pm 1, \pm e^{\pi i/3}, \pm e^{2\pi i/3}\}$.

Section VI.6

2. $\theta \sim \sum_{k=-\infty}^{\infty} c_k e^{ik\theta}$, $c_k = i(-1)^k/k$ for $k \geq 1$, $c_k = -i(-1)^k/k$ for $k \leq -1$, $c_0 = 0$. The terms of the differentiated series do not tend to 0, so the differentiated series diverges at each θ. At $\theta = \pi$, the complex Fourier series for $k \geq 1$ becomes a multiple of the harmonic series, hence the series diverges at $\theta = \pi$. The corresponding sine series is $\theta \sim \sum_{k=1}^{\infty} b_k \sin(k\theta)$, $b_k = 2(-1)^{k+1}/k$. It converges to θ if $-\pi < \theta < \pi$ and to 0 if $\theta = \pm\pi$.

3. The cosine series is $\sum a_k e^{ik\theta}$, where $a_0 = \pi^2/3$, and $a_k = 4(-1)^k/k^2$ for $k \geq 1$. By Weierstrass M-test, series converges uniformly for $|\theta| \leq \pi$.

5. Since $|c_k e^{ik\theta}| = |c_k|$, series converges absolutely for all θ. By Weierstrass M-test, convergence is uniform in θ.

Chapter VII: The Residue Calculus

Section VII.1

1. (a) $-i/4$, (b) $i/4$, (e) 0, (i) $2e^{2\pi i k/n}/n$.

2. (a) 1 at $z = 0$, (b) -1 at $z = (\pi/2) + n\pi$.

3. (a) residue at $z = 0$ is 1, integral $= 2\pi i$, (f) $z = 0$ is removable, pole inside contour at $z = \pi/2$, residue $= -2/\pi$, $\int = -4i$.

4. Apply Rule 3 to evaluate the residues.

5. The integrals along opposite sides cancel.

6. (a) pole at $z = 0$, residue $= e^{\pi i/4}/2\pi i = (1 - i)/2\sqrt{2}\pi$.

Section VII.2

1. Use semicircular contour, residue at ia is $1/2ia$.

3. Use Rule 2 to find residue $1/4i$ at $z = i$.

5. Use Rule 3 to find residues at $e^{\pi i/4}$ and $e^{3\pi i/4}$.

9. Use semicircular contour to show $\int_{-\infty}^{\infty} [e^{2ix}/(x^2+1)]dx = \pi e^{-2}$. Residue of $e^{2iz}/(z^2+1)$ at i is $e^{-2}/\pi i$. Use also $-4\sin^2 x = (e^{ix} - e^{-ix})^2 = e^{2ix} + e^{-2ix} - 2$.

10. Integral exists only for a real, and $\operatorname{Re} b \neq 0$ or $\cos(iab) = 0$. It depends analytically on b for $\operatorname{Re} b \neq 0$.

Section VII.3

1. Poles of $(z^2+1)/[z(z^2+4z+1)]$ are at 0 and $-2 \pm \sqrt{3}$. Use Rule 3 to obtain residue 0 at 1 and residue $-2 + \sqrt{3}$ at $-2\sqrt{3}/3$. Other pole is outside the unit circle.

4. Poles of $z/(z^4 - 6z^2 + 1)$ are at $z_0 = \pm\sqrt{3 - 2\sqrt{2}}$, residue by Rule 3 is $1/(4z_0^2 - 12) = -1/8\sqrt{2}$.

7. Choose branch of $\sqrt{w^2 - 1}$ on $\mathbb{C}\backslash[-1, 1]$ that is positive for $w \in (1, \infty)$. It suffices to check the identity for $w = a > 1$. Poles of $z/(z^2 + 2az + 1)^2$ are double poles at $z_\pm = -a \pm \sqrt{a^2 - 1}$. Use Rule 2 to obtain residue at z_+ of $-(z_+ + z_-)/(z_+ - z_-)^3 = a/4(a^2 - 1)^{3/2}$.

Section VII.4

2. Simple pole at $e^{i\pi/b}$ with residue $-e^{i\pi/b}/b$. Use $z = re^{2\pi i/b}$, $dz = e^{2\pi i/b}dr$, to parametrize one edge of the domain. Obtain $(1 - e^{2\pi i/b}) \int_0^\infty = -2\pi i e^{i\pi/b}/b$.

4. Integrate $z^{-a}/(1 + z)^m$ around the keyhole contour. Specify branch by $z^{-a} = r^{-a}e^{-ia\theta}$, $0 < \theta < 2\pi$. Determine the residue from power series expansion of z^{-a} about -1.

7. This is similar to Exercise 2. The integral depends analytically on the parameter a for $0 < \operatorname{Re} a < \operatorname{Re} b$.

9. Integrate $(\log z)^2/(z^3 + 1)$ around a pie-slice domain with $0 < \theta < 2\pi/3$. Specify branch by $\log z = \log r + i\theta$, $0 < \theta < 2\pi/3$. Simple pole at $e^{\pi i/3}$, residue $-\pi^2 e^{-2\pi i/3}/27$. Apply residue theorem, pass to limit, multiply by $e^{-\pi i/3}$, take imaginary parts, and substitute the values for the integrals from Exercise 8, to obtain $-\sqrt{3} \int_0^\infty +4\pi^3/81 + 12\pi^3/81 = 2\pi^3/27$. Then solve for the integral.

Section VII.5

1. Specify branch by $f(z) = r^{-a}e^{-ia\theta}(\log r + i\theta)/(z - 1)$, $z = re^{i\theta}$, $0 < \theta < 2\pi$. It is analytic on the keyhole domain. It extends analytically to $(0, \infty)$ from above, and the apparent singularity at $z = 1$ is removable. However, the extension to $(0, \infty)$ from below has a simple pole at $z = 1$, with residue $2\pi i e^{-2\pi ia}$ obtained by using $\theta = 2\pi$ in the definition of $f(z)$. Use Cauchy's theorem, multiply by $e^{2\pi ia}$, take real parts, and pass to limit using the fractional residue formula with angle $-\pi$.

3. For $a = 1$, integrate $e^{iz}/[z(\pi^2 - z^2)]$ around semicircular contour indented at 0 and $\pm\pi$. Take imaginary parts, then pass to limit using fractional residue theorem. Residue at 0 is $1/\pi^2$, at $\pm\pi$ is $1/2\pi^2$.

5. Integrate around semicircular contour indented at 0 and $\pm\pi$. Take real parts and pass to limit, using $\operatorname{Re}(e^{2ix} - 1) = -2\sin^2 x$. Residue at 0 is $2i$.

Fractional residue theorem gives limit $(-\pi i)2i = 2\pi$ for the integral over the indentation.

Section VII.6

1. $\int_0^{1-\varepsilon} + \int_{1+\varepsilon}^\infty = (1/2)\log[(2-\varepsilon)/(2+\varepsilon)] \to 0$ as $\varepsilon \to 0$.

3. By residue theorem, $\int_{-\infty}^{a-\varepsilon} + \int_{C_\varepsilon} + \int_{a+\varepsilon}^\infty = \pi/(i-a) = -\pi(i+a)/(a^2+1)$. By the fractional residue theorem, $\int_{C_\varepsilon} \to -\pi i/(a^2+1)$.

5. There are two fractional residues, at 1 and at $e^{2\pi i/b}$, each with angle $-\pi$. By Cauchy's theorem, $(1 - e^{2\pi i/b}e^{2\pi(a-1)i/b})\mathrm{PV}\int -\pi i\,\mathrm{Res}[1] - \pi i\,\mathrm{Res}[e^{2\pi i/b}] = 0$. The residue at 1 is $e^{2\pi i/b}$, at $e^{2\pi i/b}$ is $e^{2\pi i/b}e^{2\pi(a-1)i/b}/b$.

Section VII.7

2. Integrate $z^3 e^{iz}/(z^2+1)^2$ around semicircular contour of radius R in the upper half-plane. Residue at i is $1/4e$, integral $= \pi i/2e$. Let $R \to \infty$, use Jordan's lemma, and take the imaginary part.

3. The limit is πe^{-a} if $a > 0$, and $-\pi e^a$ if $a < 0$. For $a > 0$, integrate $z e^{iaz}/(z^2+1)$ around semicircular contour. Residue at i is $e^{-a}/2$. Pass to limit, use Jordan's lemma, and take the imaginary part.

Section VII.8

1. (a) -1, (b) 0, (c) -1, (d) 0, (e) $-1/(n+1)!$, (f) $\pm(b-a)/2$.

2. Follow calculation in text, use coefficient of $1/z^5$ in binomial series, residue at ∞ is $-35i/128$.

3. Converges only if $n \geq 0$, for $0 < \mathrm{Re}\,a < n+1$. Converges to $\pi/\sin(\pi a)$ if $n = 0$, to $(-1)^n[(a-1)\cdots(a-n)/n!]\pi/\sin(\pi a)$ if $n > 0$. Evaluate it for $0 < a < n+1$ by integrating $z^n/z^a(1-z)^{1-a}$ around dogbone contour.

5. Consider $\int f(z)\,dz$ around a large circle.

7. Residue at ∞ is $-3/8$, so sum of residues in the finite plane is $3/8$.

Chapter VIII: The Logarithmic Integral

Section VIII.1

2. Two in second quadrant, two in third.

5. Four zeros in open left half-plane for $\alpha < 1$ and $\alpha > 3$, two for $1 \leq \alpha \leq 3$. For $\alpha = 1$ and $\alpha = 3$ there are also two zeros on the imaginary axis.

6. Two zeros in open right half-plane for $\alpha \leq 0$, one zero for $\alpha = 1$, and three zeros for $\alpha > 0$, $\alpha \neq 1$.

8. Increase in argument of $z + \lambda - e^z$ around large rectangle with vertices $\pm iR$ and $-R \pm iR$ is 2π, so there is one zero in rectangle.

9. There is an analytic branch $\mathrm{Log}\,f(z)$ of $\log f(z)$ on D, so the increase in $\arg f(z)$ around any closed path in D is zero.

Section VIII.2

1. On circle $|z| = 1$ take BIG $= 6z$. On circle $|z| = 2$ take BIG $= 2z^5$.

2. Six.

3. Use $|p(z) - 3z^n| < e$ for $|z| = 1$.

4. Use $|e^z| \geq 1$ for $\mathrm{Re}\,z > 0$. The zeros are simple unless $\lambda = 0$.

6. (a) One in each quadrant. (b) First and fourth quadrants. (c) Compare

$p(z)$ with $z^6 + 9z^4$.

7. Neither $f(z)$ nor $g(z)$ has zeros on ∂D, so each has at most finitely many zeros in D. Estimate shows $f(z)/g(z) \notin (-\infty, 0]$ for z near ∂D, so $\mathrm{Log}(f(z)/g(z))$ is continuously defined near ∂D. Increase in $\arg f(z)/g(z)$ around any closed path near ∂D is zero.

8. They have the same number of zeros minus poles.

9. If $A = (\partial f/\partial \bar{z})(z_0) \neq 0$, consider Taylor approximation of $f(z)$ at z_0, show increase in argument of $A(\bar{z} - \overline{z_0}) + o(|z - z_0|)$ around $|z - z_0| = \varepsilon$ is < 0.

Section VIII.3

1. Take BIG $= f(z)$, little $= f_k(z) - f(z)$.

Section VIII.4

1. Apply the residue theorem. Residue at z_j is $m_j g(z_j)$.

4. Suppose that $f(z)$ is analytic across $\partial \mathbb{D}$, and consider the set of $z \in \mathbb{D}$ whose image $f(z)$ is not the image of any other point in $\mathbb{D}$.

5. (a) Apply the argument principle to $f(z) - w$ on a large disk.

6. Choose w_0 such that the number of points in $f^{-1}(w_0)$ is maximum. Then $f(z)$ attains values w near w_0 only near points in $f^{-1}(w_0)$, $1/(f(z) - w_0)$ is bounded at ∞, and $f(z)$ is meromorphic on $\mathbb{C}^*$ hence rational.

8. Let $w_0 \in \partial f(D)$, and take $z_n \in D$ with $f(z_n) \to w_0$. Assume $z_n \to z_0 \in D \cup \partial D$, then $f(z_0) = w_0$. Since $f(D)$ is open, $w_0 \notin f(D)$. Thus $z_0 \in \partial D$, and $w_0 \in f(\partial D)$.

Section VIII.5

1. Critical points ± 1, critical values ± 2.

3. Critical point 0, critical value 1.

5. Its derivative is a polynomial of degree $m - 1$, hence has $m - 1$ zeros counting multiplicity.

7. Denote the kth iterate by $f_k(z)$, a polynomial of degree m^k. By the chain rule, $f_N'(z) = f'(f_{N-1}(z))f'(f_{N-2}(z)) \cdots f'(f(z))f'(z)$. Thus the critical points of $f_N(z)$ consist of the $m^k(m-1)$ inverse images of the critical points of $f(z)$ under $f_k(z)$ for $0 \leq k < N$. Total number of critical points is $m^N - 1$.

10. Critical point at $0, \pm 1$ of order 1 and at ∞ of order 3. Critical values are $0, -1, \infty$.

13. $z_1(w) = e^{iw}$, $z_2(w) = e^{-iw}$, graphs meet at $w = \pi m$, $-\infty < m < \infty$.

Section VIII.6

1. $W(\gamma, \zeta) = 1$ for ζ in the four bounded components of $\mathbb{C} \backslash \gamma$, $W(\gamma, \zeta) = 1$ for ζ in the unbounded component.

4. (a) If γ is a closed path in D, then $W(\gamma, z)$ is constant on any connected set disjoint from γ, so $W(\gamma, z_0) = W(\gamma, z_1) = m$, and the increase in the argument of $f(z)$ is $4\pi m$. (b) The increase in the argument of $\sqrt{f(z)}$ around γ is $2\pi m$, so the analytic continuation of a branch of $\sqrt{f(z)}$ around any closed path in D returns to itself, and we can define an analytic branch of $\sqrt{f(z)}$ by analytic continuation.

5. For $n \geq 2$, $(z - z_0)^{-n}$ has primitive $(z - z_0)^{-n+1}/(1 - n)$ on D.
8. Apply the theorems in III.2 on deformation of paths.
9. The increase in $\arg f(z)$ around the circle $|z| = r$ is constant for $1 \leq r < \infty$, and it tends to 0 as $r \to \infty$, so it is identically zero. The argument principle, applied to $f(z)$ on $\mathbb{D}$, shows that $f(z)$ has no zeros in $\mathbb{D}$.
10. Suppose $0 \notin f(K)$ and $f(\infty) \neq 0$. The increase in $\arg f(z)$ around the boundary of a large square S is then 0, and $f(z)$ has only finitely many zeros. Cut S into a grid of very small squares S_j and add the increases of $\arg f(z)$ over ∂S_j.

Section VIII.7

1. As ζ crosses the curve from left to right, the Cauchy integral jumps by $-2\pi i f(\zeta)$, so the value of the integral on the lower half-plane is $g(\zeta) - 2\pi i f(\zeta)$.
2. Both sides are analytic on $\mathbb{C}\backslash[-1,1]$. They are equal for $\zeta \in (-\infty, -1)$, hence they coincide on $\mathbb{C}\backslash[-1,1]$. This does not contradict Exercise 1, since the logarithm function is not analytic on $(-1,1)$.
3. (a) $F(\zeta) = 0$ for $|\zeta| > 1$, $F(\zeta) = \zeta$ for $|\zeta| < 1$. (b) $F(\zeta) = -1/\zeta$ for $|\zeta| > 1$, $F(\zeta) = 0$ for $|\zeta| < 1$. (c) $F(\zeta) = -1/2\zeta$ for $|\zeta| > 1$, $F(\zeta) = \zeta/2$ for $|\zeta| < 1$. (d) $F(\zeta) = 1/2i\zeta$ for $|\zeta| > 1$, $F(\zeta) = \zeta/2i$ for $|\zeta| < 1$.

Section VIII.8

1. (a) and (c) are simply connected, (b) and (d) not.
3. (a) is simply connected, (b) is not.
5. If D is simply connected, and if $G(z)$ is a branch of $\log f(z)$, then $e^{G(z)/2}$ is a branch of $\sqrt{f(z)}$. If D is not simply connected, and if z_0 and K are as in the proof that (iv) $\Rightarrow$ (v), then $z - z_0$ cannot have an analytic square root $g(z)$, or else $\int_{\partial K} d \arg g(z) = \pi$ and not an integer times 2π.
9. Since $f(D)$ is simply connected, the analytic continuation of $g(w)$ is independent of path, by the monodromy theorem. Hence there is an analytic function $g(w)$ defined on $f(D)$ such that $g(f(z)) = z$ for z near z_0. By the uniqueness principle, $g(f(z)) = z$ for all $z \in D$. This implies $f(z)$ is one-to-one on D.
10. Assume $h(\zeta)$ attains only the values 0 and 1, let $E = h^{-1}(1)$, and consider $\sigma = \partial K$ as in the proof that (iv) $\Rightarrow$ (v).
11. $W(\sigma, \zeta) = (1/2\pi i) \int_{\partial U} 1/(z-\zeta) \, dz = 0$ for $\zeta \notin D$, by Cauchy's theorem.
12. Let E be a bounded component of $\mathbb{C}\backslash D$. Modify ∂K constructed in the proof that (iv) $\Rightarrow$ (v) to obtain a single closed curve γ such that $W(\gamma, \zeta) = 1$ if $\zeta \in E$ and $W(\gamma, \zeta) = 0$ if $\zeta \notin D \cup E$.

Chapter IX: The Schwarz Lemma and Hyperbolic Geometry

Section IX.1

3. Since $f(z) = 1$ has a solution $z_1 = 0$, the argument principle guarantees that $f(z) = 0$ also has a solution z_0, $|z_0| < 1$. If $\varphi(w) = M(w-1)/(M^2-w)$, then $|\varphi(f(z))| < 1$ for $|z| = 1$, and $\varphi(f(0)) = 0$. By the Schwarz lemma, $1/M = |\varphi(f(z_0))| < |z_0|$.

5. Assume $f(0) = r$, and apply the Schwarz lemma to $\psi \circ f$, where $\psi(\zeta) = (\zeta - r)/(1 - r\zeta)$. Equality holds at z_0 only when $f(z) = \lambda(\mu z + r)/(1 + r\mu z)$, where $|\lambda| = |\mu| = 1$, and $z_0 = -\bar{\mu}s$, $0 \le s < r$.

6. If $f(0)$ has distance d from ∂D, then the inverse function $f^{-1}(w)$ is analytic for $|w - f(0)| < d$, and $|f^{-1}(w)| < 1$. Apply the Schwarz lemma to obtain $|(f^{-1})'(f(0))| = 1/|f'(0)| \le 1/d$.

8. $f(z)/z$ is analytic for $|z| < 1$, and $|f(z)/z| < 1$ for $|z| < 1$, or else $f(z)/z$ would be a unimodular constant, violating $|f'(0)| < 1$. Take c to be the maximum of $|f(z)/z|$ for $|z| \le r$.

Section IX.2

1. Let $B(z)$ be the finite Blaschke product with same zeros as $f(z)$. Then $g(z) = f(z)/B(z)$ is continuous for $|z| \le 1$, analytic for $|z| < 1$, and satisfies $|g(z)| = 1$ for $|z| = 1$. Hence $|g(z)| \le 1$ for $|z| \le 1$. Since $g(z)$ has no zeros, the maximum principle applies also to $1/g(z)$, and $|g(z)| \ge 1$. Hence $g(z)$ is constant.

3. Write $f(z/3) = B(z)g(z)$, where $B(z)$ is the Blaschke product with zeros at $\pm i/3$ and $\pm 1/3$. The maximum value is $|B(0)| = 1/81$.

4. Let $d(z_0, z_1)$ denote the maximum. Since the analytic functions can be precomposed with conformal self-maps, $d(z_0, z_1) = d(0, s)$, where $s = |z_0 - z_1|/|1 - \bar{z_0}z_1|$ is the image of z_1 under conformal self-map sending z_0 to 0. Then $d(0, s) = (2/s)(1 - \sqrt{1 - s^2})$, and the maximum is attained by the conformal self-map that satisfies $g(0) = -g(s)$.

9. By Riemann's theorem on removable singularities, $f(z)$ is analytic at 0. Since $f(z)$ maps the punctured disk one-to-one onto itself, the value $f(0)$ must belong to the boundary of the punctured disk, and it satisfies $|f(0)| < 1$, so $f(0) = 0$. Thus f is a conformal self-map of $\mathbb{D}$ satisfying $f(0) = 0$, and f is a rotation.

12. For (a) there are six, for (b) there are eight, and for (c) there are two, the identity $f(z) = z$, and $f(z) = -2/z$.

15. (a) The equation $f(z) = z$ becomes $z^2 + ((\lambda - 1)/\bar{a})z - \lambda a/\bar{a} = 0$, whose solutions satisfy $|z_0 z_1| = 1$. (b) If $z_0 \in \mathbb{D}$ is fixed by f, and h is conformal self-map sending z_0 to 0, then 0 is fixed by $h \circ f \circ h^{-1}$. (d) Conjugate f by an h that maps ± 1 to the two fixed points of f. Can also replace $\mathbb{D}$ by $\mathbb{H}$, place the fixed points at 0 and ∞, and show f is conjugate to $w \to Aw$ on $\mathbb{H}$. (f) Replace $\mathbb{D}$ by $\mathbb{H}$, place the fixed point at ∞, and show that f is conjugate to the translation $w \to w + 1$ on $\mathbb{H}$.

Section IX.3

2. A hyperbolic disk centered at 0 of hyperbolic radius ρ is a Euclidean disk centered at 0 with radius r given by solving $\rho = \log[(1 + r)/(1 - r)]$. Since conformal self-maps are isometries in the hyperbolic metric, they map hyperbolic disks to hyperbolic disks of the same radius, and every hyperbolic disk is the image under a conformal self-map of a hyperbolic disk centered at 0. Conformal self-maps also map Euclidean disks to Euclidean disks.

3. Let r be the Euclidean radius of hyperbolic disk of radius ρ centered at 0. For $0 < s < 1$, this disk is mapped by $(z+s)/(1+sz)$ to the hyperbolic disk of radius ρ centered at s. The diameter $(-r, r)$ is mapped to the interval from $(s-r)/(1-sr)$ to $(s+r)/(1+sr)$. Since the Euclidean center is the midpoint of this interval, this yields explicit formulae for $c(s, \rho)$ and $r(s, \rho)$, from which the limits can be evaluated explicitly.

4. The hyperbolic circumference is $2 \int_\gamma |dz|/(1 - |z|^2)$. Set $z = re^{i\theta}$ and integrate, then substitute $r = (e^\rho - 1)/(e^\rho + 1)$.

7. Consider first an isometry $f(z)$ that fixes two points, $f(0) = 0$ and $f(r) = r$. For $z \in \mathbb{D} \backslash \mathbb{R}$, consider the two hyperbolic circles centered at 0 of radius $\rho(0, z)$ and centered at r of radius $\rho(r, z)$, and observe that they meet at only the two points z and $\bar{z}$.

8. $w = f(z)$ is an isometry if and only if $|dw|/(1 + |w|^2) = |dz|/(1 + |z|^2)$. Use $dw/dz = 1/(cz + d)^2$ to obtain equivalent equations $|a|^2 + |c|^2 = 1 = |b|^2 + |d|^2$, $a\bar{b} + c\bar{d} = 0$.

9. (a) Let γ be the geodesic from z to ζ, and estimate $\rho(z^2, \zeta^2)$ by the hyperbolic length of $f \circ \gamma$. (b) Take ζ close to z.

13. Use $g(z) = (z - i)/(z + i)$ in the definition.

14. Use $g(z) = (e^z - 1)/(e^z + 1)$ in the definition.

Chapter X: Harmonic Functions and the Reflection Principle

Section X.1
3. Use (1.11).

5. Follow the second proof of the boundary-value theorem.

6. Same as Exercise 5, except must use the symmetry of the Poisson kernel to see that each side contributes equally.

8. (a) Justify differentiating under the integral in the Poisson integral formula.

Section X.2
1. Integrate around a circle and interchange the order of integration.

2. Approximate $u(x, y)$ by a Taylor polynomial of degree two. Suffices to check formula for 1, x, y, x^2, xy, and y^2.

4. If $r_1/r_2 = \rho_1/\rho_2$, set $a = \rho_1/r_1 = \rho_2/r_2$, $f(z) = e^{iay}$, and use Exercise 3.

Section X.3
2. Express reflection in circle as composition of two fractional linear transformations and the reflection $z \to \bar{z}$ in the real axis.

3. If curves meet at angle θ, their reflections meet at angle $-\theta$.

4. (a) Use $z'(\zeta) \neq 0$ at $\zeta = 0$. (b) $\zeta(z)$ maps an interval on γ onto an interval on $\mathbb{R}$. (c) By definition, $z^* = \bar{\zeta} + ih(\bar{\zeta})$. Use $\bar{z} = \bar{\zeta} - i\overline{h(\zeta)} = \bar{\zeta} - ih(\bar{\zeta})$.

5. In Exercise 4 take $h(\zeta) = \zeta^2$, obtain $z^* = -\bar{z} - i + i\sqrt{1 - 4i\bar{z}}$. Binomial series converges for $|z| < 1/4$.

6. Divide by $(z - z_0)^n$, can assume $f(z_0) \neq 0$. The reflection formula then shows $f(z)$ has no zeros inside the circle. Use reflection formula again to see it is bounded on $\mathbb{C}$, hence constant.

8. (a) Show first that as z tends to a fixed boundary circle of the annulus, the image points tend to the the same boundary circle of the image annulus. Then apply the reflection principle. The formula shows that the reflected map maps the reflected annulus onto an annulus. Continue reflecting.

9. If $b_n = 0$ for n odd, then $\psi(z) = z - b_2 z^2 + b_4 z^4 - b_6 z^6 + \cdots$ maps the imaginary axis onto γ, so ψ^{-1} is analytic at 0 and straightens out the angle.

Chapter XI: Conformal Mapping

Section XI.1

1. $w = (z^{3/2} - s)/(z^{3/2} + s)$, for any $s > 0$. Though map is not unique, the sketch is.

2. $w = -i(\sqrt{z} - 1)/(\sqrt{z} + 1)$.

3. $w = (2A/\pi i) \operatorname{Log}((1 + iz)/(1 - iz))$.

4. $w = (1 + \zeta)/(1 - \zeta)$, where $\zeta = -e^{\pi(1+i)z/2}$. Median is mapped to top half of unit circle.

5. $w = -(z + i)^2/(z - i)^2$, $w(1) = 1$.

7. $w = (-i\zeta^2 + 2\zeta - 1)/(\zeta^2 - 2\zeta - 1)$, where $\zeta = z^{\pi/2\alpha}$ brings it to a lune.

8. $\zeta = (z - b)/(1 - bz)$, $\eta = -(\sqrt{\zeta} + i)^2/(\sqrt{\zeta} - i)^2$, $\xi = i(\eta - i)/(\eta + i)$, $w = (\xi - a)/(1 - a\xi)$, where $-1 < a < 1$.

12. $w = -(\sin z - i)/(\sin z + i)$, $w(\infty) = -1$.

Section XI.2

2. $\varphi_r(z) = z/r$.

3. Convert a connected component E of $\mathbb{C}^* \setminus D$ to a point or closed analytic curve by applying the Riemann mapping theorem to $\mathbb{C}^* \setminus E$.

Section XI.3

3. Since $1 - t/a \to 1$ uniformly on bounded sets as $a \to \infty$, $g_a(z) \to \int_0^z t^{\alpha-1}(1 - t)^{\beta-1} dt / \int_0^1 t^{\alpha-1}(1 - t)^{\beta-1} dt$ uniformly on bounded sets. This is the Schwarz-Christoffel formula in the case that $g(\infty) = w_0$.

6. $g(z) = A \int_1^z \zeta^{-1} d\zeta$.

7. $w'(z) = c(z + 1)^{-1/2}(z - 1)^{-1/2} = c/\sqrt{1 - z^2}$, $w = c \sin^{-1} z$, $z = \sin(w/c)$, $c = 2a/\pi$.

8. (a) $g(z) = \tau + A \int_0^z (t - a)^{\alpha-1} t(t - b)^{-a} dt$, where $\pi\alpha$ is the angle from the negative real axis to $[0, \tau]$. (c) The map in (b) is a special case of the map in (a) in which the z-plane is scaled so that the two vertices corresponding to 0 are mapped to $1 - \sigma$ and σ, where $0 < \sigma < 1$.

9. The Schwarz-Christoffel formula can be integrated, and $\pi g(z)$ is the appropriate branch of $\sqrt{z^2 - 1} + \log(z + \sqrt{z^2 - 1})$.

10. $g'(z) = A(z+1)^{-1}(z-1)^{-1} z^{-1/2}$, $g(z) = (2/\pi) \tan^{-1} \sqrt{z} + (1/\pi) \log((1 + \sqrt{z})/(1 - \sqrt{z}))$.

Section XI.4

3. $g'(z) = A z^{-1}(z - 1)(z + 1)$, $g(z) = (i/\pi)(z^2 - 2 \operatorname{Log} z - 1) - 1$. Stream function is $\operatorname{Arg} h(w)$.

Section XI.5

1. Use the fact that if every subsequence has a subsequence that converges to the same limit, then the sequence converges. In this case, any normally convergent subsequence has limit 0.

3. Consider the translates $f_n(z) = f(z + n)$.

4. (a) Consider the dilates $f_n(z) = f(nz)$.

6. Show first that $f_n(z) \to z_0$ uniformly on some disk centered at z_0. See Exercise IX.1.8.

7. The nth iterate $f_n(z)$ of $f(z)$ satisfies $f_n'(z_0) = f'(z_0)^n$. Since the f_n's are uniformly bounded on D, the derivatives of the f_n's are uniformly bounded on each compact subset of D.

Section XI.6

1. $g(\zeta) = (\zeta - b)/(1 - \bar{b}\zeta)$, similar formula for $f(\zeta)$, and $|(f \circ h \circ g)'(0)| = (t + 1/t)/2$ for $t = \sqrt{|b|}$.

2. Apply the Schwarz lemma to $f \circ \varphi^{-1}$.

5. Compose with conformal map of D_1 onto $\mathbb{D}$, can assume $D_n \subset \mathbb{D}$ and $w_0 = 0$. Then the g_m's are uniformly bounded. By the Hurwitz theorem, any normal limit is either constant or univalent.

7. $\rho(f(z), z_0) \approx 2|f(z) - z_0|/(1 - |z_0|^2)$ and $\rho(z, z_0) \approx 2|z - z_0|/(1 - |z_0|^2)$.

Chapter XII: Compact Families of Meromorphic Functions

Section XII.1

6. Map $\mathbb{C}^* \backslash E$ conformally onto $\mathbb{D}$, apply thesis version of Montel's theorem.

8. Show $f_n(z)$ eventually has same number of zeros as $f(z)$.

Section XII.2

3. Apply Montel's theorem.

6. (b) Assume D is a disk. Consider family of square roots with appropriate branches.

7. (a) Use Exercise 6.

8. Suppose theorem fails for $f_n(z)$ and constant $1/n$. Show first that $\{f_n(z)\}$ is not a normal family, then apply the Zalcman lemma.

Chapter XIII: Approximation Theorems

Section XIII.1

3. Exhaust D by appropriate compact sets K_n, construct $f_n(z)$ and $h_n(z)$ by induction so that $h_n(z_j) = 0$ for $1 \le j < n$, $h_n(z_n) = 1$, and $h_n(z)$ is small on K_{n-1}. Set $f_n(z) = f_{n-1}(z) + (w_n - f_{n-1}(z_n))h_n(z)$, $f(z) = \lim f_n(z)$.

Section XIII.2

9. $(\pi/3z^2)[\cot(\pi z) + \lambda \cot(\lambda \pi z) + \lambda^2 \cot(\lambda^2 \pi z)]$, where $\lambda = e^{2\pi i/3}$.

Section XIII.3

1. (a) 2, (b) 1/2, (c) 4.

9. Use the product expansion of $\sin(\pi z)$.

Section XIII.4
3. $z \prod_1^\infty (1 - z^2/\sqrt{n}) \exp(z^2/\sqrt{n} + z^4/2n)$.
6. $f(z) = c \prod_0^\infty (1 - z/2^k)$.
11. Take $g(z)$ to be analytic with simple zeros at the z_j's, take h(z) to be meromorphic with simple poles at the z_j's and residue $\lambda_j/g'(z_j)$, set $f(z) = g(z)h(z)$.

Chapter XIV: Some Special Functions

Section XIV.1
6. Combine Exercises 4 and 5.

Section XIV.2
6. (b) Substitute $s = iz$.
8. Recall Cauchy-Hadamard formula for radius of convergence.
12. Cauchy estimates show that if $f(z) = \mathcal{O}(|z|^\alpha)$ in S, then $f'(z) = \mathcal{O}(|z|^{\alpha-1})$ in S_ε.

Section XIV.4
13. Suppose $\sigma_a = 0$, and $f(s)$ extends analytically across 0. Express power series $\sum f^{(k)}(1)(s - 1)^k/k!$ as double series, substitute $s = -\delta$, justify interchanging order of summation, and sum.

Chapter XV: The Dirichlet Problem

Section XV.1
2. Use Green's first formula.
3. Use Exercise 2.
4. Use Green's first formula, with u and v interchanged.
5. Apply Exercise 4 to u and $w = v - u$.

Section XV.2
5. (a) See the Taylor series expansion in Exercise IV.8.7.

Section XV.3
5. (c) $d(z, w)$ coincides with the hyperbolic distance $\rho(z, w)$. Use Harnack's estimate to determine $d(0, r)$, then use conformal invariance.

Section XV.4
1. Try a linear combination of 1 and $\log |z|$.
2. Use $\varepsilon \log |z|$.

Section XV.5
1. Compose with Riemann maps to reduce to the case where $E_0 = \{|z| = 1\}$ and D is the domain between the unit circle and a simple closed analytic curve E_1 in $\{|z| > 1\}$. In this situation, show that $v(z)$ is strictly increasing on E_0, and choose α so that the increase of $\alpha v(z)$ around E_0 is 2π.

Section XV.6
6. Use Green's third formula.

Section XV.7

10. Express D as an increasing union of bounded domains U_m, each bounded by a finite number of piecewise smooth closed curves, let g_m be Green's function for U_m, and define $D_m = \{g_m(z) > \varepsilon_m\}$.

Chapter XVI: Riemann Surfaces

Section XVI.1

2. Use two annuli as coordinate patches.

References

Books

L. Ahlfors, Complex Analysis (3rd ed.), McGraw-Hill, 1979.

L. Ahlfors, Conformal Invariants, McGraw-Hill, 1973.

L. Carleson and T. Gamelin, Complex Dynamics, Springer-Verlag, 1993.

J.B. Conway, Functions of One Complex Variable, Springer-Verlag, 1973.

J. Dieudonné, Foundations of Modern Analysis, Academic Press, 1960.

W. Kaplan, Advanced Calculus, Addison-Wesley, 1952.

K. Knopp, Theory of Functions (Parts I and II) and Problem Book (Volumes 1 and 2), Dover, 1951.

N. Levinson and R. Redheffer, Complex Variables, Holden-Day, 1970.

Z. Nehari, Conformal Mapping, McGraw-Hill, 1952.

H.-O. Peitgen and P.H. Richter, The Beauty of Fractals, Springer-Verlag, 1986.

A. Pfluger, Theorie der Riemannschen Flächen, Springer-Verlag, 1957.

R. Remmert, Funktionentheorie (Vol. I and II), Springer-Verlag, 1984 and 1991.

J.-P. Serre, A Course in Arithmetic, Springer-Verlag, 1973.

G. Springer, Introduction to Riemann Surfaces, Addison-Wesley, 1957.

E.C. Titchmarsh, The Theory of Functions (2nd ed.), Oxford University Press, 1939.

M. Tsuji, Potential Theory in Modern Function Theory, Maruzen, 1959.

Lecture Notes and Articles

P. Chernoff, Pointwise convergence of Fourier series, Amer. Math. Monthly 87 (1980), pp. 399–400.

S. Gál, Lectures on Number Theory, Jones Letter Service, 1961.

T. Gamelin, review of *A History of Complex Dynamics, from Schröder to Fatou and Julia*, by D. Alexander, Vieweg, 1994; in Historia Math. 23 (1996), pp. 74–84.

D. Zagier, Newman's short proof of the prime number theorem, Amer. Math. Monthly 104 (1997), pp. 705–708.

L. Zalcman, Normal families: new perspectives, Bull. Amer. Math. Soc. 35 (1998), pp. 215–230.

List of Symbols

$\mathbb{Z}$	integers
$\mathbb{R}$	real line
$\mathbb{C}$	complex plane
$\mathbb{C}^*$	extended complex plane $= \mathbb{C} \cup \{\infty\}$, Riemann sphere
$\mathbb{H}$	upper half-plane, $\{\text{Re } z > 0\}$
$\mathbb{D}$	open unit disk, $\{|z| < 1\}$
$\mathcal{J}$	Julia set (Section XII.3)
$\mathcal{M}$	Mandelbrot set (Section XII.5)
$\mathcal{L}$	Laplace transform (Section XIV.2)
Re	real part of (Section I.1)
Im	imaginary part of (Section I.1)
Arg	principal branch of argument (Section I.2)
Log	principal branch of logarithm (Section I.6)
Res	residue (Section VII.1)
$\Gamma(z)$	gamma function (Section XIV.1)
$\zeta(s)$	zeta function (Section XIV.3)
$\dfrac{\partial}{\partial z}$	partial derivative with respect to z (Section IV.8)
$\dfrac{\partial}{\partial \bar{z}}$	partial derivative with respect to $\bar{z}$ (Section IV.8)
$f^\sharp$	spherical derivative (Section XII.1)

Index

absolute value of z, 2
addition formula, 7, 20
Ahlfors function, 309
algebraic function, 164
analytic boundary, 407
analytic conjugation, 325
analytic continuation, 159
analytic curve, 284
    neighborhoods of sides, 284
    reflection across, 284
analytic function, 45
    at ∞, 149
    on a Riemann surface, 421
analytic index, 332
analytic map, 422
analytic variety, 239
antiderivative, 76
argument of z
    principal value, 6
argument of z, 6
argument principle, 224, 226
Arzelà-Ascoli theorem, 307
asymptotic expansion, 368
asymptotic functions, 382
asymptotic series, 368
attains a value m times, 232
average (mean) value, 85

basin of attraction, 331
    of ∞, 327

of an attracting cycle, 331
Bernoulli numbers, 154, 369
Bessel functions, 149
Bessel's inequality, 188
beta function, 365
Bieberbach function, 323
big-oh notation, 177
binomial coefficient, 40
binomial series, 148
Blaschke product
    finite, 265
    infinite, 357
Bloch's constant, 324
Bloch's theorem, 324
boundary of a set E, 39
boundary scanning method, 328
branch cut, 16
branch point, 164

Casorati-Weierstrass theorem, 175
Cauchy criterion for convergence
    of a sequence, 36
    of a series, 133
Cauchy estimates, 118, 145
Cauchy integral, 246
Cauchy integral formula, 113
Cauchy's theorem, 110
Cauchy-Green formula, 128
Cauchy-Hadamard formula, 142

Cauchy-Riemann equations, 47
    complex form, 125
cauliflower set, 328
center of the hyperbolic component, 340
chain rule, 44
Chebyshev lemma, 384
chordal metric, 14
circle through ∞, 13
circulation around a curve, 91
closed differential, 78
closed set, 39
compact set, 39
    on a Riemann surface, 420
comparison test, 131
completely invariant set, 327
complex conjugate of z, 3
complex derivative, 42
complex number, 1
complex plane, 1
complex velocity potential, 92
conformal, 59
conformal map, 59, 289
conformal self-map
    of unit disk, 263
    of upper half-plane, 265
conformally equivalent
    domains, 295
    Riemann surfaces, 422
conjugate maps, 68, 325
    self-maps of unit disk, 266
connected set, 254
    on a Riemann surface, 420
continuous function, 37
continuum, 406
contour
    dogbone, 220
    indented, 210, 213
    keyhole, 206
convergent numerical sequence, 33, 34
convergent sequence of functions
    normally, 137
    pointwise, 133
    uniformly, 134

convergent series, 130
    absolutely, 131
    pointwise, 135
    uniformly, 135
convex function, 397
convex set, 38
coordinate disk, 419
covering map, 441
    k-sheeted covering, 240
covering transformation, 442
critical point, 413
    at ∞, 239
    of an analytic function, 236
critical strip, 375
curvature of a metric, 272
curve, 71
cycle
    integral 1-cycle, 258
    of a periodic point, 331

de Moivre's formulae, 8
deformation of paths, 81
degree of a rational function, 235
differentiable function, 42
differentiable triangle, 74
differential, 76
dilation, 63
Dirichlet convolution, 380
Dirichlet form, 54
Dirichlet integral, 393
Dirichlet problem, 98, 281
Dirichlet series, 376
    abscissa of absolute convergence, 376
    abscissa of convergence, 378
discrete set, 158
division algorithm, 5
domain, 38
    exterior domain, 219
Douady's rabbit, 329
doubly periodic function, 119

elliptic integral, 302
entire function, 118
    of finite type, 122

equicontinuous family of functions, 306
   with respect to the spherical metric, 307
essential singularity, 175
Euler numbers, 154
Euler's constant, 40
Euler's totient function, 380
exact differential, 76
extended complex plane, 11
extremal problem, 308

Fatou set, 326
Fatou-Julia theorem, 333
filled-in Julia set, 329
finite bordered Riemann surface, 423
fixed point, 68
   attracting fixed point, 331
   multiplicity, 331
   multiplier at a fixed point, 330
   repelling fixed point, 331
flux across a curve, 85
   of fluid flow, 90
Fourier coefficients, 187
Fourier series, 186
fractional linear transformation, 63
free analytic boundary arc, 286
Fresnel integrals, 219
fundamental theorem of algebra, 4
fundamental theorem of calculus, 76
   for analytic functions, 107, 108
   for line integrals, 76, 79

gamma function, 357, 361
Goursat's theorem, 123
Green's formulae, 391
Green's function, 408
   bipolar, 436
   critical point and value, 410

of a Riemann surface, 430
of general domain, 413
symmetry, 434
with pole at ∞, 416
Green's theorem, 73

Hadamard gap theorem, 163
harmonic conjugate, 55
harmonic function, 54
   on a lattice, 90
harmonic measure, 409
harmonic series, 132
Harnack's inequality, 399, 400
Hermite orthogonal functions, 111
Hilbert transform, 214
Hilbert's inequality, 112
homologous to zero, 258
homology basis for a domain, 259
Hurwitz's theorem, 231
hyperbolic
   distance, 267
   geodesics, 267
   metric, 272

identity principle, 156
imaginary axis, 1
imaginary part of z, 1
implicit function theorem, 236
increase in argument, 225, 226
independent of path, 77
infinite product, 352
   absolutely convergent, 354
   convergent, 352
interpolating sequence, 346
inverse function theorem, 234
inverse iteration method, 330
inversion, 63
irrotational, 98
irrotational flow, 91
isolated point of a set, 155
isolated singularity, 171
   at ∞, 178
   essential, 175
   removable, 172

isometric circle, 426
isoperimetric theorem, 193

Jacobian elliptic function, 302
Jensen's inequality for convex functions, 397
Jordan curve theorem, 251
    for smooth curves, 250
Jordan's lemma, 216
Julia set, 326
jump theorem for Cauchy integrals, 247
jump theorem for winding numbers, 248

Koebe's one-quarter theorem, 323

lacunary sequence, 163
Laplace transform, 365
    asymptotic appoximation theorem, 366
Laplace's equation, 54
Laplacian operator, 54
Laurent decomposition, 165
Laurent series, 168
Legendre's duplication formula, 364
length of curve, 104
    hyperbolic, 267
    spherical, 269
line integral, 71
    complex, 102
Liouville's theorem, 118
logarithmic differentiation, 355
logarithmic integral, 224
logarithmic pole, 391
lunar domain, 292

Mandelbrot set, 338
    hyperbolic component of interior, 339
    principal cardioid, 340
Marty's theorem, 318
maximum principle, 88
    for subharmonic functions, 395, 427

mean value property, 87
Mergelyan's theorem, 344
meromorphic function, 174
Mittag-Leffler theorem, 348
ML-estimate, 105
Möbius band, 249
Möbius μ-function, 379
Möbius transformation, 63
modular group, 426
modulus of z, 2
modulus of an annulus, 287
monodromy theorem, 161
Montel's theorem, 321
Morera's theorem, 119
multiplicative coefficient sequence, 379

Neumann problem, 98, 393
normal covergence of analytic functions, 137
normal covergence of meromorphic functions, 316
normal family of meromorphic functions, 317
    at a point, 324

omitted value, 321
open mapping theorem, 158, 233
open set, 37
    on a Riemann surface, 420
order, 177
    of a pole, 172
    of a zero, 154

Parseval's identity, 192
partial fractions decomposition, 180, 350
path, 70
period, 20, 182
periodic function, 20, 182
    doubly periodic, 185
    simply periodic, 185
periodic point, 331
Perron family, 402

Perron solution, 402
phase factor, 25
Picard's big theorem, 322
Picard's little theorem, 322
Pick's lemma, 264
piecewise smooth boundary, 72
piecewise smooth path, 71
Poisson integral, 277
Poisson kernel, 275
    for upper half-plane, 280
polar representation of z, 6
pole, 172
    simple, 173
Pompeiu's formula, 127
potential function, 92
power series, 138
prime number theorem, 382
primitive (complex antiderivative), 107
principal branch of square root, 17
principal part, 173
    at ∞, 178
principal value of an integral, 213
principal value of $\log z$, 22
principle of permanence of functional equations, 30, 157
purely imaginary numbers, 1

radial cluster set, 345
radial limit, 279
radius of convergence, 138
Radó's theorem, 432
ratio test, 141
real part of z, 1
regular boundary point, 403
removable singularity, 172
residue, 195
    at ∞, 221
residue theorem, 196
    for exterior domains, 221
    fractional residue theorem, 209
Riemann hypothesis, 375
Riemann map, 295

Riemann mapping theorem, 295
Riemann sphere, 315
Riemann surface, 418
    coordinate map, 419
    coordinate patch, 419
    finite bordered, 423
    hyperelliptic, 424
    of square root function, 18
    subsurface, 421
    uniformization theorem, 439
Riemann's theorem on removable singularities, 172
Robin's constant, 417
root of a polynomial, 4
    nth root of z, 8
    nth roots of unity, 9
root test, 142
Rouché's theorem, 229
Royden's theorem, 319
Runge's theorem, 342

Schlömilch formula, 170
Schottky's theorem, 323
Schröder's equation, 288
Schwarz formula, 279
Schwarz lemma, 260
Schwarz reflection principle, 282
Schwarz-Christoffel formula, 299
sharp estimate, 105
simple arc, 251
simple closed path or curve, 70, 249
simple path, 70, 249
simply connected
    domain, 252
    Riemann surface, 438
slit plane, 16
small denominator problem, 288
smooth path or curve, 71
spherical derivative, 317
spherical metric, 14, 269, 317
star-shaped domain, 39
stereographic projection, 11
Stirling's formula, 368
Stolz angle, 310

stream function of a flow, 92
streamlines of a flow, 92
strict maximum principle, 87
    for subharmonic functions,
        395, 427
subharmonic barrier, 404
subharmonic function, 394
Swiss cheese set, 344

totally disconnected, 335
trace of a path, 71, 242
translation, 63
triangulation, 74
trivial zeros, 375

uniformization theorem for Rie-
    mann surfaces, 439
uniformly convergent sequence of
    functions, 134
    with respect to the spheri-
        cal metric, 307

unimodular matrix, 425
uniqueness principle, 156
univalent function, 232
universal covering surface, 443

Wallis product, 357
Weierstrass M-test, 135
Weierstrass P-function, 349
Weierstrass product theorem, 358
winding number, 242
Wolff-Denjoy theorem, 273

Zalcman's lemma, 320
zero
    of a polynomial, 4
    of order N, 154
    of order N at ∞, 155
    simple, 155
zeta function, 371
    functional equation, 373
    trivial zeros, 375

Undergraduate Texts in Mathematics

(continued from page ii)

Frazier: An Introduction to Wavelets Through Linear Algebra

Gamelin: Complex Analysis.

Gordon: Discrete Probability.

Hairer/Wanner: Analysis by Its History. *Readings in Mathematics.*

Halmos: Finite-Dimensional Vector Spaces. Second edition.

Halmos: Naive Set Theory.

Hämmerlin/Hoffmann: Numerical Mathematics. *Readings in Mathematics.*

Harris/Hirst/Mossinghoff: Combinatorics and Graph Theory.

Hartshorne: Geometry: Euclid and Beyond.

Hijab: Introduction to Calculus and Classical Analysis.

Hilton/Holton/Pedersen: Mathematical Reflections: In a Room with Many Mirrors.

Hilton/Holton/Pedersen: Mathematical Vistas: From a Room with Many Windows.

Iooss/Joseph: Elementary Stability and Bifurcation Theory. Second edition.

Irving: Integers, Polynomials, and Rings: A Course in Algebra

Isaac: The Pleasures of Probability. *Readings in Mathematics.*

James: Topological and Uniform Spaces.

Jänich: Linear Algebra.

Jänich: Topology.

Jänich: Vector Analysis.

Kemeny/Snell: Finite Markov Chains.

Kinsey: Topology of Surfaces.

Klambauer: Aspects of Calculus.

Lang: A First Course in Calculus. Fifth edition.

Lang: Calculus of Several Variables. Third edition.

Lang: Introduction to Linear Algebra. Second edition.

Lang: Linear Algebra. Third edition.

Lang: Short Calculus: The Original Edition of "A First Course in Calculus."

Lang: Undergraduate Algebra. Third edition

Lang: Undergraduate Analysis.

Laubenbacher/Pengelley: Mathematical Expeditions.

Lax/Burstein/Lax: Calculus with Applications and Computing. Volume 1.

LeCuyer: College Mathematics with APL.

Lidl/Pilz: Applied Abstract Algebra. Second edition.

Logan: Applied Partial Differential Equations, Second edition.

Lovász/Pelikán/Vesztergombi: Discrete Mathematics.

Macki-Strauss: Introduction to Optimal Control Theory.

Malitz: Introduction to Mathematical Logic.

Marsden/Weinstein: Calculus I, II, III. Second edition.

Martin: Counting: The Art of Enumerative Combinatorics.

Martin: The Foundations of Geometry and the Non-Euclidean Plane.

Martin: Geometric Constructions.

Martin: Transformation Geometry: An Introduction to Symmetry.

Millman/Parker: Geometry: A Metric Approach with Models. Second edition.

Moschovakis: Notes on Set Theory.

Owen: A First Course in the Mathematical Foundations of Thermodynamics.

Palka: An Introduction to Complex Function Theory.

Pedrick: A First Course in Analysis.

Peressini/Sullivan/Uhl: The Mathematics of Nonlinear Programming.

Undergraduate Texts in Mathematics

Prenowitz/Jantosciak: Join Geometries.

Priestley: Calculus: A Liberal Art. Second edition.

Protter/Morrey: A First Course in Real Analysis. Second edition.

Protter/Morrey: Intermediate Calculus. Second edition.

Pugh: Real Mathematical Analysis.

Roman: An Introduction to Coding and Information Theory.

Roman: Introduction to the Mathematics of Finance: From Risk Management to Options Pricing.

Ross: Differential Equations: An Introduction with Mathematica®. Second edition.

Ross: Elementary Analysis: The Theory of Calculus.

Samuel: Projective Geometry. *Readings in Mathematics.*

Saxe: Beginning Functional Analysis

Scharlau/Opolka: From Fermat to Minkowski.

Schiff: The Laplace Transform: Theory and Applications.

Sethuraman: Rings, Fields, and Vector Spaces: An Approach to Geometric Constructability.

Sigler: Algebra.

Silverman/Tate: Rational Points on Elliptic Curves.

Simmonds: A Brief on Tensor Analysis. Second edition.

Singer: Geometry: Plane and Fancy.

Singer/Thorpe: Lecture Notes on Elementary Topology and Geometry.

Smith: Linear Algebra. Third edition.

Smith: Primer of Modern Analysis. Second edition.

Stanton/White: Constructive Combinatorics.

Stillwell: Elements of Algebra: Geometry, Numbers, Equations.

Stillwell: Elements of Number Theory.

Stillwell: Mathematics and Its History. Second edition.

Stillwell: Numbers and Geometry. *Readings in Mathematics.*

Strayer: Linear Programming and Its Applications.

Toth: Glimpses of Algebra and Geometry. Second Edition. *Readings in Mathematics.*

Troutman: Variational Calculus and Optimal Control. Second edition.

Valenza: Linear Algebra: An Introduction to Abstract Mathematics.

Whyburn/Duda: Dynamic Topology.

Wilson: Much Ado About Calculus.

Made in the USA
Lexington, KY
11 October 2013